William Boos
Metamathematics and the Philosophical Tradition

William Boos

Metamathematics and the Philosophical Tradition

Edited by
Florence S. Boos

DE GRUYTER

ISBN 978-3-11-073684-7
e-ISBN (PDF) 978-3-11-057245-2
e-ISBN (EPUB) 978-3-11-057239-1

Library of Congress Cataloging-in-Publication Data
Names: Boos, William, author.
Title: Metamathematics and the philosophical tradition / William Boos ;
    edited by Florence S. Boos.
Description: 1 [edition]. | Boston : Walter de Gruyter, 2018. | Includes
    bibliographical references and index.
Identifiers: LCCN 2018007198 (print) | LCCN 2018016157 (ebook) | ISBN
    9783110572452 ( ) | ISBN 9783110572216 (hardcover : alk. paper)
Subjects: LCSH: Philosophy. | Metamathematics. | Mathematics–Philosophy.
Classification: LCC B72 (ebook) | LCC B72 .B665 2018 (print) | DDC 121–dc23 LC record available
at https://lccn.loc.gov/2018007198

Bibliographic information published by the Deutsche Nationalbibliothek
The Deutsche Nationalbibliothek lists this publication in the Deutsche Nationalbibliografie;
detailed bibliographic data are available on the Internet at http://dnb.dnb.de.

© 2020 Walter de Gruyter GmbH, Berlin/Boston
This volume is text- and page-identical with the hardback published in 2018.
Cover image: Florence S. Boos
Typesetting: Integra Software Services Pvt. Ltd.
Printing and binding: CPI books GmbH, Leck

www.degruyter.com

Contents

Preface — IX

Editorial Remarks — XI

1 **Introduction: Boundaries of Experience — 1**
1   "Wisdom", the "Desire to Know" and the Primacy of the Practical — 1
2   "Hieratic" Inquiry and Its "Limits" — 5
3   "Zetetic" Inquiry and Its "Limits" — 7
4   "Ta Meta Ta Logika" — 9
5   "Semantic" Paradoxes — 10
6   Dialogue, "Diakritik" and "Diagonalisation" — 12
7   Interpretative Plurality and Metatheoretic Ascent — 18
8   Alternative Views of "Secular Design" — 23
9   "Transcendental" and "Proto-transcendental" Arguments — 26
10 "Proto-transcendental" Arguments and "Hieratic" "Design" — 28
11 "Coherentist Idealism" in the Hieratic Ideals of Leibniz and Berkeley — 32
12 The "Second-order Idealism" of Hume's "Empiricism" — 39
13 "Transcendental" Design — 44
14 The Zetetic *Incomplétude* of "Merely" Regulative Ideals — 52
15 Borgesian Maps — 58
16 "Philosophia" and "Scientia" — 64

2 **"Was Blind, But Now I See": Ramifications of Plato's "Line" — 68**
1   Introduction — 68
2   Metalogicians' "Intended Interpretations" — 70
3   Metaphysical "Forms" — 78
4   "The Very Idea" ("*To Eidos Auton*") of Unitary Bivalent "Truth" — 88
5   Reflective *Theoriai* — 96

3 **The Stoics, the Skeptics and Aporetic Autonomy: Is "What Is In Our Power" In Our Power? — 105**
1   Introduction — 105
2   An Early Dialectical "Refutation" — 106
3   "Semantic" Paradox(es) — 108

4	"Diagonalisation" —— 109	
5	"The World", "The Whole" and "Hypothetical Necessity" —— 115	
6	"Our" Power over (What Is In) "Our" Power —— 121	
7	Freedom, Autonomy and Aporetic "Self-Legislation" —— 125	
8	Envoi —— 129	

4 Anselm, *Fides Quaerens Interpretationem*, and *Grenzideen* as Generators of Metatheoretic Ascent —— 131

1 Outline and Introduction —— 131
2 "Gaunilonian" Relativisation(s) of "Anselmian" Argument(s) —— 134
3 Anselm of Canterbury and Georg von Halle —— 141
4 Metalogically "Intelligible" Concept-Formation —— 143
5 "Existence" and "Predication" in the First *Critique* —— 146
6 "Necessary Existence", and the "Existence" of Many "Necessities" —— 149
7 Conclusion: Anselmian "*Numina*" and Kantian "*Noumena*" —— 151

5 "Parfaits Miroirs de l'Univers": A "Virtual" Interpretation of Leibnizian Metaphysics —— 160

1 Introduction —— 160
2 "Miroirs", "Notions" et "Predicats" —— 163
3 "Necessité(s)" et "Raison(s) Suffisante(s)" —— 170
4 "Perfection" and "Plenitude" —— 173
5 Conclusion —— 180
6 "One Who Works for the People" —— 184

6 Berkeleyan Metalogical "Signs" and "Master Arguments" —— 186

1 Introduction —— 186
2 Iterated Abstraction and Metatheoretic Ascent —— 192
3 Semantic Paradox(es) and "Master Arguments" —— 197
4 Semeiotic "Coherentism"? —— 207
5 Conclusion(s) —— 223

7 The Second-order Idealism of David Hume —— 233

1 Introduction —— 233
2 Some Dogmas of Empiricism —— 235
3 "An Establish'd Maxim in Metaphysics" —— 240
4 Immaterialism and "Irrelationism" —— 252

5	"Whatever Is Clearly Conceived … —— 261
6	"[Metatheoretic] Matters of [Object-theoretic] Fact" —— 271
7	Hypotheses Non Fingimus? —— 275
8	"A Kind of Pre-established Harmony?" —— 283
9	Conclusion —— 299

8 Kantian Ethics and "the Fate of Reason" —— 306
1. Introduction —— 307
2. [Reflective Inquiry into "Ultimate" Limits of [Reflective inquiry into "Ultimate" Limits of [Reflective inquiry into "Ultimate" Limits of [….]]]] —— 312
3. Reflective Inquiry and "Transcendent" "[Self]-Validation" —— 320
4. Reflective Inquiry and *das Schicksal der Vernunft* —— 328
5. Reflective Inquiry and "Das" Reich "der" Zwecke —— 338
6. Reflective Inquiry and *Systeme von Zwecken* —— 348
7. Reflective Inquiry and Kant's Two *Unermeßlichkeiten* —— 356
8. Reflective Inquiry as a Form of Skeptical *Theoria* —— 365

9 Metamathematical Interpretations of Free Will and Determinism —— 382
1. Three Venerable Quotations —— 382
2. Preliminary Observations —— 384
3. Three Distinctions —— 386
4. "Schematic Induction" and Metalogical "Nonstandardness" —— 390
5. Gödelian "Incompleteness" and "Indeterminism" —— 392
6. Mathematically Exiguous Interpretations of Physical "Continuity" —— 396
7. The "Mystery" of Essentially Incomplete Inquiry —— 400

10 Time-Evolution in Random "Universes" —— 406
1. Introduction —— 406
2. Four Heuristic Premises —— 409
3. "Linguistic" and "Theoretical" Processes —— 409
4. "Canonical" Measures and Topologies —— 412
5. "Observational Measures" and "Observational Frames" —— 414
6. A Continuous Representation of Wave-Function "Collapse" —— 418
7. "Virtual" Set-Theoretic Universes —— 428
8. The "Direction" of "Time" —— 433

9 The Elusive Ideals of a "Closed" and "Isolated System" —— **437**
10 A Promissory Note and Postscript —— **440**

Bibliography —— 443

Index of Names —— 455

Main Index —— 459

Foreign Words Index —— 477

Preface

In this remarkable text, William Boos is seen to take on the Western philosophical tradition up to the Enlightenment in light of the twentieth-century experience in metamathematics. The stuff of incompleteness, semantic paradoxes, reflective systems, and metatheoretic ascent, as ultimately based on the Diagonal Lemma, is put into service to articulate, clarify, and cast new light on the central concerns of the philosophical tradition having to do with limits of sense, reflection, and monism. This is by no means an attempt at hermeneutic interpretation of texts, but a pursuit of a hermeneutic circle from us back to them of a deliberate sort, one that unabashedly is to inform, to systematize to some extent, and to set out anew. In other words, it is a doing of philosophy, rather than interpretation, one made possible by the experience of metamathematics. With considerable mathematical sophistication, an impressively wide range of central themes and concepts from Plato to Kant are provided searching and clarifying analysis in terms of metamathematical notions and results.

A long, long time ago, I knew Bill as an informed set theorist and logician, versed in the theory of strong axioms of infinity. As the years went by, his interest deepened in the direction of philosophy, particularly of Kant. One sees here a remarkable synthesis, one that incorporates a mathematical precisification into a sophisticated structured reading of central issues in the Western philosophical tradition. This is a work which merits sustained attention by both mathematical logicians and philosophers, and I hope it may further understanding of the deep relationship between their respective fields.

Akihiro Kanamori
February 12th, 2018
Department of Mathematics, Boston University

Editorial Remarks

This volume represents William Boos's writings over the years on metamathematics and the history of philosophy. About three years before his sudden death in 2014 he began to gather, revise, and extend essays on these topics into book form under the title, "The Boundaries of Experience." When I said to him, "Bill, no one will publish such a long manuscript!" he replied, "I want to get it into the form in which I envision it before it is cut up." The manuscript as he left it contained twenty chapters and a possible appendix, saved carefully on his computer in a folder dated the month preceding his death. The present selection represents about half of the original, weighted toward classical, medieval, and Enlightenment philosophy, with two final, more mathematical chapters offering proofs on related problems of uncertainty and time-evolution in theoretical physics.

Another unified volume could easily be constructed of the omitted chapters, most of which provide critical accounts of topics in early and mid-twentieth century philosophy of mathematics as represented in the works of Wittgenstein, Frege, Pierce, Russell, Putnam, Skolem, Dedekind, Cassirer, Cantor, Hilbert, and Gödel, as well as several proofs extending ideas of Leibniz and Gödel's successors. These have been published, however, so that an interested reader may complete the sequence (as listed in the bibliography and the bibliographical note below). For the last six years of his life he had also engaged in systematic studies on the intersection of metamathematics and physics, and two days before his death he said to me, "I believe I have a new physics proof in mind and am looking forward to returning to the cabin [in British Columbia] to write it out." He had also hoped to explicate more fully his thoughts on Wittgenstein, developed during a period of employment at the Wittgenstein Archive in Bergen, Norway, and to have elaborated thoughts on his beloved Kant in two essays on additional features of the latter's ideas as a sequel to those in "Kantian Ethics and the 'Fate of Reason'," chapter 8 in this volume.

He would have been deeply grateful to those who have been essential to making the publication of this book possible. These include Professors Victor Camillo and George Nelson of the Department of Mathematics at the University of Iowa, Professor Franklin Tall of the Department of Mathematics at the University of Toronto, Professor Akihiro Kanamori of the Department of Mathematics at Boston University, and Professor Thomas Pogge of the Department of Philosophy, Yale University. In particular, George Nelson and Frank Tall have helped determine Bill's original intent in unclear passages, and Thomas Pogge has aided in finding a suitable publisher and in making decisions regarding its contents and presentation. *Sine quibus non.*

In preparing this manuscript I also owe thanks to friends who have offered practical advice and support, among these Professor Paul Muhly of the University of Iowa, Professor Ronald Creagh and Dr. Philippe Rothstein of the Université Paul Valéry, Montpellier, France, and Dr. David Vogel of Creighton University. Christina Runnells, a Ph. D. student at the University of Iowa, provided adept mathematical typing assistance, Dr. Sean De Vega of the University of Iowa has checked the Greek quotations, and Angela Jeannette has created a useful index and glossary. I would like to acknowledge the editors of *Synthese* for their permission to include the posthumously published "Reflective Inquiry and the 'Fate of Reason'" as chapter 8 of this book. And finally, I am especially grateful to Gertrud Grünkorn and Monika Pfleghar for their continuing helpfulness in bringing *Metamathematics and the Philosophical Tradition* to completion.

Bill loved all varieties of literature, so it may be appropriate to conclude with an invocation from Chaucer, "Go [not so] little book!"

Florence S. Boos
March 16th, 2018

Bibliographical Note

Additional essay chapters include: "Theory-relative Skepticism", *Dialectica*, 41.3 (1987), 175–207; "A Self-referential 'Cogito'", *Philosophical Studies*, 44 (1983), 269–290; "'The True' in Gottlob Frege's 'Über die Grundlagen der Geometrie,'" *Archive for History of Exact Sciences* 34 (1985), 141–192; "A Metalogical Critique of Wittgenstein's 'Phenomenology'," *Quantifiers, Questions and Quantum Physics: Essays on the Philosophy of Jaakko Hintikka*, ed. Dan Kolak and J. Symons, New York: Springer, 2004, 75–99; "Limits of Inquiry," *Erkenntnis* 20 (1983), 150–194; "Consistency and *Konsistenz*," *Erkenntnis*, 26 (1987), 1–43; "Thoralf Skolem, Hermann Weyl and 'Das Gefühl der Welt als Begrenztes Ganzes,'" *Essays on the Development of the Foundations of Mathematics*, ed. Jaako Hintikka. Dordrecht: Reidel, 1995, 283–329; "The *Transzendenz* of Metamathematical 'Experience,'" *Synthese*, 114 (1998), 49–99; and "Virtual Modality," *Synthese*, 136 (2003), 435–491.

1 Introduction: Boundaries of Experience

> I am a part of all that I have met;
> Yet all experience is an arch wherethrough
> Gleams that untravelled world, whose arches fade,
> Forever and forever as I move ....
>
> We are too little, and ....
> Little remains, but every hour is saved
> From this eternal silence, something more[:]
> This grey spirit yearning in desire
> To follow knowledge like a sinking star,
> Beyond the utmost bound of human thought.
>   (Alfred Tennyson, "Ulysses")

1 "Wisdom", the "Desire to Know" and the Primacy of the Practical

A number of epistemic and normative questions might be posed about Aristotle's famous remarks in the *Metaphysics* that

1. "all human beings naturally desire to know" (*panteis anthropoi tou eidenai oregontai phusai*, 980a22 ff.); and that
2. a "sign" (*semeion*) of this desire is their "affection for the experiences of the senses (*ton aistheseon agapesis*) ... [which they] "love" for their own sakes, beyond need and use ... (*choris tes chreius agapontai di' autas*)"

Is it really "*knowledge*", for example, that "all humans" (and perhaps others) cherish and "desire"? Or is it rather the ranges of contemplative as well as sensory experiences with which those desires are intertwined?

Is the "desire to know" also a regulative property of *all* sentient or "reasonable" beings (in Kantian language), whatever their accidental qualities and particular grades of "sentience" and "reasonability" might be? Or is it simply a contingent property of *anthropoi*—a race of moderately intelligent primates, who have cherished eagerly "the experiences of the senses" in youth ("... trailing clouds of glory do they come ..."), and mourned them quietly in old age ("We are too little, and little remains ....")?

Finally, are the "signs" Aristotle observed also regulative of "sentience" and/or "reasonableness"? If so, are they signs of *value* as well? Sources of value? Should sentient or "reasonable" beings "*desire*" such signs—"love" them and seek to preserve them "... for their own sakes, beyond need and use...."?

In what follows, I will argue
3. that incomplete processes of interpretative inquiry are the principal objects of Aristotelian "*orex(e)is tou eidenai*";
4. that incompleteness of these processes is regulative of "*experience*" as well as the "*Freiheit*" Kant sought in it;
5. that "sentient" or "reasonable" beings' aptness, eagerness and openness to experience are regulative of their "reasonableness";
6. that acknowledgment of the essential incompleteness of hermeneutic inquiry and respect for the freedom and self-awareness it permits are marks of reflective "*wisdom*";
7. that "*philosophia*" (the "love of wisdom") is more accurately described as a "love" of inquiry, its objects and its apparent agents, "for their own sakes, beyond need and use ....";
8. that "dignity" ("*Würde*") is a form of solidarity among "reasonable beings" ("*das moralische Gesetz in uns*"/"the moral law within us"), in the face of vast *kosmoi* of experiences we are aware we will never have ("*der bestirnte Himmel über uns*"/"the starry heavens above us");

and finally, therefore,
9. that incomplete partial realisations of a processive "*Primat des Praktischen*" offer the only warrants of "primacy" that students of "first philosophy" should ever hope to have ("... *das moralische Gesetz in uns*")
("The journey not the arrival matters").

I could hardly claim to "prove" or "deduce" such conjectures—"transcendentally", or otherwise. I will, however, try to offer "reasonable" interpretive contexts for them.

As a first installment in this effort, I will ask the reader to consider in what follows two dialectically opposed and tacitly metatheoretic conceptions of "inquiry"—"metatheoretic" in the sense that they bear implicitly or explicitly on semantic aspects of inquiry's "course(s)": that
10. inquiry eventuates in a unique ultimate "truth" (or interpretation); and that
11. unicity of such interpretation is a *transcendental illusion*, and inquiry is a programmatic form of rational refinement which ramifies forever.

In the course of the introduction, and in more detail in the sequel, I will
12. assimilate such "course(s)" of inquiry to metalogically defined ramified processes of theory-extension and metatheoretic ascent, and

13 argue that such processes are formally *"intelligible"* miniatures of the sequences of "conditions", "grounds" or *rationales* Kant called *Reihen der Bedingungen* (series of conditions) in the first *Kritik* (introductory criteria for such "intelligibility" may be found in 6.11–6.21 below).

In the process, I will not attempt at this point to "prove" very much, but will, however,

14 offer a number of analogical conjectures derived from careful metalogical studies of the semantic limitations of formal syntactical "proof", and

15 attempt to draw from these conjectures and analogies a rationale for the pluralism implicit in this introduction's title, as well as the heuristic hypotheses which I hope will clarify it.

One of the latter is

16 that *semantic paradoxes* and metalogical theories devised to accommodate them have hermeneutic value, as formal miniatures of informal counterparts in ancient, early modern and twentieth-century metaphysics and epistemology.

Another is

17 that notions of such "existence", "design" and "experience" cannot be construed as "internally" well-defined "*Verstandes-*" or "*Inbegriffe*" (concepts of "the" understanding, or aggregates) as most ancient, early-modern and twentieth-century European and North-American metaphysicians seem tacitly or explicitly to have hoped, or simply assumed.

Another is

18 that such "criterial" notions are recurrently "metatheoretic", "problematic" (cf. *KdrV*, B 310) and processively underdetermined *Ideale der Vernunft* (ideals of reason), in search of further clarification of (and in) "courses of inquiry" which ramify "forever".

Preliminary plausibility arguments for these conjectures may be sketched as follows.

In the last sixty-odd years, mathematical logicians have drawn on precedents found in the pioneering work of Skolem, Gödel, von Neumann, Tarski and others, to explore phenomena of formal undecidability which are taught to every competent student of mathematical logic.

Many philosophical logicians and analytic philosophers of language have given little attention to such phenomena, or to the semantic interpretations of "experience" one might associate with them, but these results have proven to be extraordinarily rich and strange. "Nonstandard" interpretations of many sorts of expressible mathematical and mathematical-physical "experience" have turned out to be the rule, not the exception, in ways that can be made metamathematically precise.

It seems plausible to me therefore to conjecture that "the" theory-relative boundaries of "experience" alluded to above—which philosophers of mathematics as well as physics sometimes talk about in terms of "the" bounds of quasi-Kantian "intuition(s)"—may be comparably rich, and comparably strange.

My efforts to explore the interrelations between metalogic and metaphysics sketched above have led me to propose a number of further conclusions and working hypotheses.

One such is that

19 counterparts and analogues of semantic paradoxes and other metalogical dilemmas have pervaded metaphysical, epistemological and deontological inquiries of every variety—including reductive attempts to "dismiss" or "overcome" or "deconstruct" metaphysical speculations in the name of pragmatism, positivism, *Sprachspielstudium* and other sorts of "antimetaphysical" and "empiricist" inquiry (which seem to me as speculative as the claims they dialectically negate).

A second hypothesis is that

20 these patterns arise from common metalogical and metaphysical needs to probe limits of probation, express limits of expression, and "conceive" limits of "conception" (usually "the" limits, in each case, at least initially).

In metalogic, these recurrent "zetetic" and "hieratic" needs prompted researchers to formulate and study theories T—which I will assimilate to processive and "reflective" forms of "*synthetic a priori*" judgments—that are characterisable by their potential usefulness and near-indispensability as tacit *metatheories*—not only for other theories, but ultimately—and most intriguingly and problematically—for [*themselves*].

A third is that

21 abilities to consider, examine, and in some sense realise such theories T—in which theoretical [self]-reference is possible—may be constitutive (or at least regulative) of the sorts of [self]-consciousness Aristotle attributed to *theoria*.

If so, indefinitely ramified and graduated degrees of such theoretical "understanding" and [self]-awareness may be dimly realised in all sentient beings (Leibniz' "monads" live?).

A fourth, finally, is that
22 studies of these metalogical analogues and ectypes are unlikely to *solve* problems associated with their metaphysical counterparts. They may, however, clarify the recurrence and apparent intractability of their metaphysical and epistemological archetypes, about which we seem drawn to speculate again and again—ineluctably, perhaps, as Kant suggested in his well-known comments about the "*Schicksal der Vernunft*" (fate of reason).

If, for example, we secure and communicate processes of "understanding" in indefinitely extended hierarchies of [self]-referential theories, such "understanding" will be subject to shifts of metatheoretic perspective of the sort sketched above. Moreover, such *Verstehen* will not be able to define its own limits in determinate ways, as Kant hoped to do for the *Vermögen* (capacity) he called *Verstand* (understanding) (cf., e.g., B 692-3).

It may, however, "understand" some of its own recurrent limitations, and such "understanding", in turn, may be another embodiment of the quality Kant called "*Würde*" (dignity). "The universe", by contrast—as Pascal wryly observed— in all likelihood "understands" nothing (*n'en sait rien*).

2 "Hieratic" Inquiry and Its "Limits"

Inquiry is "hieratic", on my account (from "*hieros*", holy, priestly, sacramental), to the extent that
1 it has a uniquely directed "line" or "course";
2 this unique "line" or "course" of inquiry also converges to a unique "ultimate" limit ("that inquiry than which no greater inquiry can be conceived", so to speak);

and
3 this unique *limit of inquiry* (in Charles Sanders Peirce's phrase) may be considered its source, its origin, or (in nominally less teleological terms) its "design".

Philosophical invocations of such inquiry and its unique theological "limit(s)" or secular "designs" tend to eventuate, in turn, in a normative ideal Arthur Lovejoy once called

4   "*the monistic or pantheistic pathos*" (Lovejoy, 1965, 12–14).

Such unitary *Grenzideen* (Kant's "limit(ing) ideas") and other archetypes of this sort have appeared again and again in western philosophy and wisdom literature—in Plato's doctrine of the forms and the line- and cave-parables in Books VI and VII of the *Republic*, for example; in Aristotle's "platonic" praises of "godlike" *theoria* in the *Metaphysics* and *Nicomachean Ethics*; in the *Definitiones* and *Axiomata* Spinoza set down in the opening pages of *Ethica I*; and in the "scholastic realist" "limit of inquiry" of Peirce's essays and *Collected Papers*.

Some philosophers (Spinoza, for example, despite his intentions to "prove" the theorems in the opening pages of his *Ethica*) have simply presupposed the "existence" of such ideals. Others, such as Peirce, have explicity tried to "derive" them, as limits to which unique hierarchically ordered "lines of inquiry" (it was thought) "must" surely converge.

Aristotle, for example, sought to derive the "existence" of several such (allegedly unique) *Grenzideen*:

> (1) a *unique "first cause"* (or *"prime mover"*), for example, which initiated *"descending"* sequences of *"aitiai"* (a forensic term, like *"causae"* in Latin); and

> (2) a *unique "ascending"* convergent limit of "visual" *theoriai* which culminated in a *hieratic* ideal of (self-referential) *"thought about thought"* (*noesis noeseos*).

Early in the second millennium of the "common era"—more than fifty generations later—Anselm of Bec also sought to derive the "existence" *of* (the christian) "god" along similarly theoretical lines, as "the" maximal element of a partially ordered hierarchy of *concept formation* and *interpretation* ("*id quod maius nil cogitari posset*"), whose *unicity* he essentially never questioned.

Still later—by more than two centuries—Thomas Aquinas rang at least three expository changes on Aristotle's and Anselm's hierarchies, in his five "proofs" of the same entity's thought-experimental "existence".

Drawing on work he claimed to disdain, Descartes finally cast such patterns in the seventeenth century as semantic hierarchies of "objective" and "formal" "realities", in his *Meditationes de Prima Philosophia* III and V. Somewhat later, he attempted to reexpress these ancient scholastic motives and arguments "*more geometrico*" in the *Principia Philosophiae*, and Spinoza varied and refined such attempts in the opening pages of the *Ethica* mentioned earlier.

Other equally subtle spirits in the late seventeenth and early eighteenth centuries recognised, in effect, that such ascending hierarchies of "sense" and "reference" might not be *unique*—that they might ramify, for example, in indefinitely more complicated as well as nonlinear ways.

Leibniz and Berkeley, for instance, clearly saw that *"le" bon dieu* (or "Author of Nature") would have to sustain complex forms of synchronic as well as diachronic coordination to ensure the convergence and uniformity of different "spirits'" disparate ideas. Leibniz, however—who refused the Lutheran sacraments on his deathbed—seems also to have felt more keenly the bitter burden such "coordination" might impose.

Such a "god" (or demiurge), he saw, would have to *faire le tri* of infinitely many realisations of temporal inquiry—"*choose*", in effect, *one of* Spinoza's (internally value-free?) *deos sive naturas*. This bleak insight left him little choice—he thought—but to defend the very possibilities of epistemic nonalignment between human and "divine" "design" that Descartes had struggled most tenaciously (and unsuccessfully) to refute.

I will return later to Leibniz' apparent dilemma, and consider at this point some of the methodological complexities of *inquiries into the design(s) of inquiries into the design(s) of inquiries into* ... (an ineluctable metalogical counterpart of the "Cartesian circle").[1]

3 "Zetetic" Inquiry and Its "Limits"

Alternative views of "inquiries" and their objects—diametrically and dialectically opposed to those just sketched—might be called "zetetic" (from the Greek verb "*zetein*", to seek). "Skeptic(al)", from "*skeptesthai*", "to view" or "to look at", was a near-synonym for "*zetetic*" in later Greek philosophical usage. Pyrrhonist *skeptikoi*, for example, sometimes called themselves "*zetetikoi*", or "seekers".

Among other things, such usages also cast an interesting etymological light on the continuation of Aristotle's remarks, quoted earlier, that a "*sign*" of our natural "*orexis tou eidenai*" is (once again)

1   our affection for the experiences of the senses. These are [all] loved for their own sakes, beyond need and use, *but the experiences of sight most of all.* (emphasis mine)

---

[1] Brief accounts of mathematical and quasi-mathematical terms that may be unfamiliar may be found in "The Transzendenz of Mathematical 'Experience,'" *Synthese* 114 (1998), 49–99.

For "*eidenai*" ("to know"), was a perfect infinitive of the verb *eidein*, little used in Attic, that meant "to see" (cognate to "*videre*" in Latin).

Indeed, buried visual references seem to have underlain usages for "knowledge" in several languages—as placeholders, presumably, for cognate mental activities, but neurophysiological representations of un-"sighted" counterparts of such activities suggest that networks of interpretative "insight" (and "inquiry") may take indefinitely many cognate forms. Generalised "courses" and "designs" of such inquiry, in any event might be "zetetic", for example, to the extent they are partially rather than linearly ordered.

They might ramify, in particular, along alternative linear paths through partially ordered structures, and such paths and their limits (if they exist) might not be unique.

One might, of course, seek metatheoretic criteria to *guide* or *order* particular enquiries or "paths" of inquiry, provide bases for their comparison, and ensure that certain paths individuated by these criteria do indeed have unique limits.

Such criteria themselves, however, might not even be *metatheoretically definable* or *expressible*, much less object-theoretically ("internally") intelligible. And even if they were, the rank-orderings they imposed on inquiry's paths might not have maximal elements.

In terms relevant to Anselm's proof, for example, there might be many local or alternative "perfections"—"*ea quibus majora non cogitari possent*" ("those things than which nothing greater may be thought"), to vary a phrase. But no *one* of these paths might be intelligibly or discernibly (much less "provably") *hieros*.

None of Lovejoy's "*pathe*" quite fits my account of "zetetic" inquiry (he himself was a "critical realist", after all). But the obvious dialectical counterweight to the "monism" he characterised would be one or another form of "*pluralistic pathos*".

Such *pathe* can already be found in antiquity, though they have had a bad press in several canonical texts to which "all philosophy" has allegedly been a footnote (most conspicuously in the *Theaetetus*' putative confutation of "Protagorean relativism").

In their own texts, zetetic dialecticians cast some of their most acute analyses of interpretative plurality as elenctic responses to "dogmatic" claims about the nature and uses of reasoning or argumentation ("*logismos*") and their *techniques* ("*hai technai logikai*").

On the one hand, such techniques and lines of reasoning emerged from deeply hieratic desires (evident for example in the work of Pythagorean sages who clearly influenced Plato) to discern "hidden", "true" *eide* or *ideai* or *logoi* or other primordial structures that might underlie all experience ("... in the beginning was the 'word' ....").

On the other, other *technai logikai* offered skeptical or zetetic instruments to query the very language used to express such desires, and relativise or localise the unicity and finality of "ultimate" limits.

(What might it even "mean", for example, to talk about "all experience(s)"—as opposed, say, to "all the beakers I have just set on this table" ... ?)

4 "Ta Meta Ta Logika"

A simple but resonant example may clarify the essentially dialectic and metalogical aspects of this early Hellenistic *sic et non*.

Pioneering stoic metaphysicians elucidated most of the essentials of classical propositional logic and its "intended" interpretations before the end of the third century B.C.E. They left unresolved, however—essentially because they were not yet understood—much deeper metalogical questions about *quantificational* logic and its "intended" interpretations (quite understandably, for they remained unresolved for more than two millennia).

Working at the methodological margins of these then-recent advances, the stoics' skeptical respondents were therefore free to pose certain metalogical queries about their opponents' object-theoretic *logismoi* (whose uses and heuristic values the academic skeptics essentially accepted, despite assorted pyrrhonist disclaimers).

Stoic metaphysicians, for example, had offered semiformal formulations of what they called
1 "*criteria of truth*"—standards which would permit themselves and others to decide the veracity of arbitrary ranges of assertions.

Their skeptical opponents then asked the stoics
2 to "decide" (in effect) whether it was "criterially" legitimate to apply such "criteria" to [themselves];
3 "decide" what further "criteria" might be needed, if it was not, and
4 "decide", finally, how to respond to similar questions about *these* "criteria" in their turn.

Notice that
5 the skeptical critics in this exchange engaged their stoic opponents' claims in ways that tacitly adhered to certain common principles of eristic—"logical" or quasi-"logical"—practice;

and that

6   the skeptics' "relativisations" of stoic "truth"-definitions might also be interpreted as relational refinements of the "criteriology" they had set out to query.

In the sequel, I will argue that such interpretative divergences on common methodological ground have been typical of the most generative and creative "hieratic" and "zetetic" confrontations.

Skeptics, in any case, did not need to destroy their opponents' thought-experiments. They needed only to *relativise* and *reinterpret* them. To accomplish this, however, they had to grasp as well as query the implicit thought-experimental scope and informally "intended" semantics of their opponents' claims.

Only then could they begin to outline the needs for revision, relativisation and alternative interpretations which might belie the unicity, finality and absoluteness their opponents had claimed. And this, I believe, was a deeply generative process, part of a nascent dialectic between meta*physics* and meta*logic*—about notions such as "existence", for example—which has refined and deepened the larger dialectical confrontation between hieratic and zetetic views I wish to sketch here, and elaborate in the chapters that follow.

In the next two sections, I attempt to interpret aspects of this early dialectic as templates for more general patterns of argumentation about *semantic paradox* and *metatheoretic ascent*.

5 "Semantic" Paradoxes

"Semantic paradoxes" are mentioned in passing in a number of introductions to philosophical logic, which typically treat them as signs of some sort of formal counterpart to Kant's "*Skandal der Philosophie*"—something to be dispelled or eliminated, by any (legitimate, conceptual) means necessary.

Such approaches to these *aporiai* and *problemata*—several of which are quite ancient—seem to me quite understandable, and deeply wrong. More precisely, they seem to me to express unilaterally "hieratic" dismissals of the generative role they have played in the historical dialectic sketched above.

As ancient interrogations of "the criterion of truth" suggest, certain paradoxes seem to arise from attempts to apply linguistic or hermeneutic ("semantic") predicates reflexively (or self-referentially) to [themselves].

More precisely,

1   a number of "contradictions" or "absurdities" seem to emerge from predications of "global" alethic or interpretatively significant properties (such as "falsity") to [themselves];

2   analyses of such "reflexive" or "self-referential" *antinomies* suggest that these ostensible predications, or the "global" hermeneutic properties from which they are formed, may be indefinable;
3   partial "resolutions" of the antinomies relativise the problematic properties, or the reflexivity (or both), and reinterpret them, implicitly or explicitly, as "local" or relational notions, which depend on some sort of metatheoretic or contextually varying parameters.

In the case at hand—the *aporia* of "the criterion of truth" (the underlying ground-bass, I will argue later, of the better-known "liar paradox")—pyrrhonist and academic skeptics applied
4   reflexive or self-referential arguments (as in 1) to query
5   the definability of a "global" hermeneutic property known in English as "truth " (as in 2) (a word whose etymological origins lay in Germanic usages for "trust" or "belief", as I remark in the sequel).

These "zetetic" metaphysicians then offered refuge from such "absurdities" in variously nuanced
6   "undogmatic" avowals of the relationality, contextuality and/or probabilistic graduations of such "truth" ("undogmatic", for example, in the sense that they accommodated the metatheoretic revisions of perspective they proposed).

Relativisations and open-ended parametrisations of alethic and ontological notions are repugnant to many metaphysicians, of course. But the "reflexive" predicaments that have given rise to them have been stimulants for critical inquiry.

In Platonic terms, for example, consider
7   the Theaetetan attempt to give an account of "what it means to give an account"; or
8   the *Republic*'s attempt in the cave parable to provide a ("mere") *eikasia* (image, likeness) of "true" knowledge "outside" the "cave" that is not ("merely") *eikasia*.

In the sequel I will argue further
9   that these insights clarified needs to make local distinctions between theory and metatheory;
10 that they anticipated thought-experimental insights into the generative qualities of reflective analysis (in both senses of the word "reflective"); and

11 that these clarifications, in turn, made it possible to formulate processive and provisional resolutions of *aporiai* the stoic pioneers of logic (and most of their medieval and early-modern successors) left unresolved.

6 Dialogue, "Diakritik" and "Diagonalisation"

I have already used the criterion-problem to argue that the stoics and their academic-skeptical opponents had to care about *logike* for semantic paradoxes to become an irritant, much less a source of conceptual clarification and dialectical struggle over "true" uses of *logike/logismos*.

Similar remarks might be made about
1   the better-known *liar paradox* ("This sentence is false"—a tributary, I have suggested, of the criterion-problem), and
2   the equally venerable *sorites* or heaper paradox (A single grain is not a heap, and the addition of a single grain to a non-heap does not make a heap; therefore ....),

as well as more recent paradoxes introduced by Berry, Grelling and others (discussed briefly below, and in more detail in Boos, 1998).

(As a foretaste, consider the (potentially) "soritical" problem of determination whether a given unary predicate is (potentially) "soritical".)

In particular, it might be no accident that ancient semantic paradox arguments generated attempts to refine and recapitulate such arguments many centuries later, with the aid of "coding" procedures Leibniz envisioned more than three hundred years ago. Consciously or unconsciously following Leibniz, early twentieth-century colleagues such as Thoralf Skolem, Kurt Gödel and Alfred Tarski, among others ("fellows from another college", to borrow a phrase from John Littlewood and G. H. Hardy), developed finitary arithmetic *codes* for linguistic expressions to analyse certain self-referential arguments.

His most incisive and general observation was that for every property p(x) expressible in the languages L(T) of a wide class of formal theories T, there exists a diagonal sentence s for p(x) such that s is provably equivalent to p([s]).

One can informally sketch a construction of one such sentence in ancient dialogical (or perhaps dialectical) form, as follows
3   An ancient "stoic" claims that "apprehension" ("*katalepsis*") serves as a universal "validity"-criterion $val(\varphi)$, which determines the "validity" of every sentence φ.

4   An academic "skeptic" replies that any such "criterion" fails to determine the validity of the sentence $v = \neg val\,([\neg val])$, where
5   $[val] = s$ is a "quotation" of the stoic's original "universal" assertion (namely that "$val(\varphi)$ determines the "validity" of every sentence φ").

For
6   verification of v would falsify it, since it would show that there *is* a sentence, namely $\neg[val]$, which val does not determine; and
7   falsification of v would verify it, since it would show that there *is* a sentence, namely $\neg[val]$ once again, which val does determine).
8   Confronted with this argument, the ancient "stoic" replies (with a measure of exasperation) that the skeptic's argument is "mere sophistry," for it impugns self-referential concept formation, not the "criterion".
9   To this rejoinder the "skeptic" then responds that such a response would
(i)   shoot the messenger who brings bad news;
(ii) betray straightforward principles of "parity of reasoning";
(iii) conflict with stoic ideals of [self]-criticism and [self]-understanding; and
(iv) belie the "stoic"'s original insistence that the "criterion" be "complete" and "comprehensive".

A "formalisation" of this ancient argument in ordinary Greek is the core lemma which underlay twentieth-century "incompleteness" results.

For in 1930, the young Czech mathematician Kurt Gödel clarified the universality of the exchange outlined above in the form of a metalogical proof that every "theory-internal" "semantic" or "criterial" notion for a given "intelligible" theory (its "validity", for example) has the property that if one applies the predicate
10 [it is invalid that [it is invalid that x]] to an "intensional" counterpart of [*itself*], the resulting ("*Gödel-*)sentence" γ is provably equivalent to [its own] (invalidity).

More generally, Gödel showed in his *Diagonal Lemma*, reconsidered in more detail later, that
11 any predicate φ in a sufficiently "intelligible" formal theory T—including, a fortiori, any reconstruction of what T might think it means to "doubt"—has a "paradoxical" fixed point, in the form of an "antidiagonal" sentence $\alpha = \alpha_\varphi$ for φ such that α_φ is provably equivalent to the negation of $\varphi(\alpha_\varphi)$.

Kaplan and Montague called the special case where φ is "ignorance" the "paradox of the knower". But the scope of the diagonal lemma is in fact much wider.

Suppose, for example,

12 that a predicate φ expresses "possibility" (say), "credibility", "coherence", or "conceivability" (...).

Then one can form

13 "antidiagonal" sentences $\alpha = \alpha_\varphi$ which will be equivalent to [their own] "impossibility", "incredibility"; "incoherence", "inconceivability" (...). Along the way, as a special case Alfred Tarski observed that

14 no formalism can define a "truth"-predicate $Tr(\cdot)$ for itself such that $Tr(\theta)$ is equivalent to θ for every sentence θ. "*Truth is internally indefinable*".

To obtain a more "Cartesian" (or historically, more "stoic") example, one could apply the diagonal lemma to

15 "clarity and distinctness" once again, the notion Descartes derived in Meditation III from the stoics' alethic "*kriterion*" of "apprehensive (re)-presentation" (*kataleptike phantasia*), (paraphrased by Sextus Empiricus (cf. [9], M VII, 257) as *enarges ousa and plektike*, roughly "clear and well-struck"), and observe

16 that if there were such a formalisable predicate φ, the corresponding anti-diagonal sentence α_φ would furnish a "clear and distinct" witness to [its own] obscurity and confusion.

In more precise metamathematical detail, Gödel's definition of the generic sentence α_φ may be (re)constructed as follows.

17 Stipulate that x ranges over "codes" of predicates φ in the language of T, and let

(i) "self- application of x" be the predicate $x([x])$; and

(ii) "iterated self-application of x" the predicate $x([x])([x([x])])$

Then if T "thinks" that the predicate φ expresses inconceivability (for example), the sentence which is equivalent to [its own] inconceivability) is

(iii) $\varphi([\varphi([\varphi ([\varphi])])([\varphi[\varphi ([\varphi ([\varphi])])]])$,

or equivalently,

(iv) [self-referential application of [the assertion that [self-reference] is inconceivable]] is inconceivable.

This argument is evidently "generic", as suggested above.

If φ expresses "obscurity and confusion", for example (the negation of "clarity and distinctness"), the sentence

(v) [self-referential application of [the assertion that [self-reference] is obscure and confused] is obscure and confused.]

is provably equivalent in T to [its own] obscurity and confusion, and a T-internal truth-predicate would generate Tarski's formalisation of the liar paradox: a sentence equivalent to [its own] "falsity", a worthy descendant of the academic skeptic Carneades' counterexample to the bivalence stoic logicians claimed for their *lekta*, or propositions, sketched earlier.

Many mathematicians and metaphysicians (Gödel among them) found such arguments and constructions unsettling, and many have tried in various ways, to dismiss or "transcend" them (both deeply "hieratic" efforts, on my account).

But such efforts are ambiguous as well as problematic, for

18 the constraints on "intelligible" theories T to which they apply are actually rather minimal: namely, that

19 they be mathematically "articulate", in ways that make communication possible and (what might be called) "surveyable",

and if these constraints are satisfied,

20 the dialogical reconstructions of the sort just sketched for any T yield so-called "fixed points" for any predicates which are expressible in those theories;

and

21 reasoning about such "fixed points" does undermine "internal" attempts to say significant things about such theories' (formal or informal) attempts to ensure (or validate) [their own] validity.

(There may be a measure of historical irony in "skeptical" uses of "fixed-point" arguments, in the light of Descartes' claim to have found an "Archimedean" fixed point or *pou sto* at AT VII 24.)

In plausibly relevant theories, for example, one might formulate sentences which are provably equivalent to theory-relative miniatures (in rough temporal order) of

22 their own Stoic *akatalepsia*;

23 their own skeptical *epoche*;

24 their own Kantian *Transzendenz*;

25 their own Wittgensteinian *Unsinn* and
26 their own ("Theaetetan") *undefinability* or *inexpressibility*.

Whatever the reader's views of such reconstructions, therefore, they may at least have a certain recurrent historical resonance, especially for "transcendental" and "proto-transcendental arguments".

The last quasi-historical reconstruction, for example (cf. 22 above), turns out to have close affinities with

27 the *Berry paradox* ("Let us define n to be the least integer that cannot be "defined" in less than forty syllables in the English language ...."") (which "defines" "the least ..." in less than forty syllables),

And more indirectly, to

28 the ancient Greek *sorites*, or "heaper" argument, mentioned earlier ("One grain does not make a heap, and addition of a single grain never makes a non-heap a heap. There are therefore no heaps", in which the relevant assertion is that "x is the first grain such that it and its predecessors form a 'heap.'").

Invocations of the *sorites* go back a very long way, and the "affinities" claimed for the Berry paradox and *sorites* take the form of alternative proofs of first- and second-order incompleteness results for "intelligible" theories of the sort mentioned earlier.

Historically, it may also be worth observing that

29 the *sorites* also provided the academic skeptic Carneades with a welcome eristic generator of skeptical *epoche* (suspension of judgment),

and that

30 the *sorites* also provides a straightforward framework for formulations of "Hume's problem", and as special instances, for self-referential queries of his preferred intentional semantics of "Custom or Habit":
31 "*Une fois n'est pas coutume*" ("One time is not custom").

"Two times are also not custom".
"If *n* times are not custom, neither are n-plus-*one*".
"Therefore, ..."

(When confronted with the original *sorites*, the stoic Chrysippus is said to have abruptly called off his lecture, and Carneades later congratulated him for his shotgun conversion to skeptical *epoche*, or suspension of judgment .... "*Wovon man nicht sprechen kann, davon soll man schweigen?*"/"whereof one cannot speak, thereof shall one be silent.")

The second reconstruction, in any case (in 22 above), provides a rather clear formal miniature of the criterion-problem, outlined above. In the sequel I will gloss this problem as a generator of recurrent patterns of interpretative plurality and "metatheoretic ascent".

The third fixed point (in 23) reconstructs the ancient *aporia* of "skeptical doubt of skeptical doubt", often considered (wrongly, I believe) a decisive confutation of such doubt. In later sections, I will consider it in some detail as a Gödelian reformulation of what might be called the *pyrrhonist's paradox* ("I doubt [this sentence]").

The fourth fixed point (in 24) raises a number of potential methodological questions about Kant's *Transzendentale Analytik*, as well as other forms of "transcendental realism" (whatever may be the labels attached to such positions).

The existence of sentences "immanently" equivalent to their own "transcendence", suggests, for example, that

32 "*Bereiche der Erfahrung*" ("ranges of experience") may not be able to interpret (much less "define") unique preconditions for their own Kantian immanence; and

33 neither the *Bereiche* nor their putative *Kriterien* are uniform or univocal, much less well-defined or *bestimmbar*, in any of the ways Kant apparently sought to "deduce".

(More detailed efforts to support these claims appear in the sections on "transcendental" and "proto-transcendental" arguments in later chapters.)

The fifth ("Wittgensteinian") fixed point, finally (25 in the list above), recapitulates in many ways—and in terms less original, therefore, than many analytic metaphysicians seem to think—underlying eristic claims and counterclaims about definability and expressibility.

The existence, for example, of assertions diagonally equivalent to [their own] ineffability/unaccountability/Wittgensteinian "*Unsinnigkeit*" might call into question his claim that

34 a sufficent criterion for evaluation (and interrogation) of "*what we do*" is what "*we*" *do* (as if "we" could individuate who "we" are).

In a later section, I will argue

35 that this claim offers another self-referential redundancy-theoretic semantics, which can be interpreted

36 as a "collapsed" diagonal assertion which relies for its rhetorical force on a tacit (Kant might have said "subreptive") assimilation of the two "*we*'s."

7 Interpretative Plurality and Metatheoretic Ascent

Those who would object to introduction of the last section's "technical" arguments often claim that they impose disabling requirements on their ranges of application, and are, therefore, irrelevant to the wider purposes of philosophical debate.

They are indeed somewhat "technical", but I will argue that the sense in which they are such would apply to any effort to consider abstract ratiocinative procedures that are indifferent to their ranges of application.

More precisely, I will argue that there are four basic reasons which undermine claims that metalogic impoverishes philosophical "discourse" (These are also recapitulated at greater length in Chapter 7).

1 The first is that no one has ever proposed a consequential medium for rational communication that cannot be syntactically interpreted in relatively simple first-order theories.
2 The second is that no one has ever proposed a consequential medium for rational communication that cannot be semantically interpreted in relatively modest first-order metatheories.
3 The third is that first-order theories and metatheories cited in 1, 2 and the last section impose minimal constraints on the consequence-relations to which they give rise.
4 And the fourth is that the more "minimal" the constraints imposed on a theory's "consequence-relations" (proof-schemes), the more widely applicable are its consequences, among them the results of the last section.

In short, one cannot simply "transcend" the scope of the theories considered in the last section, or impugn their relevance by appeal to some sort of higher-order *fiat*. For the "*fiat*" would have to beg its own "intended" interpretation.

Another common objection to the sorts of arguments I have offered is that the last section's results apply "only" to formal theories, and not to "real entities" such as we conceive ourselves to be.

On the interpretation I wish to offer, however,

5 the results of the last section will apply exactly to those entities that can "conceive" themselves in consequential ways to be *anything*.

More precisely, I will argue
6   that they apply to any systems—whatever we think their physical bases may be—that evolve dynamically in time, respect certain rudimentary inferential rules, and "know" how to go on—can "anticipate the future" in inductive ways ("know", in the sense that they are able to respond to such axiomatic and inferential expectations, in simple cases of the sort they have recognised in the past.).

Such systems—let me call them "reflective"—may or may not be able to use these capacities to continue to encode or embody "internal" representations of [themselves]. Even if they do, moreover, they may or may not continue to use these capacities to reason about these "empirical selves" "correctly"—that is, their future actions may or may not continue to follow the last paragraph's axiomatic and inferential schemata.

At this point, one should perhaps acknowledge that
7   the track-record of what these "systems" have recognised in the past, and the genuineness of the "knowledge" and "continuation" just mentioned, are tacitly and processively metatheoretic, in the sense that they will have to be judged by "our" lights or those of "our" *successors*, whoever "we" and "they" may be (or become).

I will construe this acknowledgment of an apparent need for metatheoretic adjudication of object-theoretic coherence as acknowledgment of
8   syntactic sources or generators of metatheoretic ascent.

In historical terms, one might also interpret its "stages"
9   as syntactic counterparts of the "problem of the criterion of truth";
10 as a clarification of metalogically ambiguous "existence"-assertions (in the "ontological argument", for example);
11 as a formal miniature of the problem of temporal consistency of self-representation Hume brooded about in the Appendix to his *Treatise*;
12 as an analogue of Lewis Carroll's conundrum about "definitions" of material implication in "Achilles and the Tortoise"; and finally
13 as an instance of Ludwig Wittgenstein's notorious "*Problem des Regelfolgens*" ("rule-following").

On the account offered in the last section, in any case, "reflective systems" of the sort mentioned earlier which do reason "correctly" from time to time about [their own] internal representations should

14 rationally conclude that they are plurally interpretable, to the extent they think they are interpretable at all; and
15 accept therefore alternative interpretations of (or "support counterfactuals about") [their own] syntactical representations.

The results of the previous section therefore suggest that
16 recourse to metatheoretic sources will in general be needed to designate or individuate such interpretations as "intended" "working hypotheses" and to that extent (provisionally) "true" (to be "believed").

I will characterise this second recourse to metatheoretic adjudication as
17 a second, semantic source or generator of metatheoretic ascent.

One might, of course, wish to reject such relational glosses of "truth", and posit to this end
18 the 'existence of "neoplatonic", "Spinozan" or "noumenal" structures of "reality"' which are not reflective—which do not encode, define or represent [their own] consequence-relations.

Such postulations presumably have to be expressed in some communicable context, and therefore (I would claim) in some tacitly or explicitly acknowledged metatheory. One could then query this choice of (metatheoretic) context as
19 a third, metasyntactic source or generator of metatheoretic ascent.

Claims to individuate particular "true" or "intended" structures in the given metatheoretic context—make the existence-claims in 16 "constructive", in some sense—would give rise to further questions.

The results of the last section show, for example, that attempts to bar relevant forms of "infinite regress" (or "non-well-foundedness") in "the" structure of inquiry are not metatheoretically expressible. It would therefore seem legitimate to question whether such individuation is *ad hoc*, arbitrary or notional.

Such metatheoretic questions require metametatheoretic answers, and I will therefore characterise this complex of problems as a
20 a fourth, metasemantic source or generator of metatheoretic ascent.

For an "ultimate" "noumenal" theory T which has finally sorted out provisional choices of all its intermediate metatheories would not be able to

21 define "intelligibly" its own proof-theoretic completeness,
22 adjudicate "existence"-claims about [itself] in [its own] language, or
23 identify [its own] consequence-relation—its proof-theoretic "design", so to speak. ("*Es ist das Mystische*" ...)

One might be able to explain away the "malign-genius"-aspects of such misalignments in yet another metatheory ("metametatheoretically"). But "noumenality" of T would have turned out to be a "transcendental illusion", and the whole process would restart.

To me at least, it would seem reasonable to interpret this predicament as a "Pascalian" or "Plotinian" dilemma in which the "noumenon" "*n'en sait rien*" about [itself], as a
24 fifth, metasemantic source or generator of metatheoretic ascent.

Parenthetically, the arguments just developed would also seem to undermine a natural metalogical counterpart of John Stuart Mill's conjecture that one might be able to provide an "inductive proof of induction".

In any case, informal prototypes and dialectical antecedents of these formal and semiformal instruments have suggested to a small minority of working philosophers for more than two centuries that
25 everything significant may be plurally interpretable;

in the slightly modified sense that
26 "everything significant" may be plurally interpretable.

If so, and if
27 interpretation is also essential to decision, discernment and individuation,

it would seem to follow that
28 rational conflicts between alternative temporally evolving interpretations are ineluctable and never-ending,

and from *this*, finally, that
29 efforts to postulate "ultimate" interpretations which elude the essential incompleteness of interpretation are infeasible at best and incoherent at worst.

As the reader may have observed, the last three conclusions—27 through 29—parallel rather closely "absurd" doctrines of skeptical *epoche*.

Early intimations of the essential *plurality of interpretation* may perhaps be found in the presocratic Xenophanes' remark that a god of the fish would be a great fish (cf. DK 21 B 14-16), and related remarks attributed to Heraclitus, Empedocles, Leucippus and Democritus, among others. Ineluctability of such plurality would also be integral to a nontrivial interpretation of "Protagorean relativism", ostensibly "confuted" in Plato's *Theaetetus*.

It may also have been integral to the activities of Socrates himself, who seems to have woven long thought-experimental scenari—alternative interpretations, in effect—to confute *bien-pensants*' "certainty" that "standard" or "conventional" interpretations they "intended" were "right", whether they could articulate them or not.

The *cave parable*, finally—one of the more powerful images of "paradigm shift" ever presented in the history of western literature—is susceptible to deeply skeptical implications, whatever Plato's dogmatic or hieratic aims for its immediate use may have been. (All one needed to do to see this was observe once again that the parable itself is only a *parable*—and therefore a (mere) "likeness" or *eikasia*.)

Aristotle, of course, was no skeptic. But his so-called *tritos-anthropos* argument may be regarded as a forerunner of later skeptical refinements, as well as a prototype of the metalogical pattern I have called "metatheoretic ascent". Sharper anticipations of this pattern appeared in the elenctic "*tropes*" academic and pyrrhonist skeptics developed to "deconstruct" stoic views of "nature" (*physis*) and "the true" (*to alethes*), themselves early forms of the ideal of "design".

The stoic doctrine of a regulative "nature", for example—anticipated in certain respects by Heraclitus and recapitulated in very different ways by Spinoza and Kant—effectively embraced a dilemmatic alternative that
30 "everything is interpretable, except that which interprets 'everything'"

(a doctrine one might compare to Leibniz' wry rejoinder to Locke that
31 "everything in the mind was first in the senses except the mind itself").

If the arguments of the last two sections are tenable, they suggest that all one can (seem to) do is find branching paths through the *scriptoria* of metaphysics, and hope they lead to larger rooms, whose walls may be less oppressive to us. The principal aim of this work will be to add a few metalogical threads to the strings that seem to mark such branching paths.

8 Alternative Views of "Secular Design"

There are at least two obvious things one might try to do to neutralise or block the patterns of indefinitely extended metatheoretic ascent that arise from the ancient skeptics' "problem of the criterion", sketched in the last two sections, and its metalogical counterparts in the last section and later chapters.

1 One can try to "cap" the ascent at some stage (usually the first or second).
2 Or one can try to "coopt" it, in ways which implicitly or explicitly beg the unicity and linearity of order-structures to which its semantic hierarchies give rise.

In this section, I will argue

3 that both these dialectical responses give rise to (related but discernible) notions of (secular) "design";
4 that these notions may also be characterised as nominally "secular" counterparts of the "hieratic" views outlined in Section 2; and
5 that such "hieratic" notions of "design" are subject to the formal and informal analyses of metatheoretic ascent sketched above in 7.8–7.24.

In common English usage, the word "secular" has (at least) two senses.

The first derives from the familiar contrast between what is secular ("worldly"), and what is theological.

This contrast may be familiar, but it is notoriously difficult to clarify, for attempts to do so eventuate rather quickly in "reflexive" attempts to provide "rational" reconstructions of what is "rational", and "experiential" criteria for what is "experiential" (forerunners, perhaps, of Kurt Gödel's self-referential "coding" of certain theories T in T, a paradigm case of the dilemmas which drive metatheoretical ascent).

Fortunately, a few common premises about "experience" and "rationality" seem to cross a number of "world" views (itself a good skeptical *consensus gentium* argument). In considering structures and interrelations between them of the sort mentioned in 2 above and in Section 7, for example, it seems rather minimally reasonable (if not "rational") to conjecture

6 that some common attributes of these structures and interrelations may be regulative of efforts at discursive analysis;
7 that feasibly simple aspects of these common attributes may also be accessible to such discursive efforts; and
8 that these regulative notions make no appeals to Spinozan monism or "analytic" but conceptually vacuous appeals to "physical reality".

Here and in the sequel, I intend to defend the view that that feasibly simple hierarchies of recursively axiomatisable theories

9   suffice for the conduct of rational scientific inquiry;
10 are common to all efforts at discursive analysis; and
11 are *all* that *is* common to such analysis and inquiry—not of all there "is", perhaps ("it is the mystical" once again), but all that "we" and other would-be "reasonable beings" may intelligibly be able to talk about ("whereof one cannot speak," once again, "thereof shall one be silent").

This minimal view amounts, or is at least compatible with, a (metalogically inflected) "secularisation" of the zetetic ideal sketched in prior sections. It may also amount to a kind of skeptical as well as semeiotic "idealism" about "existence", "design", "experience" and "the" scope of scientific and philosophical inquiry, a point I will take up at length in later sections.

The second, "stronger" and more "hieratic" sense of "secularity" derives from other traditional senses of the Latin word "*saeculum*" (as "world"). Upholders of various forms of this view might take it as given, for example,

12 that there "exists", in some sense, an elusively structured but essentially unique physical or physical-noetic "world";
13 that this "world" provides an "ultimate" (or at least epistemically indefeasible) background-semantics for scientific as well as philosophical speculation;

and, finally,

14 that it is "obvious", therefore, or at least legitimate and uncontroversial, to talk about "truth" and "falsity" in unitary senses dictated by this otherwise unspecified semantic structure.

It is rare to credit the metalogical insights into the ambiguous stratification of "existence" which the above seem to entail, and routine to assert that notions of "truth" and "falsity" are "universal"—or at least that they transcend "naive" or overly "restrictive" model-theoretic notions of "truth-in-a given-structure". It is less routine, I believe, to acknowledge the implications of such insights for even more more naively "intended" interpretations of such transcendence itself.

Be that as it may, the assertions in 12–14 above tend to eventuate in a second, stronger, and (in my view) more "dogmatic" sense of "secularity", which I will characterise as a "secular" counterpart of the so-called argument from (or to) "design"—the most durable and "natural" (in several senses) of the "hieratic" ideals introduced in prior sections.

As a default-assumption of all but the most latitudinarian forms of (what has come to be called, in certain circles) *scientific realism*, the argument to design has provided a primary source of heuristic motivation for many working scientists. It also guided (I believe) Kant's formulation of the transcendental argument(s) of the first *Kritik*. Nevertheless, I will argue, it provides the very model of a *transcendental illusion*.

Indeed, a possible witness to the durability—and perhaps ineluctability—of this second, "stronger" ideal—both traits Kant attributed to his transcendental illusions—is the sheer variety of attempts to realise or "justify" it that have emerged in the history of western philosophy. Since they all constrain patterns of metatheoretic ascent in one fashion or another, it also seems reasonable to provide thumbnail sketches of these attempts that give some attention to a dichotomy between "capping" and "cooptation".

Realisations of attempts to "cap" the ascents, for example, might be seen in (e.g.)

15 Spinoza's conflation of stoic and neoplatonic ideals in *"deus sive natura"*;
16 Hume's "truth"-conferring "Custom or Habit" in the *Treatise* and two *Enquiries*;
17 Kant's allegedly determinate and determinable *"Bereich der Erfahrung"* ("realm of experience");
18 "early-Wittgenstein"'s elusively liminal *"Welt"*, and "late-Wittgenstein"'s elusively equivocal notions of "language game(s)" versus "our language game" (whose "world", why "games", and who are "we"?),

and a variety of other "hieratic" and quasi-hieratic assumptions which postulate a unique limit to physical and conceptual "universes".

Attempts to refine as well as *"coopt"* them, in turn, might be found (e.g.) in

19 the *regulative Ideale* of Kant's *teleologischer Urteilskraft (teleological judgment)*;
20 "Peirce's 'linear' as well as 'ultimately' convergent "limit(s) of inquiry";
21 Cassirer's more open-ended and broadly regulative *"symbolische Formen"*;
22 Wittgenstein's indefinitely defeasible but somehow definitive *Sprachspiele*;
23 Niels Bohr's elusively liminal *"komplimentaritaet"*; and
24 Thomas Kuhn's ambiguous "paradigms",

as well as a variety of other compromises between "coherentist" views of experience and "empirical realist" premises about its unicity and uniformity.

Other historical and quasi-historical allusions to thought-experimental "design" will appear from time to time in the rest of this introduction, and more systematically (or at least chronologically) in later chapters.

9 "Transcendental" and "Proto-transcendental" Arguments

Among other things, I have tried so far to convince a prospective reader
1. that "zetetic" arguments of the sort considered in Section 3 developed in "dialectical" interaction with their "hieratic" counterparts;
2. that significant aspects of "logic" as well as "philosophy" evolved in tandem from recurrent efforts to find resolutions for these disputes;

and in particular,
3. that twentieth-century metalogical inquiries consciously or unconsciously refined as well as clarified underlying aspects of this dialectic; and finally
4. that studies of these inquiries and interrelations with their historical antecedents may therefore provide significant templates for analyses of these antecedents as well as their formal miniatures.

In this section, I will explore a somewhat different aspect of this interaction, and sketch a brief preliminary study of early "transcendental" attempts by Plato, the metaphysicians of the Stoa, Augustine, Anselm, Descartes, Spinoza, Leibniz, Berkeley, Hume and Kant (among others) to "cap" and/or "coopt" the patterns of metatheoretic ascent sketched in prior sections (More carefully documented versions of these explorations appear in Chapters 2–8).

I have already argued above
5. that certain elenctic "turns" or "modes" (*tropoi*) of ancient skepticism—among them "the problem of the criterion"—anticipated patterns of "metatheoretic ascent"; and
6. that self-referential applications of some of these "modes" also gave rise rather naturally to semantic "fixed-point arguments" (such as skeptical doubt of [skeptical doubt]).

In an evolving *dialektike* between skeptical ("zetetic") and dogmatic ("hieratic") views, most "dogmatic" opponents of skepticism routinely denounced 5's indefinitely extendible hierarchies as "regressive", and 6's "paradoxical" instances of [self]-reference as "viciously circular", or simply "absurd".

Others, however, tried to turn the dialectical tables, and make "constructive" uses of such skeptical hierarchies and fixed points, in the form of "transcendental" or "proto-transcendental" arguments.

In this study, I will interpret
7   "proto-transcendental arguments" as dialectical "design-proofs"—that is, as attempts to reconstruct "experience" as a unique limit or fixed point of patterns of skeptical inquiry, of the sort just sketched.

I will call such attempts
8   "transcendental" when they purport to "prove" or "deduce" the fixity, unicity and invariance of experiential "design" (in some non-question-begging sense) from "within"; or equivalently, when they purport (in Kantian jargon) to set down conceptually immanent characterisations of "the" structure of immanence.

Almost alone before his time or since, G. W. Leibniz seems to have tacitly or explicitly acknowledged the deeply problematic or aporetic nature of such undertakings, especially in their fully "transcendental" form. In particular, he seemed to realise that each such project has at least three successive stages.

At the first (almost always overlooked),
9   one should perhaps try to establish whether such an undertaking is even intelligible or expressible.

(Descartes' *"methodus"*, Leibniz' *characteristica magna*, Berkeley's *"signs"*, Kant's *"Architektonik"* and the "coding"-schemes of Skolem and Gödel have offered partial attempts to respond to this question.)

At the second,
10 one might assume or postulate that the project is in some informal sense intelligible, and try then to "prove" in some venue that "experience" is also consistent (or interpretable).

(Aristotle and Descartes offered "prototranscendental" attempts to "derive" such a conclusion; Leibniz was content to postulate it as a "first great principle".)

At the third, finally,
11 one might assume (or postulate) the intelligibility and interpretability of experience, and try to "derive" from these premises its "determinate" or uniquely individuated interpretability, or in other words, its "truth".

(Leibniz postulated the "truth" of experience in roughly this sense, in his "second great principle ... of sufficient reason". Kant more ambitiously sought to "deduce" it in the *Transzendentale Analytik*).

In the course of this introduction, I will argue at some length that *none* of these putative "proto-transcendental arguments" can "work".

More precisely, I will argue that

12 each consistent "proto-transcendental" attempt to write out "proofs" of the sort demanded in 9–11 is itself theory-relative; and
13 each "transcendental" attempt to write out such a proof is either *inconsistent*, or equivalent to [*its own*] "transcendence".

Prompted by such considerations, I will return to the skeptical modes with which I began above. More precisely, I will attempt to coopt them, and assimilate

14 "experience" to prior sections' indefinitely ramified, "time"-evolving hierarchies of metatheoretic ascent; and
15 "things in themselves" to whatever ineffable entities may not be expressible in such indefinitely extended ramified hierarchies.

These assimilations of course, are plurally interpretable in their turn, for the patterns and ramifications in them are also sensitive to background metatheoretic assumptions. The reader may judge whether such an acknowledgment is

16 a "transcendent" counterpart of skeptical doubt of [skeptical doubt], or
17 an "experiential" acknowledgment of the processive "openness" and indeterminacy of "experience".

Or both.

10 "Proto-transcendental" Arguments and "Hieratic" "Design"

Common to most "physicalisations" of attempts to "close" or "complete" hierarchies of metatheoretic ascent are tacitly hieratic desires for unique and "universal" *design*(s) ("*kosmoi*") which comprehend "everything" but are comprehensible to us as well. Such desires are essentially theological, not philosophical.

Ancient skeptics had already recapitulated and elaborated presocratic observations to the effect that a god of the fish (say) might be a great fish, mentioned earlier (cf. Xenophanes' remarks at DK 21 B 14–16, for example, and assorted remarks of Heraclitus, Empedocles, Leucippus and Democritus). Less quaintly, such skeptics might have queried whether monotheist counterparts of the priests and scientists in Plato's cave might not interpret "god" as a *great prisoner*? ... Or the flickering shadows on its walls as models of "the" design of "the" universe?

Be that as it may, desires to find simple experimental explanations for complex thought-experimental patterns are deeply ingrained not only in hieratic interpretations of Aristotle's *orexis tou eidenai* and concomitant theological ideals, but also in early "physicalisations" of such ideals—in "linguistic" glosses of epicurean atomism and normative aspects of stoic *"physis"*, for example, and in the work of Roger Bacon, Ockham and Abelard.

They also resonated in Galilean remarks about the "book of nature" (actually "book of [natural] philosophy" in the original), and Cartesian aspirations to derive all of metaphysics and a conclusive axiomatisation of "all" physical science from early-mathematical notions of precalculus and analytical geometry. If one substitutes probability-theory, functional analysis and manifold-theory (...) for "early modern" notions of mathematical sophistication, they live on still.

The underlying methodological *Ansätze* of the last section's "transcendental" arguments proposed to make elenctic use of self-referential arguments to secure one or more fixed points of "zetesis". Typically, such efforts undertook (in language introduced earlier) to cap, coopt and/or normalise "regressive" patterns of metatheoretic ascent, and dispel in this fashion the *Skandal der Philosophie* Kant decried in the opening pages of his *Prolegomena*.

In response to ancient skeptics' embrace of relativisation of [relativism] and skeptical doubt of [skeptical doubt], what forms might antiskeptical uses of skeptical diagonalisation-arguments and fixed-point-constructions take?

A first, rather inchoate anticipation of such an argument—at least the earliest I know of—appeared in the *Theatetus*' brief and inconclusive
1   attempts to refute Protagorean relativism (cf., e.g., 171a–c, 182d, 183a–c);

and a second in the dialogue's more extended
2   attempts to explain what it is to "explain" (provide, that is, a *"logos"*, often translated in such contexts as "account", "rationale", "explanation" and/or "justification") (cf., e.g., 203e, 206d, 207c–d, 208c–e, and 210a).

Modulations of these programmatic aims also appeared in other, rather different guises in the *Physics*, *Metaphysics* and *Nicomachean Ethics* of Aristotle.

Aristotle briefly recapitulated Plato's "refutation" of "Protagorean relativism", for example (cf. *Metaphysics* IV, 1007b19ff), but he also seemed to be aware of interrelations between such arguments and *"tritos anthropos"*-arguments. I will construe the latter in Chapter 2 as early templates for Section 7's patterns of "metatheoretic ascent".

He also observed—even more aptly, I think—that relational notions of "truth" "seemed" to "talk about the indeterminate" (*to aoriston oun eoikasi legein*, 1007b 28ff), and remarked in the following lines that they undermined commitments to "apodeictic" dichotomies of "form" and "matter", "attribute" and "substance", and (what is) "actual" (*entelecheia*) as opposed to (what is "merely") "potential" (*dynamei*).

Subtler reponses to skeptical doubt of [skeptical doubt] appeared in the works of the stoics: the first in a prototypical cogito-argument, recorded several centuries later by the christian apologist Clement of Alexandria (SVF II 121); and the second in a fairly elaborate account of the rudiments of propositional logic and "comprehension" (*katalepsis*) and its modes of communication.

Augustine essentially recapitulated the first (proto-"cogito") argument in one of several "refutations" of skepticism he seems to have committed to paper in the late fourth century—arguments which his ecclesiastical eminence and respectability made more or less continuously accessible to scholars throughout the medieval and early modern periods.

Toward the end of the eleventh century, one of these scholars, Anselm of Bec, saw that one might be able to recalibrate (hitherto elenctic) skeptical arguments in "constructive" terms, as "ontological" warrants for the "existence" of a "fixed point" of a(n implicitly "metatheoretic") hierarchy he associated with "conception".

More precisely, Anselm argued

3   that we "must" be able to conceive "the" limit of "all" conception ("*id quo nihil maius concipi possit*");
4   that this limit "must" exist (for otherwise it would not be "the" limit of "all" conception); and finally,
5   that we may legitimately assimilate this ("most perfect") limit to that "which all call god" (to borrow a phrase from one of Anselm's successors).

On my account, this was the first "*transcendental argument*": that is,

6   a concise and systematic attempt to employ regulative zetetic arguments for "constitutive" hieratic ends, and do so entirely from "within" (in this case, the appropriate adverbial phrase might be "here below").

(Perhaps *this* is why the young Bertrand Russell is supposed to have shouted "Great god in boots! The ontological argument is sound!")

Five and a half centuries later, Descartes assimilated aspects of these arguments (without attribution) into the underlying structure of the *Meditationes de Prima Philosophia*.

More precisely, he

7   recapitulated Platonic, Aristotelian, stoic and Augustinian "refutations of skepticism" in his formulation of the "cogito", and

8   Aristotle's and Anselm's fixed-point arguments in his characterisations of the cogito as an Archimedean *point d'appui*.

He also
9   emulated Aristotle's identification of an archetype of this "theoretical" fixed-point with "god", and the physical universe with its intentional objects, and
10 attempted to secure the putative consistency, unicity and intelligibility of this fixed-point construction and its putative archetype in two partially complementary ways.

The first, in Meditation III,
11 recapitulated Aristotle's, Anselm's and Aquinas' attempts to "coopt" metatheoretic ascent as a hierarchy of theoretic conception which would approach the fixed point as a putatively unique "formal" limit of merely "objective" simulacra.

The second, reviewed in Meditations IV and V,
12 attempted to revive Aristotelian, Anselmian and Aquinian "*caps*" on metatheoretic ascent, in the form of a putative fixed-point "proof" *of the* consistency and semantic unicity of a limit of "all" inquiry (the "god" of the "Cartesian circle", for example).

In a later chapter, I will interpret 11 and 12 as
13 hieratic antecedents of a failed project in the history of metalogic, known as *Hilbert's Programm*, to "prove" mathematically the completeness as well as consistency of mathematical "proof" (a subtly "Cartesian" attempt to use formal methods to "prove" that mathematically "clear" and "distinct" notions "must" be "true"),

and dismiss historical prototypes of
14 Gödelian refutations of this *Programm*—which was initially focused on Russell's "type-theory", a "conceptual hierarchy" which might be regarded as a more "rigorous" metamathematical counterpart of Cartesian alternations of "formal" and "objective" (mathematical) knowledge and "reality".

Descartes' successor (and critic) Baruch Spinoza
15 employed the early modern redaction of ancient arguments sketched above as instruments to outline a nominally secular counterpart of stoic "*physis*" (in the "hieratic" senses of "secular" mentioned earlier); and

16   saw that the universality he attributed to a unique "secular" fixed point of inquiry (his *"deus sive natura"*) would be unintelligible, incomprehensible, and inaccessible to "practical" as well as "speculative" canvass from "within".

Though Spinoza essentially begged this "fixed point"'s *unicity*, and his "austere" conflation of stoicism and neoplatonism alienated him from the generations of ethicians and scientists that immediately preceded and succeeded him, I will argue in a later chapter that

17   he saw—with Plotinus and the christian fideist Pascal—that such a *"univers n'en saurait rien"*;

and that

18   "almost all" (metatheoretically) postulated "limits of inquiry" will be unintelligible (in a sense which can be made reasonably precise) *"from within"*.

11 "Coherentist Idealism" in the Hieratic Ideals of Leibniz and Berkeley

Leibniz

The word *"physis"* (like its counterpart *"natura"* in Latin) carried what we might now call "organic" associations in ancient Greek. More tellingly, perhaps, the *physis* of the ancient stoics (whose normative and alethic claims elicited the academic skeptics' "problem of the criterion") also offered an early and allegedly comprehensive version of the last section's "proto-transcendental" "design" arguments. Commentators have correctly interpreted the *tonos* the stoics attributed to this *"physis"* as a kind of proto-physical "field strength". But it is also quite transparent that the notion had strong *ethical* implications (in injunctions to "live in accordance with nature", for example), and Spinoza's *"Ethica"*, as I suggested above, can be interpreted in part as a rigorously impersonalised reading of a "fate"-driven normative ideal.

Similar speculative and normative claims could be made about the "corpuscularian" doctrines of *atomoi*-in-motion found in epicurean materialism, and only the *ad hoc* notion of a *clinamen* ("swerve") provided it with a conceptual alternative to stoic doctrines of "fate".

Almost two millenia later, the early modern heritors Descartes, Galileo, Newton and Leibniz tried to mathematise assorted fusions and distributions of these "field"- and "particle" theories, with varying degrees of success. The implicit or explicit "fatedness" they proposed for their reconstructions even reappeared, in the (begged) mathematical determinism of partial differential equations, driven by (allegedly) well-defined initial and boundary conditions.

(As most readers will be aware, early in his philosophical career the young Gottfried Wilhelm Leibniz sought out Spinoza in the Netherlands. The commentators and biographers record that Leibniz' older colleague received him politely, and that the two engaged in a conversation whose details are lost.)

Leibniz also accepted the "internal" fatalism the newly devised dynamical determinism seemed to impose "within" a given world. But he also struggled to relativise "absolute" notions of fate of the sort Spinoza had accepted, and broaden the regulative force of his metaphysical and metalogical constructions to include "ethical" decisions in humanly (and humanely) intelligible terms.

In this context, I find it interesting to observe that Leibniz—the "logician" and believer in a *characteristica magna*, did not—in his mature work—attempt to formulate "transcendental" "proofs" of the "existence", unicity and "internal" determinism of certain "metaphysically necessary" ideals and dogmas. He offered only a conditional endorsement of the ontological argument, for example, and I know of no passage in which he endorsed the alleged probative power of the Cartesian "cogito". This was a deliberate choice which his postulations of "divine justice" and "principles" of secular coherence make easy to overlook.

The reason for this abstinence, I believe, in a "proto-transcendental" metaphysician *par excellence*, was that deep if inchoate metalogical insights—about which one can speak in his case with little fear of anachronism—convinced him

1   that "transcendental" efforts to dictate "the" epistemic and normative structure of "all" possible experience tacitly begged appeals to metatheoretic hypotheses or postulates (he called them *"principes"*) which defined and localised their "intended" ranges of application;
2   that efforts to define ontological and normative parameters of such a structure "from within" ("object theoretically" in metalogical terms) would therefore be intrinsically inadequate and incomplete, however worthy and ethically as well as heuristically desirable they might be (an anti-"empiricist" view he shared, in effect, with Spinoza);

but also, finally,

3   that there might be infinitely many potential completions of any such zetetic efforts, and infinitely many internally coherent structures ("*mondes*"), therefore, of "possible experience", discernible *meta*theoretically—if at all—by a demiurge-like, artisanal entity he called "god".

In short, he realised that one could only postulate some sort of elusive alignment or interpretability of "our" normative criteria, in "framework(s)" of such metatheoretic *demiourgoi*—"benign geniuses", as it were—which might normalise those criteria, and carry out problematic triages of potential worlds.

Since he was also emphatically not the *niais* Voltaire caricatured, but a kind and deeply irenic man, he felt as well as saw a need to hope that such alignments and acts of triage might be tenuously "beneficial".

He hoped, therefore—but did not, in the end, propose to "prove"—that the insights he had helped pioneer ("simplest-path" methods, for example, called "Lagrangian" methods by later physicists) might provide glimpses—but not "proofs"—of such artisanal "designs". Awareness that such "proofs" were wanting might also provide yet another partial explanation for the absence of a grand "critical synthesis" of Leibniz' zetetic efforts at the end of his untiring philosophical career.

Leibniz' own "principles" also made the *phaenomena bene fundata* he strove to characterise and understand *intentional objects* of a sort of "god" (the "demiurge" mentioned above) whose metatheoretic "existence" he had postulated again and again (but never "proved"). On my account, this aspect of his analyses led him to formulate his evolving "system" as a kind of *coherentist idealism*. It was in this sense no accident, I believe, that he was interested in Socrates' passing interest in the *nous* of Anaxagoras, and that he commented with cautious respect toward the very end of his life and philosophical career (those being more or less coextensive) about the copy that came into his hands of George Berkeley's newly published *Principles of Human Knowledge*.

Berkeley

Many historical commentators have contrasted Berkeley's "idealism" (a term he himself never used) with Malebranche's "occasionalism". In the sequel I will assimilate, along rather different lines,

4   Leibniz' conjecture that a demiurge-like "god" conferred forms of structural coherence and uniformity on "confused" monads' perceptions of a given "world"'s *phaenomena bene fundata* to

5   Berkeley's conjecture that an "Author of Nature" confers forms of structural coherence and uniformity on lesser "spirits'" otherwise-disparate and uncoordinated "ideas" or intentional objects.

I will also argue that the interrelations Leibniz and Berkeley attributed to this "divine" artisanal orchestration of particular minds' "intentionality"
6   "capped" and "coopted" Section 7's patterns of metatheoretic ascent, and
7   eventuated in the first fully-developed *coherentist* interpretations of early modern physics.

My interpretation of the "empiricist" Berkeley as a "proto-transcendental" theorist of hieratic "design" may seem surprising here, but I would argue that he was one of western philosophy's more striking examples of a non-skeptical zetetic animated by hieratic ideals, and one of his principal declared aims was to ensure that no one should be obliged to "sit down a forlorn skeptic".

Like Leibniz, he also had a natural gift for dialogical argumentation, and he was an acute "fideist" critic of the presuppositions of his mathematical and philosophical predecessors. He also understood, I think, that "design"-arguments are elaborations of articles of internal faith, and that certain "foundational" physical arguments do not "found" themselves.

Unlike Leibniz, he was endowed with a religious faith that permitted him to "prescind" the metalogical penumbra of Leibniz' alternative "worlds", and presuppose rather than postulate a kind of congruence between "internal" structures of experience and "external" metatheoretic events.

Berkeley's "coherentism", in short, and his appeals at several points in the *Principles* and *Dialogues* to arguments-to-design, strongly suggest that his "god" was a "necessary-precondition-for-the-possibility" of something. I will characterise the underlying meta*physical* structure of this night-watchman-god (if I may borrow a phrase) in metalogical terms as
8   a—dogmatically "capped" or delimited—two-level type-theory, in which
9   finite, second-order, "active" "minds"—indefinitely many of them—intend and interpret first-order, "passive" "ideas"; and
10   "god"—an allegedly universal and self-interpreting higher-order Mind—aligns, reconciles, normalises and uniformises these lesser minds' semeiotic activities, *tota simul* (Boethius' characterization of "eternity"), and ensures that their intentions form a kind of directed hermeneutic system, with "god" at its (unique) limit.

Berkeley's "god", in short, might be identified with

11 an "ultimate" metatheoretic limit-structure of first-order inquiry—an inexpressible limit, moreover—like the "occasions", "substrates" *et alia* he so effectively mocks—in the sense that its "spirit" is beyond (first-order) ideation, in Berkeley's sense of "idea", though he insisted that we can have a (notoriously elusive) "notion" of "It".

The first steps in this formulation responded, in effect, to the exigencies of the "normalisation" and "uniformisation" mentioned above with a semeiotic doctrine of "signs", and in the chapter devoted to his metaphysics I will assimilate Berkeley's

12 "mere signs" to "syntactical" or "object-theoretic" terms and formulae of first-order theories; his
13 "ideas" to metatheoretically definable theoretical contexts and interpretation(s) for such terms and formulae, and definable elements and relations of such interpretations;

and, finally, his
14 "activities" of "perception" and "conception" to dynamic metatheoretic representation(s) of these (allegedly definable) contexts and interpretations.

Like Leibniz' *characteristica magna* (I will argue), Berkeley's recourse to
15 "signs" anticipated (needs for) metalogical completeness results (in the sense in which this expression appears in Henkin's first-order logical "completeness theorem"),

and
16 his "master argument"—which he hoped would provide "proto-transcendental" support for the "*esse* is *percipi*"-doctrine, along the lines sketched in above—

might be taken as
17 a prototype of twentieth-century semantic paradoxes and incompleteness arguments that generate the last section's patterns of metatheoretic ascent.

There are in fact two natural metalogical analogues for the "master-argument" (a phrase Berkeley himself never used), which may be interrelated in somewhat interesting ways. Briefly put, these analogues assimilated metatheoretic "definition(s)" to Berkeley's "perception(s)", and metatheoretic "interpretation(s)" to his "conception(s)".

In principle, Berkeley disjoined "perception" from "conception". In practice, however—as his commentators are well aware—he tended to conflate "perception" and "conception" in tacit ways, especially when such assimilations served immediate methodological and elenctic aims (in Philonous' more "defensive" responses in Dialogues Two and Three, for example); and disjoin them, often with an air of great rigor, when they did not (most conspicuously, perhaps, in the first Dialogue).

Perhaps all that one can reliably say about this shifting distinction is that Berkeley consistently considered "*perception*" a somewhat more "immediate", "extensional" and even eidetic activity of "spirits" (an activity which generates something roughly comparable to Hume's "impressions"), and "conception" a more "mediate", "intensional" and hermeneutic form of such ideation (which might give rise to something rather like Hume's "ideas").

Though he never quite said so explicitly, he may also have believed that one conceives something—and perhaps even that one has a "*notion*" of it—when one perceives a sign for it.

In any event, my gloss of the first, more "conceptual" analogue of Berkeley's master-argument applied to assertions (sentences) s may be sketched as follows. Given informal statements s of the form

18 "there are 'trees, for instance, in a park, or books existing in a closet but no one by to perceive them'", (a sentence s in the language of some metatheoretically consistent background-theory T),

one can assimilate such statements to *formal* assertions of the form
"s, but [s is inconsistent with T]"; i.e.,
"s, but [s holds in no interpretation of T]"; i.e.,

and formulate a metatheoretic Berkeleyan
19 "*master-argument de dicto*": that *every* "conceivable" assertion s in T must be provably interpretable in T. But this would be an "absurdity", for "… do not you yourself [interpret it] all the while?"

Results of Skolem, Gödel, Henkin and something called the "fixed-point lemma" (cf., e.g., (Bell and Machover, 1977), 117–122 and 191–202, or 1.1–1.11 of Boos, 1998 lead to a couple of relevant observations. The first is that
20 a formal counterpart of the equivocal query "… do not you yourself [interpret it] all the while?" cannot even be expressed in the theory T, if one interprets T as a stand-in for the equivocal "observer" "you."

The second is that if one assimilates
21 "you" to a metatheoretic extension U *of* T, in which the consistence of T *is* provable, the proof must be secured in a metametatheory V, whose consistency in turn would have to be secured by a metametametatheory W, whose consistency must be secured in ....,

and so on (a metalogical theorem which would have delighted ancient skeptics, who called such iterations the "*ekptosis eis apeiron*", or "lapse into the infinite"). Moreover, it would follow from
22 Gödel's and Henkin's completeness theorems and Gödel's incompleteness theorem for T in the metatheory U that there would exist uncountable many interpretations of all the axioms of T and the assertion that [[the theory T] *is* inconceivable in T].

A second, more "perceptual" metalogical counterpart of Berkeley's master argument may be outlined as follows.

If one considers terms t and existential sentences of the form [there exists an x such that p(x)] in T, and assimilates "s but [no t 'exists' such that p(t)]" to
23 "the predicate p is instantiated, but the t whose existence s postulates is imperceivable", "s holds, but [no t can be defined or individuated such that p(t)]".

One can also formulate an "existential" analogue of Berkeley's thought-experiment which I will call "*the master-argument de re*", namely
24 that every "tenable" existence-claim $s(x)$ must be witnessed by a T-definable term t.

For otherwise one could "know" that there exists a t such that $s(t)$ but never be able to define it (not an "absurdity", but an embarrassing predicament for an "empirical" philosopher).

Metalogical results of Berry, Gödel, Henkin and Thoralf Skolem establish that one can metatheoretically
25 define canonical extensions $Sk(T)$ of T in U, as well as "models" or interpretions **M** of S in which object-theoretic existence-claims s are provably "witnessed" by terms of $Sk(T)$. (These are known as "Skolem structures".)

Indeed, there are even metatheoretic and metametatheoretic senses in which "most" interpretations **M** of S have such "substitutional" characteristics. (cf., e. g., 7.9 of (Boos, 1994).

But without recourse to such "higher-order" metatheoretic constructions,

26 Berkeley's master-argument *de re* for T is not even expressible in the language of T (much less "true in T"). For "definability-*in*-T" is known to be indefinable in T (i.e., no predicate in T accurately identifies which of the "terms" in 25 *is* definable in T).

12 The "Second-order Idealism" of Hume's "Empiricism"

Obvious thought-experimental questions of the sort I have canvassed prompt queries about
1   the well-definition and unique interpretability of Leibniz' principles and Berkeley's "Author of Nature"; and therefore about
2   Leibniz and Berkeley's efforts to refine traditional theological "arguments from (or to) design" into more cogent forms of methodological and metaphysical coherentism.

David Hume, the ancient skeptics' declared but ambivalent early modern descendent, posed many of these questions with an informal but persuasive flourish in his posthumously published *Dialogues Concerning Natural Religion*.
    Or more accurately, Hume's more skeptical interlocutor "Philo" posed them, and Hume ostensibly dismissed them at the end of the work, in the voice of "Cleanthes", a name borrowed from the first stoic advocate of the regulative view of "*physis*" sketched above.
    This apparent ambivalence has often puzzled Hume's commentators, and I will argue in Chapter 7
3   that it is a natural concomitant of his un-"Humean" appeals to an elusive "preestablished harmony" and "final causation" he wryly attributed to "Custom" and "Habit" in the *Treatise* and first *Enquiry* (cf. [E] I 55);
4   that the begged qualities of these uniformities are relatively little noticed cognates of the explicit *impasse* or *aporia* of doubt about personal identity which Hume himself acknowledged in the Appendix to the *Treatise* ([T] 633–36); and finally
5   that the conflict between Philo and Cleanthes in the *Dialogues* recapitulated a deeper underlying conflict, between Hume's more "zetetic" impulses, and the "dogmatic" role he implicitly attributed to himself as the putative "Newton of the moral sciences."

I will also argue that the "skeptic" Hume's "strong" notions of "Custom" and "Habit" offered an epicurean "design"-argument, as well as a variant of

Berkeley's intentional response to the skeptical modes I have assimilated to metalogical hierarchies of metatheoretical ascent.

On my reading, therefore, Hume was what might be called a "second-order idealist": a reductive metaphysician, very like "that very ingenious author" (Hume's description of Berkeley at [E] I 155) who chose to "prescind", or at least reject, not

6   first-order non-immediate objects (which he seems to have regarded with a slightly uneasy sort of benign neglect), but rather
7   higher-order non-immediate relations ("necessity," "causation") *between such objects*,

and beg the unicity as well as temporal persistence of an ill-defined higher-order intentionality, which he believed "must" secure or preserve the intersubjective uniformity and constitutive "design" of "experience".

In the sense in which Berkeley called himself an "immaterialist," therefore, one might appropriately call Hume an "irrelationist": a tacitly hieratic metaphysician who replaced Berkeley's divine mediation and reconciliation of distinct spirits' "perceptions" with the secular mediation and reconciliation of "impressions" and "ideas" effected by the (allegedly) constitutive force of a social-psychological entity he called "Custom or Habit".

Programmatic denials of first-, second- (or higher-) order "I know not what"'s that emerge in hierarchies of metatheoretic ascent are likely to encounter sooner or later some variant of the problems of interpretive plurality that sustain such hierarchies. One aspect of Hume's "irrelationist" variant of these denials may be formulated as follows.

Hume argued brilliantly that we may consistently deny the existence of relations whose alleged "universality" we cannot divine or secure. But this would not "prove" that we "must" deny their consistency—unless, of course, one also denies the interpretative plurality of thought-experimental ratiocination altogether. For the existence of such relations might be a "matter of fact" rather than a "relation of ideas."

Metaphysical "principles" are usually interpreted as methodological premises which secure—openly and provisionally, or "subreptively" and "absolutely"—boundary conditions of philosophical "proof." Such "principles" may serve to hold some aspect of our *incomplétude* locally constant, and bridge lacunae in the recurrent hierarchies of interpretative inquiry.

Hume, however, formulated several rather high-metaphysical and "first-philosophical" "principles" (his own word)—remarkably many for a skeptic—which seem to me distinctively "subreptive", however plausible his fluent prose may make them sound.

Among the better-known are
8. a principle of the definability and well-foundedness of "sensation" and "conception" (often called "Hume's microscope");
9. a correlative principle of the identity-of-indiscernibles for "impressions" and their "corresponding" "simple ideas";
10. a clarity-and-distinctness principle for *consistency* rather than "truth": that whatever is conceivable or perceivable in clear and distinct terms is consistent (or "implies no contradiction");
11. a converse of 10's clarity-and-distinctness principle: that whatever is imperceivable and inconceivable in clear and distinct terms is inconsistent, and therefore impossible (or "a mere nothing", as Hume's colleague Berkeley sometimes put it).

From these cardinal tenets—"dogmas", if you like, of Hume's "empiricism" which flowed equably because they seemed to him "obvious"—he concluded, (among other things)
12. that first-order "impressions" and "corresponding" (also first-order?) "simple ideas" "exist" (tout court—the dogmatic assertion of such "existence" may in fact be the central dogma of empiricist metaphysics) as virtual atoms of perception and conception, which can be examined to "any" (another enigmatic boundary-condition) desired degree of precision under his "microscope or new species of optics" ([E] I 62).

From these premises, he also concluded (in effect)
13. that anything not uniformly expressible in terms derived, mediately or immediately, from "impressions" and "ideas" (notions of "necessary connexion" or "cause", for example) is "inconceivable" .... and therefore "impossible" .... and therefore to be "consign[ed] to the flames" (an enthymeme sometimes called "Hume's Fork").

"Examined", any ancient skeptic might ask, by what—or by whom, in what circumstances, and with the aid of what instruments? Berkeley may have been closer here to the ground-motives of "skeptical" inquiry attributed to Hume in his meditations on the noninvariance, under different perceptual resolutions, of a "mite's foot" ([Pr] I, 60–61); cf. Hume's reflections on a cognate problem at [T] 189–90.

Hume's apparently unconscious ambiguity about "the" boundary-conditions of "causation" and "experience" also seem to reflect

14 unwarranted confidence—recapitulated by twentieth-century "logical positivists"—that one can distinguish unequivocally between
(a) "evident" (first-order) experiential *phaenomena* (the pyrrhonists' "appearances;" compare his "impressions" and "corresponding simple ideas"), which Hume trusted; and
(b) "inevident" (second- or higher-order *noumena* (literally "things thought", or "thought-objects"; compare again empiricists' "ideas of reflexion"), which he did not.

They also reflected, I believe.
15 a deeper unawareness that such confidence is difficult to reconcile with recurrent equivocation about "the" metatheoretic boundaries between "factual" and "counterfactual" modalities and their implication(s).

For these boundaries can be ignored only if one denies the very patterns of metatheoretic ascent, counterfactual modality and interpretative plurality which underwrite the contingency of "experience":

> Cf. [E] I, 76: ... we may define a cause to be an object, followed by another, and where all the objects similar to the first are followed by objects similar to the second. Or in other words where, if the first object *had not been* the second *never had* existed.

Correlatively, they also begged, rather than "proved", a deferred but remarkably high-metaphysical "design" argument or presupposition:
16 that one can "ground" a first- and second-order instrumentalism about physical relations ("cause"), in an irrefragable higher-order essentialism about ill-defined mental or intentional relations ("force", "vivacity") and their alleged "moral" archetypes in his sense of "*mos*" ("custom", "habit", "human nature"), in a kind of apodictic apotheosis of these words' original Latin senses.

Hume never really explained, for example, why we should not query the metatheoretic "force" and "uniformity" of "Custom", in the same terms as the ones he invoked to question both "necessity" and "causality."

Indeed, one of the principal motives of my arguments here and in the sections which follow is a zetetic suspicion: that Hume's anacoluthic decision to hold harmless the "moral" from his judgment of *non liquet* against the physical effectively made his "custom" as dogmatically immune to criticism as Berkeley's "god."

There is evidence that Hume was aware of some of the dogmatic and hieratic aspects of the un-"empirically" uniform and integrative higher-order mental and "moral" entities he presupposed. The best known may be the well-known afterthoughts and reservations he expressed about the *Treatise*'s self-referential limit of the mental in its appendix (cf. [T] 633–636). But equally interesting, I believe, is the wryly high-metaphysical cast of his praise for this uniformity and integrality in the full text of [E] I 55, mentioned earlier:

> Here, then, is a kind of *preestablished harmony* between the course of nature and the succession of our ideas, and although the powers and forces by which the former is governed be wholly unknown to us; yet our thoughts and conceptions have still, we find, gone on in the same train with the other works of nature. Custom is that principle, by which this correspondence has been effected; so necessary to the subsistence of our species, and the regulation of our conduct, in every circumstance and occurence of human life. Had not the presence of an object instantly excited the idea of those objects, commonly conjoined with it, all our knowledge must have been limited to the narrow sphere of our memory and sense; and we should never have been able to adjust means to ends, or employ our natural powers, either to the producing of good, or avoiding of evil. Those, who delight in the discovery and contemplation of *final causes*, have here ample subject to employ their wonder and admiration. (The second emphasis was Hume's.)

As for "modality", I would also ask the reader to consider the potential metalogical and metaphysical resonance of the counterfactual subjectives which appear midway through this long passage ("Had not the presence ... we should never have been able ...."). In effect, Hume is essentially arguing in such passages that "Custom" is the "cause" of "cause".

Finally, I will argue at greater length in Chapter 7 that Hume's uses of these "principles" posited a tacit but fallacious *quantifier-shift* (an instance of something logicians sometimes call a "uniformisation-argument"): from assertions that

17 "Every "chain of argument" has a beginning (which we may denominate its "experiential" "evidence"); to
18 "There exists a single uniform beginning ("Experience," "the" source of "Evidence") for every such "chain of argument."

The first of these assertions may well be plausible, though it may simply beg a response to skeptical queries about pragmatic well-foundedness and "metatheoretic ascent".

The second, however plausible it may sometimes look in Hume's fluent prose, does not follow from it, unless one again invokes auxiliary assumptions—traditionally invoked in "hieratic" "first-cause" proofs of the existence of "god"—to provide some sort of warrant for its quantifier-shift, and ensure "the" intersubjective and structural uniformity of "all" (?) human experience.

On the interpretation I have offered, in short, Hume's claim to have carried through a consequent form of "empiricism" may rest on a reading of his arguments that wryly recalls in some respects Newton's famous dictum that "*Hypotheses non fingo*" ("I do not frame hypotheses"). For his apparent uniformisation-arguments do seem to postulate, at least tacitly (or "subreptively," again, in Kant's language) a uniform and essentially nonempirical ("absolute?") reference-frame—"Experience"—for "moral" ("matter-of-factual"), rather than "physical" ("mind-independent") phenomena.

Instead of "absolute space," in other words, the Newton of "Moral" Science, who averred that

> ... there is no reason to despair of equal success in our enquiries concerning the mental powers and economy, if prosecuted with equal capacity and caution .... (cf. [E] I 14)

effectively "framed" a thoroughly "hypothetical" and equally *non*-experiential entity which one might call—in honor of Hume's great predecessor—*absolute Experience*.

In his role as "Cleanthes", for example—the genial interlocutor who accepted secularised "design"-arguments in the *Dialogues*—Hume was willing to accept such hypotheses—even posit them, even though he rejected their analogues as claims about relational structures in a physical external world. He only required that we express them in lower case, as it were, as assertions about "Custom", "Habit" and other uniform structures of "the mind".

If so, Hume may have appealed to a form of "proto-transcendental idealism" to resolve the tensions of his "dogmatic empiricism".

It may then have fallen then to Kant to formulate

19 an "apodictic", "synthetic a priori" "(psychological) deduction" or judgment that the "deductions" and judgments of ratiocinative experience are "apodictic" and "synthetic a priori",

and modulate "rationalist" phrases like "absolute experience" into different but more or less neutral German counterparts, such as "*Erfahrung überhaupt*" ... .

Perhaps "*die kopernikanische Wende*" had already taken place. Berkeley ("*der schwärmende Idealist*"), not Kant, was Copernicus. Hume was Kepler. And Kant was Laplace.

13 "Transcendental" Design

The foregoing reading of the dogmatic aspects of Hume's empiricism gives rise to a correlative reading of Kant's "constitutive" claims in the *Kritik der reinen Vernunft*. In particular, I have already argued

1   that Berkeley and Hume, in different but complementary ways, had already negotiated the outer shoals of Kant's "*kopernikanische Wende*"); and in particular
2   that the putative ontological closure and self-sufficiency of the "second-order idealism" and social-psychological "design"-arguments Hume developed in the *Treatise* and first *Enquiry* closely anticipated essential aspects of Kant's "deduction(s)" in the *Transzendentaler Analytik*.

In the sequel, I will also argue
3   that the theory-extensions of "metatheoretic ascent" and Kant's *Reihen der Bedingungen* eventuate alike in an indefinite continuum of underdetermined but dialectically opposed *Grenzideen* or regulative ideals;
4   that certain *Sätze* and *Gegensätze* of this dialectic have historically had deeply heuristic as well as normative values for their adherents, and therefore for the evolution of western philosophy;
5   that cognate antinomies which have cut across conventional philosophical boundaries have emerged again and again from the evolution of this dialectic;
6   that "ultimate" limiting resolutions of these antinomies are not to be found, and "reasonable" partial resolutions of these antinomies are therefore zetetic, rather than hieratic; and finally,
7   that the incompleteness of these partial resolutions' normative as well as discursive force underlies the "metaphysical pathos" of Kant's "fate of reason", the *Schicksal der Vernunft* he invoked with rare eloquence in the opening lines of the first *Critique*.

German equivalents of the word "custom" ("*Brauch(tum)*," sometimes "*Sitte*") play no significant role whatsoever in Immanuel Kant's metaphysical writings. Furthermore, Kant explicitly identified Hume's "habit" ("*Gewohnheit*") with (merely) "*subjektive Notwendigkeit*" ("subjective necessity") at [Pr] 258.

With respect to the first claim above, that Berkeley effectively ascribed to "god" and Hume to "custom" and "habit" most of the attributes of *notwendige Allgemeingültigkeit* and *objektive Notwendigkeit* Kant later associated with "*Wahrheit*" and the "synthetic *a priori*", it may be of some interest
8   that one notion which does bear a significant functional resemblance to Hume's "Custom" is Kant's "*Natur*," which the latter explicitly characterized at B164 of the first *Critique* as "*das verknüpfende Vermögen*" ("the connecting faculty").

At [Pr] 319, Kant also referred to *"Natur überhaupt"* ("nature in general") as *"die Gesetzmässigkeit in Verknüpfung der Erscheinungen"* ("lawlikeness in the connection of appearances;" compare again Berkeley's coherentism, and "Custom"'s role as the "faculty" which "produces" "necessary connexions").

Also reminiscent of "Custom"'s capacity to order "possible experience" is [Pr] 36's claim that *"Natur und mögliche Erfahrung"* ("nature and possible experience") are *"ganz und gar einerlei"* ("utterly the same (thing)"), and Kant's characterizations at [Pr] 318 of *natura formaliter* (as opposed to *materialiter*) *spectata*, as

> ... *der Inbegriff der Regeln, unter denen alle Erscheinungen stehen müssen, wenn sie in einer Erfahrung als verknüpft gedacht werden sollen.*
>
> ... the aggregate of rules [cf. the "general rule" at the bottom of [T] 141], *under which all appearances must stand, if [or when] they are to be thought of as connected in an experience.* (italics mine)

The originality of Kant's argumentation therefore lay in his endeavor to "deduce" the conclusions about the ideality and structural invariance of "experience" he clearly shared with Leibniz and Berkeley as well as Hume, an undertaking no one had seriously attempted since Descartes. On the evidence of the *Transzendentale Aesthetik*, Kant proposed to "deduce" the existence of a complete theory axiomatised by "constitutive" "Euclidean" "forms" of "inner" and "outer sense", and a complete list of "synthetic" "categories of the understanding". Sustained by such "completeness"-assumptions, he also

9 believed that he had shown the way to a unique physical, conceptual and mathematical axiomatisation of "the starry heavens above us" (if not "the moral law within us") which was uniquely interpretable up to isomorphism, and

10 proposed (in effect) to call such a theory "transcendentally ideal", and identify this, conjecturally at least, as an axiomatisation of something which might justifiably be called a "common core" of *"Erfahrung überhaupt"*.

Nomenclature aside, these considerations gave Kant good reasons to believe that he had provided a genuinely mathematical-physical "theoretical core" of Berkeley and Hume's dogmatic coherentism. Nothing available to him in his time suggested that "the" form of "inner sense" (roughly speaking, arithmetic and the rational numbers) and "the" form of "outer sense" (again roughly speaking, "the" geometry of set-theoretic continua which underlay Newtonian space-time) might have myriads of "standard" and "nonstandard" interpretations.

Nor did he take into account, at least in the *Tranzendentaler Analytik*, that
11 the "common core" mentioned above might be syntactically and semantically exiguous—preliminary "conceptual analyses" to the contrary (Who are "we"? How do "we" know whether "we" have canvassed "all" of "us"?).

And finally,
12 The "common core" might be both syntactically significant and semantically consistent, but syntactically incomplete, in the sense that there might be many assertions in the common language that this core-theory would not decide. Equivalently, the putative "core"-theory might have many incompatible interpretations, and "Euclidean" as well as "constitutive" assertions made in it might therefore fail to be *"durchgängig bestimmt"* ("thoroughgoingly coherent"), a phrase Kant often used.

These predicaments suggest that there is nothing particularly "transcendental", much less probative, about the attempts to identify a "constitutive" theoretical core of discursive reasoning I have sketched above.

They also suggest that such attempts—which resemble certain aspects of his "psychological deduction"—may be informative, but do not secure the universal intersubjective validity and *"notwendige Allgemeingültigkeit"* ("necessary generality") he identified with "truth" and the "systematic unity of apperception".

More "transcendental" as well as "dialectical" attempts to relativise or mitigate the force of the thought-experiments in 11 and 12 might roughly proceed as follows.

In the case of 11, one might respond
13 that there "must" be an incomplete directed system of interpretative translations between disparate agents' "languages" or semeiotic systems (whatever they might be); for we would otherwise be unable to communicate to each other our skeptical reservations about such systems—or anything else, for that matter.

In the case of 12, similarly, one might argue
14 that syntactical inconsistency of an otherwise "constitutive" linguistic or information-theoretic core-theory would vitiate all "internal" argumentation whatsoever in that core—including any attempts opponents of "proto-transcendental" arguments might make to evaluate those arguments' tenability.

A specious ineffective response to the problem posed in 12, by contrast, might tacitly or explicitly (and mistakenly)

15 conflate incompleteness of "constitutive" inquiry with *inconsistency* of such inquiry, and attempt to "derive" a "proof" of the completeness of the experience associated with delimited forms of such inquiry from a fixed point- or generalised-"cogito"-argument which allegedly yields 13.

In my view, such a response would parallel significant aspects of the *reductio*-arguments Kant tried to employ in his attempts to formulate "transcendental" refinements of the "psychological" and "metaphysical deduction(s)" of the "existence" of "synthetic *a priori* judgments", in which

16 Kant conflated a "relative consistency proof" of a reduct of "*Vernunft*" (which he called "*Verstand*") with a "deduction" that a consistency-and-completeness-proof of a sufficiently fine-tuned such reduct would be carried through [*in itself*] (as he might himself have characterised, in other contexts, as a "*Deduktion an sich*").

Along the way, I believe that he also conflated—again and again in the *Transzendentale Analytik*, and in thoroughly understandable but metatheoretically untenable ways—

17 provisionally "unified", "proto-transcendental" metatheoretic interpretations and reinterpretations of "experience", with
18 provably "unified", "transcendental" and object-theoretic interpretations and reinterpretations of "experience" "from within".

Many of Kant's better arguments have purchase, I think, because they remind us that we need to postulate ramified hierarchies of metatheoretic "way stations" of the sort just considered, in order to "go on".

But a need should not be confused with a necessity, diodorean or otherwise. Such needs and forms of inquiry carry the zetetic force of Kantian regulative ideals, and should not be violated by attempts to collapse these hierarchies, or beg the "internal" finality of their temporal substructures. For zetetic readings of the "paradoxes" generated by Gödel's diagonalisation-lemma, sketched earlier, undermine such attempts, in the sense that

19 expressible attempts to "*close*" hierarchies of metatheoretic ascent are precisely what generate—or witness the need to generate—ever more of them.

For
20 they do not (and cannot) "close" or "bound" metatheoretic ascents: they bear witness to particular instances of semantic paradox which sustain and generate them, in the sense that
21 (putatively "presuppositionless") "proofs" *within* an "intelligible" theory that it is consistent, complete or "*durchgängig bestimmt*" ... (and therefore *immanent*) are metatheoretically equivalent to their own *Transzendenz*, and witness the *incoherence* of the (meta)theories in which they are formulated.

It would be wrong, moreover, to interpret these arguments as *"unsinnig"*, in (roughly) Wittgensteinian senses, or "deconstructions" of the regulative ideal sketched above. It would be fairer and more respectful, I believe, to interpret them as
22 sources of guidance for avoidance of "transcendental illusion";
23 confirmations of the ancient "zetetic" need for provisional theory/metatheory distinctions;
24 marks of the definitive (if unacknowledged) need for consideration of such distinctions in the history of "hieratic" metaphysics;
25 evidence for the relevance of ancient hieratic and zetetic disputes, and the force of "skeptical doubt of skeptical doubt";
26 anticipations of the relevance of metalogical arguments which give rise to such paradoxes in renewed and recurrent forms of "eternal return";
27 refutations of Kant's claims to "define" or "determine" ("*bestimmen*") "the" boundaries between "immanence" and "transcendence" and appeal to such "determination" to "deduce" the structural unicity of a *Bereich der Erfahrung*.

These predicaments give rise to a natural question: do any patterns of "judgment" accompany the forms of zetetic inquiry I have considered in ways that might—in weak and relativisable but "regulative" senses—*be* "synthetic *a priori*"?

There are, in fact, metalogical candidates for such forms of "judgment", but they are metaphysically rather modest. Among them are
28 quantifier rules and induction or replacement schemes—closely related to "number"-orderings of the sort Kant associated with "the" form of "inner sense"—which permit encoding, retrieval and reproduction of language and information,

to the extent that such rules and schemes are

29 "synthetic", in the sense that they permit us (for example) to "put together" (*syn-tithesthai*) data in ways which appear to be relatively independent of their physical support; and "*a priori*", in the sense that any modes of conception, inscription, storage, communication and review of efforts at discursive reasoning we can understand seem to "presuppose" them.

But
30 exactly because such theories are "intelligible", Gödel's original incompleteness results establish that they can decide little or nothing about [their own interpretations], much less [interpretations of wider theories which employ them as "regulative" presuppositions].

To the extent therefore that "contingency", "physicality", "physical support" and "physical evidence" all seem to involve "intelligible" but indefinitely complex (and nominally semantic) interpretations (or "realisations"), therefore,
31 theory-relative "*synthetic a priori*" judgments seem to "generate" their own "contingency", and in that sense make "necessary" the hierarchies of "metatheoretic ascent" to which I have repeatedly appealed.

Prompted by such observations, I have attempted to assimilate
32 such incompletely and indefinitely generated hierarchies to processive modes of "experience"; and
33 stages of the hierarchies themselves to "local", "synthetic" and "experiential" attempts to resolve pressing choices such ("experiential") contingencies pose.

A kind of pragmatically necessary precondition argument for these assimilations may be sketched as follows.

Expression and consequent articulation seem to be prerequisite for reflective inquiry, and reflective inquiry (broadly conceived to embrace many forms of sentience, of the sort considered in the next section) distinguishes "experience" from information processing.

So potential expressibility in some finitary context seems to be prerequisite for "experience".

Since, moreover,
34 finitude of expression in "reflective" theories also gives rise to *undecidability*, and
35 undecidability, in turn, to generation of indefinitely extended hierarchies of (metatheoretic, ...) (re)expression and (re)interpretation,

it seems reasonable to construe

36 *stages* of such hierarchies as "local" aspects of thought-experimental "immanence" or "experience."

It should immediately be acknowledged, of course, that "experience" so conceived would include many Kantian *Grenz-* and *Vernunftideen*, provisionally considered and acknowledged as such (as a matter of intellectual honesty) in ramifying stages of zetetic inquiry.

Transitions between such stages, in particular, may involve

37 indefinitely many "local" stages of provisional reexpression and reinterpretation—new forms, perhaps, of Wittgensteinian *Übersicht* (overview)—none of which provide "ultimate" ranges of such quantification over their *ree*xpressions and *re*interpretations.

For "all object-theoretic interpretation" is an ineluctably theory-relative notion (a consequence of Gödel's work), and so zetetic inquiry is as well.

In this sense, therefore, the frameworks of "zetetic inquiry" and "experience" I have postulated are zetetic in their turn (another fixed point, in effect, and a zetetic counterpart, once again, of skeptical doubt of skeptical doubt).

To put a point on it: I have informally

38 attempted to "coopt" such frameworks for zetetic structure for "zetetic" purposes, and identified them with "experience."

More precisely, I have proposed to identify

39 "experience in (a given background-theory) T" with "zetetic inquiry" *in* T (and therefore with local, intentional and hierarchically organised attempts to decide the contingencies of such "experience in T"—another zetetic fixed point).

This "local" argument would seem to have a very obvious companion: to identify

40 "*Dinge an sich* in (or for) T", with entities which may elude "all" stages of thought-experimental inquiry in *T*, but "exist" for T in the sense(s) of ramified temporal hierarchies of wider metatheories which assert T's consistency.

Such zetetic explorations of frameworks for "experience" and zetesis would obviously continue *ad indefinitum* (rather than "*ad infinitum*"). As parts of their own contingency and locality in turn, they would become instances of the ineluctability of Kant's "fate of reason".

But they might also bear witness to an essentially normative conjecture I will take up again in the next section: that the *journey—not the indefinitely deferred arrival—matters*.

14 The Zetetic *Incomplétude* of "Merely" Regulative Ideals

In this section, I will argue that that the real "*kopernikanische Wende*" of Kant's lifework lay in
1. his exposition of the *Antinomien* (as he himself suggested with modest pride), and that
2. the subtlety of his *transzendentale Dialektik* bore witness to this deeper "zetetic" undercurrent in his work.

I have not attempted yet to discuss in any sustained way "*the primacy of the practical*", as I promised to do. I have, however, suggested at length that discursive reasoning is all we have, and that such reasoning is potentially communicable, and therefore formalisable, and therefore incomplete.

This view may not be bereft of modest practical implications, for almost any medium for communication of information might be "discursive", on my account, and complex forms of such "communication" are essential to the sustenance of life, as well as the "life of the mind".

I have also attempted to assimilate "incompleteness" to the "reedlike" quality we recognise in Pascal's elegiac remarks about the "thinking reed" (*Pensées* 6.347), and suggested that
3. "zetetic" forms of *dialektike* eventuate in forms of moral as well as metaphysical pathos which monists typically find troubling, and believe must be overcome (hence the disciplinary recurrence of "last" philosophers).

More precisely, I have argued that
4. "zetetic" attempts to answer emergent inquiries naturally tend to ramify in branching patterns (or hierarchies) of further inquiries, each of which is incomplete in its turn; and
5. ethical, aesthetic and emotional responses which guide such patterns (give "regulative" value to them) naturally eventuate in recurrent forms of "pluralistic pathos" (a variant of the other metaphysical *pathe* mentioned earlier, which Arthur Lovejoy introduced and considered in (Lovejoy, 1965), 10–20).

14 The Zetetic *Incomplétude* of "Merely" Regulative Ideals — 53

In the sequel, I will draw on the original sense of the Latin word *fatum* (an obsolete synonym for "what is spoken") to assimilate in greater detail

6   the inherence of this incompleteness—which impels us to seek "local" resolutions to "globally" ill-defined problems in metatheoretic hierarchies—to Kant's "fate of reason",

and

7   recourses to these hierarchies as "constructive" as well as "regulative" realisations of "skeptical doubt of skeptical doubt".

Are there any invariantly "practical" considerations (much less "imperatives") in all this?

The answer I will offer—a good dialectical one—is "yes and no" ("*sic et non*").

"No", in the sense that every practical or casuistic decision in evolving structures of "zetetic" inquiry would be theory-relative (in the usual formulation "context-relative"), and subject to the extensive reinterpretations such inquiry might bring.

But "yes", in the sense that the characterisations I have offered for zetetic inquiry itself would recur in internally probative forms at every stage and in every form of such inquiry. And "practical" extensions of current "speculative" inquiry attributed recurrent if theory-relative value to these characterisations.

In particular, I will attempt to elicit certain attributions of value from "constructive" interpretations of the "pluralistic pathos" I have attributed to "zetetic" inquiry. Among these are

8   contemplative as well as "communicative" notions of "freedom" and "ethical agency"; and

9   "zetetic" reconstructions of "dignity", "respect", and (a counterpart of) Kant's notion of the "purely good will".

I will, for example, interpret (or reconstruct)

10 "freedom" as an unconstrained, "action-theoretic" counterpart—call it *incomplétude*—of the recurrent incompleteness of "experience", which I have assimilated to metatheoretic ascent;

11 "ethical agents" as entities which understand that they have limits and act on that understanding, at least "in thought" (there may be more forms of cognitive agency than are dreamt of in our philosophies); and

12 "reasonable" ethical agents as ones which can also understand that reasons may have limits ("reasons" of state, for example), and act on that understanding, at least "in thought" (there may also be more "reasonable" forms of *ethical* agency than are dreamt of in our philosophies).

One of the deepest "hieratic" tenets in metaphysics is that there are, "ultimately", no ontological middle-grounds between (useless) "inconsistency" and (vacuous) "truth"—or, equivalently, no metaphysical distinction between "consistency" and "truth".

This amounts to the view that "contingency" is a makeshift or illusion, and physical "truth" is one, apodictic (in some equally elusive sense) and determinative of "experience" (which in some latitudinarian sense converges to it or aligns with it).

One of the deepest "hieratic" tenets in moral philosophy is that there "ultimately" are, should and "must be" no axiological middle-ground between "inconsistency" and "truth"—or, equivalently, no "ultimate" practical or normative distinction between moral "consistency" and normative "truth".

This would amount—on my account, at least—to a view that
13   moral "contingency" is a makeshift or illusion, and that
14   moral "truth" is one, apodictic (in some equally elusive sense) and determinative of "all" praxis or practical "experience" (which in some latitudinarian sense converges to or aligns with it).

Conversely, anyone who would accept this dichotomic view but deny the "existence" of such a unitary realm of ends or moral truth, would therefore be committed to the view that "anything goes".

I believe that the moral as well as metaphysical variants of this view are "transcendental illusions", in roughly Kant's senses of the phrase, and have already argued to this end in earlier sections
15   that incompleteness is regulative of discursive rationality (and that *this* is the "fate of reason"); and
16   that semantic paradoxes provide formal miniatures or analogies for larger dialectical clashes between the two principal *pathe* I have explored (the "hieratic" and the "zetetic").

More particularly, I will argue
17   that variants of these paradoxes provide analogies for dialectical tensions that hover at the margins and presuppositions of hieratic views of "practical philosophy";
18   that these dialectical tensions are generative rather than "frustrating", *à la limite*(s), and reflect the "dignity" of Kantian "freedom" as well as its "fate"; and finally,
19   that efforts to understand "freedom" and "fate" in such double aspects offer a stable "middle ground" of the sort sought above, as well as moments of

practical as well as speculative *ataraxia* (a stoic as well as pyrrhonist-skeptical term).

I will also propose two "practical" constraints on "normative" and "evidentiary" theories in stages of "enlightened" inquiry and metatheoretic ascent:
20 that normative theories accepted at such stages extend as well as interpret their current evidentiary counterparts; and
21 that normative theories at "higher" stages constrain ranges of evidentiary interpretation at "lower" ones, and conversely.

These metatheoretic conditions and constraints, in effect, embody attempts to
22 localise Kant's observation that "speculative" evidence underdetermines "practical" criteria, and
23 articulate processive, zetetic and evolving forms of Kant's "*Primat des Praktischen*".

The constraints or conditions in 20 and 21 are proposed as minimal, processively normative assertions about initial stages of "enlightened" inquiries.

They do not, for example, characterise attempts to engage in zetetic inquiry in "constitutive" ways. They simply propose incomplete regulative constraints on enlightened inquiry, which many potential courses of such inquiry might satisfy. In particular, extensions of these constraints might therefore evolve together with the attempts at zetetic inquiry they "regulate", along many divergent paths.

In the minimal situation, in other words, these constraints are weak, in metalogical and metaphysical terms. Nevertheless, I will argue for a number of normative conjectures and working hypotheses which resonate "zetetically" with them in interesting and suggestive ways.

I have already suggested earlier
24 that brief and unfinished eudaimonic realisations of the "desires" and "regrets" of "sentient beings" must formulate their own regulative ideals, and project or outline in the process their own "eternal rewards"; and
25 that incomplete "zetetic" attempts to sustain and regulate (apparent preconditions of) such consciousness and an enlightened "life of the mind" offer the only "*a priori*" language for expression of such ideals we will ever have.

In this sense, conscious moral theories that would regulate consciousness are inherently "self-referential", in ways which resonate with "the golden rule" and allegedly more precise counterparts thereof (in one of Kant's versions of the "categorical imperative", for example).

It may be no accident, for example, that one of the more intractable problems of eudaimonic, stoic, Humean and Kantian ethics has always been "the" nature of "the" elusive moral "self" to which they were supposed to apply, for the analyses I sketched suggest

26 that whatever is capable of [self]-reference—whatever can engage in processes of (partial) [self]-observation—is inherently liminal and in flux, and

27 that what seeks or is sought *"in itself"*, "for its own sake", *en soi, an sich*, is not absolute or invariant. It is rather what varies and "absolves".

For

28 what can recognise [itself] is no longer [what it recognised]. It has relativised in transit what it observed and "identified" (individuated); and therewith [its own observations]; and therewith [itself].

I also believe that "what recognises" also effectuates quantum-theoretic *Schnitte* ("cuts" of the sort which occur when something is "measured"), large and small, which set aleatory initial conditions for further branching paths of observation and [self]-observation.

Be that as it may, Kant's notion of the *"schlechterdings gute Wille"* ("utterly good will") may serve as a test case for the conjectures I have offered so far about elusiveness of normative [self]-reference and "practical" implications of *"incomplétude"*.

The *"schlechterdings gute Wille"* is elusive exactly because it is "merely" dispositional and incomplete. It cannot "know" the limitations and boundary-conditions of its continued existence (if it has any). It cannot, therefore, "know" how fragile and tenuous it is, or by what it might be overwhelmed.

What might it "know" (or "reasonably" conjecture)?

It might "know", first (or reasonably conjecture),

29 that freedom and autonomy, minimally characterised as in 10 above, are regulative of inquiry, and therefore of moral inquiry, and therefore of moral identity;

30 that other properties obviously constrain as well as regulate the boundaries of inquiry, moral inquiry and moral identity in indefinitely complex and contingent ways; and finally, as I remarked just above,

31 that *"schlechterdings gute Wille"* can*not*, therefore, identify precise initial or boundary-conditions of [its own] freedom and autonomy, and does not, to that extent, "know", [the boundaries of its moral identity].

From 26, in particular, "purely good will" might infer that "knowledge" is not enough. It might also accept Kantian analyses of "dutifulness" and the actions of "*kaltblütige Bösewichte*" ("coldblooded villains") (cf. *Gr* 394), and conclude that casuistical canvasses cannot characterise "goodness", and that one might "know" and "know" and be a villain.

Such analyses leave the way open to postulate or conjecture

32 that potential moral agents should value freedom and autonomy more highly than other, more constraining aspects of moral identity;
33 that such agents should also value these aspects of moral identity "for their own sakes", even though discernment of their precise boundaries is infeasible.

In the sequel, I will also interpret the liminality and indefinition of identity as a personal "veil-of-ignorance"-argument, and attempt to develop (not "derive") from it an ethic of mutual "valuation" and "respect" (the root-meanings of Kant's "*Achtung*" distribute over both).

This "development" is obviously couched in quasi-Kantian language, as I have sketched it, but one could formulate a variant of it from "Leibnizian" premises about monadic "confusion", the mirroring-principle, and the consequently elusive as well as liminal nature of personal identity.

Hume, in any case, was clearly right that one cannot deduce ethical precepts from neutral experiential premises. But I believe this "development" becomes an argument with a certain probative force, if one ackowledges two auxiliary hypotheses:

34 that free inquiry has an intrinsic moral value (cf. 33 above); and
35 that this value can guide action in provisional and incomplete but potentially decisive ways.

In particular, and in more obviously affective language, I believe zetetic moral agents have reason to "respect" their fellow-inquirers, their fragility and their incompleteness, and reach out to what they value and respect. This "reaching out", I believe, is also one of the deepest sources of *sympatheia/sympathie/ Mitgefühl*—the moral sense, as it were, of Hume's relatively hidebound "Moral Sense".

To me at least, the depth and recurrence of these sources offer a rationale for the fact that many have found Kantian ethics of mutual *Achtung* and *Würde* deeply poignant, even though their applications will inevitably remain culture-bound, and they are casuistically of little use *in extremis*. For theory-relative variants of these sources enjoin us consciously to respect and enlarge boundaries of consciousness neither they nor we can define. They also urge us, I believe, to respect the capacity for the reflective and contemplative *dynamis* of indefinitely extended inquiry and "metatheoretic ascent", in ways that transcend the penury of any list of casebook examples.

15 Borgesian Maps

"Gott" Schuf Inferenz; alles Andere ist Erfahrung

"God" Created Inference; Everything Else is Experience.
(With Apologies to Leopold Kronecker)[2]

I have already argued, in effect, that "the" "*argument from design*" was a tacitly presupposed premise before anyone ever conferred on it the dignity of an "argument"—a premise which expressed a world-view that might be called "the monotheism of the scientists" (in rough analogy with "the god of the philosophers"), and a premise which begged (in its subtler forms) a methodological assumption in lieu of an ontological one in the form of ever-more elusive regulative ideals of scientific "design".

On the assumption that common vulnerability flows from a common "proof"-theoretic structure, I have also argued

1. that metalogical analogies provide extensive support for Hume's metaphysical critiques of design-arguments in the *Dialogues Concerning Natural Religion*;
2. that these analogies also support Kant's slightly different analyses of "physiological proofs" in the *Transzendentale Dialektik*; but
3. that they strongly suggest that the "psychological deduction" of the *Transzendentale Analytik* may also be analysed, in very similar terms, as an (inconclusive) "physiological proof".

On this analysis,

4. convergence arguments and appeals to experiential adequacy typically invoked to warrant ancient, medieval and early modern design presuppositions are essentially variants of their theological counterparts, with different notational and nomenclatural key-signatures; and

---

[2] The German mathematician Leopold Kronecker (1823–91) is quoted as having said, "Die ganzen Zahlen hat der liebe Gott gemacht, alles andere ist Menschenwerk" ("God made the integers; all else is the work of man").

5   analogues of design-arguments which fail to secure the unicity and well-definition of an intelligible deist "god" may also fail to secure corresponding attributes of an intelligible agnostic "universe", however "obvious" the "existence" of the latter may seem; for they may withstand skeptical scrutiny no better than their theological ancestors.

For impeccable metalogical arguments seem to ensure the plural interpretability of such arguments, in senses which can be made precise, and suggest therefore that "[d]er Begriff des Ganzen" ("concept of the whole") may be "nur relativ zu verstehen" (only relatively to be understood"), to quote Hans-Georg Gadamer, one of such arguments' more dismissive post-Kantian critics.

If one declines to embrace unique-convergence claims and their associated heiratic ideals as articles of unknowable and perhaps uncommunicable faith, such claims may rest on "subreptively" begged questions and equivocations ("*Amphibolien*"), of the sort Kant identified in other contexts with great acuity in the *Transzendentale Dialektik*.

At the risk of *lèse majesté*, for example, one might cite "arguments" that look suspiciously like simple quantifier-mistakes ("… if everything has a design [e.g.], there "must" surely be a design for everything" … ; or: "… if everything has a cause, there "must" surely be a cause of everything" …).

Cognate iterations of alternating quantifiers still pose the single greatest barrier to students' understanding of "rigorous" university level mathematics. More tellingly, perhaps, they seem to have been rather ill-understood for two millenia after Aristotle and others sorted out their singly quantified counterparts.

Metalogical "choice"-, "uniformisation"- and "well-foundedness"-principles may sweep objections to such quantifier inversions off the table, at least rhetorically ("… but this would be a *regressus in infinitum*, which is manifestly absurd …").

Logicians, however, have extensively studied formal miniatures of such "uniformisation"- and well-foundedness principles. They turn out to be theory-relative and eminently defeasible, and uncritical invocation of them may simply replace a mistake with a *petitio principii*.

Similar remarks apply in somewhat different ways to traditional rejections of contextual and relational notions of "truth in (a structure)", or "truth with respect to (wider theoretical or hermeneutic reference-frames)". Such notions have always seemed unsatisfying, disturbing or worse to more hieratically inclined metaphysicians in the history of western philosophy, "analytic" defenders of various "realist" doctrines among them.

Unrelational "truth", however—even when defined, by fiat, in the form of so-called redundancy theories ("'Snow is white' is true if and only if snow is white") —is a vacuous second-order template or "convention", as Donald Davidson, in effect, never tired of pointing out (in different language, of course, and with other aims in mind).

Attempts to make do with such jejune "definitions" therefore have a tendency to suppress debatable background contexts and venues for them (cf. Kant's "subreption", once again), and assimilate them to not-so-universal counterparts which "we" may regard as "obvious" (whoever "we" are, once again).

Like cognate justifications of arguments from design, such "subreptive" assimilations also identify "our" counterparts of semantic notions ("design", "truth") with "universal" or "absolute" counterparts of them. In various chapters of this book as well as this introduction, I argue that such counterparts elevate them to the level of their methodological incompetence.

For if expressibility is relational (to syntactical interpretative schemes) and local (in metatheoretic hierarchies), "global" and "irrelational" notions of truth would have to be

6 an empty template (a "redundancy-theory", in search of branching metatheoretic clarifications); or
7 an inexpressible "hieratic" ideal, access to which would "pass all understanding", and therefore "all" discursive relevance for concept-formation and guidance of belief (cf. again the Germanic etymology of "truth" sketched above).

In the second chapter's examination of Platonic "forms", for example, and *passim* thereafter, I interpret the *eide*, along the lines of 7, as ghosts of departed "idealist" appeals to "intended" interpretations, which (allegedly) have no "intender" and have become therewith "absolute".

The transcendentally "hieratic" early twentieth-century early-Wittgensteinian notion of "the" "*Welt*" is another, much later "hieratic" variant, and the very model of such a subreptively "universal" interpretation which passes *use* as well as understanding, as Wittgenstein himself later realised (cf. Boos, 2004, and 14.26–14.28 above).

Ontological realisations of hieratic ideals, as I remarked earlier, tend to come in two forms, each of which can be reconstructed as a denial or repudiation of indefinitely extended zetetic patterns of inquiry and metatheoretic ascent.

However different they may be in other respects, for example, Kant's Bereich der Erfahrung and Wittgenstein's Welt both "close" or "cap" "experiential" inquiry in a single "universal" interpetation, (allegedly) exempt by "transcendental analysis" or mystical fiat from further reexamination.

Peirce's "limit of inquiry", by contrast (cf. Chapter 8), does not "close" inquiry "vertically", but "narrows" or aligns it "horizontally", into a single linearly ordered "course", whose directedness or linearity ensures that it has a unique infinitary but directed limit.

The end result in both cases, however, is the same: a single "universal" interpretation, once again, (allegedly) exempt by "transcendental analysis" or "scholastic realisms" or mystical fiat from further scrutiny and reexamination. The elusive "existence" and unicity as well as the internal intelligibility of these nominally "universal" interpretations are also begged in both cases.

One way to explicate briefly my argument that metatheoretic ascent is essential to "experience" and cannot be begged into insignificance may be to invoke the notion of a *Borgesian* or *Funesian map*—a map that is so precise that it is structurally isomorphic to what it "maps". In terms introduced above in Sections 4–6, such a map may be identified with a theory that interprets [itself].

Such maps are not in themselves paradoxical. Any Henkin interpretation of a formal theory, for example, is Borgesian in the sense that it is [self]-interpreting with respect to (metatheoretically defined notions of) satisfaction. But a few "Russellian" thought-experiments with such notions as

8 "the design of all and only all those designs that are not designs of [themselves]", or
9 "the theory that interprets all and only all those theories that do not interpret [themselves]", or
10 "the (Peircean) interpretant which interprets all and only all those interpretants which do not interpret [themselves]",

may convince the reader that "absolutely intended" interpretation(s) narrowly avert absurdity in ways that require that they be "absolutely" [self]-interpreting.

To gain a sense of the interrelations between less controversial notions of "Borgesianness" and patterns of metatheoretic ascent, it turns out to be instructive to examine Borgesian cartography for what might be called its non-self-similarity, or non-alignment of "global" designs and patterns with their "local" counterparts, seen in such structures "from within". For it turns out that several aspects of such misalignments bear on Hume's and others' "internal" critiques of traditional "arguments from (or to) design."

The Platonic cave parable's aporetic use of (mere) *eikasia* to denounce (mere) "*eikasia*" pointed to one underlying aspect of this problem. A millenium and a half later, Anselm's respondent Gaunilo correctly observed another: that one cannot expect to discern within a "divinely perfect"

Borgesian theory that it *is* Borgesian, an essentially zetetic insight that also reappeared in several of "Philo"'s counterexamples in Hume's *Dialogues Concerning Natural Religion*.

The problem is not just the obvious "extensional" one—that Borgesian attributes would presumably be infinitary, and that physical canvasses of them would therefore be infeasible. A deeper, more "epistemic" or "intentional" refinement of this problem, I believe, is

11   that there may indeed be no way to recognise a Borgesian theory as a "map", for very general Gödelian arguments establish that a subtheory of it cannot be both finitary and complete.

A first indication of this appears in attempts to talk (say) about
12   "'all' the theories that are consistent (or interpretable)", or
13   "'all' the theories that are definable (or expressible)".

For Gödelian arguments yield that the universal quantifiers in these phrases are themselves intelligible only in theory-relative terms.

Another manifestation of the problem emerges from natural attempts to "solve" it with the aid of semantic forms of induction or "model-theoretic forcing", reviewed in several succeeding chapters, which generate "generic" structures as minimal extensions of evidence collected "from within".

Such "minimal" closures of partial inductive "evidence" turn out—unexpectedly perhaps—to be very extensively random, in ways that contravene Hume's dismissals of randomness in the *Treatise* and first *Enquiry*, and bear out quite well the reservations he (or "Philo") formulated in the *Dialogues Concerning Natural Religion* (perhaps one should talk about an "early" and "late" Hume?).

Genericity or inductive generation of such structures can be shown, in particular, to efface [forms of internal self-recognition] in interesting ways. Most such structures are generic, for example, and each such structure is Borgesian, but each generic structure (once again) will also satisfy many sentences of the form [s and [s is not interpretable]], the sort of conjunction Berkeley's "master argument" labelled "absurd".

Finally, canonical notions of "inductiveness" or "randomness"—borrowed from insights of Kolmogorov, Solovay, Chaitin and other pioneering investigators—seem to be deeply relational and sensitive to their situation in hierarchies of metatheoretic ascent. These are ineluctable notions, as far as we can tell, but they seem to elude or "transcend" [themselves], in something like Kant's sense of "transcendence" in the *Dialektik*.

As I read this evidence, to say that "the" structure of "reality" is "Borgesian" therefore opens it, *en abyme*, to questions that are "*unabweis[bar]*" (as Kant put it):

> denn sie sind [der Vernunft] durch die Natur der Vernunft selbst aufgegeben, die sie aber nicht beantworten kann; denn sie übersteigen alles Vermögen der menschlichen Vernunft. [*KdrV* A VII]
>
> not to be dismissed, for they are posed by the nature of reason itself, but cannot be answered, for they would transcend every capacity of reason itself.

Other, equally "*aporetic*" but generative questions arise from study of the liminal nature of "all (provably) consistent theories", mentioned earlier, or attempts to express "the" range of "all" expression, conceive "the" limits of "all" conception, or interpret "the" scope of "all" interpretation ....

For to me at least, such thought-experimentation suggests that

14 methodological arguments can only be *a priori* in "merely regulative" senses which would frame or "regulate" incomplete ramifying hierarchies of the sort I have assimilated to "experience"; that

15 "practical" as well as "speculative" efforts to discern heuristic boundaries between stages of these hierarchies can only have *a priori* value in the modest sense that they regulate conscious inquiry; and that

16 no stage, threshold or horizon of experiential inquiry, however refined it may be, will ever "determine" what it endeavors to "interpret", "... *denn [solche Streben] überstiegen alles Vermögen der menschlichen Vernunft*" ("... for such aspirations would transcend every capacity of human reason").

16 "Philosophia" and "Scientia"

I have tried in this introduction to make qualitative use of metalogical ideas to outline a number of "zetetic" analyses of traditional arguments from the history of ethics, metaphysics and epistemology.

The central chapters of the work itself offer more detailed attempts to apply these analyses to texts of Plato, Aristotle, Sextus Empiricus, Descartes, Leibniz, Spinoza, Berkeley, Hume, Kant, Weyl, Skolem and Wittgenstein, among others,[3]

---

3 For Descartes, Weyl, Skolem and Wittgenstein, see Boos 1983, 1994, 1995 and 2004.

prompted in part by my admiration for Ernst Cassirer's multi-volume work *Das Erkenntnisproblem in der Philosophie und Wissenschaft der Neueren Zeit*.

As part of these analyses, I have already argued in preliminary ways that

1. free use of every finitary instrument and conceptual resource, and concomitant suspension of judgment about "whatever else" there may be, are "locally" and "provisionally" "synthetic a priori" and regulative of "experience".

I have also suggested, more controversially, that

2. displacement of love of inquiry into love of "the" (alleged) ultimate *terminus ad quem* of "all" inquiry (which I characterised at one point as "the monotheism of the scientists") may be the real "scandal" of philosophy Kant decried in the opening pages of the *Prolegomena*.

Such language is clearly normative—as it was, of course, in Kant's Vorrede. As I see it, desires for assurance of a unique "ultimate" ("hieratic") knowledge/theory/understanding would be comparable to desires for eternal life—understandable, but hubristic all the same.

Drawing on common premises about experience and rationality that seem to cross a number of "world"-views (as mentioned, itself a skeptical consensus gentium argument), I have also to sketch some "reasonable" reconstructions of what is "rational", and "zetetic" criteria for what is "experiential".

It seems rather minimally reasonable (if not "rational"), for example, to conjecture

3. that some common attributes of these structures and interrelations may be regulative of efforts at discursive analysis;
4. that feasibly simple aspects of these common attributes may also be accessible to such discursive efforts; and
5. that these simpler aspects of discursive analysis are *never* "bivalent," in the sense in which this adjective is thought to characterise "ultimate truth".

Working with the texts mentioned above, it seems to me

6. that feasibly simple hierarchies of concept formation and theory construction are indeed common to efforts at discursive analysis along the lines just sketched;

but

7. that such "feasibly simple" hierarchies of axiomatisable theories may be all that is common to such analyses—not all there "is", perhaps, but all we and other would-be "reasonable beings" can intelligibly talk about.

This minimal view amounts, or is at least compatible with, (metalogically inflected) "secularisations" of the zetetic ideals sketched earlier, and with concomitant forms of semeiotic or linguistic skepsis about "experience" and "the" scope of scientific and philosophical inquiry.

It is commonplace, for example, to observe that no stable demarcation has been or is likely to be found between theory and observation. But it seems to be less commonplace to conjecture that there may also be no "ultimate" time- and situation-invariant distinction between *theory* and *evidence*.

In at least one obvious sense, for example, "theories" seems to be one provisional type-level higher than "evidence" for them, to the extent that

8   theories quantify over the entities they are "about", and one typically claims that they are "about" certain sorts of object-theoretic "evidentiary" data.

In another sense, however, "theories" seems to be at least one provisional type-level lower than "evidence" for them, to the extent that

9   evidence for theories is tested in semantic interpretations for these theories, and
10 acceptability of semantic "evidence" as such is a matter of metatheoretic judgment, which creates its own conceptual event-horizons.

The latter reconstruction also provides background interpretations of a straightforward sort for

11 standard arguments that "observation is theory-laden", as well as
12 presuppositions that one can formulate syntactical and semantic interpretations, at least "in principle", for the theories in question (for experimental "isolation" and "localisation" of conceptual "boundary conditions" would otherwise be impossible).

The local and provisional metatheoretic "type"-distinctions evoked in 9–12 seem to me more persuasive than their "object-theoretic" counterpart(s) in 8, but

13 the independence, undecidability, and plurality of ("evidentiary", "experimental") interpretation generates needs for
14 ever more (meta ...)theoretic decisions (about experimental "outcomes" and their "measurement"-readings, for example), and
15 "types" (or the more flexible theory/metatheory-distinctions that replace them) may be reinterpreted as "sorts" at subsequent stages of metatheoretic ascent.

In either event, such ramified time-indexed courses of concept-formation, reinterpretation and "evidential" verification in "experience" might be construed as temporal realisations of metatheoretic ascent.

The recurrent incompleteness (and concomitant methodological "contingency") of scientific inquiry, so conceived, suggests that one might find a "perennial" role for philosophy as the *recurrent liminal margin of such inquiry* ("... whose arches fade/Forever and forever as I move ..." ["Ulysses"]).

More precisely, a complementary rationale for the underdetermination I attributed to "experience" in Section 2 may emerge if one views philosophy, in historical and processive terms, as a study of questions that remain undecidable by a given period"s "science". This characterisation would suggest

16 that "philosophy", so conceived, may not be a "science" (or even a *Wissenschaft*); and
17 that metalogic and metamathematics, formal disciplines devoted to the study of questions that are contextually undecidable (neither provable nor refutable from a given contextual set of theoretical premises) may indeed provide hermeneutic and heuristic insight into the history of philosophy— and conversely.

One might elaborate this view as follows.

Historically, most forms of modern "science" began as species of "(natural) philosophy"—consider, for example, the title of Newton's *Principia Mathematica Philosophiae Naturalis*, or William James' ambiguous role as "psychologist" and "philosopher" two centuries later. From time to time, consensus grew that theoretical inquiry into particular areas of "experience" had become more or less "decidable", within carefully demarcated and "feasibly finite" initial and boundary conditions, often translated into stipulations of "tolerance" and experimental "isolation". Such theories then became "scientific"—itself an early modern usage (compare also Thomas Kuhn's notion of a "normal science").

In prior articles, I have tried to work with contextual and theory-relative interpretations of such *scientiae*, and attempted to assimilate the undecidability that demarcates "philosophy" from them to metalogical phenomena of formal undecidability.

The divisions of labor have historically been provisional, of course, in varying degrees (indeed, this is one way to interpret "Hume's problem" in the philosophy of science). And current penumbrae of undecidable methodological or metaphysical questions have always remained, *and always will*.

For

18 object-theoretical notions of "boundedness" and "initiality" and "feasible finitude" are neither feasibly finite, "physically" bounded nor "temporally" well-founded,

and
19 consistency and "intelligibility" of these notions' object-theoretic *definitions* will ensure that the cardinality of their semantic interpretations is the power of the continuum.

In seems to me therefore that
20 notions of "feasible finitude", "experimental isolation" and initial and boundary conditions have traditionally been relegated to
21 de facto metatheoretic examinations by practitioners of the evolving discipline(s) of "philosophies of ..." (mathematics, physics, science, language, mind....).

For such examination may be characterised as an
22 attempt to define, determine, characterize, individuate and adjudicate what may be metatheoretically decidable (or "experienceable"), and what may not,

and a source of
23 heuristic evidence that "immanence" is intrinsically emergent may be the recurrent underlying source of transcendence, and that no theory of "experience" can define the structure (or preconditions) of its own "immanence".

2 "Was Blind, But Now I See": Ramifications of Plato's "Line"

1 Introduction

In this chapter I will argue for a conjectural assimilation of middle- and late-platonic *eide* or "forms" to the idealised and extratheoretic notions which twentieth-century mathematical logicians call "intended interpretations". Among other things, this assimilation may provide a quasi-rigorous analogical approach to certain notorious obscurities of Plato's usage—defensible metalogical senses for his talk about "partaking of" or "participating in" a form, for example. It may also help localise with some precision those aspects of nineteenth- and twentieth-century mathematical practice which can most reasonably be called "platonist", a term philosophers of mathematics sometimes use rather loosely.

1.1 In Section 2, I will outline in some detail what mathematical logicians' "intended interpretations" of certain archetypes of set- and number-theory "are"—or at least what explanatory role(s) they are thought to serve. I will also sketch several specific ways in which attempts to postulate such interpretations may clarify *obscura per obscuriora* (appropriately enough, perhaps, if I would compare them with Plato's invocations of the forms).

Mathematicians' attempts to express claims that such "intended interpretations" "exist", for example, are recurrently extratheoretic, or at least marginal to the ("object"-) theories to which they are intended to apply. Results of Thoralf Skolem, Kurt Gödel and others have shown that such interpretations are subject to graduation and indefinite shifts of perspective which I call *metatheoretic ascent* in 1.3.3 and Section 2, in which I argue for a theory-relativism which accepts such shifts, and assimilate such acceptance to a form of mitigated ontological skepticism.

In more familiar twentieth-century language, such acceptance might also be phrased as a

1.1.1 ramified context principle: that a sentence makes sense only in the context of an interpretation, which must, in turn, be secured in a background-metatheory.

I will also suggest in this chapter and the next (cf. 1.4) that such shifts may be "transcendental" preconditions for refinements of Aristotelian "theoretical" inquiry, and assimilate them to ramified and iterated escapes from metamathematical counterparts of platonic "caves".

1.2   In Section 3, I will review certain aspects of Plato's original doctrine(s) about the "forms", and argue that metamathematical postulation of such "intended interpretations", as sketched in Section 2, does, indeed, bear significant comparisons with metaphysical postulation of the forms.

These comparisons will include "positive" conjectures about structural similarities, as well as "negative" claims that the two notions are vulnerable to cognate *elenchoi* and skeptical critiques.

The "positive" conjectures will focus briefly on Plato's well-known desires to "mathematise" the forms, but more specifically on an analogy between Plato's "dichotomisation" of a particular form in the *Parmenides* (136a–d), and metalogical characterisation of a particular interpretation of a given theory T in T's Stone space, or space of interpretations (cf. 3.4–3.6).

The "negative" arguments will explore analogies between modes of transcendence and inexpressibility that the "forms" and "intended interpretations" arguably share (cf. the *obscura per obscuriora* mentioned above) and parallels between *elenchoi* and skeptical critiques to which they may be subject.

1.3   Several of these analogies will draw on results of twentieth-century metamathematics that I sketch at some length in Section 2, and gloss further in Section 4.

In later passages, for example, I call variants of a natural conceptual framework, in which one might posit the existence of form-like entities and deny the implications of such critiques, "semantic monism".

In its most uncompromising and deeply problematic lines of argument (I will also argue), semantic monism essentially requires postulation of a hypostatic or extra-theoretic "last metatheory", which would have to be inexpressible to itself, by the results reviewed in Section 2 (cf. 1.3, and 2.36–2.42). The analogies of Section 2 may therefore suggest a potential gradient for the historical convergence of Platonism toward its Neoplatonic successors.

In Section 4, I will identify certain common metatheoretical assumptions that I believe underlie "Socrates"' well-known rebuttals of "Protagoras" in the *Theaetetus*, construed here as refutations of a mitigated relativism to which plurality of interpretations at every temporal stage gives rise.

Two subsidiary arguments of Socrates' *elenchos* are generally considered persuasive, if not decisive. These are

1.3.1   his proto-"transcendental" claim that Protagorean semantic relativism (which I will assimilate to the theory-relativism of Sections 2 and 3)

"must" eventuate in an incoherent elision of "all" feasible semantic criteria; and

1.3.2 his subtler, proto-Cartesian assertion that reflexive applications of such semantic relativism "must" eventuate in a form of skeptical doubt which elides or confutes [itself].

I will suggest in Section 4 that

1.3.3 the claim in 1.3.1 rests on a mixture of tacit metatheoretic quantifier-error, and begged denial of the patterns of metatheoretical ascent alluded to above in Section 1.1; and

1.3.4 that the assertion in 1.3.2 can be countered by metamathematical reconstructions of skeptical [self]-doubt which follow directly from Gödel's diagonal lemma, a ground base in Section 6 of this book's introduction.

1.4 In Section 5, the chapter's conclusion, I develop a contrast between Section 3's "semantic monism" (a methodological cousin, perhaps, of Arthur Lovejoy's "monistic pathos"), and processive (and quite "rational") forms of "theory-relativism" or "semantic pluralism".

On this view, once again, variants of Section 4's *epoche* or *ataraxia*—and a concomitant openness of the paths of inquiry to which they give rise—might be assimilated to a mitigated-skeptical form of Aristotelian *theoria* ("contemplation"), or at least "transcendental" preconditions for the active [self]-awareness Aristotle associated with such "theoretic" *energeia*.

2 Metalogicians' "Intended Interpretations"

Mathematical logicians typically invoke "intended interpretations" to normalise, in indefinitely idealised ways, what they "mean" by certain especially problematic assertions about intensionally infinite entities in number theory, analysis, set-theory and model theory.

A partial canvass of such notions—in roughly ascending order of set-theoretic *Transzendenz* and *Inkonsistenz* (cf., e.g., (Cantor), 443–444)—would include the following:

2.1 "the" hereditarily finite sets HF, and their numerical counterparts, "the" set N of "all" natural numbers;

2.2 "the" continuum, construed either as:

2.3 "the" collection of "all" subsets of the natural numbers; or

2.4 "the" collection of "all" sequences of countably consecutive coin-tosses; or

2.5 "the" collection (or Stone space) of "all" countable models or interpretations of Peano arithmetic (the usual formal axiomatisation of number theory);
2.6 "the" smallest "natural" model of Zermelo set theory;
2.7 "the" class of "all" constructible sets;
2.8 "the" class of "all" sets ("the" set-theoretic universe);
2.9 "the" (hyper)class of "all" classes.

All of these notions are nominally "extensional". A bridge to more obviously "intensional"—and perhaps "metaphysical"—notions may be provided by Cantor's original example of "intentional" *Inkonsistenz*, in his letter to Hilbert, cited above:
2.10 "the" "*Inbegriff* "*alles*" *Denkbaren*" ("aggregate of everything thinkable").

The "aggregate" invoked above is a rather expansive example of Aristotelian "divine thought" ("*noesis theia*"), but Cantor seems to have thought of it in more or less Kantian terms, as a kind of Kantian *Grenzidee* (cf., e. g., *KrdrV*, B 310–311), a notion Hilbert himself later invoked in a reflective passage of his essay "*Über das Unendliche*" ((Hilbert), 288; cf. also a related sense of "*ideale*" Objekte, discussed earlier in the same essay).

Typically, one secures the nominal "extensionality" of notions such as the ones cited in 1–9 by metatheoretic appeals to echeloned "intensions" and "intentions"' notions which become interwoven with them in indefinitely complex ways.

Theory-relative "existence" of all these notions, for example, can be posited in a wide assortment of "stronger" finite or recursively axiomatisable theories, which become "intelligible" if they can also "talk about" (or "encode") their own intensions, in ways outlined below.

Most thoughtful adherents of various forms of mathematical "*realism*" or "Platonism" would consider such existence-assumptions—which I assimilated to "mere" Platonic *dianoia* above in 1.2—inadequate to the task.

But a growing instrumentarium of "semantic paradoxes" and well-accepted dialectical and limitative results have shown that such semantic notions—if they are consistent—require further postulation of theory-marginal "intended interpretations" in their turn. And "all" of them, *ad indefinitum*—together with "all" the metatheoretic contexts one posits to formulate them—can then be (re)interpreted in "nonstandard" ways.

Among these more obviously semantic notions and intensional constructs, one might cite the following—in roughly ascending order of likely appearance, once again, as one invokes them to secure the "extensional" unicity of notions such as those listed in 2.1–2.9:

2.11 "consistency";
2.12 "interpretability";
2.13 "existence";
2.14 "definability";
2.15 "truth" ("satisfaction"); and
2.16 "standardness" (or "intendedness"; or "well-foundedness").

I will focus primarily on these notions in the remainder of this section.

Consider, for example, the result of examining the notions of consistency, interpretability, existence, definability and "truth", in theoretical contexts in which relevant "coding" (or "quoting") arguments make this possible.

We begin, therefore, with a particular "intelligible" theory T, fixed until further notice, which is "parsable", in the sense that it is recursively axiomatisable, and "autological", in the sense that we can "talk about T" in encoded ("quoted") intensions [p] of formulas p in the language L(T) of T, and prove things inductively about the structure of these intensions. Hilbert, Skolem, Gödel and others observed that relatively weak set theories or number theories do permit us to carry out such (information-theoretically indispensable) procedures, in systematic and natural ways.

Working along such lines, it then becomes possible (*via* straightforward but tedious coding-arguments, which I will here suppress) to define, for each expressible property $p(\_)$ in the language L(T) of T, a function-symbol, which I will call

2.17 self-referential application of (_), or self-reference (_),

and prove the following, for each such formal property $p(\_)$,

2.18 **Gödel's Diagonal Lemma**

For each expressible property $p(\_)$,

(i) self-referential application of $([p(\_)])$ equals $[p([p(\_)])]$, the code or intension of $p(\_)$ applied to [itself]; and

(ii) the sentence δ = self-referential application of $([p(\_)])$ has the property that δ is equivalent to $p([\delta])$.

Gödel himself used this lemma to prove that the encoded assertion that [T is inconsistent] is metatheoretically consistent with T, for theories T "strong" enough to insure that this assertion makes sense, but other intriguing paradoxes abound.

Suppose, for example, that "rigorous philosophical analysis" had led the "late" Wittgenstein, say—who sarcastically contemned Gödel's arguments—to formulate a property $p(\_)$ which asserts that "$(\_)$ is 'philosophically irrelevant'". Then the sentence

2.19 δ = (self-referential application of ([the assertion that self-reference $(\_)$ is philosophically irrelevant]) *is* philosophically irrelevant

would be provably equivalent to [its own] "philosophical irrelevance". Similar arguments would yield sentences provably equivalent to (putative formalisations of):
2.20 [their own] Stoic *akatalepsia*;
2.21 [their own] skeptical *epoche*;
2.22 [their own] Kantian *Transzendenz*;
2.23 [their own] Wittgensteinian *Unsinn*; ....

For any first-order "intelligible" theory—systematically axiomatised, and "strong" enough to talk about [its own syntax and semantics]—Gödel also proved the following results;
2.24 The "Internal" ("Object-Theoretic") Completeness Theorem
[T is consistent] if and only if [T has a model or interpretation]; and
2.25 The "External" ("Metatheoretic") Completeness Theorem (without the "[...]"s) T is consistent if and only if T has a model or interpretation.

Consistency of T, therefore (cf. 2.8) is metatheoretically equivalent to T's interpretability (cf. 2.9), and thus to T's semantic "existence" in senses introduced by David Hilbert in (Hilbert 1935), 300–301, and (Hilbert 1930), 257–258 (= (Hilbert 1905), 182–183).

Inconsistency of a given T—with or without the '[...]'s—may be interpreted in the light of Gödel's observations in two equivalent ways: that
2.26 T *is* absurd; and that
2.27 T *is* uninterpretable.

Gödel's incompleteness theorems give rise to an interesting philosophical question, examined in detail in the sequel: what "would it be like" to "live" in a semantic interpretation **M** of T in which [T is inconsistent] happened to hold.

It would mean the following.
2.28 No evidence will ever be found to undermine or refute T in **M**. But neither will any concrete realisation of T ever be encountered in **M**. For in any metametatheoretic venue in which one can "identify" it, none exists in **M**.

The most natural semantic interpretation of this situation is that T would remain forever globally "[nonconstructive]" and "[abstract]" in a potentially illimitable ramified hierarchy of ever-more capacious Platonic "caves" **M**, **M'**, **M''**, ..., even though it is locally—sentence by sentence—correct in **M**.

2.29   If interpretability were construed as conceivability, for example, [uninterpretability of S] in **M** would mean that

[s is "the case", even though s can never be "conceived" or "perceived" to be the case] in **M** (an observation which might have interested Bishop Berkeley).

Tangentially, such thought-experiments involving metatheoretic interpretation **M** of S in which T is interpretable but [s is not] thus provide a reasonably natural interpretation of nonconstructive concept formation, and a rebuttal—as I point out in Chapter 6—of a rather strong form (one that does not rest on an equivocation between "conception" and "perception") of George Berkeley's so-called "master-argument" (cf. (Berkeley), I, 50–51).

In Chapter 6, I also characterise [s is consistent with T] as conception "*de dicto*" of s in T—a relatively weak nonconstructive notion—and consider also a supplementary, stronger notion of conception "*de re*": s is conceivable *de re* in T if and only if there is a definable interpretation **M** of T in which s holds.

Plato, of course, explicitly considered problems of "definition" and "(in) definability", in connection with the forms and elsewhere (cf., e.g., *Republic*, 531e and 534b, *Sophist*, 238c and 259e, *Parmenides* 133d, 134b–c and 135a, and *Theaetetus*, 207c–d, 208c–e, and 210a), and I will comment on some of these problems in the next section.

At this point, I would simply like to observe that "definability" and "conception *de re*" *for* a given T are not even metatheoretically expressible by any predicate *in* the language of T. If it were, one could use arguments that parallel Cantor's original "diagonal" anticipations of Russell's paradox to enumerate the definitions in the language of T, and "define" an (*ex hypothesi*) "indefinable" interpretation.

The conclusion of this metatheoretic reductio-argument—an obvious cousin of the more familiar Berry paradox (which asks us to consider, e. g., "the least natural number not definable in the English language in less than 100 syllables")—is therefore that "definability" is "internally" indefinable.

Less provocatively, the argument yields

2.30   that definability (by Russellian definite descriptions) is a contextual and theory-relative notion (so there are many "definabilities") and

2.31   that these contextual definability-predicates require implicit but recurrent appeals to metatheories, which cannot, in turn, provide for such definitions for [themselves].

If one considers seriously the (indefinite) boundaries of human finitude—adumbrated so vividly in the metaphor of the cave—these analogical conjectures and results, and variants of them sketched below, may seem less surprising.

They would simply confirm—under the aspect of dialectical eternity—a not unfamiliar hypothesis:

2.32 that significant hermeneutic and alethic notions such as interpretability, definability, and (*a fortiori*) "truth" and the platonic "forms", may—rather like Kantian *Grenzideen*, perhaps, as in 2.10—be as dialectically elusive as Stoic claims to identify an irrefragable criterion of "truth" exercised by a nonexistent "sage".

Or at least

2.33 that such hermeneutic and alethic notions are relational—"merely" *dianoetic*, in platonic terms: conjectural and schematic templates, which our finite intellects are fortunate enough to be able to project for further investigation and dialectical and practical inquiry.

"Truth" of a theory T, for example, is a formally rather jejune second-order notion, a kind of blank template ("Convention T"), as one learns in (Tarski). Even interesting second-order extensions of Tarski's original ideas developed by the authors of (Kripke) and (Hintikka) (among others) remain sophisticated "redundancy"-theories—parodied, in effect, by the philosophers' mantra that "'Snow is white' is true if and only if snow is white".

Little substantive emerges from such "disquotational" analyses (in my view), unless we refine them to relational notions, of truth in a structure, viewed in the context of an appropriate metatheory (we have to be able identify a given "cave", if you wish, by the light of a given contextual "sun").

And even then, there will usually be infinitely many branching, non-isomorphic candidates for such structures, and these may be considered in infinitely many branching, nonequivalent candidates for appropriate metatheories.

Correlatively, the common practice of philosophers of language—to refer to such transitions as "mere" changes from object-language to metalanguage—is wrong. If one wants to prove that alethic predicates "mean" what we want them to "mean", we must enlarge and reinterpret the theories we formulate in such languages as well.

Tarski further showed, in effect, that substantive individuation of such structures can only be undertaken in metatheoretic contexts, that cannot be both unique and internally expressible (cf. the remarks about "complementarity" in 2.36).

2.34 One obvious way to cope with the situations such metalogical results identify—as Tarski himself recognised—would simply be to acknowledge a recurrent need for recourse to wider and wider metatheories, whenever we wish to ground various object-theories' "intended" semantics. Here and elsewhere I refer to this response as admission of a recurrent need for metatheoretic ascent.

For some reason, most (but not all) philosophers have seemed to find the need for such "ascent" unsatisfactory.

2.35 Typically, such philosophers have sought or simply postulated a kind of last or global metatheory with some special properties—metaphysical, epistemic, "practical", "pragmatic"—that would allegedly render that theory [self]-sufficient—or [self]-"grounding", or [self]-"illuminating" if you like (cf. Sections 4 and 5).

Postulation of an "intended interpretation" in this putative context, for example, would serve to interpret Plato's most ambitious, semantic monist desiderata for the forms.

Indeed, one might assimilate this Last Metatheory itself to Plato's allegedly universal, ontologically generative and "truth"-conferring Form of the Good. The reflexive, [self]-constituting aspects of this Metatheory might also recall Aristotle's characterisation of "god", as that Entity which "grounds", "moves" and "thinks about" everything that does not "ground", "move" or "think about" itself.

If, indeed, "all" [self]-referential acts and intentions were "intended" by the universally [self]-actuating *energeia* Aristotle imputes to *ho theos* and *to theion*, such an Aristotelian Entity would have to "ground", "move" and "think" about [itself], by one half of the argument Russell used to develop his paradox. (Theory-relative counterparts of this *energeia*—whose "thought(s)" would not be uniquely determinate—need not undertake such [self]-creation; cf. Section 5.)

2.36 Both "speculatively" and "practically", however—if I may resort to Kantian language once again—attempts to formulate such "global", [self]-grounding theories cannot escape a kind of complementarity between the intentional finitude of our conceptual horizons, and lack of closure of the concept-formation(s) we undertake within them.

For such "global" theories face a choice between two complementary alternatives: between undecidability and inexpressibility.

This (literal) dilemma—a simple consequence, in this case, of Gödel"s observations about [internal] diagonalisation, sketched above—may be posed as follows.

2.37 An expressible (internally encodable) theory—of the integral "self" of the Stoics, say; or "The Form of The Good"; or Aristotelian [self]-regarding divinity; or "that than which nothing greater can be conceived" (cf. also Kant's later canvasses of "*psychologische, kosmologische und theologische Ideen*")—must be incomplete. It can formulate, but never prove, [its own] existence (consistency / interpretability).

By simple contraposition, then:

2.38 Any complete limiting (meta)theory of such metaphysically *inkonsistente Ideen* could not formulate or express (encode) [its own] existence.

To put a mildly provocative point on it:

Even if "the" Form of the Good, or *katalepsis* of "the" Stoic *sophos* (or "*die*" *Selbstgesetzgebung* "*des*" *heiligen Willens*; or ....) were

2.39 provably "complete"—in ways I will assimilate in the next section to The Sophist's "dichotomisation"-arguments—

and therefore

2.40 "existent" and [self]-interpreting, in some tacitly metatheoretic venue (presumably given by the notions themselves),

then:

2.41 such a "Form" or "*katalepsis*" (or *Wille*; or ...) could be relegated, in this tacit metatheoretic context, to the role of a mere object-theoretic *demiurge* (merely empirical / hypothetical simulacrum of [itself]);

and moreover,

2.42 such a "Form" or "*katalepsis*" (or *Selbstgesetzgebung*; or....) could not, after all, even in this metatheory, [internally] define [itself] (cf. *Grundlegung*, 463).

This recurrent complementarity—between expressibility and metatheoretic universality—seems to me to take many cognate forms in the history of philosophy, in ways which form this monograph's basic *Ansatz*.

More precisely, I have argued that we cannot ensure, or even canonically express, certain ideals and *Grenzideen*, with anything like the constitutive *Vollständigkeit* and semantic unicity that most classical metaphysicians (and most analytic philosophers) demanded of (and implicitly attributed to) "truth". I will return to variants of this argument in Sections 3 and 4.

3 Metaphysical "Forms"

Plato's well-known parable of the cave is set out with great eloquence and mordant political irony at the beginning of Book VII of the *Republic* (514a–517b). The historical, literary and political resonance of this parable may exceed that of any other thought-experiment or "mere likeness" (*eikasia*) in the history of western philosophy.

Somewhat less attractive to poets and other *littérateurs*, however, has been the model of the "divided line", introduced just before, at the end of Book VI (509d–511e), which the parable allegedly illustrates.

A certain diminution of poetic interest may follow not only from the model's literal "linearity", but also from its "pythagorean" obsession with apparently gratuitous numerology (in this case, the equiproportionality—cf. 508b–c and 509e—of the line's various internal divisions; cf. the notorious "nuptial number" for the ages at which guardians may marry and give birth, at 546b–c). We will return briefly to Plato's meticulous concern with the line's internal proportions below.

Plato's fascination with mathematics, at any rate, may help legitimate the following (obviously anachronistic) questions, for which I have already proposed tentative answers.

3.1   Do mathematicians' "intended interpretations", first, provide "forms"—or at least formal miniatures or informative metamathematical counterparts—for Plato's "forms" (or conversely)?

3.2   Do the patterns of metamathematical perspective-shifts I have called metatheoretic ascent in the last section reflect or interpret critiques of the forms in potentially significant ways?

3.3   Can one draw from these and other metamathematical analogies any significant implications for the parable of the cave, or for the characteristic two-stage linearity of Plato's "line"-model for noetic enlightenment?

Most of this section will be devoted to 3.1 and 3.2; 3.3 will be taken up at greater length in Section 5.

I will begin with a brief review of some aspects of the "forms", and an attempt to offer some of the positive arguments for my assimilation of such forms to "intended interpretations", alluded to above in 1.2.

In Plato's metaphysics, the "forms" obviously served as idealised exemplars of some sort. But it is otherwise rather ambiguous—notoriously so, in fact—what metaphysical or ontological roles Plato may have had in mind for them, in the dialogues in which they most prominently appear.

Referring to the cascade-of-imitations argument in Book X of the *Republic*, the editor of one recent handbook suggests indirectly that the "form" of the bed (or, presumably, "the" Good; or ....) may be identified with "what it means *precisely* to be the bed" (or "the" Good; or ....) and adds that "it is only the Form that is *really* real (596e–597a)" ((Kraut), 11; the emphases here are mine). This obviously expresses something of the "exemplary" or "paradigmatic" quality of the forms, alluded to above. But what might it "mean" to "mean *precisely*", and to be "*really* real"?

On the analogy between "forms" and "intended interpretations" I have suggested, it might "mean" that forms are idealised archetypal semantic entities, adumbrated by assorted syntactical, theory-internal ectypes, but never adequately represented by them (in some Spinozan sense of the word "adequate").

Notoriously, Plato also employed several different verbs to express relationships that might be said to obtain between such ectypes and their "formal" originals: among the usages that appear are *mimein* (imitate); *koinonein* (share) and *metechein* (participate in). What might these relationships "mean", on the analogy I wish to propose?

To answer this, it may be helpful to recall the dichotomising construal of the forms Plato offered in several passages of the *Sophist* and the *Parmenides*, among them the following (*Parmenides* 136a and 136b–c):

> ... you must examine not only the consequences of a hypothesis if each proposed (hypothesised) thing *is*, but also the consequences of hypothesising that that same thing is *not*....

> ... in a word, about whatever you may hypothesise as being or not being or undergoing any influence whatever, you must examine the consequences with respect to it and each other thing, whatever you may choose, and with respect to more and even all things in the same way. And the other things, in turn, you must examine, with respect to themselves and whatever else you may choose, whether you hypothesise what you hypothesise as being or as not being, if you wish in a completely trained and comprehensive way to see (through) the truth.

Many of these passages are standardly construed as taxonomic analyses of forms as "*infimae species*". More adequate and suggestive metalogical counterparts for them, however, might be found in Marshall Stone's topological analysis of a space $S(T)$ of interpretations of a given theory T, now called the Stone space of T (cf. (Bell and Machover), 143), for two reasons.

The first is that such interpretations are transparently semantic entities, as any respectable metalogical counterparts of the forms should be.

The second is that S(T) will include—for a wide variety of interesting theories T—"incorrect", "nonstandard" or "unintended" interpretations **M** of T, as well as "correct" "standard" or "intended" ones (forms?), **F**.

Stone proved the following pair of results more than fifty years ago.

3.4   Each interpretation **M** of T may be represented in a natural way by a unique list of decisions about assertions of L(T) that are undecidable (neither provable nor refutable) in T.

3.5   Furthermore, this representation maps the Stone space S(T) of interpretations of T homeomorphically onto the Cantor space, 2^N, of countably many binary choices (as "being or as not being")—in this case, choices about sentences s that are undecidable in T.

Such an infinite canvass of undecidable assertions in L(T) corresponding to a given "form" ("intended interpretation") **F** = **F**(T) for T ("the form of T") is then my metamathematical analogue for Plato's interpellation of claims and counterclaims about **F**. A particular sentence or description s in the language L(T) of T is "opinable" if and only if it is undecidable in T, and such an s partakes of or participates in the form **F** if and only if p holds in **F**.

On this interpretation, moreover, a "truly" exhaustive, potentially infinite canvass of each such "merely opinable" sentence s about a given form **F** = **F**(T) —each s

> ... which partakes of both, being and non-being, but is called neither absolutely
>
> (*Republic*, 478e)

would (counterfactually) yield not "mere opinion" about **F** = **F**(T), but "knowledge" about **F**.

In mathematical logical terms, once again, **F** might be individuated as the "standard" or (extratheoretically) "intended" interpretation of T.

There will also, however, in all the interesting cases mentioned in Section 2, be many other (imperfect?) models or interpretations **M** of T (badly designed "beds", *et al.*).

3.6   Such "nonstandard", "imperfect" interpretations **M**, moreover, might also be said to "partake of" **F** in a partial and graduated sense, as follows.

Each undecidable s of the form considered above can be identified with an element **b**(s) of a boolean algebra **B**(T) of "truth-values", defined in a natural way from S(T). Then the (boolean) truth-value [[**M** partakes of **F**]] of the assertion that "**M** partakes of *F*" is the supremum in **B**(T) of the values **b**(s) such that **M** and **F** both satisfy s.

If this "truth"-value is **0** (the least element of **B**(T)), then **M** is utterly discrepant from the "form" **F** (a very inadequate "bed"; or mere "*eikasia*" ...).

If that value is a **b** which is intermediate between 0 and 1 (the greatest element of **B**(T), then **M** might be regarded as a "mere image" or "copy", which "partakes of" **F** in a correspondingly intermediate way.

If the value is 1, however, then **M** is "elementarily equivalent" to **F**: it cannot be discerned from **F** by sentences in L(T). One might reasonably paraphrase this by saying that **M** "partakes of" **F** as fully as can be expressed in the language L(T).

If, indeed, the original theory T is rich enough to "witness" in some nominal way every existential claim made in its language L(T), then **M** essentially *is* **F**, up to isomorphism.

3.7 All of the assertions just made in 3.4–3.6 are to be understood in the sense of some fixed but tacit metatheory, in which "one" can decide whether they (seem to) make good mathematical sense.

Whether and where an "ultimately" extratheoretic or "final"-metatheoretic venue might "exist", however—an ultimate "one" in which a definitively "true" canvass of these sentences *s* which converge to the "intended" **F** could be canonically discerned from others which converge to mere simulacra—is another and much more elusive question, whose *aporiai* lead naturally into the "negative" aspect of these analogical arguments, mentioned above.

For we could query "the" "true" canvass of this venue"s "form" (intended interpetation) **F** in its turn (another instance of the "metatheoretic ascent" of 2.34; cf. also 3.13 and 3.15).

The "negative" evidence for my assimilation of forms to intended interpretations, referred to above in 1.2, accordingly, will focus on two related clusters of properties that the forms and intended interpretations share:

3.8 their apparent vulnerability to "regress"- ("ascent"-) arguments, and
3.9 their ultimate inexpressibility, which emerges when one tries to "cap" such regresses.

In both cases, I will argue, the problems of 3.8 (which seem to me less damaging than they did to Aristotle) can only be mitigated at the price of accepting 3.9 (which seems to me much more serious).

The extratheoretic nature of existence-claims about "intended interpretations", first—mentioned above in 1.1—might seem an aspect of the assimilation that twentieth-century mathematical counterparts of the Socrates of *Republic* Books VI and VII would have reason to welcome.

For *aporiai* of reductive attempts to express certain limiting mathematical ideas in merely formal terms would then become readily understandable. They would simply be instances of mathematical "*dianoia*"—intellection that takes place somewhere in the "lower" part of the line's upper, "intelligible" section.

"True" understanding of such ideas, by contrast, would inevitably require the deeper, "higher" insight of mathematical *noesis* (which Gödel, in a tacit inversion of Kant's architectonic, called "intuition"; for more about mathematical appeals to "intuition" see 3.26).

Other aspects of the assimilation, however, have consequences that twentieth century mathematical realists or "platonists" would prefer to explain away or avert.

There is, first, a sense in which natural metamathematical critiques of mathematicians' tendencies to posit intended interpretations parallel or reflect certain aspects of Parmenides metaphysical critiques of the forms, as well as Aristotle's subsequent allusions to *tritos anthropos*-arguments, in his review and endorsement of such critiques (cf. *Parmenides*, 133a and *Metaphysics* 990b18).

Indeed, the hypostatic quality of a mathematical "intended interpretation" recalls that of

3.10 the putatively unique extratheoretic exemplar Aristotle called in the *Metaphysics* (cf. 990b18; also the *Parmenides*, 133a) a "*tritos anthropos*" ("third human")—something that is

3.11 neither an x (particular human), nor a (quasi-Aristotelian) form of x (formal predication of x's "humanness" or "humanity", in some appropriate metatheoretic context); but an ambiguously situated "third x" ("third human").

Such a "third x"—postulated in Plato's theory of forms, according to Aristotle—would indeed be deeply problematic, though not quite for the reasons Aristotle thought. Recurrent appeals to metatheoretic contexts to interpret alethic notions may be quite coherent, *contra* Aristotle. Indeed, I will argue below that they are regulative of conscious inquiry.

But Plato's exemplars would have to do more than merely guide our inquiry into an ultimate *eidos* or form in heuristic ways. Guides would usually be expected to admit that there might be other, perhaps better guides.

Such infallible guidance, I will argue, would require

3.12 that there "must" exist one and exactly one "universal" interpretation of sufficiently comprehensive (consistent) theory T (which must, then, be the "intended" interpretation of T, if there is one);

or equivalently,

3.13 that there "must" also "exist" a last metatheory for T, that uniquely interprets [itself], and therefore "proves" or "grounds" [its own existence] (with or without the "[....]"s; at this lofty level, the distinction is usually blurred).

On the analogy I am suggesting, then, a "third *T*" for an ambitious but incomplete theory T—such as Peano arithmetic, for example, or ZF—would be provided by

3.14   its (Platonic, not Aristotelian) "form", an (allegedly) unique "intended" interpretation of T.

As such, this "intended interpretation" would mediate between the abstraction (the syntactical/theoretic "form" of x) and particular partial realisations of it in particular x's, but not (it is claimed) in any "merely" psychological or "dianoetic" way.

One might, of course, ("merely") "intend" such interpretation(s) in the sense of appropriate background theories (in which they might turn out to be multiply interpreted in their turn).

But this alternative (proposed originally in this century by Thoralf Skolem) would lead to an indefinite hierarchy of metatheoretic "intentions"—2.34's "metatheoretic ascent"—and most philosophers of mathematics who take mathematical-realist ontological positions about infinitary notions seem to find this no more satisfactory to their philosophical temperaments than Aristotle did, even though their own formal inquiries tell them that such notions can never be encompassed in any "final" (meta)theoretical context.

This view then leads such mathematical realists, in effect, to posit a similar sort of "final" metatheoretic "*tritos anthropos*", or perhaps some sort of extra-theoretic "*tritos nous*".

To "ground" any sort of [self]-constituting Platonic form or Anselmian maximum or Leibnizian limit of such metatheoretic positing(s), however, the mathematical realist or metaphysical "platonist" would also have to posit 3.13's Last Metatheory, or great Metatheoretical Subject—"*die*" *Grenze seiner metatheoretischen Welt(en)*, as it were—a close analogue, I believe, of the "final", regress-grounding "formal reality" of an Aristotelian *prote aitia* [first cause], and a correspondingly begged solution to the supposed problem of "one over many", with a vengeance.

Following Aristotle, again, or at any rate 3.11, we might observe that such a putatively universal Metatheory, for an already rather comprehensive theory T, would have to be

3.15   neither the (quasi-Aristotelian) "form" (syntactical theory) of T (T itself); nor any one of the many particular interpretations of T, in any particular (specifiable) metatheory; but "the" third T: an "intended" interpretation of

T (and perhaps of more as well) that "must" somehow provide [its own] metatheoretic ground.

If there is any message that emerges from the practice of twentieth-century metalogic, in my view, it is that there is no such complete, [self]-interpreting "third T".

For such an entity would not only have to have the "unconditioned completeness" and *durchgängige Bestimmung* Kant associated with *transzendentale Ideale*, in another analogue of the "dichotomisation"-analysis reviewed above (cf. 3.4–3.6 and *KrdrV*, B599 ff). It would have a deeper problem: it would be, as I've already remarked, inexpressible to [itself]. And it is certainly arguable that nothing can provably "ground" or "interpret" what it cannot express.

3.16 Correlatively, it seems to me that attempts to "ground" (end) metatheoretic ascent in [self]-interpretation—allegedly secured in the sort of "last" metatheory considered above—are parade-examples (*pace* Spinoza) of what Kant called "*transzendent(al)er Schein*" (cf. *KrdrV*, B 352 ff.)—no less than attempts to ground the god(s) of Aristotle, Anselm and others in [self]-causation, [self]-interpretation, [self]-origination, [self]-organisation, or [self]-design.

This comparison, in turn, suggests again that "the" "intended interpretation" of infinitary theoretical entities may be the very model of a Kantian idea-of-reason, as suggested above in 1.3.4. Study of this quasi-metalogical "idea", in turn, may yield further insights into common underlying structure(s) that are implicit in Kant's three overarching *Vernunftideen*—the "psychological", "cosmological", and "theological" ideas.

For an attempt to "intend" and secure a canonical extratheoretic interpretation of a theoretical [self], first, would seem to parallel rather closely attempts to render some sort of "psychological" [self] an "archimedean point" for assorted forms of relativism and skeptical doubt, in ways outlined in (Boos, 1983b), and reviewed briefly in Section 4.

Attempts to find 3.9's last, "perfect", self-realising metatheory, moreover—for theoretical notions that are putatively open-ended as well as object-theoretically universal ("the" set-theoretic universe, once again; or "the" hereditarily finite sets; or thoughts about "divine" thought; or attempts to conceive "the" limits of conception, at "the" top of Plato's line) would aspire to complete [themselves], in ways that suggest both the ontological argument, and other—Aristotelian as well as Aquinian and Spinozan—regress-ending theological arguments, mentioned above.

3.17 The deeper underlying problem of the "thirdness" Aristotle identified, finally, is not that it gives rise to "insupportable" iterations of metatheoretic regress *ad indefinitum*, though Aristotle makes no attempt to disguise his distaste for such regresses (or ascents) and presumably, therefore, for the ramifying semantic pluralities of 2.34 and 2.37–2.42.

The problem is rather that the mathematical and metaphysical realist emphatically does share this distaste, and more specifically, that

3.18 any such iterations would be insupportable for Plato's semantic monism, for they would conflict with the metaphysical closure he clearly intended the forms to provide; and

3.19 the only way to cap such iterations would make his alethic exemplars inexpressible and unintelligible, in ways that Plato (unlike Plotinus) apparently did not wish to accept.

A natural metamathematical analogue of the "regress" is the process of metatheoretic ascent, mentioned above in 1.3.3, which emerges naturally from the intrinsically aporetic nature of searches for "the" intended interpretation. For such searches may never come to rest in any particular metatheory U, since U itself will in general be plurally interpretable.

In this and subsequent chapters, I assimilate the open-ended iterations of "metatheoretical ascent" to the patterns identified in the skeptical "problem of the criterion", and more particularly the Agrippan "mode" of "*ekptosis eis apeiron*", or "lapse into infinite(s)" (cf. also the hierarchies of "abstraction" Leibniz wryly contemplated in, e. g., (Leibniz-Couturat), 512–513). I will return to this analogy below in Section 4.

3.20 Here—prompted by such assimilations, and by the expressibility-problems sketched above—I would simply argue for provisional acceptance of ramified "Peircean" limits of inquiry generated by such interpolated "thirds", and indefinite ramifications, therefore, of the "forms" that go with them.

In effect, then, I am suggesting that Parmenides' and Aristotle's critiques are apt, but only decisive if one shares Aristotle's abhorrence of such indefinite metatheoretic ascent. What is regressive in the mind's eye of one philosophical beholder may be progressive in that of another. I will also return to this point in subsequent sections.

3.21 There is little doubt in my mind, in any case, that we need to refer to the semantic notions mentioned above in (e.g.) 2.11–2.16. Such appeals, if you like, are transcendental preconditions of the possibility of inquiry. We make them (I believe) from earliest childhood in tacit ways.

If they are either theory-relative or ineffable, therefore, then so much the worse (I would argue) for ineffability. For such appeals are also preconditions for the possibility of communication. And the latter is intertwined with inquiry in ways that we will never unravel.

The consequence of this acceptance is not, as "Socrates" claims in the *Theaetetus*, some sort of incoherent relativism, but simply an acceptance of an emergent plurality of metatheoretic interpretations for the notions adumbrated above in 2.11–2.16.

By this I mean, once again, that they are only expressible in recurrent metatheoretic contexts, in which their interpretations will fail, in general, by the Gödelian analyses cited above, to be unique.

3.22 On the assimilation I offer, therefore, the unicity Plato demanded of his ultimate forms or semantic exemplars—and the unicity mathematical "platonists" demand of their intended interpretations—could coherently be maintained only at the expense of their (metatheoretic, or metametatheoretic, or ....) expressibility or definability.

Genuinely "noetic" archetypes of "merely" dianoetic ectypes in both cases, therefore, would have to be ineffable, in senses that can be made metamathematically precise. To the extent one grants the metaphysical relevance of the analogical framework, moreover, one would also expect that a consequential "Platonism", so construed, would eventuate in something very like the "Neoplatonism" associated with Plotinus and his followers.

Twentieth and twenty-first century mathematical platonists, then, may face a similar dilemma.

3.23 I would therefore like to close this section with some idiosyncratic observations and conjectures about the interrelations between 2.1–2.9 and 2.11–2.16 that have emerged in the divergent *praxeis* of mathematical and philosophical logic.

The plausibility, first, of some sort of analogical relation between Platonic forms and metamathematical *Gedankengut* is already suggested by the commonplace nature of talk about "Platonism" in twentieth- and twenty-first-century philosophy of mathematics.

This usage—popularised especially by Paul Bernays—is admittedly very loose. People who uphold other views, for example, often employ it as an elenctic synomym for various forms of mathematical "realism" which they find objectionable.

3.24 Defenders of the *reine Lehren* of intuitionism and other forms of constructivism, in particular, sometimes use "Platonism" as a pejorative tag for any talk about countably or uncountably infinite entities, however provisional, nominal or hypothetical such talk might be.

Constructivists as well as other philosophers of mathematics, on the other hand, often dismiss attempts to interpret infinite entities in theory-internal ways as "(mere) formalism" (another term popularised in the philosophy of mathematics by Bernays).

3.25 Both these dismissive usages, in my view, amount to metamathematical *ignorationes elenchi*. For they ignore approaches to infinity, pioneered by the theory-relativist Thoralf Skolem, which have been enormously productive in the history of twentieth-century mathematical logic.

Practitioners of these approaches typically relativise particular ontological (semantical) assertions about infinite entities to particular (meta)theoretical contexts, in case-by-case ("processive") ways.

They do not, however, treat them as mere uninterpreted symbology. They work with them hermeneutically, as (syntactical) interpretations—"conceptual schemes", if you like—that can be (plurally) "intended" in immanent and provisional ways.

Most mathematical logicians do not do this as a matter of ideology, moreover, but rather as an straightforward information-generating aspect of their metamathematical practice. Set- and model-theorists, for example, typically exploit the theory-relativity of 2.11–2.16 in intricate ways, to clarify aspects of the fine structure of 2.1–2.9.

3.26 Philosophical logicians, by contrast, often propose elaborate epicyclic adjustments to the notions mentioned in 2.11–2.16, refinements and interpolations which promise to minimise or suppress such theory-relativity, if only in rather nominal ways.

In this sense, standard allegations of set-theoretic "Platonism" often distort the actual normative shadings of much mathematical practice, and attribute ontological views to set theorists that most of the latter would find gratuitously naive.

Set- and model-theorists, for example, often try to prove independence-results of great subtlety, and these efforts require delicate respect for the relativity ("non-absoluteness", in set theorists' terminology) of various sorts of mathematical claims. These efforts prompt them to seek multiplicity and ramified complexity of results-to-be-proved, rather than the canonical unities ("*henades*" and "*monades*" in the *Philebus*, for example) that Plato sought in the forms.

Many philosophical logicians, by contrast (among them many professed intuitionists), tirelessly seek unitary "forms" under the supposed welter of mathematical "appearances." To this end, they search dutifully for new epicycles which might suppress the regresses of metatheoretic ascent, and "save" (in effect) some metamathematical counterpart of the semantic monist hypothesis that there is, at the level of such "forms" and "intended interpretations", a mathematical analogue of Plato's *anupotheton*.

Such philosophical logicians, I will suggest in the sections that follow, are the true heritors of "Platonism" in the philosophy of mathematics.

4 "The Very Idea" ("*To Eidos Auton*") of Unitary Bivalent "Truth"

In section 2 and the "negative" arguments offered in Section 3 to support my assimilation of metamathematical "intended" interpretations to Platonic forms, I have outlined a critique of the position I called "semantic monism", and sketched an alternative semantic pluralism or theory-relativism based on the metalogical process I called metatheoretic ascent.

In this section, I will try to respond to two aspects of "Socrates"'s well-known "refutations" in the *Theaetetus* of the mitigated relativism to which such plurality of interpretations gives rise. I will then respond in the next and final section to the position that such theory-relativism is dialectically clever (perhaps) but ethically as well as metaphysically and epistemologically jejune.

As I remarked above in 1.3.1–1.3.4, this *elenchos* holds that such relativism would inevitably

4.1 undermine "all" feasible semantic criteria; and in particular,
4.2 undermine or elide "itself".

For reasons I will elaborate below, let me call the claim made in 4.1 Plato's "transcendental argument", and the subtler dialectical assertion in 4.2 his "Stoic" or "Cartesian refutation".

I will attempt to argue

4.3 that the claim in 4.1 rests on a tacit metatheoretic quantifier-reversal which begs the position it is supposed to "prove";

and that

4.4 the assertion in 4.2 is undermined by

4.5 the indefinitely ramified structures of metatheoretic revision associated with the patterns I called in Section 2.34 metatheoretic ascent; and by
4.6 a "diagonal" (Gödelian) construal of mitigated-skeptical *epoche*.

Counterparts of 4.3 and 4.4 also seem to me to apply to the anachronistic name-givers of 4.1 and 4.2 (cf. below).

Let us examine, then, the justification for 4.1, a recurrent *Deduktion* that has often been informally alleged to secure, in some elusively extratheoretic sense, the "existence" of a definitive "criterion of truth", ultimate "intended interpretation", or other form of semantic closure of the sort considered in Sections 2 and 3.

I will call this argument-type, for lack of any other name, the metatheoretic **EA**-argument, from a standard usage for a sentence with an **EA**, existential-universal ("there exists-for all") quantifier-structure: a sentence, in other words, of the form "there exists an x such that for all y (s(x,y))." **AE**- or universal-existential sentences, by contrast, have the form "for all x there exists a y such that (s(x,y))."

A metatheoretic **EA**-argument does clearly play a prominent role in one of Plato's "refutations" of the semantic relativity "Socrates" attributes to Protagoras in the *Theaetetus*, when Socrates fallaciously claims (in effect) that anyone who rejects the **EA**-claim,

4.7 that there must exist a single "best" objective or intersubjective standard for all human judgments, must also reject a cognate, but quantifier-reversed and weaker, **AE**-claim,
4.8 that for all human judgments, "better" contextual standards exist for evaluation of those judgments' truth.

A similarly begged metatheoretic **EA**-argument—to the effect that (not (there exists an x such that for all y s(x,y)) would "imply" (not (for all x there exists a y such that s(x,y))–also seems to me to play a strong but tacit role in Kant's "*transzendentale Deduktion*", but I will not try to elaborate this view here, beyond the remarks in 4.9–4.12.

At this point, I would like to emphasise a point I have argued in an examination of Gottlob Frege's dogmatic rejection of David Hilbert's multiple interpretations of "neutral geometry" in the *Grundlagen der Geometrie* (Boos, 1985): that

4.9 the specious attractiveness of semantic **EA**-arguments is typically begged on the basis of a vigorous denial of the patterns of metatheoretical reinterpretation I have called metatheoretical ascent; and that
4.10 this denial, in turn, is essentially equivalent to a reassertion of the central point at issue, the extratheoretic position I have called "semantic monism" in 1.3 and Section 3.

This view, in turn, is often begged on the basis of the very "comprehensiveness" of the theory, or at least of the theory's ontological aspirations. This is an attractive but thoroughly dubious enthymeme—as Cantor saw (when he called such theories "*inkonsistent*"), Frege later learned, and Skolem and Gödel clarified.

Sometimes, however, as in Kant's *Deduktion*, this view is begged on the basis of a slightly subtler metatheoretic argument, one that is actually related to the metalogical "cogito" discussed in the next section:

4.11 that the position I called semantic monism above must be begged, or at least assented to; for if it is not, then
4.12 incoherence would be the consequence, since we would not be able coherently to query the claim

(cf. the interpretation of the *Theaetetus*-argument above, and of the related claim that semantic relativism is "self-refuting", and therefore incoherent).

This meta-argument, dialectically attractive though it may seem, is again really little more than a kind of "transcendental" second-order *petitio* and metatheoretic reassertion of the semantic-monist claim that a theory without a unique interpretation is inconsistent.

Not only, in short, is the implication (4.9, 4.11–4.12) now known to be refutable, for relevant theories T, in any theoretical venue in which it can be expressed. It is also essentially a type-raised equivalent of what is being "deduced", as in 4.8.

The basic unsupported premise of "Socrates'" *petitio*, then, amounts to

4.13 Plato's dogmatic refusal—"supported" in the *Theaetetus* by a metatheoretic quantifier-mistake—to countenance ramifications or iterations of the "cave"-escaping' process—a pattern I call "metatheoretic ascent" in Section 2.

From this double refusal, "Socrates" implicitly derives

4.14 the capped, two-stage structure of the "line", and explicitly deduces the *Theaetetus*' well-known conceptual-schematic denial of conceptual schemes, which I have assimilated to the "semantic monism" of Section 3.

Against this background, I also wish to consider a somewhat more substantial argument "Socrates" offers against semantic relativism in the *Theaetetus*: his claim that

4.15 plurality of conceptual schemes and forms of *epoche* that may arise from clashes between such schemes are "self-refuting".

This claim ironically resembles rather closely an *elenchos* applied somewhat later by the Stoics, Augustine and Descartes, among others, to Socrates' disclaimers of "knowledge." Compare, for example, the passages from Augustine's writings quoted in (Gilson, 1976), 295, and the Stoic criticisms recorded about two centuries earlier by Clement of Alexandria (SVF 2.121):

> If (pyrrhonian) suspension of judgment says that nothing is confirmed, then it is clear that proceeding from itself, it first disconfirms itself. ... (For) either (such suspension) is itself truly asserted or not. If it is, it involuntarily grants that something is true. If not, it leaves standing as true those things it wished to confute. (But) to the extent the confuting suspension is shown false, to that extent the things confuted are shown true, *as (in the case of) the dream that says that all dreams are false. For if it confutes itself it becomes confirmatory of all the others* (cf. Descartes' "Archimedean point" at AT VII, 24). *But if it is true, in general, it will make a beginning from itself, and not be suspension of something else, but of itself first. For if someone grasps that he (or she) is human, or suspends judgment, he (or she) clearly does not suspend judgment. For how would the person reach the beginning of the dispute, suspending judgment about all things? How would the person reply to what was asked? About this very thing the person clearly does not suspend judgment, even while declaring that he (or she) does.* And if we are to be persuaded by the pyrrhonians' arguments to suspend judgment (cf. "... si quid mihi persuasi", AT VII, 25), we will first suspend judgment about suspension of judgment, whether it is to be trusted or not. And if that itself is true that we do not know the truth, then nothing is given by that (as) true at the beginning. But if one says that even *that* is disputable, that we do not know the truth, then one grants that the truth is knowable, by the very fact that one clearly does not confirm suspension of judgment about *that*.

(I have italicised some of the passage's "Cartesian" remarks.)

Augustine readily acknowledged that he did not read Greek, and there is no reason to impugn this admission. He may of course have meant that he "read" Greek slowly and laboriously (a state with which I am well familiar), and analogues or counterparts of the points made in the italicised passages may well have been available in Latin sources as well as Greek compendia.

Suggestive anticipations in these passages of the so-called "Augustinian cogito", in any case, seem fairly clear, and I will return briefly to these interrelations below.

In response to claims such as 4.14—to which many twentieth-century analytic philosophers also seem to subscribe, for reasons they seem to consider independently rigorous (cf., e.g., (Davidson) and the comments on (Putnam, 1981), 7–8 in (Boos, 1983a))—I would respond that we should suspend judgment about "ultimate" metaphysical closure and criteria of "truth". I base this recommendation on a straightforward reading of the metalogical evidence of Section 2.

More precisely, I argue that we should

4.16 assimilate pyrrhonist and academic-skeptical *epoche* and *ataraxia* to formal undecidability; and

4.17 respond to the traditional dogmatic challenge to explain skeptical doubt of skeptical doubt (the other aspect of Socrates' "refutation" of Protagorean relativism in the *Theaetetus*) with an interpretation of such doubt as

4.18 acceptance of the ramifications of potential syntactical as well as metatheoretical (re)interpretation to which any claim about alethic predicates may be subject; and in some cases as

4.19 a theory-internal diagonal [fixed point] for itself (the notation "[ ]", once again, is a variant of a traditional metalogical notation for theory-internal "intensions").

The Gödelian "dialectic" of previous sections, moreover, provides (cf. Boos, 1983b; "Theory-relative Skepticism," 1987b) a more detailed construal of 4.19 as (what might be called)

4.20 the "doubter's paradox" ([I doubt [this sentence]) (as opposed to the liar paradox—"[this sentence] is 'false'"—which is essentially inexpressible, in the formal venues to which it applies).

Along similar lines, one can also make use of Gödelian diagonalisation (cf., again, Boos, 1983b), to clarify and interpret

4.21 the wryly inconclusive, self-questioning tone "Socrates" sometimes seems to adopt in some (but emphatically not all) of the dialogues in which he appears (cf., e. g., *Apology* 21b and 29b; other references may be found in (Woodruff), 62 ff.) and rebut

4.22 Stoic and Augustinian "refutations" of such self-doubt and self-questioning, along the lines alluded to in 4.15 and 4.17, along with the "cogito"-arguments Descartes derived as epicycles from them.

Indeed, skeptical and theory-relativist forms of [self]-doubt have a "positive", "constructive" aspect as well. If one interprets

4.23 "knowledge" in "my" theory T (or, more precisely, "knowability") as (encoded) [provability] in T, and

4.24 "doubt(ability)" as [unprovability], it turns out that

4.25 the supposedly "self-refuting" doubter's paradox is provably equivalent, as Gödel showed, to the claim that [T is interpretable]; or [T exists]; or equivalently (in T) that [T is consistent].

4 "The Very Idea" ("*To Eidos Auton*") of Unitary Bivalent "Truth" — 93

Even if "knowledge" is taken to be some more general "internal" predicate that is simply implied by [provability], and "doubt" as the negation of "knowledge", a fixed point of such doubt would still imply the [interpretability of T].

If "knowledge", for example, is construed as [provability(_)] in some finite or recursively enumerable extension K = K(T), 2.18's diagonalisation of 4.24's "doubt" is the assertion that

4.26 [self-referential application of [the assertion that [self-reference (⌐)] is unknowable] is unknowable.

This can reasonably be read, in turn, as the assertion that

4.27 [self-referential application *of* [skeptical self-doubt]] is unknowable.

Since the platonic (Stoic, Augustinian, Cartesian) rebutter of "skeptical relativism" typically asserts that such [self-referential application of [skeptical self-doubt]] is exactly what the skeptic "must" know, such a rebutter would assert either

4.28 metatheoretic possession of a proof of such [self-referential application of [skeptical self-doubt]]; or

4.29 "internal" negation of the sentence in 4.27.

The metatheoretic claim represented in 4.28 would convict the rebutter (or the rebutter's epistemic theory K of 4.25) of incoherence, for Gödel showed no such metatheoretic proof can exist from a metatheoretically consistent theory K.

The "internal" claim made in 4.29, on the other hand, would assert the sentence

4.30 [K is incoherent/uninterpretable/"nonexistent"] in K.

In either case, the skeptic's allegedly "self-refuting" answer is not only coherent. It offers the only consequential way to preserve certain object- and metatheoretic claims that the rebutter might *a fortiori* wish to preserve, but has unwittingly undermined.

To the extent, then, that we might wish to uphold the metatheoretic coherence of our expressible epistemic theories, or the "internal" [coherence] of [their internal counterparts], we should not only acknowledge that skeptical [self-]doubt is theoretically coherent. We should accept that certain forms of denial of skeptical [self-]doubt are not.

As I initially developed this interpretation in a preliminary way in (Boos 1983b), the implication (skeptical [self]-doubt implies [existence]

(or [consistency], or [interpretability]) is provable in K. But the consequent of this implication is not, unless K itself is incoherent, and neither, therefore, is the antecedent (focus on this contrast in Cartesian contexts is one of several points of difference between (Boos 1983b) and (Slezak 1983)).

4.31 Ironically, perhaps, the Stoics' (and Augustine's, and Descartes') characteristic claim in the passages I have cited—that skeptical-Socrates's position is in some sense self-eliding, or "self-refuting"—also resembles rather closely dogmatic-Socrates's dismissal of Protagoraean relativism in the *Theaetetus*, as I remarked above in 4.15. Certain aspects of this claim also anticipate Stoic, Augustinian and Cartesian "cogito"-arguments: the claim that I could coherently undertake to doubt "everything", except that I (can) doubt.

This would hardly be surprising, of course, since the Stoics and Augustine explicitly cast their arguments as confutations of skepticism, and Descartes, who drew heavily on his predecessors (to put it politely), cast his as an "archimedean point" for destabilisation of "hyperbolic doubt".

4.32 I have already argued that postulation of such regress-halting "archimedean points" seems *haltlos* (as the Germans would put it), in the case of metatheoretic ascent. One step of such an ascent, indeed, is marked by the observations made above about 4.28. It should perhaps be observed that the Diagonal Lemma strongly calls into question attempts to halt such "regresses" object-theoretically, as well—"in one theory", as it were.

For consider the fixed point of 4.25's "doubt". Iterated applications of the Diagonal Lemma yield (as they would, indeed, with any particular instance of the lemma, which is really a scheme of lemmas) an indefinitely extended imploding series of sentences, each of which is provably equivalent to the original fixed point:

4.33 "It is not knowable whether
[[self-referential application of [skeptical self-doubt]] is unknowable];

4.34 "It is not knowable whether
[It is not knowable whether
[[self-referential application of [skeptical self-doubt]] is unknowable]];

4.35 "It is not knowable whether
[It is not knowable whether
[...
[self-referential application of [skeptical self-doubt is unknowable]...]];

If one could form the infinite conjunction of all these sentences, it would be semantically equivalent to any one of them. It might be paraphrased as

4.36 "It is not knowable whether

[It is not knowable whether

[...]]].

In any theory T, or K, therefore, in which one can talk about T or K in information-theoretically useful "intensional" (encoded) ways, one either encounters the metatheoretic ascents considered earlier, or the object-theoretic descents of 4.33–4.36. By contraposition, therefore, any theory in which one did not encounter this (literal) dilemma would be inexpressible to itself—one in which it could not talk about [itself] in these information-theoretically useful ways.

"*Es ist das Mystische*" (?).

I would like to close this section with two loosely related historical remarks.

4.37 It may be interesting, first, to observe that Descartes specifically derogated Socrates' (intermittent) diffidence about what he "knows", in the thirteenth of his *Regulae ad Directionem Ingenii* (AT X, 432; cf. also the *Recherche de la Vérité par La Lumière Naturelle*, X, 505–506), and his language in this passage strongly suggests that Descartes did indeed draw very heavily on this elenctic tradition, as he formulated his variant of the Socratic/Stoic/Augustinian "confutation" of skeptical [self]-doubt.

In my view, the ardent admiration of "minor Socratics" such as the Cynics and Cyrenaics, and middle-academic skeptics such as Arcesilaus and Carneades, strongly suggests that an open, [self]-questioning demeanor may indeed have been closer to the *habitus* of the historical figure than the rather dogmatic "Socratic" *elenchoi* we have been examining in this section.

4.38 In later chapters, I argue that if Descartes did try to use the Socratic/Stoic/Augustinian "confutation" of skeptical [self]-doubt to "prove" (or "know", in 4.25's K), that [he exists], as I believe he did, from assumptions that did not augment [his] theory K with the conclusion to be "proved", then he would catch himself on the horns of this dilemma:

either:

4.39 Descartes' [self]-referential epistemic theory K would be "internally" inexpressible in K (in which case he could not have been talking object-theoretically about [himself] at all—exactly what he fears in Meditation III;

or:

4.40   any "proof" he might claim to provide in K that [he exists] would actually yield conclusive evidence that K is metatheoretically incoherent.

In a sense, then, Descartes' appeal to the ontological argument was not so much circular as epicyclic: in support of the cogito, an attractive but specious derivation of the conclusion that [he exists], he adduced another argument, that essentially reran a variant of that attractive but specious derivation for an allegedly [self]-constituting platonic form he called "god".

Indeed, the fact that the cogito and ontological argument both seek to secure such speciously extratheoretic [self]-constitution for conjoint limiting exemplars of "pure" theoretical "existence", "knowledge", and (in Anselm's case, "perfection") may lend a certain oblique plausibility, even a sort of poignance, to Descartes' strange worry in the Third and Fourth Meditations: that he hasn't yet established "he" is not "god".

5 Reflective *Theoriai*

5.1   In anticipation of the next chapter, I wish to suggest that variants of Section 4's *epoche* or *ataraxia*—and a concomitant openness of the paths of inquiry to which they give rise—might also be characteristic of a mitigated skeptical form of Aristotelian *theoria* ("contemplation"), or preconditions for the active [self]-awareness Aristotle considered characteristic of it.

By this I do not mean, of course, that any Bekker numbers can be found to support such a heterodox as well as anachronistic conjecture. But I will suggest however that Aristotle's "theoretical" ideal of "thought about thought" may be assimilated to forms of [self]-reference which eventuate in metalogical *epoche* and theoretical undecidability of the sort considered earlier.

5.2   Aristotle's "*theoria*" originally embodied a visual metaphor (from "*theorein*", to view, observe), as did Plato's "*eide*" and "*ideai*" from "*idein*" to *see*, in (among other things) the sense of John Newton's and William Walker's poignant hymn.

Kant derived his "*Ideen*" and "*Ideale*" via assorted intermediaries in part from Plato's usage, of course, and his "*Anschauung*" was (among other things) a direct translation of the Latin "*intuitio*", another word with strong visual associations (the related verb "*intueri*" meant "to look into"). Such semantic associations surely hovered over some of these usages for Greek and Roman readers (for a parallel, compare our "insight").

Plato was presumably aware, for example, that his references to "*ideai*" and "*eide*" (or "forms") sounded like allusions to "vision(s)". And Aristotle's appeals to "*theoria*" probably carried a suggestive resonance and metaphysical pathos that are absent from the word "theory" in twentieth and twenty-first-century English.

I would like, in what follows, for something of this resonance and pathos to hover here as well.

5.3 It is easy to forget that Aristotle's *theoria*—the standard translation for which is "contemplation"—appears most prominently in an ethical context: Book X of the *Nicomachean Ethics*. But it also makes a guest appearance in the *Metaphysics* (cf. 1072b24 ff.), and it has clear associations in both these texts with "reflexivity" and [self]-reference.

At NE 1177a20 and 1177a23, for example, Aristotle simply praises *theoria*, as the most "powerful" ("*kratiste*") and "continuous" ("*synechestate*") of human endeavors.

At NE 1177b1, however, he observes more strongly that theoria "alone is thought to be loved for itself" ("*doxai t' an aute mone di' auten agapasthai*") and is "worthy in itself" ("*kath' auten timia*") (NE 1178b32).

One should consider these remarks in conjunction with the almost "Platonic" assertions at NE 1177b1, 1178b7–9, and 1178b22–25 that the "theoretic life" is not only the sole human activity that is loved (*agapasthai*) for [its own] sake, but also the best approximation to "divine" *energeia* and *eudaimonia*. For these passages in EN 1178–1179 closely parallel Aristotle's well-known characterisation of "divine" thought in the *Metaphysics* as (explicitly [self]-referential) *noesis noeseos*: "thought about thought", which "intentional object" I will assimilate to thought focused on incomplete but unending inquiry.

I will return to these reflexive and reflective analogies again in subsequent chapters, and to a reconsideration of some of the potential interrelations Aristotle drew between *theoria* and *praxis* (action).

5.4 "Skepticism", once again, derived from a visual metaphor: the word *skeptesthai* meant to "look at" or "examine". The skeptics also liked to call themselves *zetetikoi*, according to Sextus Empiricus—"seekers" or "inquirers". It is this *search* and *inquiry* that I have tried to assimilate to a skeptical variant of Aristotle's *theoria*, and wish to contrast here with Plato's monistic/unitary "vision" of the Form of the Good.

Searching normally presupposes some sort of yearning or desire, or at least of incompleteness, which the searcher typically wishes to assuage. Mathematical logicians, in their admittedly narrow senses of these words, assimilate incompleteness to undecidability: that a question is apparently unanswerable, from whatever premises, "facts" or information we may have. We are uncertain which of several branching paths to take. We search, therefore, for wider ranges of alternative interpretations which may literally suggest "how to go on".

If such *epoche* / undecidability, in turn, is assimilated to freedom of (dia)‑noetic inquiry, or at least to a "precondition" of such freedom (as well as the drossy contingency of *doxa* and *eikasia* Plato so deprecated), then desirable forms of contemplative openness might indeed require as well as presuppose such "freedom".

But this, in turn, would require that individual and collective forms of "theoretical" [self]-definition be indeterminate.

5.5 In quasi-Kantian terms, such openness, freedom and conceptual "eternal recurrence" might be regulative, perhaps even constitutive, of some of the "higher" forms of sentient [self]-awareness and [self]-determination that we typically—and ideally—wish to attribute to [ourselves], and whose *energeia* Aristotle associated with *theoria*.

It is in this sense that I have contrasted Section 3's semantic monism (a methodological cousin, perhaps, of Arthur Lovejoy's "monistic pathos") with an antithetical, processive view which I call "theory-relativism" or "semantic pluralism".

5.6 It is also in this sense that it is our queries, moreover—the metalogical evidence suggests—that are robustly [self]-generating of the allegedly definitive "answers" we attempt to impose on them. Such "answers" are provisional conjectures, little more.

Or rather, close examination seems to suggest that most such "answers"—Plato's invocations of the forms, for example; or Aristotle's ultimately vacuous postulations of a *"prote philosophia"*; or the Stoics' claims to "know" what the knowledge and wisdom of the *sophos* "must" be—tacitly embed [self]-referential premises that are really disguised queries, and stopgap-solutions to conceptual *cris de coeur*. Our descendants (if we are lucky) may be expected to recognise them as such (along with their modern and "postmodern" counterparts), if we do not.

To paraphrase Pascal (and Kant; see below) our minds (and hearts) seem to pose questions that "reason" cannot answer.

5.7 Why does this simple observation seem so deeply unsatisfying to so many people?

Perhaps because we have what the Germans (borrowing from the French) call *Angst vor der eigenen Courage* (fear of one's own courage). We seem to fear the very potential for perspective-shifts that is a regulative precondition for our own "freedom".

Consider the example of Plato himself, who provided one of the most eloquent parables of the need for such perspective shifts ever written, but insisted (in effect) that emergence from that parable's cave can only take place once, and in one (prescribed) direction, and cannot be iterated.

Or that of Pascal, who resorted to a manifestly threadbare "wager" to convince himself and others that the groundless religious monism his "heart" demanded was "necessary".

5.8   Sometimes it seems to me that we have a kind of horror-of-undecidability, rather like an (alleged) *horror vacui*—as though non-unicity of interpretation implied inconsistency.

If so, this (fallacious) implication would itself be an instance of the begged semantic monism I have tried to critique. It would also be (I believe) the fundamental premise begged in most applications of Kant's "transcendental argument".

5.9   Whether or not we have such a horror, at any rate, such hermeneutic perspective-shifts seem to me regulative of the very possibility of theoretical inquiry, as I suggested above in 5.5.

For one of the principal messages I have tried to convey in this chapter is that any global theory reflective (or reflexive) enough to pose encoded, internal questions about [itself] will eventually have to recognise that [it is plurally interpretable], if [it is interpretable at all].

5.10   Correlatively, it would seem that to recognise the very force and eloquence of the Platonic cave-parable in the first place (or of any other metaphysically or epistemically relevant shift of metatheoretic perspective) we must be able to countenance—genuinely and without equivocation—[our own] incompleteness, and acknowledge, at any conceivable stage of inquiry ("enlightenment") we may have reached to date, that plural interpretation of prior inquiry is possible, even desirable, and in any case inevitable, if the pleasures Aristotle (correctly) attributed to discovery and the [self]-respecting "theoretic" life are to continue. If the *noesis noeseos* of 5.3 is to be expressible, it must be incomplete.

A plurality of ramifying ways "up" from "the" cave, indeed, might be more appropriate than Plato's capped, two-stage linear model of noetic enlightenment to the "dialogical" senses of the Greek words *dialogos* and *dialektike*, as well some later senses of *"Dialektik"* in the history of philosophy. There is surely a modicum of irony in the effective narrowing of *dialogos* to *monologos* at the "end" of *dialektike*.

It does seem to me ironic, in any case, that such multiplicity—whose meta-logical counterparts are provable—arises in the specific case of entities whose reason for being is their alleged unchanging universality and uniqueness.

It also seems to me even more ironic that philosophers—not only Plato, but Aristotle, in this case—who admire the infinities implicit in eidetic "visions" and near-"divine" *theoria*, are so quick to reject critical refinements of interpretation(s) which are such infinities' natural and recurrent mark.

5.11   This elenctic argument aside, I have always been genuinely puzzled at most philosophers' attraction to an allegedly stringent and austere Ideal which Lovejoy called the "monistic pathos" (cf. Lovejoy, 13), e. g.:
5.12   the Parmenidean One;
5.13   the *eide, ideai, henades* and *monades* of Platonic dialogues;
5.14   the One of Plotinus' *Enneads*;
5.15   the *deus sive natura* of Spinoza's *Ethics*;
5.16   the ineffable unicity of Kantian *Erfahrung* and *Pflicht*; and all the other, allegedly probative, subtler forms of "apodicticity" and "transcendental unity", demanded under various ontological guises by "major" metaphysicians and ethicians in the history of western philosophy.

(Leibniz, I believe, was a distinguished partial exception. But his principal concession toward such monism—the unicity of the "best"—is, by Voltaire's and most other commentators' accounts, the weakest and least persuasive aspect of his system.)

One unstated premise that seems to have motivated many such demands is a presupposition (which I believe is deeply mistaken),
5.17   that the more "holist" one's ("intended") metaphysics, the more unitary will be one's "intended" universe of metaphysical discourse.

On one rather straightforward reading of the metalogical analogues I have sketched, stringent and rigorous lines of argument lead quite clearly in the opposite direction, if they lead in any direction at all.
5.18   This is not—or is at least not "intended" to be—some sort(s) of facile claim(s) about what "we" can "know" about our "condition" (whatever that is). It is, rather—or is "intended" to be—a considered judgment

about thoroughly natural attempts to "conceive" ramifying limit(s) of "conception"—natural consequences, if you like, of Kant's *Schicksal der Vernunft* (cf. *KrdrV*, A VII).

Such global interpretations would have to be provably theory-marginal, and perhaps "*transzendent*" as well, in the sense that one can prove—in any metalogical venue in which one can raise the issue—that they can only be "intended" in wider and wider metatheories. For if Plato (or Aristotle; or Anselm; or Descartes; or Leibniz; or….) could "prove" the [(unique) interpretability] of any (suitably anthropomorphised and encoded) theory of divine or ultimate attributes, *within* such a theory, then such a "proof" might actually witness that theory's incoherence.

Even the nonskeptical dialecticians Zeno and Parmenides, for example (or at least their counterparts in Plato's *Parmenides*), suggested indirectly that the entire scheme of Plato's divided line, and therefore the place of his *eide* in that scheme, might itself be just an [*eikasia*], a [likeness]—an instance of the lowest order of conjecture in that line. The cave-parable is, after all, a parable.

Such *eide*—forms—would then become subject to further relativisations and reexaminations in their turn. And—more importantly, perhaps—so also would the entire scheme in which they were introduced.

(An almost exactly parallel observation could be made, I believe, about Kant's claims to demarcate from "within" "*der*" *Grenzbestimmung zwischen Immanenz und Transzendenz* (boundary-demarcation between immanence and transcendence); cf., e. g., *KrdrV*, B 786, *Prolegomena* 350–362).

For the more global one's theoretical aspirations—and the more holist one's *theoria*—the more open to metatheoretic ascent and emergent reinterpretation(s) such aspirations and *theoria* will have to be.

5.19 These tensions of our (regulatively useful and desirable) attempts to "conceive" (emergent) limits of "our" "conception" are generative and heuristic if they forego illusions of ultimate unicity.

They already appear, for example, in Socrates' intermittent moods of epistemic modesty (cf. 4.21), whose ambiguities may derive in part from the evanescent mysticism of his appeals to an intellectual and emotional *daimon*), and in the arguments of minor Socratics and academic and pyrrhonist skeptics. Variants of them also reappeared in the writings of Nikolaus of Cusa, and other medieval, Renaissance and early modern writers who are usually called "skeptics" of one sort or another (Nicolas of Autrecourt; Francisco Sanches; Michel de Montaigne; Pierre Gassendi; Blaise Pascal).

Indeed, such tensions provided much of the dialectical basis of Pascal's deepest (and in his case paradoxical) insights in passages such as the following, perhaps the most eloquent evocation ever written of the sort of courage Kierkegaard later attributed (dismissively) to his agnostic "knight of infinite resignation" ("*Uendelighedens Ridder*"):

> *La dernière démarche de la raison est de reconnaitre qu'il y a une infinité des choses qui la surpassent ...*
>
> *L'homme n'est qu'un roseau, le plus faible de la nature, mais c'est un roseau pensant. Il ne faut pas que l'univers entier s'arme pour l'écraser; une vapeur, une goutte d'eau suffit pur le tuer. Mais quand l'univers l'écraserait, l'homme serait encore plus noble que ce qui le tue, puisqu'il sait qu'il meurt et l'avantage que l'univers a sur lui. L'univers n'en sait rien.*
>
> *Toute notre dignité consiste donc en la pensée. C'est de là qu'il nous faut relever et non de l'espace et de la dureé, que nous ne saurions remplir. Travaillons donc à bien penser: voila le principe de la morale.*
>
> The last step of reason is to recognise that there are an infinity of things that surpass it.
>
> A human being is a reed, the weakest of nature, but it is a thinking reed. It isn't necessary for the whole universe to take arms to crush it; a vapor, a waterdrop is enough to kill it. But when the universe crushes it, the human being is still nobler than what kills it, for it knows that it is dying and the advantage the universe has over it. The universe knows nothing.
>
> All our dignity consists therefore in thought. It's from there we should take our orientation, and not from space and time, which we cannot fill. So let us work to think well: this is the principle of morals. (*OC*, 418)

Finally, they resonate in a somewhat different way in David Hume's basic insight in the *Dialogues Concerning Natural Religion*—that human notions of [design] might provide a very dubious template for "ultimate" extrapolation(s) of [themselves]:

> What particular privilege has this little Agitation of the Brain which we call Thought, that we must make it the Model of the whole Universe? (168)

For whatever it is worth, I find "Philo"'s observations in this passage and most of the rest of the *Dialogues* (apart from his notoriously implausible "reconciliation" with Cleanthes in Book XII) much more acute than most of Hume's better-known arguments in the *Treatise* and first *Enquiry*.

The latter, after all, are ultimately devoted to Hume's dogmatic aspirations to provide the "Principles" of an essentially foundational "moral" science or science of epistemically constitutive *"mos"*—"Custom", construed wryly as a "final cause" or "pre-established harmony" in paragraph 44 of the first *Enquiry*.

Pascal, at any rate—the "fideist" skeptic—effectively answers Hume, the "consequent" one (and cogent critic, I believe—at least in the *Dialogues*—of "transcendental deductions"), in the only way that seems open to us: "regulatively", with an essentially ethical invocation of our "dignity" and fragility. This way, I believe, lies a more open interpretation of the *primacy of the practical*, as well as a "merely regulative" interpretation of "transcendental arguments".

5.20   There are, of course, many other ethical and "practical" counterparts of this dilemma as well—as there should be, indeed, if one wishes to honor and extend the original double aspect—ethical as well as epistemic/metaphysical—of the cave-parable.

Many of them are clearly reflected in the blend of Stoic insight and Kantian *Autonomie* expressed in Pascal's *défi*. The same contrast, moreover—between the universe "without" and the universe "within"—appears in the famous remark from the end of the second *Critique*, engraved on Kant's cenotaph.

> *Zwei Dinge erfüllen das Gemüt mit immer neuer und zunehmender Ehrfurcht, je öfter und anhaltender sich das Nachdenken damit beschäftigt: der bestirnte Himmel über mir, und das moralische Gesetz in mir.*

> Two things fill the sensibility with ever new and growing admiration and awe, the more often and persistently reflection occupies itself with them: the starred firmament above me, and the moral law within me.   *(KdpV, 288)*

To his credit, Kant essentially acknowledged in the *Grundlegung der Metaphysik der Sitten*, that a "purely good will" is indefinable and indiscernible in human experience (cf. 459–463). This seems to me methodologically right, on the metalogical analogies I have sketched. But it does suggest that no internal "transcendental deduction" of such a will's existence—even in very broad and extended senses of *"Deduktion"*—could really be formulated. Postulated, yes. "Deduced", no.

It is, in any case, the recurrent contrasts between *eikasia(i)* and original(s) whose discoveries move and awe us, I believe: the vulnerable flare of Pascal's *pensée(s)*, not the *univers* that ends them; the indefinite yet-more of Kant's *bestirnter Himmel*, not the (specious) fixity of its *Fixsterne*; the eloquence of

Wittgenstein's sternly youthful and multiply-violated injunctions to "silence", not the truly unitary silence of the grave.

We yearn, in processive alternation, for fleeting "invariant" symmetries in the flux, and for reinterpretations which "break" those symmetries. We understand, therefore, the anonymous homilist in *Ecclesiastes* who tells us that there are time(s) to cast away stones, and time(s) to bring stones together.

We yearn, in short, for yet-more journeys, from yet-more "caves". But we also understand Leonard Woolf's reminder—that the *journeys* not the arrivals matter.

In this and other writings, I have sought to articulate such contrasts and interrelations between the potential *aporiai* of iterated [self]-referential application of alethic thought experiments, and the "dignity" of "reasonable beings" *capable* of such iterations.

5.21 Several concluding unscientific thought experiments may serve here to illustrate some of the ethical and epistemic consequences of such a pluralism or theory-relativism, and adumbrate the more "open", "theoretical" view of dialogical or dialectical enlightenment I have tried to propose.

5.22 The first of these is that "enlightenment"—emotional, ethical, noetic—may be a complexly relational notion; in particular, it may make sense only when understood as something which "emanates" from a(nother) (metatheoretical) source.

5.23 The second is that we may, indeed, find ourselves in a "cave", of the sort Plato described, in his famous parable of epistemic oppression and illumination—a parable, if you wish, of "Amazing Grace" ("... I once was lost, but now I'm found .../Was blind, but now I see").

5.24 The third is that we may, in temporal succession, find ourselves in many "caves", and find many paths that branch out of such "caves", in diverging directions. Some may lead "downward", into epistemic or moral confusion. Others may lead "upward", into "freer" and more expansive outer "worlds"—illuminated, perhaps, by initially blinding "suns" that fade in their turn.

("Who knows the spirit ... that goeth up, and the spirit ... that goeth down to the earth?"—*Ecclesiastes* 4.21)

5.25 There may, in the end, be many forms of mutually incomparable recurrent "grace" (and many concomitant forms of dis-"grace")—more, indeed, than were dreamt of in Plato's philosophy.

Or in ours.

3 The Stoics, the Skeptics and Aporetic Autonomy: Is "What Is In Our Power" In Our Power?

1 Introduction

If one can credit the late-second-century commentator Clement of Alexandria (*SVF* II, 121; see Section 2), the ancient stoics anticipated essential aspects of Descartes' "cogito"-argument and its alleged refutation of skepticism long before Augustine and Descartes took it up. More generally, a defensible case can be made that the stoics' "logical", "ethical" and "physical" *ordines saeculorum* influenced significant aspects of Descartes' "*methodus*" and "circle" as well—so "clearly and distinctly", I believe, that one might construe them as attempts to
1   set aside the stoics' monist "ηθική"; and
2   Spinoza's *Ethica* as an attempt to restore it.

A number of students of stoic arguments' ancient antecedents have credited the stoics' "logical" insights, but queried the ethical and physical determinism of their *ananke* (necessity) and *heimarmene* (fate). In this chapter I will argue that metalogical analyses and insights—traceable historically to subtle dialectical exchanges between the stoics and their skeptical opponents—generate methodological critiques of stoic "dogmatism" in all its forms. More precisely, I will interpret
3   the stoics' original "refutation" of their opponents' skepticism, as a failed "fixed-point" argument;
4   their more innovative "logical" inquiries as propaedeutic rationales ("*méthodes*") for the existence of a canonical ethical (or "metaethical") self; and
5   their early "physics" as an attempt (among other things) to embed that self in a "hieratic" *phusis* "in accordance with" which the counterfactual stoic "sage" (*sophos*) enjoined us to live.

I will also appeal to metalogical arguments to argue that
6   "what is in our power" is not epistemically definable by "us", and not in that sense in "our" power;
7   draw on this *aporia* to query stoic unicity, determinacy of *tonos*, *phusis*, *ananke* and *systema* of "the" *cosmos* as well as "the" *moral self*;
8   interpret stoic claims of such unitarity and determination as attempts to develop a theodicy of cognitive *episteme* and quasi-hieratic design-arguments;

and finally,

9   sketch alternative ideals of incomplete inquiry, and derive from them skeptical counterparts of stoic *sophia*, Kantian *Würde*, and other forms of solidarity among "reasonable beings" ("… *das moralische Gesetz in uns*"), in the face of vast *kosmoi* of experiences we will never have ("… *der bestirnte Himmel über uns*").

2 An Early Dialectical "Refutation"

I will begin with several passages from von Arnim's *Stoicorum Veterum Fragmenta*, among them the following *elenchos*, attributed to the stoics in the late second or early third century of the common era by Clement of Alexandria, and later applied by Augustine and Descartes as "refutations" of Socratic as well as skeptical disclaimers of "knowledge" (cf., *Apology* 21b and 29b; Augustine's repudiation of such views in (Gilson, 1976), 295, and Descartes' Second Meditation (AT VII, 24–25); the italics are mine):

> If (skeptical) suspension of judgment says that nothing is confirmed (*bebaion*), then it is clear that proceeding from itself, it first disconfirms (*akurosei*) itself. … (For) either (such suspension) is itself truly asserted (*aletheuai*) or not. If it is, it involuntarily grants that something is true. If not, it leaves standing as true (*alethe apoleipei*) those things it wished to confute (*anelein*). (But) to the extent the confuting suspension is shown false, to that extent the things confuted are shown true, *as (in the case of) the dream that says that all dreams are false. For if it confutes itself* (*autos gar anairetike*) *it becomes confirmatory* (*kurotike*) *of all the others* [cf. the liar paradox below, and by way of dialectical inversion, Descartes' "Archimedean point" at AT VII, 24].
>
> *But if it is true, in general, it will make a beginning from itself, and not be suspension of something else, but of itself first. For if someone grasps that he (or she) is human, or suspends judgment, he (or she) clearly does not suspend judgment. For how would the person reach the beginning of the dispute, suspending judgment about all things? How would the person reply to what was asked? About this very thing the person clearly does not suspend judgment, even while declaring that he (or she) does.* And if we are to be persuaded by the pyrrhonists' arguments to suspend judgment (cf. Descartes' "… *si quid mihi persuasi*", AT VII, 25), we will first suspend judgment about suspension of judgment, whether it is to be trusted or not. And if that itself is true that we do not know the truth, then nothing is given by that (as) true at the beginning. But if one says that even *that* is disputable (*amphisbetesimon*), that we do not know the truth, then one grants that the truth is knowable, by the very fact that one clearly does not confirm suspension of judgment about *that*. (SVF 2.121)

Since Descartes' alleged innovation in this context was to recast such self-referential "confutations of skepticism" as a confirmation of reference to "the" self, it may also be interesting to conjoin these passages with the following passage, by Tertullian (of all people), a religious despiser of all things philosophical (except perhaps when it served his purposes):

> Since therefore the soul provides the ability to sense all things, and even sense the senses of all things, how likely is it that it is not granted a sense of itself from the beginning? ... . Indeed, how can it be rational which renders the human a rational animal, if it does not know its own rationale (*rationem*), ignorant thus of itself? (Tertullian, *SVF* 2.845)

Given Descartes' well-known recourse to ancient denials of "infinite regress" as guywires for his (rhetorically) powerful arguments in Meditation II, it may also be of interest to consider the following ancient instance of non-skeptical tolerance of temporal regresses and ascents):

> How is it *not* absurd to say that causes extend to infinity, and the order and succession of them (*heirmon auton kai ten episyndesin*) (is) such that nothing is either first or last? ... . Comprehensive knowledge (*episteme*) would be eliminated according to that view, if such knowledge is indeed primarily an overview (*gnosis*) of first causes, and there is in their conception no first among causes. *But not every departure (parabasis) from overall structure (taxeos) is destructive of the things in it* ... . Nor if such a thing were the case does it wholly dissolve the well-being of the world (*luei ten eudaimonian tou kosmou*). (Alexander Aphrodisias, *SVF* 2.949; the English italics are mine)

In effect, Alexander correctly identified as invalid an inference from existence of "locally" adequate designs, to existence of a "globally" adequate design—a logical quantifier-mistake.

I will consider "partially ordered" refutations of such "inferences" later, and consider first a dialectical "fixed-point-argument" which anticipated the "Augustinian cogito" and its better known Cartesian seventeenth-century successor.

Stoic metaphysicians had claimed to offer semiformal formulations of what they called
1 "criteria of truth"—standards which would permit themselves and others to decide the veracity of arbitrary ranges of assertions.

In response, their skeptical opponents asked them to
2 "decide" whether such "criteria" applied to [their criteria],
3 "decide" what further "(meta)criteria" might be needed, if it was not, and
4 "decide" how to respond to similar questions about these "(meta)criteria" in their turn.

Notice that

5 the skeptics engaged their stoic opponents' claims in ways which tacitly adhered to shared principles of *eristic*—"logical" or quasi-"logical" practice, in the sense that
6 they did not reduce alethic thought-experimentation to "absurdity". They simply observed that its conceptual boundary conditions are subject to refinement, ramification and regeneration without end.

In the next two sections, I attempt to interpret aspects of this early dialectical exchange as templates for more general patterns of argumentation which may be characterised as semantic paradox and metatheoretic ascent.

3 "Semantic" Paradox(es)

"Semantic paradoxes" are mentioned in passing in most introductions to philosophical logic, often to treat them as formal counterparts of Kant's "*Skandal der Philosophie*", to be dispelled or eliminated by any conceptual means necessary. Such approaches to traditional *aporiai* and *problemata*—several of which are quite ancient—seem to me understandable but deeply wrong. More precisely, they seem to me to express unilaterally "hieratic" dismissals of the generative role such paradoxes have played in the historical dialectic sketched above.

As the last section's interrogations of "the criterion of truth" suggest, certain paradoxes arose from attempts to apply linguistic or hermeneutic ("semantic") predicates reflexively (or self-referentially) to [themselves].

More precisely, as I have sketched in Chapter 1,

1 a number of "contradictions" or "absurdities" have often seemed to emerge from predications of "global" alethic or interpretatively significant properties (such as "falsity") to [themselves];
2 analyses of such "reflexive" or "self-referential" antinomies suggested that these predications, or the "globally" hermeneutic properties from which they were formed, might be locally indefinable; and
3 "resolutions" of the antinomies relativised the problematic properties, or their reflexivity, (or both), and reinterpreted them, implicitly or explicitly, as "local" or relational notions, which depended on metatheoretic or contextually varying parameter(s).

In the last section's case at hand—the *aporia* of the "the criterion of truth" (the underlying ground-bass, I will argue later, of the better-known "liar paradox")—pyrrhonist and academic skeptics applied

4   reflexive or self-referential arguments (as in 1) to query
5   the definability of one of 2's "global" hermeneutic properties, "truth" (a word which, as mentioned, we have in English from older northern European usages for "trust" or "belief").

Typically, skeptical metaphysicians then offered refuge from such "absurdities" in variously nuanced
6   "undogmatic" avowals of the relationality, contextuality and/or probabilistic graduations of such predicates ("undogmatic", for example, in the sense that they accommodated metatheoretic revisions of the perspective they proposed).

Relativisations and open-ended parametrisations of alethic, ontological and methodologial notions have been repugnant to many metaphysicians, then and now. But such reflexive (or [self]-referential) predicaments to which these have given rise are not confined to "truth"-criteria.
    In Platonic terms, for example, consider
7   the Theaetetan attempt to give an account of "what it means to give an account"; or
8   the *Republic*'s attempt to provide an *eikasia* in Book VII of "true" knowledge that is not (merely) *eikasia*.

In the sequel, I will therefore argue quite generally
9   that "locally" analytical thought-experiments are quite compatible with "global" underdetermination of their systematic counterparts;
10 that this observation clarifies a recurrent need to make local conceptual distinctions between theory and what would now be called metatheory;

and in particular
11 that such clarifications have made it possible to formulate provisional resolutions of *aporiai* the stoic pioneers of logic (and most of their medieval and early-modern successors) left unresolved.

4 "Diagonalisation"

I have already used the criterion-problem to argue that the stoics and their academic skeptical opponents had to care enough about *logike* for semantic paradoxes to become sources of conceptual clarification of *logike*.
    Similar remarks might be made about
1   the better-known liar paradox ("This sentence is false"—a tributary, I have suggested, of the criterion-problem), and

2   the equally venerable *sorites* or paradox of the heap ("One grain is not a heap, and the addition of a single grain to a non-heap does not make a heap. There are therefore no heaps."),

as well as more recent paradoxes, introduced by Berry, Grelling and others (see Cantini, 2014).

In particular, it might be no accident that ancient semantic paradox arguments which had slumbered for centuries generated later attempts to refine and recapitulate them many centuries later, aided still later by "coding"-procedures Leibniz envisioned more than three hundred years ago. Consciously or unconsciously following Leibniz, early twentieth-century colleagues such as Thoralf Skolem, Kurt Gödel and Alfred Tarski, among others ("fellows from another college", a phrase of John Littlewood brought into circulation by G. H. Hardy), devised finitary arithmetic codes for linguistic expressions to analyse certain self-referential arguments.

Accessible expositions of these ideas can be facilitated if one considers assorted theories T as "object-theories", in tacit metatheories T', T" ... , and abbreviates long "intensional" assertions "..." in T, by intentions [...]. These will typically have the form of semiformal English assertions "..." isolated in ("intentional") ordinary brackets [...]. Such enclosures may be iterated, or nested, as will become clear in subsequent sections. Especially interesting for dialectical purposes will be situations—commonplace, I will argue, in philosophical speculation—in which one considers a particular T as a potential metatheory for *itself* (or more properly [itself]).

In many such situations—formal miniatures, in a sense, of certain forms of linguistic "oblique contexts"—one can introduce recursively "encoded" "intensions" [for expressions "..." of L(T) within T itself (for a more careful exposition of this idea, see (Smorynski, 1977), 826–829). Such "internally intensional" coding can take many forms, and I have suggested that it may be regulative of various forms of [self]-awareness in ways Leibniz anticipated quite explicitly in the seventeenth century. Its pioneering investigators in the twentieth century were Gödel, and before him the Norwegian logician Thoralf Skolem, who used an arithmetic form of it to demonstrate the contextuality and theory-relativity of certain "higher" infinities ("Skolem's paradox").

Gödel's most incisive and general innovation was to employ such coding in a different way to "encode" the argument of Russell's paradox, and show that for every property $p(x)$ expressible in the languages L(T) of a wide class of formal theories T, there exists a diagonal sentence s for $p(x)$ such that s is provably equivalent to $p([s])$.

One can actually sketch the construction of this s in dialogical (or perhaps dialectical) form, as follows. A "dogmatic" metaphysician claims that a

3 "semantic" property **valid** (x) ("*katalepsia*"; "accordance with nature"; "what is in our power" . ...) confers an "ultimate" "*validity*" on x.
4 A "skeptical" opponent asks whether the property **valid** (x) confers the alleged "*validity*" ("existence"; "truth"; "homologia", "autonomy", ...) **valid** (x) on [*itself*].
5 The "dogmatist" responds that such self-referential concept-formation is "*invalid*", by the lights of **valid** (x), that is, that **not-valid** ([**valid** (x)])
6 The ancient skeptic usually argued at this point that such a bar on [self]-reference was a crippling admission of methodological failure, and there the conversation would end, typically with some sort of "dogmatic" *argumentum ad hominem*.

In the wake of Russell's and others' paradoxes, the Czech logician Kurt Gödel examined (in effect) the result of taking this ancient exchange a deep and remarkable step further, and proved that

7 the assertion **diag(valid)** = $_{Df}$ **not-valid** ([**not-valid** ([**valid** (x)]]) is "**valid**" if and only if [it] is "**invalid**": i.e, that **diag(valid)** *is* equivalent to [**not-valid (valid)**].

(Notice that this does not establish that any given "validity"-predicate **valid** is "false", or "worthless"—simply that the scope of its applicability lies beyond its ken.) Gödel (who was no skeptic)

8 applied his original "diagonal" analysis to a property normal provability in formal theories;
9 showed that the resulting diagonal sentence is equivalent to an assertion that [the theory makes sense (is consistent)]; and
10 observed that this sentence is therefore undecidable—neither provable nor refutable—in the theory in and for which it is formulated.

Many mathematicians and metaphysicians (Gödel among them) found these constructions unsettling, and many have tried, then and since and in various ways, to dismiss or "transcend" them (both deeply "hieratic" efforts, on my account). But as I have argued in Chapter 1 (6.18–31),

11 constraints on the theories to which they apply are actually rather minimal (the theories have to be "articulate" and "inductive" in ways that make communication possible and reproducible; and

12 such diagonal arguments clearly yield "fixed points" of such v (a somewhat ironic expression, in the light of Descartes' claim (AT VII, 24) to have found, in effect, a single "Archimedean" fixed point or "*pou sto*" for any such predicates in any such theories; and
13 such reconstructions do therefore undermine "internal" attempts to say significant things about any such theory's (formal or informal) semantics.

In plausibly relevant theories, for example, one might formulate sentences which are provably equivalent to (theory-relative miniatures) of
14 [their own] Theaetetan unaccountability, undefinability and/or inexpressibility;
15 [their own] stoic *akatalepsia*;
16 [their own] skeptical *epoche*;
17 [their own] Kantian *Transzendenz*; and
18 [their own] Wittgensteinian *Unsinn* ....

Such reconstructions may have a certain recurrent historical resonance, especially for "transcendental" and "proto-transcendental arguments" (cf., e.g., 22 and 23 below, Section 8 *passim*, and Chapters 4–6). The first quasi-historical reconstruction (in 14 above), for example, turns out to have close affinities with
19 the Berry paradox ("Let us define n to be the least integer that cannot be defined in less than forty syllables in the English language ... ."),

and more historically, to
20 the *sorites* or "heaper"-argument, mentioned earlier ("The number one is not infeasibly large, and addition by one does not render feasibly large numbers infeasible. Therefore there are no infeasibly large numbers.").

Invocations of the *sorites* go back a very long way, and the affinities claimed for the Berry paradox and *sorites* take the form of alternative proofs of first- and second-order incompleteness results for "inductive" theories of the sort mentioned above (informal sketches of these proofs appear in Chapter 1). Historically, it may also be worth observing that
21 the *sorites* also provided the academic skeptic Carneades with a welcome eristic generator of stoic *epoche* (suspension of judgment), or at least of unwonted stoic "silence", a kind of tacitly acknowledged counterpart thereof.
("*Wovon man nicht sprechen kann, davon soll man schweigen? ... .*")

Whatever the merits of silence, I would respond to the stoic hopes for complete apodeixis that we should suspend judgment about "ultimate" metaphysical closure and criteria of "truth", or the tenability of claims to "comprehensive representation" (*kataleptike fantasia*) Chrysippus and other stoics derived from them, evinced in the vigorous passage from *SVF* 2.121, quoted above, and base this recommendation on a straightforward reading of the metalogical evidence I have sketched.

More precisely, I have argued at various points that we might counter the *Theaetetus*-like "refutations" of skeptical doubt and "protagorean relativism" with

22 assimilation of pyrrhonist and academic-skeptical *epoche* and *ataraxia* to formal undecidability,
23 honest acknowledgments that inductive forms of discursive rationality are plurally interpretable, and
24 interpretations of such acknowledgments as diagonal [fixed points] for [themselves]

(where the notation "[]", once again, abbreviates certain theory-internal "intensions"). Along similar lines, one might interpret *SVF* 1.121 and its Augustinian and Cartesian variants as incredulous formulations of what might be called

25 the "doubter's paradox" ("I doubt [this sentence]" (as opposed to the liar paradox—"[this sentence] is 'false'")—a claim that is essentially inexpressible in formal venues to which it applies.

To see that skeptical and theory-relativist forms of [self]-doubt may have "positive", even "constructive" aspects as well, it will be useful to interpret

26 ["knowledge"] (or, more precisely, ["knowability"]) in "my" theory T as (encoded) [provability in T], and
27 ["doubt(ability)"] in T as [unprovability in T].

For if one does this, it turns out that

28 the supposedly self-refuting doubter's "paradox" is provably equivalent, as Gödel showed, to the claim that [T is consistent]; or equivalently (in reasonably sophisticated theories T) that [T is interpretable]; or [T *exists*].

Even if ["knowledge"], moreover, were taken to be some more general "internal" predicate that is simply implied by [provability], and "doubt" as the negation of "knowledge", a fixed point of such "doubt" would still imply the [interpretability of T].

Nor is all this so abstruse that one cannot say it, or at least recapitulate it, in a slightly formalised version of ordinary English, with the aid of the "intentional" brackets introduced above.

For if "knowledge" or "knowability" is construed "internally" as in 26 and 27, the diagonal sentence for this predicate—the last formal claim generated in 3 through 7 above—asserts

29 that [self-referential application of [the assertion that [self-reference] is unknowable] is unknowable]; or equivalently,
30 that [self-referential application of [skeptical self-doubt] is unknowable]; and in either case, therefore,
31 that [self-referential application of [skeptical self-doubt] is subject to [skeptical self doubt]].

In their dialectical examinations of such predicaments, platonic, stoic, Augustinian and Cartesian "confuters" of "skeptical relativism" simply assumed (as "obvious")

32 that "serious" skeptics could not suspend judgment about [self-referential application of [skeptical self-doubt]].

They therefore asserted (as equally "obvious")

33 that "serious" skeptics, who acknowledged such [self-referential application of [skeptical self-doubt]], would therefore "reduce their skepticism to "absurdity".

From this putative *reductio*, the "confuter of skepticism" then "proved the 'knowability' of [skeptical doubt of [skeptical doubt]]" (Descartes' "fixed point"), and from this the "knowability" of other claims the skeptic professed to hold in suspension as well.

Consider, however, what happens when one actually takes the care to formalise "provability"-claims about [self-referential application of [skeptical self-doubt].

If one construes "provability" object-theoretically in T (as what I have called "[provability]"), the "obvious" acknowledgement in 32 implies

34 that [the theory T is incoherent/uninterpretable/"nonexistent"], as an "internal" or "intensional" assertion in T, and therefore that [[everything] is [provable] in T].

If one construes the "provability"-claim metatheoretically, the "obvious" assumption in 31 becomes even more problematic. For then

35 the claim would have as one of its consequences the outright metatheoretic incoherence and uninterpretability of T, and ensure therefore that everything "really" *is* provable in T.

For one of the best-known consequences of Gödel's work is that no consistent metatheoretic proof of such a claim—equivalent to a metatheoretic proof in T of T's internal consistency—can be found.

On either construction, in short, the skeptic's allegedly "self-refuting" answer is not only coherent. It offers the only consequent way to preserve local forms of object- and metatheoretic coherence the "confuter" or rebutter has globally and unwittingly undermined. To the extent we wish to uphold the metatheoretic coherence of our expressible epistemic theories, or the internal [coherence] of [their internal counterparts], we should therefore

36 acknowledge that skeptical [self]-doubt is theoretically coherent; and
37 accept, in fact, that certain denials of skeptical [self]-doubt are not.

5 "The World", "The Whole" and "Hypothetical Necessity"

My intention in this section will be to examine with some care the somewhat ambiguous nature of stoic metaphysical and ethical "monism", and its interconnections with the tripartite division of philosophical inquiry mentioned in the chapter's introduction. Chrysippus and other stoics often took care to distinguish between

1 *"to holon"* ("the whole", "the universe"), a comparatively amorphous notion they sometimes characterised as "to ti" (the "something" or "what(ever)"); and
2 *"ho kosmos"* ("the world"), a notion that suggested order, structure and beauty in ways that elude its English counterpart.

Consider, for example, passages such as the the following, which seem to construe "nature" (*physis*) as a kind of metatheoretic entity which interprets the world—provides a semantic structure for it—but is not *"in"* it.

> The most general (*katholikotata*) terms [are] three: one, being and something (what (ever)), [and] they are applied in the same manner to all of the things that are—the *one* (*to hen*) according to Plato, *being* (*to on*) according to Aristotle, and the *something* (the *what*(ever)) (*to ti*) according to the stoics. (A comment by an anonymous Aristotelian scholiast, *SVF* 2.333)

> [The world] is not infinite but finite (or bounded, *peperasmenos*), as is evident from its administration (*dioikeisthei*) by nature. There cannot be a nature of anything infinite (unbounded), for nature must have in its power that whose nature it is. (*Apeirou men gar oudinos physin einai dynaton dei gar katakratein physin outinou estin.*) That (the world) has

a nature that administers it is known, first, from the arrangement (*taxeos*) of its properties and substructures (*meron*), then from the arrangement of the things in it (also *meron*), and thirdly from the sympathy (*sympatheia*) of those substructures with each other. (Cleomedes, *SVF* 2.534)

The stoics conjectured that the cosmos or "the all" (*to pan*) was structured by a kind of "tension" or *tonos*, (sometimes translated "tenor"), which Samuel Sambursky once assimilated rather aptly, I think, to a kind of spatially distributed field-strength. Their conjectures about this "field" were no less "physical" than the epicureans' pre-mathematical accounts of atomic trajectories and swerves, and their "material" ontology embraced "waves" as well as "particles" (cf, *SVF* 2.871–872).

Stoic ideology also had a number of distinctly idealistic, dualistic and even Platonist aspects, however—not surprisingly, perhaps, given that it was a collective construct that extended over several centuries. Consider, for example, the following remarks.

> ... [Zeno] says that the world ("*to pan*", "the all") is the most beautiful work wrought according to nature, and on the most likely account a living being, ensouled, intelligent and rational. (Sextus Empiricus, *SVF* 1.110)

and

> Chrysippus ... constituted *the highest good* [of a human being] in such a way that it seemed not [to be] to excel in mind, but *to be nothing but mind* (Cicero, *SVF* 3.20; the italics are mine)

Such passages suggest that one might distinguish two contrasting views of stoic "nature" and "the world" it "administered", and characterise them in somewhat nonstandard terms. The following dichotomous account of the contrast is by Plotinus, though my anachronistic characterisations of it obviously are not.

3 [A quasi-"Leibnizian" view]

Some, when they come to the principle (*archen*) of the all (*tou pantou*), derive (*katagousi*) all things from it and say that it pervades (*phoitasasan*) everything as a cause, not only as the moving cause but as a producing cause of each thing, and this they call fate and the overarching cause (*heimarmenen tauten kai kuriotaten aitian themenoi*) producing (*poiousan*) everything [they say], not only [that] all other things that come into being, but even our thoughts proceed [*ienai*] from its motions ... .

4 [A quasi-"Spinozan" or –"Newtonian" alternative]

> And others ... [say that] the superposition (*epiploken*) of causes with each other in a descending sequence (*anothen heirmon*), that the consequents always follow the antecedents, that the former refer back to the latter (*ep' ekeina anienai*) through which they come into being and without which they would not come into being, and that they are subject to [*douleuein*] them. *Anyone who says these things is clearly introducing fate in a different manner.* [So] *one would not be far from the truth if one put these people in two* [groups]: *some link* [*anartosin*] *all things to one* [principle], *and others do not.* (Plotinus, SVF 2.946; the italics are mine)

On the "*Leibnizian*" view in 3, in short, a world-marginal "nature", "god" or otherwise adumbrated "principle" (*arche*) individuates a particular *kosmos* or "world", and well-orders (or minimally ill-orders) it.

The "Spinozan"/"Newtonian" interpretation in 4, by contrast, would identify "god"/"nature"/"the world"/"the" *principium et causa sui* from "within", as *deus sive natura*, and interpret "fate" as the deterministic internal structure of that entity, whatever it is called.

In principle, the "Leibnizian" view could therefore distinguish two levels of potential "physical" determinism, and associate each with a prototype of metalogicians' completeness, and philosophical logicians' "bivalence"—whose importance Chrysippus well-understood:

> Chrysippus the stoic [said] that the necessitated (*katenankasmenon*) does not differ from the fated (*eimarmenou*), for the fated is an eternal, continuous and structured motion (*kinesin aidion syneche kai tctagmenon*) (Theodoretus, SVF 2.916a)

> ... so Chrysippus fears that if he fails to maintain that everything which might be enunciated is either true or false, he will not be able to hold that all things are brought about (*fieri*) by fate and eternal causes of future events. (Cicero, *De fato*, SVF 2.952)

A "Leibnizian" determinist, on this account, could therefore postulate—as Leibniz explicitly did—two sorts of "completeness" or "unicity" and monistic integrality:

5   unicity of "the" higher-order or metatheoretic "*physis*" or "god" or "*natura*"; and
6   unicity of "the" lower-order or object-theoretic world (*kosmos*) this super-ordinated empyrean entity would "chose", "regulate" or (as the stoics sometimes put it) "administer" (the verb that recurs is *dioikeo*).

An object-theoretic, "Spinozan" conflated version of *to tiv* and *to pan*, by contrast —*deus sive natura*—would somehow have to "choose", "regulate" or "administer" [itself], and it was no accident therefore that Spinoza himself made such

[self]-realisation or [self]-interpretation the defining characteristic of "substance".

It may also bear mention here that Leibniz himself stratified "necessity" along essentially similar lines, and his language and background explanations for such "necessities" suggested he had two complementary background aspects for them in mind.

On the one hand, he called the unicity and "necessity" of 3 "absolute", and that of 4 (merely) "hypothetical", in terms which recalled the stoic or quasi-stoic distinction sketched above.

On the other, he assimilated

7   the plurality of 3's "divine" "choices" to "logical possibility"—whose natural metalogical counterpart, on the evidence of assorted completness theorems, would be consistency; and
8   4's "internal" monism and "fate" to ("Laplacean") deterministic time-evolution, governed by certain then-recently discovered (and uncritically extrapolated) differential equations. In the case of the stoics, both forms of determinism sketched above in 3 through 6 left little room for forms of indeterminism or underdetermination that might be needed for unillusioned moral autonomy. Plutarch formulated a version of this notorious problem as the following trichotomy:

> ... [I]f representations do not arise from fate, [then what is the cause] of assent? If on the other hand, [fate] makes representations inducing assent, and assents are said to occur by fate, how does fate not conflict with itself ... ? ... Yet one of these three must hold: not every representation is the work of fate; or not every access and assent to representation is unerring (*anamarteton*); or fate itself is subject to reproach. (Plutarch, *De stoicorum repugnantiis*, SVF 2.993)

Much later, Hume developed a more nuanced and rhetorically powerful canvass of these dilemmatic alternatives in his *Dialogues Concerning Natural Religion*.

Chrysippus himself had also formulated an early response to this problem, in the form of yet another dichotomy. In the rest of this section, I will propose and explore some analogies and interrelations between this dichotomy and

9   the one Plotinus outlined above (cf. 3 and 4);
10   the early-modern variants of it sketched in 5 through 8;
11   the hierarchy of stoic *ordres de raisons* (*logike*, *ethike* and *physike*);
12   local or contextual metalogical distinctions between theory and metatheory;
13   local or contextual mathematical distinctions between deterministic time-evolution of a given system, and its more contingent initial and boundary conditions.

In Cicero's transmission, the original dichotomy went as follows.

> Chrysippus ... distinguishes two kinds of causes, so that he may escape necessity and retain fate. For, he says, some causes are *perfect* and *principal*, and others are *auxiliary* and *proximate*. So when we say that everything is brought about by fate and antecedent causes, we do not mean (*non ... intelligi volumus*) perfect and principal ones, but auxiliary and proximate ... . [T]herefore, ... if everything is brought about by fate, it follows that everything happens by prior causes, but not in fact by *principal* and *perfect* ones, only *auxiliary* and *proximate*. And if these [causes] are *not in our power*, it does not follow that our appetites are not ... . (Cicero, *De fato, SVF* 2.974; the italics are mine)

I would first like to argue that a two-stage hierarchy of the sort Plotinus considered in 3 and 4 may in fact have provided a rationale for (and even motivated) the dichotomy of "causation" Cicero glossed. In this hierarchy, for example, forms of personal identity at "higher" levels of "final" causation could transcend "lower" levels of rigid "moving" causation (recall Kant's "*intelligibele Ursachen*"), or at least seek to align themselves in spirit with such a realm.

I would also like to suggest

14 that the priority it seemed to give to "formal" and "final" causes, and the explanatory accounts associated with them (latent in the original etymology of "*aitiai*") may in fact have anticipated significant aspects of the metalogical distinction between object-theory and metatheory.

15 that other aspects of Chrysippus' dichotomy are also strongly suggestive of now standard mathematical-physical distinctions between system-under-evolution, and initial and boundary conditions for such evolution, for the latter commonly account for chaotic and stochastic departures from the otherwise deterministic trajectories of such systems.

More vulnerably, I would finally like to propose a conjectural framework in which one might interrelate these analogies.

In "Virtual Modality" ([Boos] 2003), I developed a "Leibnizian" metatheoretic semantics for modal extensions of first-order theories T, and in Chapter 5, "*Parfaits miroirs de l'univers*': A 'Virtual' Interpretation of Leibnizian Metaphysics." I offer a historical and metalogical rationale for this label. In part at least, my purpose in this chapter is to suggest that certain aspects of this framework of "virtual" modality may also

16 provide a common speculative framework for these interrelations;

17 relativise as well as reconstruct the "Leibnizian" reconstruction of stoic "nature"; and

18. clarify a common *aporia* Chrysippus, Spinoza, Leibniz and Kant have all faced in their attempts to explain their way out of the "thoroughgoing determination(s)" of what Kant called *Naturnotwendigkeit*.

In essence, "virtual" interpretations of modal extensions of T appear in boolean-valued extensions of a set-theoretic universe, and "truth" of model assertions in such extensions is not bivalent, but distributed in graduated and partially-ordered ways over a space of interpretations (called the Stone space) of T.

More precisely, such "virtual" "world"-extensions can be specified by particular boolean-valued interpretations **v** for T, which may be individuated as definite objects in semantic terms given in later metatheoretic contexts, though observers in such contexts will be unable to discern whether such individuations are "really" definite or "merely" virtual in their turn.

In the more historically oriented Chapter 5, "'*Parfaits miroirs de l'univers*' . ....", I offer textual analogues in Leibniz' published writings for much of this, and assimilate, in particular,

19. **v**'s semantic structure to a semantic rationale for Leibniz' "principle of sufficient reason", which secured a counterpart of 6's "internal", "hypothetical" necessity;
20. the allegedly invariant metatheoretic "world"-structure **v**'s "virtuality" to the recurrent graduation and "confusion" of Leibniz' monadic "mirroring", which he sometimes likened to the superposition of "waves heard at the seashore".

In the present context, I would draw similar parallels between
21. the structures of particular structures **v** and those of alternate stoic *kosmoi*, and
22. the metatheoretic individuations of such structures to Chrysippian perfect and principal "causes" of them, and internal relations modeled within them to their auxiliary and proximate counterparts.

If these reconstructions are correct, finally, they might suggest
23. that one should reorder the last two terms in Chrysippus' original ascending "order of reasons" (logic/ethics/physics); and
24. that the Leibnizian variant of stoic ethics Plotinus and Cicero sketched may have been closer to its Kantian successor than most Kant scholars think. (Perhaps Antisthenes or Diogenes or Epictetus were the original "grinches who didn't steal Christmas"; cf. *Grundlegung*, 398).

What these reconstructions and analogies do not suggest, on my view, is that Kant came significantly closer to a resolution of the deeper underlying problems than Chrysippus or Leibniz had done. For they all

25 begged the question of higher-order metatheoretically determinate and determining designs; and, more seriously,
26 postulated views of moral identity that look suspiciously like varieties of higher-order predestinarianism, however "successfully" questions of their metatheoretic unicity and determinacy might be begged.

What, for example, if the possession or attainment of an intrinsically "good will" is not "in the power" of a would-be "intrinsically good will"?

I will consider this question in a preliminary way in the next section, and examine it more carefully in a study of Kant's second and third *Critiques* in Chapter 8.

6 "Our" Power over (What Is In) "Our" Power

In this section and the chapter's conclusion, I will argue
1   that the stratification examined in the last section ineluctably leads into a conceptual pattern I have elsewhere called metatheoretic ascent, and
2   that a concomitant ideal of indefinite inquiry, and not in any unitary physis or *Reich der Zwecke*, is the only escape we are ever likely to find from the anachronistic Wittgensteinian *Fliegenglas* of stoic fate.

Consider first, by way of brief review, some classical expressions of the stoic ideal of "concordance" *or* "concurrence" (*homologia*) with "nature".

> Zeno ... set forth that the end of goods is to live honorably, and this is drawn from the conciliation of nature (*ducatur conciliatione naturae*). (Cicero, *SVF* 1.181)
>
> ... a duty (*officium*) ... is what is so performed (*factum*) that a reasonable account (*probabilis ratio*) can be given of it (Cicero, *SVF* 1.230).
>
> ... virtue is a harmonised disposition (*diathesin homologoumenon*), chosen for itself and not from hope or fear of any external things; in this is spiritual enlightenment (*eudaimonia*) since in this state the soul is directed toward concordance (*homologia*) of all life. (Diogenes Laertius, *SVF* 3.39)
>
> ... when the mind ascends with the concurrence of reason from the things in accordance with nature, then finally it arrives at the notion of the good. (Cicero, *SVF* 3.72)
>
> ... herein lies the chief good ... in what the stoics term *homologia* ... that good to which all things are to be referred ... the sole thing that is sought for its own strength and worth *(vi sua et dignitate)*; none of the first objects of nature, by contrast, is sought for itself. (Cicero, *SVF* 3.188) Whatever is carried out by a wise person (*a sapiente*) must be continuously fulfilled (*expletum*) in all its parts; for what we say is sought lies in this; in the same way, things carried out virtuously ... are to be judged right in their first inception, and not in their final completion (*perfectione*). (Cicero, *SVF* 3.504)

> An action will not be right unless the will has been right, for the action derives from this. Nor will the will be right unless the state of the mind (*habitus animi*) is right, for the will derives from this. (Cicero, *SVF* 3.517)

The first observation I would offer is that action in concordance with a "perfect" higher order stoic realm of "nature" of the sort sketched in the last section may be inexpressible, and in that sense irreconcilable with the more discursive notion of "duty" Cicero referred to in SVF 1.230.

Common to most hieratic attempts to "close" or "complete" hierarchies of metatheoretic ascent, by contrast, are tacitly hieratic desires to "hypostasise" them in "unconditioned" forms of "absolute" existence and *durchgängige Bestimmung* (Kant's "thoroughgoing definition" or "determination"). But the best metalogical evidence—drawn, I have tried to suggest, from dialectical refinements of the early stoic and skeptical debates outlined in earlier sections—suggests that efforts to prove alethic predicates "mean" what we want them to "mean" require

3   that we enlarge and reinterpret the theories we formulate in hierarchies of metatheoretic inquiry, and
4   that we enlarge and reinterpret *ourselves* in the process.

For the metalogical insights just mentioned remind us that one should not only distinguish consistency from "truth", in object- as well as metatheoretic contexts, but also

5   proofs that one must postulate ever more metatheoretic reinterpretations to "save" consistency for "constitutive" theories, from
6   ultimately incoherent "proofs" that we "must" consolidate these partial postulations in a *single* constitutive theory that proves its own "truth".

Indeed, such insights suggest that there will usually be infinitely many branching, nonisomorphic candidates for interpretation of "intelligible" concept-formation, and that these may be considered in infinitely many branching, nonequivalent candidates for appropriate metatheories.

Expressibility, in short, is local. But aspirations to comprehensiveness and perfection are not. Truth without expressibility is an empty template, but "truth" which "passes all understanding" passes all relevance. For such "universal" limits of reason-giving and design ("*kosmoi*") would (ostensibly) comprehend everything which continues to perplex us, yet be comprehensible to us as well.

As mentioned in chapter 1, ancient skeptics had already recapitulated and elaborated presocratic observations to the effect that a god of the fish (say) might be a great fish (cf. Xenophanes' remarks at DK 21 B1416, for example, and assorted remarks of Heraclitus, Empedocles, Leucippus and Democritus). Less amusingly, such skeptics might have queried whether the priests and savants in Plato's cave might not have had "good reasons" to interpret "god" as "the" "great prisoner (or warden)"? ... . Or the flickering shadows on its walls as models of "the" design of "the" universe? ... .

(This is the essential problem of Leibniz' theodicy, of course. Has no one ever inquired why "the sun" did not illuminate that Orphic "cave"?)

Appeals to such hieratic ideals as an intelligible guide to "experiential" action and reflection are therefore speculatively untenable, and the very notion that they are accessible to "us" is a normative claim that may rise to the level of its incompetence. It may also express a form of normative arrogation—a belief that "we" are not so simple after all; that our "stimuli" are not so impoverished; and that "we" are the chosen people of *physis*, as it were, or at least at the head of its class (the fish, by contrast, are just fish).

Be that as it may, desires to find "ultimate" exemplars for complex but incomplete thought-experimental patterns remain deeply ingrained in hieratic aspects of stoic *"physis"* and other theological ideals. And nominally secular hieratic ideals—of an allegedly unique and uniform *Bereich der Erfahrung*, perhaps, or "the universe of the philosophers", or "the god of the scientists"— remain remarkably robust, for reasons that may have something to do with the human need for heuristic encouragement, and other psychological and temperamental aspects of "reason's fate."

For some find *ataraxia* in a welcome rest, but others find it (or aspire to find it) in "the peace that passes all understanding". This has metaphysical as well as normative implications in Chrysippus' original ordering, which essentially reversed in its last two stages Kant's "primacy of the practical". For scientists who are temperamental *croyants*, some counterpart of stoic *physis* may well seem to be "god", or at least one of "its" principal manifestations.

In this and other chapters, I have tried to offer a historical and metalogical outline of a kindred dialectical opposition betweem "hieratic" and "zetetic" approaches to philosophical inquiry. Among other things, the opposition may be cast in terms of antinomially opposed views of "wisdom" and its embodiments (such as the stoic sophos).

On the "hieratic" view, inquiry is valuable because it converges to wisdom. Ataraxia is then to be found in the quiet authority of that wisdom, which decides and guarantees comprehensiveness and unicity of interpretation, with whose (alleged) personifications one might be drawn to identify. On the "zetetic" view, wisdom is valuable because it fosters and generates inquiry. Nothing is expressible or discernible in "hieratic" formulations that could not be reconstrued in alternative interpretations of inexhaustible beauty and contemplative refinement.

And nothing in them provides probative warrants for the non-theory-relativity they claim for [themselves].

The metalogical evidence sketched above also seems to me lead to a natural "zetetic" assimilation of "experience" to such hierarchies of metatheoretical ascent. For it strongly suggests that

7   expressible attempts to "close" transcendentally hierarchies of metatheoretic ascent are precisely what generate—or witness the need to generate—ever more of them.

For if indefinitely iterable processes of encoding, interpretation and reinterpretation of information are inherent to life, for example, such processes might plausibly be immanent to "experience". In more affectively intense forms, they might also be inherent to the *eudaimonia* of humans and other sentient beings. In such processes, moreover, we seem to pass almost immediately to interpretation (or ranges of interpretation) of such processes. It may, therefore, be pragmatically as well as zetetically unwise to try to delimit our quantification over them, and over their interpretations, and … . So construed, our experiences would come to us embedded in processes of analysis *ad indefinitum*. This would not mean, however, that *we* are infinite, or that we transcend analysis, or that we have "comprehensive presentations" of anything that does.

On this view of "experience", moreover, "ultimate" forms of unicity and individuation, if they "existed" in any sense at all, would do so by definition in a sense that would be inaccessible to us. And if there were anything that eluded "all" "our" demands for expressibility and communicability, these elusive entities—not the *Noumena* or "objects of thought" Kant wrote about—might well be described—by some other sentient beings, perhaps—as "our" *Dinge an sich*.

In temporal concomitants of metatheoretic ascent, finally, whatever can recognise [itself] is no longer [what it recognised], and an infallible judgment which purported to provide a guide for judgment of actions would therefore not be an action, and to that extent not a judgment either.

Indeed, this follows arguments cognate to those sketched above in 4.29 through 4.33, which give rise to a kind of metalogical anti-"cogito". For a claim to provide an infallible guide … of the sort just sketched would be a claim to "internally" provable completeness, which would imply its own [inconsistency]; and metatheoretic "perfection" could only be interpreted in a metatheory that could not express or "conceive" it. In more metaphorical terms, how would we "know", once again, that the light from the sun outside the cave does not emanate from a *demiourgos* rather than "god"? How—more painfully—could we discern "the" distinction, if "it" "crafted" us?

I believe the counter-"cogito" argument just sketched comes very close to a refutation of certain forms of "hieratic" postulation. But a hieratic metaphysician has a carefully nuanced open option, which I have considered elsewhere (Chapter 8): acceptance of a theory that is, roughly speaking, inconceivable or incomprehensible to [itself]. ("*Es ist das Mystische* . ....") Indeed, I do not see a way to formulate a consequential elaboration of the hieratic view which does not

8   employ indefinitely many stages of a zetetic's metatheoretic hierarchy—which would seem to offer a de facto concession to the philosophical opposition; or
9   eventuate in some variant of the alternative—a form of mysticism (or mystification), perhaps—just sketched.

To put a more precise point on it: hieratic postulation of an "intended" interpretation or "horizon" of interpretation that is beyond intention (and therefore in certain senses beyond postulation) could not even be formulated as a discursive candidate for acceptance or rejection unless it accepted some counterpart of the disjunction in 8 and 9.

Since some aspects of the "mystical" but metatheoretically coherent alternative in 9 recall the dictum "*credo quia absurdum est*", it may not attract a very wide following. There is, however, substantial evidence that attempts to halt or arrest the metatheoretical hierarchies sketched earlier encounter such predicaments.

This brief argument, finally—one I've sketched in greater detail elsewhere—also amounts to a reversal of one of Davidson's more influential claims. For it amounts to a metalogical analysis of "the very idea of a nonconceptual scheme".

The "zetetic" view, in conclusion, also has at least the merit that it is rather readily expressible, and open therefore to discursive (re)examinations of [itself]. In responses to dubious "hieratic" uses of zetetic self-referential/semantic-paradox arguments to "end" zetesis, for example (the "absurdity" of skeptical doubt of skeptical doubt; the cogito; the "master-argument"; the "transcendental deduction"; ...), the zetetic metaphysicians can respond that they do not need to "end" hieratic argumentation—only reinterpret it (one more time).

7 Freedom, Autonomy and Aporetic "Self-Legislation"

Prompted by the arguments in the last two sections, I have tried elsewhere (Chapter 1, 14.1–12) to elicit attributions of value from "constructive" but essentially incomplete interpretations of the "pluralistic pathos" one might attribute to zetetic inquiry, in opposition to hieratic metaphysicians' "monistic pathos" (a phrase lifted from Lovejoy's *Great Chain of Being*).

These seem to me "zetetic" in the sense that they draw on assumptions of incompleteness and plurality of interpretation, rather than the presuppositions of completeness and hermeneutic unicity (often conflated with "universality") that seem to underlie their hieratic counterparts. One of the deepest hieratic tenets in metaphysics, I have argued, is a "completeness"-assumption that there is no "ultimate" ontological or methodological middle ground between "inconsistency" and "truth"—or, equivalently, no "ultimate" metaphysical distinction between "consistency" and "truth".[1]

This amounts to the view that "contingency" is a makeshift or illusion, and physical "truth" is one, apodictic (in some equally elusive sense) and determinative of "experience" (which in some latitudinarian sense converges to it or aligns with it). Correlatively, one of the deepest hieratic tenets in moral philosophy would be a concomitant or analogous claim that there is no "legitimate" practical or axiological middle-ground between "inconsistency" and "truth"—or, equivalently, no "ultimate" normative distinction between "consistency" and "truth".

And this would amount—on my account, at least—to the view that moral "contingency" is a makeshift or illusion, and moral "truth" is one, apodictic (in some equally elusive sense) and determinative of praxis or practical "experience" which "must" converge to or align with it.

According to this latter, "axiological monist" or "realist" view, in particular, anyone who denied the "existence" of such a unitary realm of ends or moral truth, would inexorably be committed to the view that "anything goes", a dialogical fallacy which underlies sooner or later every special plea for such a realm's "necessary existence".

In reply to this view, I would argue that the moral as well as metaphysical variants of this view are "transcendental illusions", in roughly Kant's senses of the phrase, and respond (as outlined in more detail in Chapter 1, 14.15–19)

1   that these dialectical tensions are generative rather than "frustrating", *à la limite(s)*, and reflect the "dignity" of Kantian "freedom" as well as its "fate"; and finally,
2   that efforts to understand "freedom" and "fate" in such double aspects offer a stable middle ground of the sort sought above, as well as moments of practical as well as speculative *ataraxia* (a stoic as well as pyrrhonist-skeptical term).

---

[1] This section includes a few sentences from 1.14, but seems required for the sense here as well; I have recast the order slightly to minimize repetition. Ed.

In the rest of this section, I will comment briefly on three rubrics—of particular relevance to the preoccupations of classical stoicism—that appear in the last two of these assertions.

The first is

3   "Freedom and Incompleteness"

Early critics of stoic teaching were right, I believe, that a unitary "truth" that encompassed temporal experience would seem to entail some sort of diodorean determinism, and right again that zetetic freedom and plurality of temporal outcomes are essential to the ideals of autonomy the stoics cherished.

As a zetetic response to what might be called "the stoic's paradox"—that "I am free to accept my unfreedom"—I would therefore offer a less provocative assertion (or acknowledgment): that "I am free to accept the fragility, uncertainty and incompleteness that accompany that freedom."

Temporal as well as conceptual incompleteness would also seem to me to be a precondition of the coherence of talk about "will", good or otherwise, and as such, part of our strength and our "fate".

The second rubric is

4   Autonomy, "Reasonable Faith" *and "Reine Selbstätigkeit"* ("pure self-determination and/or actuation")

The stoics' critics were also right, I believe, that injunctions to "submit" to the ostensible stasis and invariance of stoic *physis* would be a violation of stoic ideals of "autonomy", not fulfilment of them.

It also seems strange, more generally, that so many who inquire, or at least claim to value deeply certain sorts of inquiry, seem to believe

5   that inquiry's "end" is its end—its termination;
6   that its "spirit"—its dynamical ability to generate new, not-yet-unswered, "interesting" questions—can only be found in something that is bereft of such spirit; and
7   that saving words can be found only in something that is beyond interpretive plurality, and therefore beyond dialogical language.

Arguments that draw in part on desires for lucidity and mutual expressibility, but also on tacit acknowledgment that aspirations to autonomy may otherwise be doomed, suggest that

8   more "reasonable" counterparts of stoic *physis* and Kantian "reasonable faith" may be found in the indefinite ascents of metatheoretic inquiry, and

9   more "reasonable" ideals of autonomy and *"reine Selbsttätigkeit"* in stage-by-stage efforts to refine moral and physical hypothesis-formation.

Processive ideals of this sort might also provide constructive alternatives to the circularity of contentless [self]-reference, and the vacuous ideals of completeness and uniformity that pass all understanding.

Autonomy in the literal sense of Kantian [self]-legislation, for example, may be a "discursive" and dispositional undertaking. For "legislation" itself is.

And *"reine Selbsttätigkeit"*, finally—a "practical" counterpart of Aristotle's *noesis noeseos* Kant attributed to the "self-legislating will"—may be as ethically significant as Kant thought it was, but it is also provisional, transitory (as are we), and inherently incomplete.

A final rubric is

10   Solidarity, Moral "Ataraxia" and Primacy of the Practical

One of my motives in this chapter has been to offer a zetetic alternative to Kant's individualisation (as I see it) of stoic notions of *"autonomia"* and *"homologia"*. Kant, of course, appealed to the *"regulativ"/"konstitutiv"*-distinctions and rather *ad hoc* notions of *"intelligibele Ursachen"* to exempt moral agency from Laplace's mechanistic counterparts of stoic "fate".

I have replaced them with an incomplete but open interpretation of "experience" which construes metatheoretic ascent as a more open and processive alternative to stoic *physis*, and accepts recurrent conceptual limitations as part of a less rigid notion of zetetic *homologia*. (It is morally significant, on this account, that "Now I've seen everything" can never be said.) An ideal of mutual intelligibility would be part of such provisional and latitudinarian counterpart of stoic "concordance". Since "god" doesn't provide it for us, moreover, we must try to realise it —as incomplete forms of reflective equilibrium, perhaps—in our incomplete and provisional ways.

A "zetetic" imperative, therefore—not "categorical", but very deeply "hypothetical"—might enjoin us to (try to) improve our own and others' interpretive abilities. This might be the only way to help each other escape from Wittgenstein's *Fliegengläser* without *petitio*. For our abilities to envision indefinite alternative hierarchies in other *Gläser* may be what enable us to "see" ourselves, and to that extent be, something other than *Fliegen* (admittedly not the sense of *"Gläser"* Wittgenstein had in mind).

We have failed, of course, and will fail—spectacularly. But our guilt (our moral responsibility) will be diminished to the extent we "blunder" with respect to what we cannot interpret. Correlatively, we merit no praise if we act in alignment with what we cannot interpret (an instance of metalogical "heteronomy").

I would hardly efface normative distinctions between such "success" and "failure", in short—only "localise" them. There will be many local and potential criteria for "success" of such efforts, in the reconstructions of "experience" I have offered—particular criteria for particular kinds of understanding, and others for interpretations that underlie understanding of such understanding.

In keeping therefore with the imperative sketched earlier, "zetetic" inquiry
16 should therefore give stage-by-stage priority to the "practical"; and
17 should endeavor to make those stages both *responsable* and "accountable"—subject, that is, to egalitarian constraints of mutual interpretability (consider the dialogical etymologies of these two words).

To the extent "practical" reflections, finally, become part of semantic interpretation of "empirical" problems they tried to resolve, suffusion of first-level "observation" with second-level ("theoretical") "values" becomes "practically" useful as well as "speculatively" significant. So conceived, in short, the primacy of the practical is a graduated and processive as well as counterfactual ideal, no less elusive perhaps than Kant's *Reich der Zwecke*. But its primary imperative seems to me straightforward enough: respect observant consciousness wherever it may be found.

8 Envoi

Among other things, in this chapter I have tried to offer a skeptically inflected answer to the hauntingly recurrent question David Hume asked in his posthumous *Dialogues concerning Natural Religion*:

> What particular privilege has this little Agitation of the Brain which we call Thought, that we must make it the Model of the whole Universe?

"None", would be the response, if one really tried to propose such "agitation" as a model for "the whole universe". But "every", of course, if one offered it as a model for practical action, humane behavior and for those forms of normative evaluation of which we may be humanly capable.

What is "in our power", in particular, is not in our power. But one consequence of such practical *incomplétude* is plurality of practical action, as plurality of interpretation is a consequence (or at least a concomitant) of its epistemic and metaphysical counterparts. Explicit or implicit self-referential *aporiai* are therefore marks of "reason"'s *honor* as well as its "fate". And "subreptive" *petitiones* abridge that honor.

This "abridgement", in my view, is marked by assorted dialectic ironies: the *praxeis* and *energeiai* of inquiry, for example, are intrinsically skeptical. But the motivating ideals of many who pursue various sorts of inquiry for other ends are not.

In response to such a *de facto* "two-truths"-theory, I have tried to argue that dogma is both heteronomous and conceptually self-defeating, but that inquiry— which is deeply "skeptical", in original senses of the word—is neither. Indeed, inquiry is regulative of the "life of the mind", or of whatever confers the dignity of such "life" on minds—or processors—in whatever material supports they may be found.

In the end, therefore, I have argued

1. that what is "*hieros*", is not "god", but a good inquiring will,
2. that eagerness to inquire—or at least willingness—is regulative of that goodness,
3. that this eagerness should not be valued as a means to some other end, but as an end in itself, and
4. that freedom, autonomy, solidarity and mutual aid are limiting and processive ideals of such "goodness", and
5. that small epiphanies of "*homologia*", or "life in accordance with" these ideals can provide partial measures of *ataraxia* and peace.

Eternal, invariant and uniform "peace", however, comes only in death. The journeys, not the arrivals, matter.

I will finish with a comment about "*aporetic*" aspects of the view of autonomy I have sketched. Lexically, of course, *aporia* meant something like: no-way-out; *huis clos; sans issue; ohne Aussicht; kein Ausweg*; … . an end, as it were, of a journey. But not "*the*" end, perhaps, of *the* journey. Others will continue, for a while, and new doors will be open as long as "we" are, to untold rooms of great potential horror, and great potential beauty—more in each case than we may be able to bear or comprehend.

And our very awareness of our vulnerability to all this may give us glimpses of deeper variants of mutual aid and stoic *sympatheia*—deeper, at least, than could be found in most versions of their *kosmoi*.

4 Anselm, *Fides Quaerens Interpretationem*, and *Grenzideen* as Generators of Metatheoretic Ascent

1 Outline and Introduction

The principal aims of this chapter are to
1. examine good-faith alternatives to metaphysical monist "faith", and
2. interpret "confutations" of skeptical critiques of such monism as *ignorationes elenchi* which dismiss the implications of "impredicativity" and semantic paradox.

More precisely, I will argue that dialectical, empirical and "transcendental" variants of such "confutation" have taken many forms, among them
3. ancient stoic arguments that skeptics cannot doubt the "doubter's paradox" (that "I doubt [this sentence]");
4. cognate stoic, Augustinian and Cartesian "cogito"-arguments (that I may "doubt", but cannot doubt that "I" doubt);
5. Berkeleyan and quasi-Berkeleyan "master arguments" (that I must "conceive" "what" transcends my "conception" to "conceive" "that" something transcends "it");
6. Humean "empiricist" *petitiones* (that it "must", e.g., be a "matter of fact" or "relation of ideas" whether a given assertion is a "matter of fact" or "relation of ideas");

and
7. Kantian "transcendental" "deductions" (e.g., of "synthetic a priori" uniformity of "experience" from (alleged) "synthetic a priori" uniformity of "the" faculties which "constitute" "it".

An ancillary aim will be to draw on metalogical analyses (usually but not always Gödelian) to argue that such attempts have blurred theory and metatheory in ways which tacitly
8. relativised their authors' intended conclusions, in (implicitly skeptical) hierarchies of metatheoretical ascent, or
9. begged these conclusions, by exempting their methodological premises from (their own standards of) critical scrutiny.

In particular, I will interpret Anselm of Bec's "ontological argument"—which appealed explicitly to "our" alleged ability to "conceive" a "unique" limit of "our" "conception"—along similar lines: as
10 a refinement of stoic "refutations of skepticism", on the one hand; and
11 a template and precursor of various "critical arguments" and "transcendental deduction(s)" on the other.

I will also argue that
12 the ontological argument's underlying fallacy lies in its (implicit) attempts to "define" "the" limit(s) of definition and "conceive" "the" limit(s) of conception;
13 consequent analyses of such attempts inevitably give rise to "locally critical" hierarchies of object-theoretic conjecture and metatheoretic (re)-interpretation;
14 theory-relative, processive, stage-by-stage appeals to such hierarchies are heuristic preconditions and "infinitesimal generators" of "consciousness" and reflective inquiry;
15 provisional attempts to "conceive limits of conception" become specious (only) when they claim to "bound" "all" ramifying paths of conception and inquiry, in a unique limit which allegedly transcends such paths; and finally that
16 conceptual hierarchies of the sort mentioned above in 1.3–1.7 provide thoroughly "natural" but also processive, skeptical, "merely regulative" and non-uniform interpretations of phenomenal "experience".

Anselm was right, on this account, to derive semantic significance (however elusive) from conjectures about "conception of limit(s) of conception" (an *Ansatz* which Kant later refined). For processive variants of such noetic activity are deeply regulative of the "reasonableness" of "reasonable beings", as well as the "understanding" (*intellectus*) Anselm hoped would ground or confirm his faith.

But he was wrong (and many other metaphysicians with him) to assume that we "*homunciones*" ("human-lings") "must" presuppose the coherence (much less "existence") of a "unique" hypostatic limit for boundless ramifications of conceptual inquiry. For straightforward metalogical arguments strongly suggest that such (pre)suppositions are not only unsecured by refinements of the discursive analysis he hoped would strengthen his religious convictions: they are incompatible with them.

More explicitly, such metamathematical arguments indicate that
17 attempts to "define" as well as "prove" *the* "existence" of "that than which nothing greater can be conceived" ("*id quo majus nihil cogitari potest*") do not give rise to "proofs", but to semantic paradoxes; and that

18 these paradoxes become outright contradictions (*"credo quia absurdum est ... ?"*), unless one clarifies them with the aid of not-so-transcendental (and less-than-numinous) relativisations of "proof", "definition", "conception" and the claims of "unicity" implicit in the grammatical singularity of Anselm's *"id quo"*.

Still more explicitly, I believe they also yield that
19 in any metatheory T which can "talk" about "existence" of such (hypothetically) unique limits for [itself], [provability of such existence] is provably equivalent (following Gödel) to [provabilility of everything] (and therefore to the [inconsistency of T]);

and therefore that
20 "proofs" of such a limit's "existence" would therefore have to be more than "nonconstructive" (a generative as well as harmless trait of abstract mental activity); they would have to elude scrutiny in the very metatheories one might invoke to "prove" that [such "proofs" "exist"] ... .

To me, finally, such reflections (which are quite straightforward in the metalogical ambits to which they are normally confined) suggest that
21 theory-relative attempts to "define" limits of "definition", "conceive" limits of "conception" and "demarcate" limits of "demarcation" are regulative aspects of reason's office (*officium, duty*) as well as its "dignity" (Kant's *"Würde"*);

but also the following ineluctable dilemma (which runs like a ground base through these chapters): that
22 "ultimate" projections of such attempts are "subreptive" aspects of reason's capacity for self-deception and reluctance to accept its "fate": that

> ... sie durch Fragen belästigt wird, die sie nicht abweisen kann; denn sie sind ihr durch die Natur der Vernunft selbst aufgegeben, die sie aber auch nicht beantworten kann; denn sie übersteigen alles Vermögen der menschlichen Vernunft. (KdrV, A VI) ... . they are harassed by questions they cannot dismiss[,] for they are posed by the nature of reason itself[;] but which they also cannot answer[,] for they surpass every capacity of human reason.

Such "impredicative" dilemmas suggest an idiosyncratic variant (and oblique critique) of Wittgenstein's notorious injunction to "silence":[1]

---

[1] *Logisch-Philosophische Abhandlung*, 1984, Proposition 7, "Wovon man nicht sprechen kann, darüber muss man schweigen."

> *Wovon man nicht sprechen kann, davon soll man absehen ...,*
> (Whereof one cannot speak, thereof should one suspend judgment ... .),

for skeptical inquiry seems to me a better response to such dilemmas ("*von etwas absehen*" can mean—among other things—to "refrain from" something; to "abstract from" it; or to "withhold judgment about" it).

In the first section of this chapter, I will briefly review Anselm's original argument and Gaunilo of Marmoutiers' critique, and sketch preliminary interpretations of both in the language of 1.1–1.4.

In the second, I will examine the striking similarity between Anselm's well-known formula ("*id quo majus* ... ") and George Cantor's paradigmatic instantiation of a "liminal" property he called "*Inkonsistenz.*"

In the third, I will review some metalogical warrants for skeptical (or "locally transcendental") examinations of such "*Inkonsistenz*".

In the fourth section, I will reconsider Kant's early modern critique of Anselm's *Beweisführung* in the light of such metatheoretic analyses, and then apply them to more recent efforts to provide "modal" reconstructions of Anselm's argument in the fifth.

In the sixth section, finally, I will appeal to the indefinitely ramifying hierarchies evoked in 1.12–1.14 above to propose

23 more skeptical variants of Kant's "*vernünftigen Glauben*" ("reasonable faith"); more provisional forms of "*Vertrauen in die Vernunft*" ("trust in reason"); and more "locally critical" counterparts of the untenable *fides* Anselm sought to secure.

2 "Gaunilonian" Relativisation(s) of "Anselmian" Argument(s)

In his *Proslogion*, Anselm of Bec proposed a well-known compact dialectical refinement of the more diffuse and conventionally apologetic arguments he had set forth in his *Monologion*.

> I do not attempt, Lord, to penetrate your vastness .... For indeed *I do not seek to understand in order that I might believe; but I believe in order that I might understand* .... Therefore, Lord, grant to me to understand, since you are as we believe, [and] we believe you to be *something than which nothing greater could be conceived*....
>
> And certainly "*that than which nothing greater can be thought*" cannot be in the mind alone. For ... [then something] "*than which nothing greater could be thought*" would be the very same as "*that than which [something] greater could be thought*"; but certainly that cannot be. (The italics, ellipses and translation are mine.)

The originality of these arguments lay in Anslem's attempts to
1. "diagonalise" over conceptual scholastic "grades" or "degrees" of explanation and "reality" ("formal causation", "that through which", ...), and
2. "define" an elusively "conceptual" (and speciously "unique") fixed point or limit of "all" such "epistemic" as well as "ontological" gradations.

Whatever its other merits, this assessment seems to have ordered a plausible historical framework for Bertrand Russell's youthful exclamation, "Great god in boots! The ontological argument is sound!" For such a "revelation" might well have come naturally to someone about to
3. exploit a diagonal argument to devise a self-referential paradox, and
4. "prorogue" such paradoxes in an infinitely ascending formalism of linearly ordered "types", a *de facto* logicist recapitulation of the great medieval chain of "formal" and "(merely) objective" "reality".

Put somewhat differently: Anselm was not the first dogmatic philosopher (consider, for example, Aristotle's "thought about thought", as glossed below) to
5. acknowledge and attempt at least implicitly to exploit the self-referential infinity, or at least indefiniteness (if not indefinition), of such conceptual ascents, and
6. beg the linearity of the indefinitely extendible hierarchy to which they gave rise, and the consequent unicity of "the" limit to which they might converge.

But he may well have been the first such philosopher (with the exception of Aristotle, and perhaps Plotinus) to "diagonalise" over them, in an attempt to identify a unique noumenal "fixed point" and attribute "unicity" as well as "existence" to it.

The "exception" I have in mind is Aristotle's well-known characterisations of
7. "divine" *noesis* as "conception of conception" (*noesis noeseos*), and
8. "first philosophy", in effect, as "that philosophy than which none greater can be conceived", as a
9. refutation of elenctic arguments (offered by the academic skeptics, for example) that
10. such "fixed points" cannot be well-defined, much less "uniquely" or "ultimately" "fixed", and that
11. there might perhaps be many "local" as well as incompatible fixed points for different provisional ranges of metatheoretic "conception".

A variant of such conjectures, cast as *reductio*-arguments, quickly (re)appeared in Gaunilo of Marmoutiers' implicitly skeptical (and proto-Humean) evocation of the "*praestantissima insula*", in his "*Pro Insipiente*" (Gaunilo, "On Behalf of the Fool"):

> For example: … you cannot doubt that the island more impressive than [others] which you do not doubt to exist in your mind truly *is* [exists] somewhere, in reality; and inasmuch as you have no doubt that [it is] more impressive not only in [your] mind but in reality, that it *is* [exists] *necessarily*[;] for if it were not, any other real island would be more impressive, and the island understood by you to be more impressive would not be.

Gaunilo (who for some reason is almost always described rather dismissively as an "obscure monk") offered quite a bit more than a countermodel for the ontological argument's aspirations to ultimate validity. He also furnished persuasive thought-experimental evidence that Anselm had begged certain auxiliary assumptions about "the" nature of "divinity" he needed to make the argument "go through".

In so doing, Gaunilo—who was, after all, a monk who presumably had his own notions of noumenality—had to tread lightly, for it was presumably clear to him and his interlocutor that one might vary his *reductio* to furnish "manichean" ontological arguments for the "existence" of other ultimate conceptual limits—the *praestantissimus genius malignus*, say. In the event, Anselm sternly (and in my opinion, humorlessly) called his obscure monachal opponent to theological order in his reply. But Gaunilo's proto-"Humean" point about the multiplicity of "design(s)" and "*kosmoi*" had been made.

For the protean, *portrait-robot*-like qualities of Gaunilo's countermodel have interesting historical and methodological implications for theodicies as well as the interrelations between ontological arguments and their "cosmological" and "physico-theological" ("design-based") counterparts, in Kant's language. They gently recapitulated, for example, Xenophanic as well as Humean conjectures that the god of the fish (for example) might be a *praestantissimus piscis*. More problematically, they indirectly anticipated certain aspects of Descartes' (slightly bizarre) need to establish in Meditatio III that he had not "proved" he was "god".

Be that as it may, Anselm's attempt to free his argument from the obvious implications of Gaunilo's counterexample amounted to a series of strenuous assertions that "conception" of a "maximum" for "conception" is "different" from "conception" of other (putative) "maxima". In metalogical terms, one can outline an argument that he was right, but not in senses which would have served his apologetic purposes.

"Conception", for example—unlike "island-hood"—is a relational notion, in the sense that it tacitly posits the "existence" of some sort of "intentional" agent—or metatheoretically definable "system"—which "interprets" what it "conceives".

"Perfection" too, I would argue, is relational as well as "intentional", in subtler ways which may not be so readily apparent. For something is "perfectum" in Latin if and only if it is "finished" or "accomplished"—a usage which prompts certain natural questions: "finished" or "accomplished" by what, for example, by what measure, and to what end? "Perfection", in short, may be simply a notion of "completion" or "accomplishment" raised to the level of its incompetence—an observation which suggests that the "ontological" argument may be more "cosmological" (and even "physico-theological") than initially meets the eye.

For since "conception" and "perfection" are relational notions, one is also free to conjoin them, concatenate them and iterate them—"conceive" ways to "perfect" one's "conception", for example, or "conceive" more "perfect" ways to "conceive" "perfection" (cf. 16–19 and 7.24 below). Such alternations and concatenations may well seem artificial. But I would argue that they are closely related to hierarchies of alternating "formal" and "objective" "reality", reconfigured as hierarchies of concept-formation of the sort Descartes invoked in the *Meditations*. More of this later.

Gaunilo, in any case, might well have acknowledged that such constructions seem to be artificial. But who are "we" (he might have asked) to "define" a "unique" limit of "all" such "intentional" and "systematic" hierarchies, or discern which of them might be "ultimately" (much less uniquely) convergent? (Which "*limes*"? What "*natura*"?). If we acknowledge, in other words (as Anselm professed to do), that sufficiently complex systems and conceptual constructions lie beyond "our" conceptual horizons, who are "we" to individuate them, decide whether they are compatible with each other, and assert that they "must" "converge" to a "unique" "ultimate" limit?

All we can do is work with certain fine-structural aspects of such iterations which seem to lie within our ken. Anselm has offered no "proof", for example, that "conception" of "(greater) perfection" can be assimilated to "(greater) perfection" of "conception"—a tacit quantifier-mistake, and a gulf Descartes sought unsuccessfully to bridge over a *pons eruditorum* called the "Cartesian circle".

Latent in Gaunilo's counterexample, in brief, were

12 a cogent critique of the tacit contextuality of Anselm's implicit and explicit appeals to notions of "perfection"; and
13 a correlative insight into the theory-relativity (or at least "relationality") of "impredicative" attempts to refute theory-relativity.

A number of relevant metalogical analyses arguably have their origins in such insights, and I will close the section with an informal account of one of their ancestors: the ancient skeptics' "problem of the criterion".

Attributions of "existence" and conceptions of "more perfect than" tacitly presuppose some sort of noetic "leverage" or "semantic" vantage-point, as pyrrhonist skeptics observed, and Descartes tacitly acknowledged in his metaphorical allusion to the "*Archimedean point*" in the opening lines of the First Meditation.

For "parity-of-reasoning" arguments would seem to suggest

14 that provisional fixations of such a semantic "position" (or "postulation"—"*positio*" could mean both) can be queried in their turn, and
15 that there may be many ramifying "vantage-points" or *positiones* for each such attribution (alethic, deontological, teleological ...).

"All" "ultimate" fixed points of concept-formation might therefore be more or less equivalently "*amphibolic*" (to borrow Kant's usage), in the sense (cf., once again, 7.24 below) that

16 one could "perfect" a "true" notion of "truth" if and only if
17 one could make "ultimate" sense of what is "ultimate" if and only if
18 one could make sense of a "final" notion of "finality", if and only if
19 one could "conceive" a "perfect" notion of "perfection", if and only iff ... .

But a modest measure of religious as well as non-religious humility would seem to suggest that one can do no such thing.

One simple way to see this is to observe that if one could do any of these things, one could, presumably—at the very least—"identify" who or what "oneself" is. But there are good reasons to conclude—as did Hume in the appendix to his *Treatise*—that "one" can't even do that. And one of those good reasons, I believe, is that the attempts to "define", "conceive" or "perfect" the "ultimate" fixed points sound like equivocation because they—literally—are.

For they cease, I believe, to be "equivocal" when we acknowledge the relationality ("binarity", "ternarity", ...) of their "liminal" postulations of yet-higher *points d'appui*. For "perfect" as well as unique "conceptions" of "perfection" and unicity (much less forms of "divinity" associated with them) could not—by stipulation—be relativised in ways which would permit them to be individuated "from beneath".

If, therefore,

20 "we" "*insipientes*" and "*homunciones*" acknowledged—as did Anselm in his rhetorical invocations of "our" finitude—that "we" cannot discern "true" "perfection", then
21 "we" should also acknowledge—with Gaunilo—that "we" have no "definitive" or "*insipiens*-proof" criteria for discernment of "god" from (e.g.) "the" "most

perfect and beneficial demiurge", much less "the" "most perfect and beneficial notion of divinity"—an argument Leibniz took up six centuries later.

The problematically "ultimate" (claims of) individuation and "maximality" are the underlying points at issue here—or more precisely, a tension between the "maximality" and the "individuation".

Suppose, for example, one postulated that "that than which nothing more complex could be conceived" "must" exist. Could one "conceive" "it"? Express "it"? Interpret "it"? Could one formulate criteria for discernment of its "unicity" and "maximality" which would not be more "complex" than "it" is? Could one, in short, "prove" that "it" "existed" in ways which would not either relativise "its" "existence", or over-"simplify" "its" "complexity"? Or both?

In subsequent sections, I will adduce metalogical arguments which suggest that the answer to these questions is "no"—and more precisely, that

22 the more "elevated" the conception, the more tenuous will be our hold on its "unique" instantiation, and that
23 whoever "we" may be, our conception of "unique" as well as "ultimate" "perfection" is one of the least "perfect" conceptions we will ever have.

A few stray observations—prompted in part by William James' remarks about "philosophical temperaments"—will bring this section to its end.

Limitless ramified interpretative hierarchies of the sort just sketched have often evoked a kind of existential shudder. Why? Why not existential admiration? One reason, I believe, is that these question are not about metaphysics, but about a sense of emotional as well as *noetic* security.

Anselm, for example, sought to "prove" the existence of an "authoritative" vantage-point, toward which "everything must" rise and converge. But the attraction of his proof, I believe, has never been its aspirations to rigor, but its expression of what Arthur Lovejoy once called the "monistic pathos".

Two enduring ideals of this "pathos" (which has a "pluralistic" counterpart) —and more generally of Lovejoy's "great chain of being"—have always been

24 its appeals to a kind of authoritative "unicity"; and more deeply,
25 (begged) assurances that inquiry and "reality" "must" be linear, or at least directed,

for these appeals and assurances assuage a widely felt and fundamentally affective *horror pluralitatis* (or perhaps *complicationis*).

Seen through this lens, such ideals and assurances are not objects of conceptual "proof", but of affective as well as thoroughly understandable noetic desires for something like a (perhaps non-unique) "eternal father, strong to save". (Or "mother" ... . What we yearn for are the "strength" and the "salvation". We wouldn't recognise the "eternity" if we "saw" it.)

As Kant essentially acknowledged, appeals to such ideals and assurances have also had a kind of personal as well as "regulative" force for many mathematicians, as well as philosophers and theologians, since the time of Pythagoras and Plato. Drawn to these ideals, some of these savants—Anselm among them—sought, quite naturally, to grace their "normative" assurance with the dignity of some sort of "proof" (or at least "*probatio*": cf. Anselm's "*credo, ut intelligam*").

The most venerable figure who might come most naturally to mind in this context—other than Anselm himself—would probably be Spinoza. For Spinoza, like Descartes, explicitly designed his *axiomata* to accommodate variants of the ontological argument, as well as its "cosmological" and "physicotheological" variants. But he also did something else—something which might well have horrified Anselm, and clearly scandalised those of Spinoza's contemporaries and immediate successors who denounced him not only as an "atheist" but as a "skeptic". He took Gaunilo's generic counterexample *seriously*.

For it is conventional in such contexts to characterise Spinoza's *deus sive natura* as a particular austere "god of the philosophers", rather than "god of faith". But Spinoza's estrangement from his predecessors' tacit presuppositions went deeper than that.

For what his "proofs" undermined was (among other things)

26 the groundless supposition Descartes had strenuously "saved" (*à la* Duhem) with the aid of his "circle", and Leibniz later "derived" from his wistfully hopeful meliorism, namely:

27 the epistemic intelligibility of "*dei sive natura*" to "us", "us" to "*deo sive natura*" and "*dei sive natura*" to "itself".

For it would follow from later arguments in the *Ethica* that his own "proof"—once he had begged all the essential questions in the *axiomata*—was utterly, completely and inexpressibly "nonconstructive", in the strong sense that no "intelligible" theory of the sort considered in the next section could consistently secure the unicity and individuation of *his own axiomata*, much less a "unique" "intended" interpretation for them.

For Gaunilo had suggested (in effect) that Anselm was asking his readers to accept

28 a "proof" of the "unique" "interpretability" of an assertion whose "intentional" ambiguities might seem to generate indefinitely many "non-standard"

interpretations, whose interpretations would transcend those of his "most perfect island", (derived in all likelihood from wistful ideals of an undiscovered "island of the blessed").

Spinoza, by contrast, incorporated the "impredicativity" and petitional self-referentiality of Anselm's formulation into his axiomata, and tacitly accepted an "austere" (and quasi-Neoplatonic) conclusion: that
29 a genuine "proof" of the "unique" interpretability of his axiomata would have to be inexpressible in any immanent (or "intelligible") framework to which they might apply (i.e., not only would his *deus sive natura* be unintelligible, but any "proof" of its existence would be as well).

In the following sections, I will attempt to
30 compare certain ideas of Anselm, Spinoza and Georg Cantor,
31 explain what I mean by an "intelligible" theoretical framework, and
32 argue that Spinoza's (essentially neoplatonic) conclusion that an Anselmian *deus* would be unintelligible "from beneath" was essentially correct, though his assertion that it would be unique was not.

3 Anselm of Canterbury and Georg von Halle

In a well-known 1899 letter, Georg Cantor wrote the following lines to his old friend Richard Dedekind (Cantor 1966, 443–444; cf. also van Heijenoort 1967, 144):

> [The] notion of a definite multiplicity (a system, an aggregate) ... can be so constituted that the assumption of a "being-together" of *all* its elements leads to a contradiction, so that it is impossible to construe the multiplicity as a unity, as "*a complete [finished] thing*". Such multiplicities I call *absolutely infinite* or "*inconsistent*" *multiplicities* . ... [T]he "aggregate of everything thinkable" is such a multiplicity . ...
>
> If [however] ... a multiplicity can be thought of "being together", *without contradiction* ... . I [would] call [it] a "*consistent*" *multiplicity* or a "*set*".

Cantor's insight is conventionally interpreted as the *Leitmotiv* for "schematic" (as well as "impredicative") "definitions" of "sets" in Gödel-Bernays or Zermelo-Fraenkel set-theory (e.g.): that is, as
1   "classes" "definable" from "schemes" of formally "definable" "predicates", as sub-"classes" of other, simpler "sets".

The "liminal" (or at least problematic) role of "definability" in such "definitions" is more deeply aporetic, for carefully encoded formalisations of the Berry

paradox establish (with cornered brackets [.] serving as theory-internal definitions) that
2   "[definability] is not definable", "[univocality] is not univocal", and "[well-definition] is not well-defined",

which may be taken quite reasonably to establish that the (implicitly *relational*) relations of "definability" and "univocality" are undefinable and equivocal—in short, "*inkonsistent*".

In ways which refine "Lewis Carroll"'s whimsical resort to hierarchies of metatheoretic ascent to "define" "material implication", therefore, metalogical arguments do not resolve metaphysical dilemmas. What they can do—fairly well, I think—is
3   coopt them, and refine them, in carefully "controlled" heuristic (thought-) experiments whose analogical "outcomes" offer persuasive evidence that they are irresolvable.

Consider, for example, the case at hand. Cantor—a devout Lutheran—called his "'inconsistent' multiplicities" "absolutely infinite"—precisely the same formulation ("*absolute infinitum*") which Nicholas of Cusa and Spinoza had employed in their "definitions" of "god".

Consider also the fact that Cantor's initial example of "*Inkonsistenz*" in this context was not "the" ordinal class of "all" ordinals (the proximal object of his remarks to Dedekind)—or any other "formal" notion, for that matter. It was the "ontologically" deeper and historically more resonant notion of "the" "*Inbegriff alles Denkbaren*" ("aggregate of everything thinkable").

My point here is that Anselm had set out to "prove" (in effect) not only that
4   his medieval precursor of Cantor's "*Inbegriff alles Denkbaren*" ("*id quo majus non cogitari potest*") "must" be *konsistent* (or Cantor's "*zusammenseiend*"); but also
5   that such a scholastic notion—whose "*Inkonsistenz*", Cantor seemed to suggest would be clear—"must" have a unique "definable" "realisation"—namely "god".

To me at least, this assimilation—despite its anachronism—bears out several of the "skeptical" implications of Gaunilo's counterexample I outlined earlier. But it also seems to cast Cantor's remarks to Dedekind in an interesting "theological" light, and give an unexpectedly literal resonance to David Hilbert's oft-quoted mathematical-logical rallying cry, some years later—that

6 "[a]us dem Paradies, das Cantor uns geschaffen, soll uns niemand vertreiben können"[2]

("[f]rom the paradise Cantor has created for us, shall no one have the power to expel us") ... .

4 Metalogically "Intelligible" Concept-Formation

I have drawn very briefly several historical comparisons from the eleventh century, from the fifteenth and seventeenth centuries, and more extensively from the dawn of late nineteenth and early twentieth-century metalogic. To refine and clarify these comparisons, it will be helpful to recur once again to a semi-formal notion of theoretical "intelligibility".

A formal theory T (or n-type, logical shorthand for an n-fold predicate in such a theory)—often, but not always, first-order—is intelligible in this sense if and only if it is

1 "parsable" (its language is countable, and the metatheoretically defined set of Gödel-codes of its axioms and their consequences is recursive); and
2 "autological": (it syntactically interprets a theory of finite sets or theory of arithmetic "strong" enough to include a mathematical-induction scheme which permits (arithmetic) encoding of its own syntax),

where the philosophical relevance of

3 "parsability" is that a recursive programmed "device" could "recognise" the theory's premises and trace out, step by step, its consequence-relations;

and of

4 "autologicality" that the theory can pose (but not, Gödel discovered, answer) "internally" coded (or represented) counterparts of semantic questions about "itself" ("Am 'I' 'consistent'"? "Do 'I' 'exist'"?) (cf. Kant's *KdrV*, A VI)

It is more or less generally acknowledged in computer science as well as other logical and metalogical disciplines that such

5 "intelligible" first-order theories and their n-types provide tentative frameworks for communication and systematic interpretation of information;

---

[2] David Hilbert, *Mathematische Annale* 95(1926), 170.

6   physical "inquiry"—including "inquiry" in systems which are initially not first-order—may "eventually" be interpreted (or "represented") in such theories; and
7   systematic interpretations of such theories provide "regulative", "locally critical" and "synthetic *a priori*" accounts of "experience" in all its problematic complexity.

Such theories may also be
8   "merely regulative" (to borrow a Kantian phrase), in the sense that they may ramify in alternative hierarchies of "outcomes" whose "initial" and "boundary" conditions are not controlled;
9   "locally critical", in that "local" distinctions between intelligible theories and metatheories may clarify "semantic paradoxes", and discern them from outright contradictions;
10 "synthetic", in that "experiential" distinctions between "object"- and "concept-formation" may be adjudicated in "contingent" ramified hierarchies of such metatheoretic interpretations; and
11 "*a priori*", in the sense that the syntactic and semantic interpretations which "represent" such distinctions suggest "preconditions" for graduated forms of "pattern-recognition", "information-processing" and "discursive reasoning".

Less obviously, studies of ramified hierarchies of "intelligible" theories also give rise to graduated notions of "noumenality" (an elusive predicate whose etymological origins are sketched below), namely
12 an intelligible predicate or sentence $\varsigma = \varsigma_{vou\mu}$ is noumenal with respect to an intelligible T, if and only if it "trancends" *T*,

by which I mean (once again following Gödel) that T cannot prove that [$\varsigma_{vou\mu}$ is consistent], or equivalently that
13 no such predicate $\varsigma_{vou\mu}$ in the language of an intelligible theory T has the property that σ semantics interprets (is a "model" or "realisation" of) T.

(Notice also once again that ς may be a provably intelligible theory or "type" in the sense of some yet-"stronger" (or more "complex") intelligible theory U which lies outside of the semantic horizon of T.)

One might also try (following Kant) to ask whether there exist syntactically expressible *noumenal* "ς *an sich*" which are uninterpretable in "all" intelligible theories T, and conjecture that an ineffable "god of the philosophers" might "realise" such a notion. Metalogically, however—as Cantor suggested—such questions cannot even be "intelligibly" *posed*.

For it follows once again from Gödel's results—which are provable in relatively "weak" intelligible theories (Peano arithmetic, for example), which can syntactically "encode" but not semantically interpret much "stronger" theories—that
14 every consistent "intelligible" theory T can formulate (or interpret syntactically) notions which it cannot interpret semantically; that
15 no consistent intelligible theory T can make consistent (semantically interpretable) sense of "all" consistent intelligible theories; and finally that
16 no consistent intelligible theory U is a "fixed" point for the intelligible theories it provably interprets—i.e., such that U semantically interprets [itself].

In no intelligible theory, therefore (contra Anselm and Spinoza), can one
17 syntactically or semantically interpret "everything" one "might" intelligibly "conceive" in indefinitely "stronger" and more complex counterparts.

And in no intelligible theory, *a fortiori*, can one "prove" the "existence" of a "unique" semantic interpretation or "ultimate" realisation for
18 Anselm's syntactically as well as semantically equivocal "*id quo majus ...* ";
19 Spinoza's "unique" "substantia" (or *id quod "in se est, et in per se concipi"*); or
20 Cantor's ("impredicative" as well as) *inkonsistenter "Inbegriff alles Denkbaren"*.

It may be informative, finally, to contrast these "ultimate" formulae's un-localisable *Inkonsistenz* with
21 metalogical localisations of Nicholas of Cusa's "*coincidentia oppositorum*"— a comparably mystical notion, with which Anselm's and Spinoza's "definitions" have sometimes been compared.

For given an arbitrary intelligible theory T in one of the metalogical frameworks I have sketched, let
22 T's "coincidentia oppositorum" be the collection C_T of "all" the continuum-many consistent but mutually inconsistent consistent intelligible subtheories S which are semantically interpretable in T.

Then
23 C_T might be viewed as one of T's conceptual "horizons" in some "stronger" intelligible metatheory U;

and
24 each such C_T would include, for each of its elements S, an uncountable family of "coincident" but mutually inconsistent "alternatives" (or "opposites") S_n.

In brief: each such C_T might be construed on the one hand as a local "Archimedean point" or "place to stand" ("*pou stō*"), and on the other as a sandbar in one of Heraclitus' great conceptual "rivers" ... .

5 "Existence" and "Predication" in the First *Critique*

Immanuel Kant—the most searching early modern critic of Anselm's ontological argument and its "cosmological" and "physico-theological" counterparts—focused his deconstructive analyses of such "*theologische Ideen*" on two premises of the sort he considered "subreptive": that

1   "*Sein*" is "*ein reales Prädikat, d. i., ein Begriff von irgendetwas, was zu dem Begriff eines Dinges hinzukommen könnte*" (that "existence" is "a real predicate, that is, a concept of something which could adjoin itself to the concept of a thing" (*KdrV*, B626);

and

2   the maxim that "*alles Existierende ist durchgängig bestimmt ... .*" (that "everything existing is exhaustively determined"), i.e., that "*um ein Ding vollständig zu erkennen, muß man alles Mögliche erkennen, und es dadurch, es sei bejahend oder verneinend, bestimmen*" (that "in order to understand a thing completely, one must recognise everything possible, and thereby determine [or define] it, either [by] affirming or denying [every predicate of it]"). (*KdrV*, B601).

The purpose of this section is

3   to examine metalogical analogues of such "definitions" and characterisation of "absolute" "existence"; and more particularly,
4   to explore metamathematical counterparts of Kant's suggestive "critical" remark that

> [e]xhaustive determination is therefore a concept which we never can represent in its totality *in concreto*, and grounds itself in consequence on an idea which has its seat entirely in reason, which prescribes the rule of complete use to understanding. (B601)

Consider, for example, its adjectival counterpart "*durchgängig bestimmt*" ("exhaustively determined"), usually translated as "completely determined". In ordinary German, first, "*bestimmt*" means "defined" or "definite", and would normally be so translated in most everyday speech. Secondly, Kant's uses of the word "*durchgängig*" (from "*durchgehen*", to "go through", "canvass" or "traverse") evoked infinitary limits of taxonomic classification, but also

5 the Parmenides' "exhaustive" canvass of "all" predicates which might be asserted of a given Platonic *eidos* (cf. Kant's choice of the word "*Idee*"), and
6 the "*analyses infinitorum*" which underlay Leibniz' definition of "complete (individual) notions" (*notiones completae*) and suggestions that every monad "*conçoit tout, mais confus[é]ment*".

In both cases, I believe, these verbal constructs anticipated metalogical counterparts of such analyses in
7 semantic studies of proof-theoretic "completeness" and intelligible theories' continuous "Stone spaces"; and
8 metatheoretic studies of (complete) "types" over a given (intelligible) first-order theory T, which provide "virtual" "objects" for a rigorous semantics of modal quantification.

Kant's suggestions, finally, that suppositious appeals to "*durchgängige Bestimmung*" might themselves be "underdetermined", and his uses of this observation to characterise "*Vernunftideen*" ("ideas-of-reason") also anticipated metalogically valid observations that
9 "most" such "types" (in well-studied topological and measure-theoretic senses of the word "most") are not "realised" in most semantic interpretations of a given intelligible theory.

For a cardinal tenet of Kant's "transcendental idealism", reflected in the second passage from *KdrV*, B601, quoted above, was that
10 "*Verstandesbegriffe*" ("concepts of the understanding") (such as "force") are "constitutive"; characterised by finitary forms of "*durchgängige Bestimmung*"; and realised in a unique structure of "*Erfahrung*" which "*constitutes*" them; but
11 "*Vernunfideen*" ("ideas of reason") (such as "god") are ("merely") "regulative"; characterised by infinitary (or at least indefinitely extendible) forms of *durchgängige Bestimmung*; and not realised or "constituted" in the unique structure of such "experience".

If one keeps these views and Kant's near-assimilations of "*Sein*", "*Existenz*", "*Erfahrung*" and "*Konstitution*" in mind, in fact, Kant's observations that "*Sein ist kein reales Prädikat*" suggest cognate assertions—that
12 "*Konstitution*" *ist kein* "*konstitutives*" *Prädikat*, for example; or that
13 "*Erfahrung*" *ist kein* "*erfahrungsmäßiges*" *Prädikat*; or even that
14 "*Existenz*" "*besteht*" ("*existiert*") *nicht*.

Critiques that self-applications of such "probative" or "normative" predicates are circular and mere logical tricks seem to me mistaken, at best, and acts of methodological bad faith at their worst. For such "probative" predicates have typically introduced themselves into ordinary discourse as well as philosophical argumentation to impose certain sorts of authority, and validate assertions to which they apply. Critiques of such "self-validation", therefore, lay at the heart of

15 ancient "zetetic" responses to stoic, "rationalist" and dogmatic-"empiricist" claims; and
16 skeptical observations that attempt to make proto-"critical" distinctions unwind these claims in criterial hierarchies which do not close.

What is the point of these observations for Anselm's ontological argument, and Kant's "critical" analyses of it?

The first is that Kant's objection to the ontological argument would lose much of its force if one asserts (as Kant effectively did in the *Analytic*, but not in the *Dialectic* or the second and third *Critiques*)

17 that "existence" can (or "must") be relativised to a conceptually "closed" "system" of "*Erfahrung*".

For then the "transcendental" framework in which Kant effected this relativisation would not, by his own stricter criterion, "exist", and might be as "merely regulative" as Anselm's "god".

One more hermeneutically generous way to interpret Kant's assertion that "existence" is "not a predicate" would be to stratify or echelon "existence" as well as "predication" processively, and assimilate

18 "critical" distinctions between "constitutive" and "regulative" predication to "local" metalogical distinctions between theory and metatheory (and similarly with types over such theories).

For then one could also assimilate Kant's assertion that "existence is not a predicate" to the metalogically tenable observation that

19 "existence" of a predicate or theoretical claim is metatheoretic predication of the "existence" of a semantic interpretation for it,

and make graduated or processive sense of "schematic" observations that

20 "*Sein*", "*Existenz*", "*Erfahrung*" and "*Konstitution*" are "relational" notions, whose open metatheoretic parameters "define", "determine", or "individuate" alternate "*realia*" in ramifying metalogical hierarchies which do not "close".

The limitless ramifications and lack of closure of such hierarchies may then be viewed
21 as alms (or cradle-gifts) for their "intelligibility"; or more "aporetically"
22 as schemes of conceptual "complementarity"-principles, which weigh intelligibility against the (putatively) "ultimate" [self]-interpretation of Aristotle's "godlike" *theoria*.

I will consider "modal" extensions of such "complementarity"-relations between "finality" and "intelligibility" in the next section.

6 "Necessary Existence", and the "Existence" of Many "Necessities"

In a number of essays and monographs, Alvin Plantinga and other authors have argued that there "must" be a definable entity whose "possibility" entails its "necessary" "existence". In earlier sections, I considered aporetic aspects of "ultimate conception" in terms which suggest that indefinite hierarchies of "counterfactual" concept-formation may have something to do with the elusive purchase of Anselm's argument. In the present one, I will briefly compare Anselm's tendency to conflate "*concipi potest*" ("can be conceived") to "*concipi posset*" ("might be conceived") with a similar modal ambiguity in David Hume's definition of "Cause" in paragraph 60 of the first *Enquiry*, and argue that
1   such ambiguities are characteristic of "ultimate" modal usages as well as "ultimate" causation, "ultimate" interpretation and "ultimate" concept-formation;

and therefore that
2   appeals to adequate semantic frameworks for alethic modality are in order, but generate more elenctic questions about such "ultimate" "limits" than they can possibly answer.

I have already argued, in effect, that traditional beliefs that "liminal" notions such as "existence" and "conception" "must" have unique "intended" interpretations raised these notions' etymological origins to the level of their incompetence. ("*Concipere*", for example, simply meant to "grasp firmly" or "draw together".)

Less evident may be the extent to which such observations apply to ancient, medieval and early modern as well as "ordinary-language" uses of "modal" expressions such as the Latin "*potest*", "*posset*", "*possibile*" and "*necesse* (*est*)". ("*Necesse*", for example, originally meant something like "unceasing(ly)"—a "diodorean" usage—and "*possibile*" "what one can do"—a "potential" or "conative" one). One might infer from such observations that appeals to the full apparatus of modal quantificational logic may overinterpret ancient, medieval and early modern uses of such expressions—or at least that such apparatus may be more relevant to later refinements of Anselm's argument. This argument seems to me warranted, to the extent one applies it evenhandedly to "ultimate" "hypostaseis" of "conception", "existence", "perfection", "necessity" and "possibility" alike.

For my point is
3   that such notions are contextual (or relational), and etymological origins are hermeneutic contexts,

but also
4   that first-order logical semantic hierarchies offer as "neutral" (and un-"loaded") a venue for relational studies of such contexts (including [themselves]) as one is likely to find.

I will therefore continue to work with (and in) such hierarchies in what follows.

A second, technically more vexing problem for anyone who wishes to undertake a serious examination of such assertions is to make relational metalogical sense of abstract modal "entities" and their "necessary existence".

(Such problems should be "technically more vexing" if the conjecture about "neutrality" just made is tenable. For "neutral" efforts to find resolutions for them must question the [self]-evidence of [their own] premises.)

In any inquiry into "necessary existence", for example, it would seem relevant to consider
5   semantic (or "ontological") interpretations of modal quantification and its "intended objects".

In "Virtual Modality," ([Boos], 2003) I offered a "canonical" semantic interpretation of "necessity" and "possibility" in which
6   modal quantification is not "sharp", but irreducibly graduated and distributive, in the sense that
7   modal "truth"-values do not range over the simplest two-valued boolean algebra which ranges over "truth" and "falsity", but over
8   more complex boolean algebras of which a the Stone space of T has the power of "the" set-theoretic continuum (whatever "that" is),

and argued that

9 "graduation" and "distributivity" cannot be eliminated from the modal semantics which characterise such algebras;

that

10 probative recourses to such semantics are subject to "appeals" *ad indefinitum* to ever more graduated and distributive adjudication in ever higher "courts";

and that

11 ever more nuanced "adjudications" in such "courts" will not "hand down" "ultimate" *adequationes* for the conceptual "*res*" they may generate,

much less

12 "close" the conceptual hierarchies implicit in Anselm's argument, much less "prove" the "necessary existence" of a unique as well as "ultimate" Anselmian "*god*".

For such ideals (for that is what they are) will forever

13 ride the wavefronts of the hierarchies of intelligible theories they clarify,
14 reflect as well as recapitulate intermediate theories' "essential incompleteness", and
15 bear witness to conceptual boundary conditions which require recourse to indefinitely graduated refinements of conceptual "truth" (a word which, as mentioned, originally meant "trust" as well as "belief").

Compare Leibniz' suggestion that

16 ... the Soul itself only knows that of which it has perception ... *according to the measure of its distinct perceptions. Every soul knows the infinite, knows all, but confusedly*, as in walking at the seashore ..... I hear the particular noise[s] of each wave, of which the total noise is composed, but without discerning them; *our confused perceptions are the result of the impressions the whole universe makes on us.* ("Principles of Nature and of Grace" [1714], section 13, *G* VI 604, *AG* 211)

7 Conclusion: Anselmian "*Numina*" and Kantian "*Noumena*"

Everyday professions of faith in the "existence" of an "utterly" transcendent "god" typically fall back on a deeply equivocal assertion: that

1   an agent or entity which "transcends" "all" "our" experience can be "evinced" or "witnessed" within "our" experience as "its" "source" or "cause".

On the analysis I have offered, the ontological argument draws on a thought-experimental analogue of this equivocation: an assertion that
2   an agent or entity which "transcends" "all" "our" experience can be "defined" or "proved to exist" within "our" experience as its conceptual "limit" or "fixed point".

The second assertion, I have argued—Anselm's "proof"—abstracted and sublimated the contradiction implicit in the first, and begged the "existence"- and "uniqueness"-assertions it was supposed to "prove".

If we do not want our arguments to eventuate in such *petitiones*, what can we do? We can work and communicate in (modest, feasible fragments of) "intelligible" theories, in which we can identify portable and provisional templates of "conception", "interpretation" *et al*, and their ethical counterparts. What we cannot do is instantiate it (prove "consistency") "on site"—"from within", as it were—much less instantiate it "canonically" (in a "true" notion of "intended" interpretation, or "truth", much less "necessary" truth).

In the spirit of 1.3–1.6's "confutations of skepticism", one might also construe the methodological circularity of 1 and 2 above as a "fixed-point" argument, in which Anselm
3   "defined" "god" as a (putatively unique, and therefore "necessary") hermeneutic fixed point and "intended" interpretation for conceptual refinements of discursive speculation about [itself],

in his effort to "confute" skepticism about the "existence" of such a "god". If this analysis is correct, it also suggests that there is little which is intrinsically "theological" (as opposed, say, to "cosmological", or "epistemological") about this argument.

One might, for example, construe George Berkeley's "immaterialism" as an argument for the [self]-sufficiency of "perception" (or "conception"), in which he
4   "defined" "perception" (or "conception") as a (putatively unique, and therefore "necessary") interpretive fixed point for discursive speculation about [*itself*];

and Kant's "metaphysical deduction' as a claim that

5   "'the' structure of 'experience' ("*in spekulativer Absicht*") is a (putatively unique, and therefore "necessary") fixed point for discursive analysis of [itself])";

and Kant's "transcendental deduction" as a claim that

6   "'the' structure of 'experience' ("*in spekulativer Absicht*") is a (putatively unique, and therefore 'necessary') condition for the possibility of [itself]".

Parenthetically, it seems to me no accident in this connection that

7   Berkeley did not characterise "god" as an ontological "limit", but a master-"designer", who assured the "existence" and intersubjective uniformity of lesser "spirits"' "perceptions",

and that

8   Kant did not characterise "god" (or the "*theologische Idee*") as an ontological "limit", but as a "merely" "*regulatives Ideal*", whose "systematic unity" furnished a "design"-like "*Analogon*" of the (allegedly) "constitutive" unity of "our" "understanding".

For neither Berkeley nor Kant needed to identify the "existence" of an alleged "fixed point" of conception with "god". Both of them were content, in effect, to postulate a

9   "regulative" guarantor of the intersubjective communicability and mutual intelligibility of "experience",

for they both believed (mistakenly) that such "experience", like Anselm's *deus*, could semantically "fix" or interpret [itself].

Be that as it may, such assimilations

10   suggest the existence of common underlying patterns in otherwise disparate attempts to prove "secular" and "theological" conclusions, and

11   confirm Section 5's suggestion that Kant's "*Deduktion*" of his "*empirischer Realismus*" was a fixed-point argument devised to "save" an exquisitely interpolated argument from design.

The movability and "modularity" of "conception", "interpretation" and informal "fixed-point arguments" which employed them may also have provided a template for Kurt Gödel's remarkable metalogical observation that

12   formal fixed-points provably abound in every "intelligible" theory—i.e.,

13 every theory which is "inductive" enough (and therefore "reflexive" enough) to encode [its own syntax] and formulate conjectures about [its own semantics].

For a bit of reflection also suggests a sense in which

14 informal antecedents of such "reflexive" results "should" have emerged in an ostensibly analytical discipline committed to an echeloned imperative to try to "know [thyself]—an injunction whose formal counterpart gave rise to the "critical" practice and discipline of "metamathematics".

Fixed points, in short, open interrogations, and generate forms of "eternal renewal" (if you wish, they are "self-similar"). But Berkeley and Kant were mistaken because they do not "close" them.

They do not, that is, individuate unique "intended interpretations" for [themselves]. They have no "ultimate" semantic purchase, except as recurrent sources of elenctic counterexamples for "probative" claims about "ultimate" forms of semantic purchase.

On the evidence of such metalogical arguments, therefore, what is problematic in philosophical employment of fixed points is not their reflexivity, however "paradoxical" or "viciously circular" it may seem.

(As a matter of historical fact, "dogmatic" (i.e., "anti-skeptical") philosophers have seldom found self-referential argumentation "absurd" or "disabling" when it served their elenctic aims. Aristotle's "thought about thought", for example, was deep. But skeptical doubt about skeptical doubt was deeply disturbing, if not "absurd"). What is problematic about fixed-point arguments and their iterations arises from dogmatic efforts to "ground" or "close" or "terminate" them—either in "experience", or in "god(s)", or in other "ultimate" "intended interpretations"—and with them the tacit or explicit recourses to wider metatheoretic contexts to which they give rise.

Sometimes such efforts take the form of table-thumping claims—easily locatable, for example, in Aristotelian and Aquinian texts—that patterns of "infinite regress" ("open" topological evolutes of conceptual "fixed points") are "absurd". More subtle, but equally problematic, are efforts to secure individuation, "thoroughgoing determination" (Kant's "*durchgängige Bestimmung*"), and allegedly "unique" uniform convergence to "ultimate" semantic limits of what it is that fixed-point arguments allegedly "fix".

For the "semantic paradoxes" which arise from fixed-point arguments are

7 Conclusion: Anselmian *"Numina"* and Kantian *"Noumena"* — 155

15 "object-theoretic" traces of underdetermined "metatheoretic" notions such as "consistency" and "definability", as well as "design", "existence" (or "being") and "truth".

And they bear witness to

16 multiplicity of interpretation of any "intelligible" counterparts of such liminal notions in any metatheoretic contexts in which they are "intelligibly" interpretable at all.

Kant's original employments of the phrase *"durchgängige Bestimmung"*, for example, considered earlier, reflected such a tension in interesting ways. On the one hand he associated it with

17 the alleged completeness and determinacy of "the" forms of intuition and categories of the understanding, and on the other with

18 "subreptive" forms of "completeness of the conditions" (*"Vollständigkeit der Bedingungen"*) he identified with experientially underdetermined "ideas of pure reason".

Such ambiguities seem to me to become more understandable when one considers that a common "subreptive" premise employed in classical arguments to "prove" (i.e., "prove" the "necessary existence" of) certain forms of "(necessary) existence"—

19 the "(necessary) existence" of an "ultimate" "divinity";
   the "(necessary) existence" of an "ultimate" *"causa sui"*; and
   the "(necessary) existence" of an "ultimate" "criterion of truth", for example—

is also

20 the "necessary existence" (or at least well-definition) of an "ultimate" "predicate" (*pace* Kant) of "necessary existence",

of the sort I queried in the last section.

Dismissive responses to formally analogical arguments of the sort I have just made typically hold that they are anachronistic at best and gross "category-mistakes" at worst. A case can obviously be made for their anachronism, though some of the case's points of evidence might just as readily be interpreted as points of vindication. Historically, for example, the "subreptive premise(s)" mentioned above—present in arguments of Aristotle, Aquinas and others—often rested on an elementary quantifier-mistake: that "infinite regress" is an "absurdity" (in effect: that it is an "absurdity" to posit that "for all x there exists a y such that xRy", but deny that "there exists a y such that xRy for every x"). (It is probably no accident in this context that systematic studies of arbitrary quantifier-alternation first appeared in print in the nineteenth century, more than two millenia after the death of Aristotle. Alternations of *three* such quantifiers—in the formal definitions of "continuity" and "differentiability"—still trouble many students of elementary calculus.)

What relevance do such formal observations have for the ramifying hierarchies I have argued attempts to analyse "conception" and what "it" "conceives" naturally generate? The answer I have to offer is that quantifier-alternations within first-order logic are no longer abstruse, but metalogical counterparts of them are.

Otherwise it would seem as straightforward to
21 accept that "for every intelligible conceptual framework x there is another such framework y which provides a 'criterion of truth' for x",

but
22 deny that "there is an intelligible conceptual framework y which provides a 'criterion of truth' for every such framework x",

as it is to
23 accept that "for all x there exists a y such that xRy", but
    deny that "there exists a y such that xRy for every x".

Another thoughtful and sophisticated critique of skeptical uses of metalogical arguments, common among philosophical logicians, is that such arguments are "merely formal" and should be subordinated to higher philosophical imperatives. In response to such critiques, I have tried to reply that an extratheoretic commitment may be a bar, but it is not an argument, and that an intelligible theory may be as deep, rich and reliable as we (or our successors, as self-nominated "reasonable beings") will ever be. It will simply not be complete.

More significantly, it will be "essentially incomplete": its intelligible extensions will pose inexhaustible hierarchies of questions they will be unable to answer, and reproduce *ad indefinitum* the dilemmata of Kant's *Schicksal der Vernunft* (which are closely interrelated with the *Freiheit* which "regulated" reasonable beings' *Würde*).

Intelligible theories "exist", at least syntactically (they are recursively axiomatisable, and their consequences are recursively enumerable). But they cannot recognise or define, much less prove their semantic "existence" (interpretability, "conceivability", or consistency). Even their notions of

7 Conclusion: Anselmian *"Numina"* and Kantian *"Noumena"* — 157

"finitude" and "infinity" may be nonstandard, in ways which reflect forms of randomness which lie beyond the horizon of localised attempts to characterise them.

They do not, therefore—on pain of metatheoretic inconsistency—"determine" intelligible metatheoretical interpretations which would answer "every" question they might express (one characterisation of metatheoretic completeness).

Ontological arguments identify (at some indistinct "limit") "conception" of "perfection" with "perfection" of "conception" ... (a "parade-example" of Cantorian "*Inkonsistenz*").

(To see that there is something "indistinct" as well as "impredicative" about [self]-application of quasi-Kantian "*Grenzideen*", consider the following (not entirely parodic) permutations:

24   "conception" of "existence" of "perfection";
"existence" of "perfection" of "conception";
"perfection" of "conception" of "existence";
"existence" of "conception" of "perfection";
"conception" of "perfection" of "existence";
"perfection" of "existence" of "conception"; ... .)

I have argued, by contrast, that we can "conceive" things that we cannot "comprehend", all right (graduated notions of "existence" and "perfections", for example). But we can do so only in "nonconstructive", theory-relative terms, which

25   secure the "clarity" of what we conceive at the expense of its "distinctness" ("clarify" it in precise but provisional object-theoretic terms which are metatheoretically reinterpretable *ad indefinitum*).

"Ideas" of "maximally" ("formally") "real" "ideas", for example, need not be "maximally" ("formally") "real". And claims that they "must" be such amount to *petitiones* of a quasi-platonic "form" of "divinity", not "proofs" of "its" "existence". In more Cartesian terms: Anselm's *petitio*—and Descartes' "circle"—was to assume that "the" "idea" of a "maximally" "clear and distinct" "idea" must be "maximally" "clear and distinct" (and therefore "the" "unique" such "idea"). On the evidence of the metalogical analogies I have tried to adduce, they are relatively—if not "maximally" obscure and diffuse . ...

At the end of this chapter's introduction I varied Wittgenstein's notorious injunction to "silence" to enjoin adherence to an ideal of intelligible inquiry, tempered by conceptual suspension of judgment in the face of indefinite pluralities of alternative interpretation. At the end of its conclusion, I will vary a rather different author's equally famous phrase to express a mark of metalogical reason's "fate": Wittgenstein's "*nicht weiter Wissen*" (lack of "knowledge" how to "go on").

The sigil of a university where I anxiously began graduate study of mathematics many years ago was (and presumably still is) "*Numen Lumen*" ("God [is] Light"). *Numen* and its adjective *numinosum* probably derived from "*nuere*" (to nod), a derivation sometimes thought to have reflected notions of divinity as a tentative source of *arbitrium* as well as arbitration.

Both Anselm and his "design-theoretic" successors, I have tried to suggest, sought to identify "divine" forms of *arbitrium* with forms of "thought" ("*noumena*") they hoped would both underlie and "understand" them. But fossilised metaphors embedded in "underlie" and "understand" point to deep tensions in alternative initial and boundary conditions of such searches. For *arbitrium* (judgment) would seem to presuppose widening wavefronts of intentionality; and efforts to "understand" such judgment(s) new forms of "second-intentional" metatheoretic frames for them.

Assumption of the burdens of incomplete and non-"ultimate" *arbitrium* might therefore be regulative of ideals of inquiry which are not "arbitrary". And "free" attempts to continue such inquiry would therefore suspend judgment about the alleged finality or "ultimateness" (or "perfection") of any forms such numinous or "ultimate" *arbitrium* might take. (There is a sense, I believe, in which one of the few assertions Descartes successfully defended in the *Meditations* was his wry insight, mentioned earlier, that the absence of discernible "limits" for sentient beings' "wills" does not render them "divine".)

To put a point on it: if an "intelligible" *numen* were a "*noumenon*" (a Greek neuter middle participle which could refer either to an agent of thought, or to one of its objects), it would not be "ultimate".

For
26 if such a "*noumenon*" were a speculative or practical agent, first, the intelligibility of its (idealised) agency and praxis would bear witness to its incompleteness.

And
27 if, on the other hand, it were an object of action or contemplation, what "contemplated" it or "acted upon" it would arguably be more "numinous" than "it" is.

Anselm's *insight*, I suggested earlier, was to realise that one might try to "diagonalise" indefinitely ramifying hierarchies of "judgment" and "understanding". But such diagonalisations are not marks or guarantors of "ultimate" unicity and "determination". They are way stations of further inquiry, and insights into the hermeneutic plurality to which such inquiries give rise.

"Thought about thought", for example, is not an "ultimate" form of contemplation, as Aristotle believed. It is a precondition of every stage of "consciousness". As such, it may be "*theios*" ("divine"), in the liminal as well as "contemplative" sense that there is no "ultimate" conceptual boundary to its capacities for renewal. It may even be generative, in the sense that it can sometimes understand itself in the poignant spirit of Pascal's *roseau pensant* (thinking reed).

It may, therefore—for a time, and collectively, if not individually—"go on": reemerge, in new, more comprehensive and therefore more "sympathetic" forms (the French sense of the word "*compréhensif*"). If so, such *compréhension* might help it (and "us") see that

28   we cannot "perfectly" understand "perfection", any more than we can "define" the "existence" of something that is "maximally" "definable"; and that
29   we cannot even "define" what it would mean to be "definable" to "us" (essentially because there might be many alternative as well metatheoretically tenable "definitions" of "us").

In metalogical paraphrase:
30   there may be many non-"ultimate" (and incompatible) "fixed points" and limits of "directed systems". But no inquiry can "prove" that "all" (potential) inquiry is directed toward a "unique" limit.

For
31   ascending paths of "proof", "validation" and "fixed-point" identification open out into practical, speculative, "ontological" and "teleological" hierarchies which ramify without end.

What more should we want? What more would we "understand"?

"We" don't know, of course—any more than we know "where all past years are, Or who cleft the devil's foot". But that—in a sense—may be the point. In this chapter, I have argued that

32   no intelligible theory T can "conceive" "all" and only "all" those intelligible theories which do not "conceive" [themselves]; and that
33   if Anselm's "god" were "defined" as "the" intelligible theory which proved the consistency (or "existence") of "all" intelligible theories, such a "god" would not exist.

(To paraphrase Samuel Beckett: "I *must* go on. I don't *know how* to go on. I *will* (to) go on" … .)

5 "Parfaits Miroirs de l'Univers": A "Virtual" Interpretation of Leibnizian Metaphysics

> Mir scheint daher, daß in der Leibniz-Forschung mit Recht wiederholt der Versuch von Leibniz kritisiert worden ist, das mathematische Konvergenzmodell auch auf Begriffe zu übertragen.
>
> ([Schneider], 302, quoted in [Mates], 112)

> Mea Principia talia sunt, ut vix a se invicem divelli possint.
> Qui unum bene novit, omnia novit (*G* II, 412)
>
> (My principles are such that one could hardly disjoin them.
> Who knows one well, knows them all.)

1 Introduction

In Boos (2003b)—a long technical argument—I develop an adequate metalogical semantics for modal extensions of a recursively axiomatisable first-order theory T, and call the motivations for this semantic framework "Leibnizian". The purpose of this somewhat less formal discussion will be to offer a partial justification for my invocation of one of the more venerable names in the history of western philosophy.

To this end, let T be such an incomplete axiomatisable theory, and **C** the completion of T's Lindenbaum algebra—the (Dedekind-MacNeille) completion of the boolean algebra of formulas of T modulo T-provable equivalence.

In the present context, **C** may be concretely realised as the Stone algebra, or boolean algebra of regular open sets in T's Stone space, where the latter is construed as the (metatheoretically defined) space of Henkin interpretations of T. Detailed information about Stone algebras and Stone spaces (so named for their twentieth-century discoverer and investigator, Marshall Stone), as well as the notions of boolean-valued "randomness" which underlie these results, may be found in (Bell, 1985) and (Bell and Machover, 1977), (Jech, 2002) and other articles, textbooks and monographs cited in the bibliography to this volume.

Briefly, the semantic framework set out in "Virtual Modality" ([Boos], 2003b) may be characterised as follows.

1. Modal "intensions" *or* "intensional objects" in this framework are topologically "random" or "virtual" (**C**-valued) 1-types over T (maximal consistent sets of properties in the language of T (I will assimilate such "types" in the sequel to Leibniz' "complete individual notions").

2   Modal "truth" **u** is a canonically defined "random" (**C**-valued) Henkin interpretation of T (**u** will also provide a metalogical gloss of Leibniz' elusive notion of "metaphysical necessity" for T).
3   Modal valuations of the names or constants of T are "random" (**C**-valued) valuations of the constants of T in the "virtual" model **u** (and therefore particular instances of 1's "modal intensions").
4   Valuations of modal "necessity" are "random" (**C**-valued) theoretical extensions **N** of T in the language of T. Such **N** may be also construed as "virtual" "accessibility"-relations between the **C**-valued Henkin interpretations **w** of T.

Finally,
5   "Leibnizian" valuations of modal necessity are particular "random" (**C**-valued) extensions of T which are **C**-valued intersections of the vacuously "true" theory **u** for T, and one other, "actual" random Henkin model **w** of T. I will construe such **w** as virtual sources of "sufficient reason" and "hypothetical necessity" for T below.

The principal theorem of "Virtual Modality" is a completeness result, in which I derived the existence of "countermodels" **N** for unprovable modal assertions s from the bridge between virtual theories and accessibility-relations mentioned above in 4. The existence of simpler "Leibnizian" countermodels **N** for s then followed as a corollary.

I will not recapitulate these results in detail here, but sketch instead some of their potential implications, and construe the structures they employ as metalogical ectypes of metaphysical archetypes in Leibniz' "*principes*".

Particular metamathematical glosses, for example, emerge for
6   Leibniz' appeals to "complete individual notions";
7   his "predicate-in-notion"-principle;
8   the "principle of the identity of indiscernibles";
9   his "holist" "mirroring principle";
10 the elusive distinction(s) he tried to draw between "metaphysical" and "hypothetical" *necessity*;
11 his "*grand principe ... de la raison suffisante*";
12 his many attempts to characterise "perfection";
13 his "principle of plenitude"; and finally
14 his persistent invocations of "virtual identity", and many allusions to continually (and continuously) graduated forms of monadic "confusion".

I will provide more detailed metalogical accounts of 6 through 9 in Section 2 below, turn to 10 and 11 in Section 3, consider 12 and 13 in Section 4, and assimilate 14's boolean-valued "virtuality" to Leibnizian "confusion" in Section 5.

I will also offer detailed textual rationales for these reconstructions and assimilations, and make a final effort to draw together these conjectural interrelations between my metalogical miniatures and Leibniz' original *principes* in the chapter's conclusion.

All of these metamathematical glosses, it should be mentioned, will also be theory-relative, in a dual sense: they are formulated in tacit metatheoretic venues U, in which one can define canonical spaces of interpretations for particular theories T—the Stone spaces mentioned above. I will therefore conclude this introductory section with a brief preliminary rationale (with apologies to Wittgenstein) for my introduction of such *(meta)logische Räume*.

In a very terse but systematic and suggestive note, reprinted as (Couturat, 1–3), Leibniz tried to outline a "numerical" analogue of "truth"(s) of "reason" and of "fact" (cf. also other, kindred texts such as *FdC* 187ff.), and I will assimilate these explicitly analogical arguments—arrayed in parallel columns—to a protoanalysis of an unspecified theory's Stone space.

In the original fragment, of course (also translated in (*AG* 98–101), as "The Source of Contingent Truths"), Leibniz did not talk about any formal background-theory T for his reflections, much less observe that such a theory might be incomplete. He was, however, aware that "contingency" required some sort of incompleteness or underdetermination of the "merely logical" *characteristica(e)* he had elsewhere posited.

In any event, Leibniz' still-inchoate analogies in their original form led him to distinguish between

15 (binary?) decimals, which he assimilated (in running entries in two adjacent columns) to assertions that are decidable (provable or refutable), in ways one might now call "finitary" (from finitely many premises); and

16 infinite (binary?) decimals, which he aligned with assertions that could not be finitarily decided, and would, in that sense, be contingent.

It is now quite standard to observe

17 that Stone spaces of incomplete theories T are homeomorphic to (Cantor-) spaces of infinite binary (rational as well as irrational) decimal expansions; and

18 that such spaces, in turn, are homeomorphic to "random" spaces of countably many independent coin tosses.

In ways I will try to elaborate further below, my introduction of Stone spaces of theories and their free-variable counterparts may therefore be defensible attempts to elaborate an incomplete but deeply original Leibnizian insight into the semantics of the *characteristicae magnae* he so cherished.

2 "Miroirs", "Notions" et "Predicats"

In this section, I will begin to sketch the interrelations promised above between the metalogically random formal semantics of 1–5 and Leibniz'
1   ontology of "complete individual notions"; his
2   "predicate-in-notion"-principle; his
3   "principle of the identity of indiscernibles"; and, finally, his
4   (holist, "flower-in-the-crannied-wall"-, or) "mirroring"-principle.

I will also begin to examine
5   the graduated states of "confusion", "potentiality" and relational underdetermination Leibniz attributed to monadic "perception".

Consider first the following well-known passages, from Sections 8, 9 and 13 of the text we call the "Discourse on Metaphysics" (1686), from *Primary Truths* (1686), and from *Principles of Nature and of Grace* (1714). I have cross-referenced them here for convenience with the translations of [*AG*]:

6   Complete individual notions and the predicate-in-notion-principle

("Discourse on Metaphysics" (1686), Section 8, *G* IV 432–433, *AG* 40–41)

... puisque les actions et passions appartiennent proprement aux substances individuelles (actiones sunt suppositorum), il seroit necessaire d'expliquer ce que c'est qu'une telle substance. Il est bien vray, que lorsque plusieurs prédicats s'attribuent à un même sujet, et que ce sujet ne s'attribue à aucun autre, on l'appelle substance individuelle; mais cela n'est pas assez, et une telle explication n'est que nominale ...

... lors qu'une proposition n'est pas indentique, c'est à dire lorsque le predicat n'est pas compris expressement dans le sujet, il faut qu'il soit compris virtuellement, et c'est que les philosophes appellent *in-esse*. ... en sorte que celuy qui entendroit parfaitement la notion du sujet, jugeroit aussi que le predicat lui appartient.

... la nature d'une substance individuelle ou d'un estre complet, est d'avoir une notion si accomplie qu'elle soit suffisante à comprendre et à en faire deduire tous les predicats du sujet à qui cette notion est attribuée.

(And since actions and passions properly belong to individual substances ... it will be necessary to explain what such an individual substance is. It is indeed true that when several predicates are attributed to a single subject and this subject is attributed to no other, it is called an individual substance; but this is not sufficient, and such an explanation is merely nominal.... and when a proposition is not an identity, that is, when the predicate is not explicitly contained in the subject, it must be contained in it virtually. That is what the philosophers call *in-esse*, when they say that the predicate is in the subject. Thus the subject term must always contain the predicate term, so that one who understands perfectly the notion of the subject would also know that the predicate belongs to it.... we can say that the nature of an individual substance or of a complete being is to have a notion so complete that it is sufficient to contain, and to allow us to deduce from it all the predicates of the subject to which this notion is attributed.)

7 Indiscernibility

("Primae Veritates" (1686), C 519, AG 32)

Sequitur etiam hinc, non dari posse in natura duas res singulares solo numero differentes.

(From these considerations it also follows that, in nature, there cannot be two individual things that differ in number alone.)

("Discourse on Metaphysics" (1686), Section 9, G IV 433, AG 41–42)

Il s'ensuit de cela plusieurs paradoxes considerables, comme entre autres qu'il n'est pas vray que deux substances se ressemblent entierement, et soyent differentes solo numero ....

(Several notable paradoxes follow from this among others, it follows that it is not true that two substances can resemble each other completely and differ only in number.)

8 Mirroring and *Confusion*

("Discourse on Metaphysics" (1686), Section 9, G IV 434, AG 42)

De plus toute substance est comme un monde entier et comme miroir de Dieu ou bien de tout l'univers, qu'elle exprime chacune à sa façon, à peu près comme une même ville est diversement représentée selon les différentes situations de celuy qui la regarde . ... Car elle exprime quoyque confusement tout ce qui arrive dans l'univers, passé, present ou avenir, ce qui a quelque ressemblance à une perception ou une connoissance infinie.

(Moreover, every substance is like a complete world and like a mirror of God or of the whole universe, which each one expresses in its own way, somewhat as the same city is variously represented depending upon the different positions from which it is viewed ... For [each substance] expresses, however confusedly, everything that happens in the universe—whether past, present, or future—this has some resemblance to an infinite perception or knowledge.)

("Principles of Nature and of Grace" (1714), Section 3, *G* VI, 599, *AG* 207)

Il s'ensuit que chaque Monade est un miroir vivant, ou doué d'action interne, representatif de l'univers, suivant son point de vue, et aussi reglé que l'univers lui-même.

(It follows that each monad is a living mirror, or a mirror endowed with an internal action, which resepresents the universe from its own point of view and is as ordered as the universe itself.)

9 "Absolute" and "Metaphysical" Necessity

("Discourse on Metaphysics" (1686), Section 13, *G* IV 436–437, *AG* 45)

... il semble que par là la difference des verités contingentes et necessaires sera détruite, ... et qu'une la fatalité absolue regnera .... A quoy je réponds ... que la connexion ou consécution est de deux sortes, l'une est absolument necessaire, dont le contraire implique contradiction ...; l'autre n'est necessaire qu'ex hypothesi, et pour ainsi dire par accident, et elle est contingente en elle même .... Et cette connexion est fondée ... sur la suite de l'univers.

(We must distinguish between what is certain and what is necessary ... future contingents are certain, since God foresees them, but we do not concede that they are necessary on that account .... But (someone will say) if a conclusion can be deduced infallibly from a definition or notion, it is necessary .... To address it firmly, I assert that connection or following is of two kinds. The one whose contrary implies a contradiction is absolutely necessary .... The other is necessary only *ex hypothesi* and, so to speak, accidentally ... And this connection is based ... on the sequence of the universe.

10 Compossiblity and Graduated "Perfection"

(Letter to Bourguet, 1714, *G* III 573, *L* 662)

Vous y ajoutés ces paroles: Si l'on regarde l'univers comme une collection, on ne peut pas dire qu'il puisse y en avoir plusieurs. Cela seroit vray, si l'univers etoit la collection de tous les possibles; mais cela n'est point, parce que tous les possibles ne sont point *compossibles*; .... et l'Univers actuel est la collection de tous les possibles existans, c'est à dire de ceux qui forment le plus riche composé. Et comme il y a de differentes combinaisons des possibles, les unes meilleurs que les autres, il y en a *plusieurs Univers possibles, chaque collection de compossibles en faisant un.*

(You add to this the words: "If one considers the universe as a collection, one cannot say that there could be many worlds in it." This would be true if the universe were a collection of all possibles, but it is not, since all possibles are not compossible .... and the actual universe is a collection of all the possibles which exist, that is to say, those which form the richest composite. And since there are different combinations of possibilities, some of them better than others, there are many possible universes, each collection of compossibles making up one of them.)

11 "Chaque Ame connoit l'infini, connoit tout, mais confusement; comme en me promenant sur le rivage de la mer . . . ."

("... through a glass, darkly. ...")

("Principles of Nature and of Grace" (1714), Section 13, *G* VI 604, *AG* 211)

On pourroit connoitre la beauté de l'univers cans chaque ame, si l'on pouvoit deplier tous ses replis, que ne se developpent sensiblement qu'avec le temps. Mais comme chaque perception distincte de l'Ame comprend une infinité de perceptions confuses, qui enveloppent tout l'univers, l'Ame même ne connoit dont elle a perception, qu'autant qu'elle en a des perceptions distinctes et revelées; et elle a de la perfection, à mesure de ses perceptions distinctes. Chaque Ame connoit l'infini, connoit tout, mais confusement; comme en me promenant sur le rivage de la mer, et entendant le grand bruit qu'elle fait, j'entends les bruits particuliers de chaque vague, dont le bruit total est composé, mais sans les discerner; nos perceptions confuses sont le resultat des impressions que tout l'univers fait sur nous. Il en est de même dans chaque Monade. Dieu seul a une connoissance distincte de tout, car il en est la source. On a fort bien dit, qu'il est comme centre partout; mais sa circomference n'est nulle part, tout luy étant present immediatement, sans aucun eloignement de ce Centre.

(One could know the beauty of the universe in each soul, if one could unfold all its folds, which only open perceptibly with time. But since each distinct perception of the soul includes an infinity of confused perceptions which embrace the whole universe, the soul itself knows the things it perceives only so far as it has distinct and heightened perceptions; and it has perfection to the extent that it has distinct perceptions. Each soul knows the infinite—knows all—but confusedly. It is like walking on the seashore and hearing the great noise of the sea: I hear the particular noises of each wave, of which the whole noise is composed, but without distinguishing them. But confused perceptions are the result of impressions that the whole universe makes upon us; it is the same for each monad. God alone has distinct knowledge of the whole, for he is its source. It has been said quite nicely that he is like a center that is everywhere, but that his circumference is nowhere, since all is present to him immediately, without any distance from this center.)

On my metalogical reading of these middle- and late-Leibnizian passages, they reflected Leibniz' decision to identify (at least "*virtuellement*"),

12 "complete individual notions", "monads" and / or "metaphysical points" with
13 maximally consistent classes of theoretical predicates that might "correctly" be attributed to them, in the sense of a proto-metatheoretical *characteristica*.

I have therefore proposed to assimilate these classes to
14 metalogical types—1-types, more precisely—which appear naturally in the semantics of first-order theories (cf., e.g., [Bell and Machover, 1977], 205),

that is,

15 maximally consistent sets **m** of formulas in a single free variable in the language *of* T.

This assimilation also fits precisely into a wider pattern of "Stone duality", which may be sketched as follows.

If a theory T is given as above, let S(T) be the set of all maximally consistent sets of formulae in the language of T (or equivalently, using Henkin constants to emulate free variables, Henkin interpretations of T).

One may then say of a property or "predicate" $s(x_1)$ in T that

16 **m** verifies s, or s *holds in* **m**, if and only if [s] is *in* **m**—is a set-theoretic member of **m**; or equivalently, if and only if

17 the sentence obtained by substituting the first Henkin constant for the variable x_1 in s holds, in Tarski's sense, in the Henkin model which corresponds to **m**.

One can also define the Stone topology for S(T). Each formula s in the language of T corresponds to a basic open (and closed) set [s]: [s] consists of the "notions" **m** which verify s—equivalently, are such that the predicate s is in the notion **m**.

Let us quickly check that this assimilation bears out analogues of Leibniz' original claims.

It does, first, closely parallel Leibniz' attempts to anchor substantiality in some sort of metatheoretic completeness, as in 1 above, *via* a conception of such "completeness" as a systematic canvass of "all" the predicates one might attribute to a given "substance".

Leibniz, moreover, took pains at several points to express his admiration for Plato, and hypothetically "complete" canvasses of predicates had already appeared in the *Parmenides* as glosses of the forms.

Integration of time-evolution and dynamics of infinitesimally-generated "well-founded properties" into the formalism of T, finally, can accommodate quite readily the "New System" account of substantiality Leibniz developed in the 1690s (cf., e.g., (Rutherford, 1995a), 145–158, and (Rutherford, 1995b), 124–132).

Such "complete" canvasses also surfaced later in the metaphysics of Kant, most conspicuously in the "completeness" or *Vollständigkeit der Bedingungen* that characterised "transcendent" *Vernunftideen*. Kant also argued that the *Bereich der Erscheinungen* is "complete", and alleged to "deduce" this in some elusive "transcendental" sense, but struggled for much of the rest of his career to discern such "constitutive" completeness from "merely regulative" counterparts.

Be that as it may—the metalogical reconstruction of Leibniz' predicate-in-notion-principle, in the framework sketched above, is extensionally correct. All one need do is construe "in" as "is a member of", as in 10 above.

The principle of the identity of indiscernibles is also immediate, in the framework of the identification sketched above in 6 through 9. "Complete notions", identified as types in L(T), are simply sets of predicates in L(T). By completeness and the axiom of extensionality, they can therefore be discerned in a set-metatheory for T if and only if the negation of a predicate "in" one is "in" the other.

The mirroring principle, finally, holds in the following qualified, theory-relative sense. A type or "complete notion" t trivially includes many predicates of the form "s and $x = x$", where s is a sentence in the language L(T) of T. The set of such sentences s forms a complete theory in L(T), and this theory is in fact the reduct to L(T) of any complete Henkin theory m which realises t.

It would seem reasonable, therefore, to argue that "complete" knowledge of the type t would "express" everything that could be said in the original language of T about the "world" determined by **m**.

In a number of passages, Leibniz also defended, or seemed to defend, a stronger "compossibilist" formulation of the mirroring principle, loosely associated with a relational property of complete notions he called "compossibility" (cf. e.g., the remarks in (G III 573), (L 662), the letter to Bourguet quoted above).

"Compossibility" is a somewhat vexed notion, historically and metalogically. There is an obvious if somewhat simplistic sense in which every Henkin structure or complete Henkin extension of T may be regarded as a collection of "compossibles"—its so-called elementary diagram—in ways that seem to fit the passage in the letter to Bourguet.

Other passages, however, suggest a need to search for stronger or more carefully differentiated glosses. According to one of these, defended by Mates, Rescher and others,

18   every complete notion **t** would express its binary, tertiary. . . . (inter)relations with every other complete individual notion **t***; and

19   a (maximal consistent) aggregate of such complete notions (and therefore of the interrelations between them) would consitute a "world".

As I indicate in Boos (2003b), the metalogical counterparts **t** of Leibniz' complete notions are "one-types over T"—maximally consistent collections of predicates in the original language L(T) of T.

In this context, a rough but natural counterpart of 19's "aggregate" can be elicited from the "Leibnizian" structure **w** mentioned in 1.5—namely the **C**-valued collection of one-types (in the language L(H) of the Henkin closure H of T) that are realised in **w**.

In the same context, one could also sketch a metalinguistically stratified partial gloss of 18. For each random **C**-valued 1-type **t** over a Henkin-closed extension H *of* T uniquely defines a random Henkin structure **w**, essentially as in the interpretation of the "mirroring principle" above.

And each pair (u,w), in turn, would then determine a "Leibnizian" interpretation of T-expressible modal "relations" between virtual 1-types.

Some attempt at metalinguistic disambiguation (here between "type over T" and "type over H") may be needed to "save" the informal claim made in 18. For straightforward metalogical interpretations of it would be untenable.

Such disambiguations, in turn, typically give rise to metatheoretic hierarchies, and problematic attempts to "collapse" such hierarchies seem to underlie "vexed" aspects of many classical metapysical thought experiments—Leibniz' "compossibility", perhaps, among them. Similar patterns of metalinguistic and metatheoretic ascent also emerge, for example, when one tries to "collapse" hierarchies which disambiguate notions of "definability" (cf., e.g., Boos, 1998, Section 3, and Chapter 4, Section 3). Both "definability" and type-formation, of course, are also forms of (meta)logical individuation—pointwise definability in one case, and "complete" specification in the other.

More generally, as argued in (Boos, 1998), "all" disambiguations of significant semantic or "alethic" notions give rise to such ascents. To the extent that one can generalise or philosophise about such notions, they seem to be relational, or perhaps liminal: expressible only in ways that generate (and demand) new threshholds of reinterpretation. In this sense, they may all be spiritual descendents of Parmenidean and Aristotlelian "tritos-anthropos"-arguments, or ancient skeptical modes of "lapse into infinites".

More concretely, I wish to argue here—in support of the particular claims made in Section 2 above—that the metalogical reconstructions in "Virtual Modality" ([Boos], 2003b), do in fact yield a kind of modal ontology.

(In the light of Willard van Orman Quine's opposition to modal ontologies of any sort, it may also be slightly ironic that I make this assertion in the sense of his well-known dictum that "to be is to be the value of a bound variable"). For in "Virtual Modality"'s formal semantics for model extensions of T, outlined above, the entities one metatheoretically and ("virtually") quantifies over are, in fact, the 1-types I have assimilated to Leibniz' "complete individual notions". In the next section, I will examine the tenability of "Virtual Modality"'s semantics as an analogical reconstruction for Leibniz' "principle of sufficient reason", and his somewhat elusive distinction between "metaphysical" and "hypothetical" necessity.

3 "Necessité(s)" et "Raison(s) Suffisante(s)"

It may be relevant, first, that Leibniz explicitly introduced his second "great principle", in the *Monadology* (G VI, 612), as a semantic refinement of its first (and, implicitly, weaker) counterpart—the Aristotelian principle of "(non)contradiction" (consistency). For consistency of a first-order theory T is a precondition for nontriviality of T's Stone space in one sense (the space is nonempty), and incompleteness in another (the space is larger than $\{0,1\}$).

Indeed, one may gloss assorted semantic paradoxes and results of Skolem, Gödel, Chaitin and others as refinements of
1. Leibniz' attempts to discern consistency (or interpretability) from "truth"; and
2. his remarks that acknowledge the elusiveness and liminality of these and other criterial, alethic notions, viewed "from within".

I will return to such glosses at the end of this section, and in the chapter's conclusion. For now, I wish to argue that Leibniz' "second great principle"—that of "sufficient reason"—fulfilled the following offices in his system. It
3. postulated forms of metamathematical coherence, in the form of initial and boundary conditions, for the illimitably complex "*raisons particulières*" of his "monads"' graduated sentience and experience. In the process, it also
4. re-secularised—or at least re-mathematised—Descartes' appeals to the veracity of "god" as the source of a quasi-stoic "criterion of truth", including mathematical truth.

Finally, the principle of sufficient reason anticipated, in my view—or at least provided some carefully-considered prototypes for—Kant's claims (in very different language)
5. that the "forms" and "categories" of "*Erfahrung*" are uniquely "constitutive" of our experience,

and
6. that the latter therefore has some sort of "transcendentally" identifiable as well as determinate structure.

Along the way, it also preempted the views expressed in two of Albert Einstein*s well known quasi-eschatological dicta:
7. that the most inexplicable datum about "the universe" is that "it" is (he firmly believed) explicable; and

8   that *"der Alte"*—Einstein's "old man", who is *"raffiniert" aber nicht "boshaft" —nicht "würfelt"* (a wry but thoroughly "Leibnizian" remark).

From the 1670s on, Leibniz also believed quite clearly
9   that both of his "two great principles"—of consistency and sufficient reason —and
10 most of the ancillary *"principes"* he adduced to clarify and elaborate them, were "logical", in some deep sense, rather than theological; and
11 that they were integrally related in some way to an as yet unrealised *"characteristica universalis"*.

It is no accident, I believe, that twentieth-century metalogicians have derived enormous benefit from Thoralf Skolem's and Kurt Gödel's insights into the limitations of such just *characteristicae*, elaborated as precisely expressible schemes for coding and "arithmetising" finitary proof-systems of the sort introduced and annotated by Frege, Peano, Russell and others (reviewed, e. g., in [Boos], 2003b).

It would seem appropriate, therefore, to look for explicit *Rückbeziehungen* between Skolem and Gödel's ground-breaking metalogical ideas about the limitations of metalogical ideas, and the insights and expressions of hope Leibniz cast in the form of his allegedly constitutive *"principes"*.

Finally, it also seems appropriate historically, as well as methodologically, to adduce Stone's metatheoretically definable spaces of interpretations, and look for interrelations not in explicitly modal formalisms—for these (I believe) were derivative in Leibniz' metaphysics—but in the not-so-universal *characteristicae* and theory-relative ontologies provided by axiomatic number- and set-metatheories. For it is precisely within these rather vacuously abstract metatheoretic contexts that one might try to impose more concrete formal constraints, which might serve, in turn—in various alternative ways—as conjectural counterparts for Leibniz' *"principles"*.

Indeed, one of Leibniz' greatest virtues as a metaphysician was his amiable willingness to (try to) identify—as such *"principes"*—the assumptions he knew he had to beg—a practice most of his predecessors, contemporaries and successors (including some who notoriously claimed to present their more "systematic" conclusions *more geometrico*) honored only in the breach.

In any case, the "principle of sufficient reason"—in my reconstruction, at least—is an intrinsically metatheoretic existence-claim. It posits the existence of a unique, divinely-"intended" structure or ratio for T's "experience(s)". In effect, I am assimilating such a ratio to an element **w** of a "virtual" space St(T) of interpretations for a given theory T. Such a model, Leibniz observed, could only be specified "completely" by what he often called an *"analysis infinitorum"*—which we human beings, in our finitude, can only adumbrate with the aid of mathematical "analysis" and indefinite extrapolation.

In effect, then, I have

12 assimilated Leibniz' *"analyses infinitorum"* to Bolzano-Weierstrass-like canvasses which converge to element **w** of St(T); but also
13 observed that our "finitude" may also induce us to demand "absolutely" "sharp" interpretations of experiential evidence, and of modal assertions about such evidence, when only "relatively" ("hypothetically"?) "sharp" counterparts of such interpretations are to be had.

Arguments cognate to these apply, finally, to one of Leibniz' most notoriously vexed and elusive distinctions—the one he struggled to draw between "absolute" and "hypothetical" necessity.

In his study of the dual distinction, I would argue—the distinction between "absolute" and "hypothetical" possibility—Leibniz effectively outlined an inchoate but well thought through attempt to discern

14 the existence of an interpretation for T—a (Henkin) structure which models a given (incomplete) theory T (any structure at all which models T)—as a guarantor of "absolute" *or* "metaphysical" possibility; and
15 the existence of an "intended" interpretation **w** for T ("discerned" or individuated, if at all, only by (a) "god").

The latter—the fixed structure of **w**, including its internal "time-evolution"—would then determine the boundary conditions, as it were, for what might be called actuality, regarded as a kind of "hypothetical possibility".

The language and substance of Leibniz' arguments was greatly complicated, I believe, by the fact that he also believed the divinely "intended" structure **w** would lock into place a kind of **w**-internal "hypothetical" "necessity"—which one could identify with some appropriate sort of "Laplacean determinism": the notion that time-evolution of **w**-internal events is subject to some sort of universal (differential) equation.

Much metaphysical and eschatological *Unfug*, of course—brilliantly parodied by Voltaire—resulted from Leibniz' attempts to reconcile his temperamental meliorism with this then-paradigmatically-"scientific", "determinist" view.

In any event, I believe that Leibniz tried to base an inchoate but well-grounded modal semantics for "universal" but expressible theories T in which "contingencies" (undecidabilities) arise on some kind of superposition of two determining structures—**u** and **w**—for such T.

These are, once again:

16 a comparably "universal" but rather vacuous notion **u** of "truth" for T; and
17 a second structure **w** for T, which provides a "sufficient reason"—that is,
18 a "complete" and particular account (*logos, ratio, raison*) for "all" of our (past, present and potential future) "experience(s)".

The first structure would impose a kind of initial condition of "absolute necessity", given by "merely logical" inference (and in the tension between "absolute" and "merely", one might see the source of Wittgenstein's "*Sinnlosigkeit*").

The second structure **w** would "virtually" decide—"analyse"—the boundary-conditions of "metaphysical necessity"—the illimitably nuanced complexities and interrelations of monadic "perceptions". Such a structure "must" exist, Leibniz believed, or postulated, even though we finite monadic minds might "perceive" its patterns and superpositions only in "confused" ways.

4 "Perfection" and "Plenitude"

The last section's metalogical reconstructions of Leibnizian "actuality" and "sufficient reason" provide complementary interpretations of his notion(s) of perfection and plenitude, if one construes the latter as stipulations imposed on "actual" worlds or interpretations **w** for T, and types t over T.

Consider first the following middle- and late-Leibnizian characterisations of (moral and metaphysical) "perfection".

1   **("Discourse on Metaphysics" 6, *G* IV 431, *AG* 39)**

>   ... Dieu a choisi celuy [le monde] qui est le plus parfait, c'est à dire celuy qui est en même temps le plus simple en hypotheses et le plus riche en phenomenes, ...

>   (But God has chosen the most perfect world, that is, the one which is at the same time the simplest in hypotheses and the richest in phenomena.)

2   **("De rerum originatione radicali", *G* VII 304, *AG* 151)**

>   Et ut possibilitas est principium Essentiae, ita perfectio seu Essentiae gradus (per quem plurima sunt compossibilia) principium existentiae.

>   (And just as possibility is the foundation of essence, so perfection or degree of essence (through which the greatest number of things are compossible) is the foundation of existence.)

3 **("Monadology" 50, *G* VII 615, *AG* 219)**

Et une créature est plus parfait qu'une autre en ce qu'on trouve en celle ce qui sert à rendre raison a priori de ce qui se passe dans l'autre, et c'est par là qu'on dit, qu'elle agit sur l'autre.

(And one creature is more perfect than another insofar as one finds in it that which provides an *a priori* reason for what happens in the other; and this is why we say that it acts on the other.)

4 **(Letter to Wolff, *LW* 161, *AG* 230)**

Perfectio, de qua quaeris, est gradus realitatis positivae, vel quod eodem redit, intelligibilitatis affirmativae, ut illud sit perfectius, in quo plura reperiuntur notatu digne.

(The perfection about which you ask is the degree of positive reality, or what comes to the same thing, the degree of affirmative intelligibility, so that something more perfect is something in which more things worthy of observation are found.)

5 **(Letter to Wolff, *LW* 170, *AG* 233)**

Plus observabilitatis est in re, est plures in ea proprietates universales, plus harmoniae; ergo idem est perfectionem quaerere in essentia, et quaerere in proprietatibus quae ex essentia fluunt.

(The more there is worthy of observation in a thing, the more general properties, the more harmony it contains; therefore it is the same to look for perfection in an essence and in the properties that flow from the essence.)

6 **(Letter to Wolff, *LW* 171, *AG* 233)**

Consensus quaeritur in varietate, hic placet eo magis, quo facilius observatur, et in hoc consistit sensus perfectionis. Perfectio autem in re ipsa est tanto major, quanto major est consensus in majore varietate, sive a nobis observatur vel non.

(Agreement is sought in variety, and the more easily it is observed there, the more it pleases; and in this consists the sense of perfection. Moreover, the perfection a thing has is greater, to the extent that there is more agreement in greater variety, whether we observe it or not.)

7 **(Letter to Wolff, *LW* 172, *AG* 233–234)**

Perfectio est harmonia rerum, vel observabilitas universalium, seu consensus vel identitas in varietate, posses etiam dicere gradum considerabilitatis.

(Perfection is the harmony of things, or the state where everything is worthy of being observed, that is, the state of agreement or identity in variety; you can even say that it is the degree of contemplatibility.)

Leibniz also used *"perfectum"* from time to time as a near-synonym for *"completum"* (cf., e.g., "... in perfecta notione cujusque substantiae individualis continientur omnia ejus praedicata tam necessaria quam contingentia ....", *G* VII 311)).

In passages such as the following, Leibniz interwove claims about "perfection" with similar-sounding but metalogically distinct ontological assertions (often implicit) about "plenitude".

8 (Letter to Malebranche (1679), *G* I 331, *L* 211)

Quicquid agit, quatenus agit, liberum est. Il faut dire aussi que Dieu fait le plus de choses qu'il peut, et ce qui l'oblie à chercher des loix simples, c'est à fin de trouver place pour tout autant de choses qu'il est possible de placer ensemble.

(Whatever acts is free insofar as it acts. We must also say that God makes the maximum of things he can, and what obliges him to seek simple laws is precisely the necessity to find place for as many things as can be put together.)

9 ("De rerum originatione radicali" (1697), *G* VII 303, *AG* 150)

Ut autem paulo distinctius explicemus, quomodo ex veritatibus aeternis sive essentialibus vel metaphysicis oriantur veritates temporales, contingentes sive physicae, primum agnoscere debemus eo ipso, quod aliquid potius existere quam nihil, aliquam in rebus possibilibus seu in ipsa possibilitate vel essentia esse exigentiam existentiae, vel (ut sic dicam) praetensionem ad existendum et, ut verbo complectar, essentiam per se tendere ad existentiam. Unde porro sequitur, omnia possibilia, seu essentiam vel realitatem possibilem exprimentia, pari jure ad essentiam tendere pro quantitate essentiae seu realitatis, vel pro gradu perfectionis quem involvunt, est enim perfectio nihil aliud quam essentiae quantitas.

Hinc vero manifestissime intelligitur infinitis possibilium combinationibus seriebusque possibilibus existere eam, per quam plurimum essentiae seu possibilitatis perducitur ad existendum.

(Furthermore, in order to explain a bit more distinctly how temporal, contingent, or physical truths arise from eternal, essential or metaphysical truths, we must first acknowledge that since something rather than nothing exists, there is a certain urge for existence or (so to speak) a straining toward existence in possible things or in possibility or essence itself; in a word, essence in and of itself strives for existence. Furthermore, it follows from this that all possibles, that is, everything that expresses essence or possible reality, strives with equal right for existence in proportion to the amount of essence or reality or the degree of perfection they contain, for perfection is nothing but the amount of essence.)

Moderately complex interrelations can be traced between metalogical analogues of these notions, and I will review several of these below.

Preliminary attempts to disambiguate metalogical counterparts of Leibnizian "*perfectio*" might distinguish "perfection" of a theory from "perfection" of its (Henkin-) interpretations, seen as particular extensions of T.

Considering theories first, and abusing for a moment the distinction between theories and their axiomatisations, one might, for example, introduce an early modern descendant of Occam's razor, and define "perfection" *of* T as

10   independence of T's axiomatisation (cf. 4.1 above); or, more sharply, as
11   recursive axiomatisability of T.

On the simpler reconstruction in 10, a given axiomatisation of a theory T would be more "perfect"—more streamlined, one might say—if it is minimally redundant.

Given appropriate metatheoretic premises, it is not hard to establish that every first-order theory may be "perfected" in this relatively straightforward way. But the axiomatisations that witness this "perfection" would not in general be unique, and such "perfection" would not change the consequences of T.

The second candidate for theory "perfection"—in 11—might seem to offer a more promising realisation for Occam's basic idea. For it would require that T have an epistemically accessible and communicable axiomatisation—if not finite, then at least expressible, in "principle", in a prescriptive or algorithmic form. As the Germans put it, such a condition might require that "god"'s design for T be "*nachvollziehbar*".

It would follow however from Gödel's results that 11's accessibility-, communicability- and *Nachvollziehbarkeits*-condition conflicts, in general, with another metalogical reconstruction of "perfection" Leibniz evidently valued, namely

12   completeness of T (cf. 2, 5 and 6 above).

Indeed, Gödel's arguments suggest a kind of complementarity relation may exist between the "epistemic" desiderata in 11 and their "ontological" counterparts in 12.

For a theory is "epistemically" more "perfect" if it "abstracts" elegantly from more structures in its language, considered as sources of "particular", merely empirical data. But it is "ontologically" more "perfect" if it accounts— metaphorically speaking—for every sparrow that falls. And Gödel's results pointed to an inherent tension between these two conditions.

Historical antecedents of metalogical completeness notions had already appeared in Parmenidean infinite-canvass reconstructions of the Platonic forms, mentioned earlier, and in Leibniz' own *notiones completae*. And metalogical descendents of these notions lie at the heart of proofs of the completeness theorem, and of the Stone duality properties briefly invoked in Section 2.

Every Henkin interpretation of T, in particular, is "ontologically perfect" in the sense of 12.

This points, in turn, to another difficulty with 12—that "ontologically" more "perfect" extensions of "epistemically" more "perfect" theories are anything but unique. There are, for example, continuum-many such extensions of any theories that inductively recapitulate ("encode") their own syntax (itself a reasonable criterion of "epistemic perfection"), among them the theories in 11. "Ontological arguments" focused on the criterion in 12 would therefore have to consider whole *pandaimonia* of such "godlike" extensions.

In response to these persistent problems, one might shift one's attention to "perfection" of interpretations **w** for a given theory T, and try to elicit from more nuanced semantic conditions on such interpretations notions of

13 "epistemic perfection" of such **w** as their inductive accessibility "from within" (cf. also 5.1 and 5.2 below); and
14 "ontological perfection" as their "*richesse*", "plenitude" or realisation of maximally many types over T (or *notiones completae* for T).

Interestingly enough, these notions—more latitudinarian forms of 11 and 12, which also reflect Leibnizian desiderata of the sort quoted earlier—are also incompatible. For the natural metalogical counterpart of 13—called "model-theoretic genericity" of such **w**—is equivalent to the assertion that they omit—fail to realise—every type they possibly can.

Also interesting, perhaps, is that the metalogical counterpart of the condition in 14—various forms of which are called "saturation"—does yield a kind of unicity-result for such **w** (up to isomorphism)—when it holds. But such validity is itself contingent on acceptance of detailed (and defeasible) background set-theoretic assumptions.

Roughly speaking, then, one way to formulate persistent "complementarities" between 11/13 and 12/14 might be to suggest

15 that if a "god" (or demiurge) were to devise a **w** that mirrored "all" the "actual" variety it could, we would be unable to gain an inductive epistemic overview of that ontological variety "from within".

This might provide warrants for a (controversial) "theodicy", perhaps; but not a "theology".

Such formulations also suggest that Leibniz may have been "right" to postulate some sort of "plenitude" (or "richness", or "continuity") as a guarantor of relevant forms of metalogical unicity. Monotheists who wish to formalise modal or metalogical counterparts of ontological arguments might keep this mind.

But Leibniz may have been "wrong", once again, if he believed that the "existence" of such "rich" structures could unproblematically be "proved"; or that such "plenitude" could be anticipated or characterised processively "from within". Indeed, the "complementarity"-relations just sketched suggest that "ontological arguments from within" might not even be (expressible, communicable, intelligible) arguments.

(By way of technical side comment: the "complementarity"-relations between 13 and 14 also suggest that the modal Barcan formula-scheme may reflect an "epistemic" constraint on modal quantification. For it holds in the "Leibnizian" semantics associated with **w** if and only if **w** satisfies a very strong boolean-valued form of 13.)

Such considerations of conflicting "epistemic" and "ontological" alternatives have not exhausted the range of Leibnizian attempts to provide proto-metalogical reconstructions of "perfection" (perhaps it is not exhaustible).

Consider once again, for example, the passages in which he considered

16 eam [combinationem possibilium seriemque possibilem], per quam plurimum essentiae seu possibilitatis perducitur ad existendum (*G* VII 303, *AG* 150) (... of the infinite combinations of possibilities and possible series, the one that exists is the one through which the most essence or possibility is brought into existence.);

and the

17 "[c]reature ... plus parfait qu'une autre", in which one finds "ce qui sert à rendre raison a priori de ce qui se passe dans l'autre" (*G* VII 615, *AG* 219). (One creature is more perfect than another insofar as one finds in it that which provides an a priori reason for what happens in the other.)

If one counts theories, for example, among Leibniz' "creatures" (his spirit might be willing), one might assimilate

18 the "essence" of a class of structures for a given theory T to the theoretical extension of T common to (satisfied in) all of them;

19 "existence" of such an "essence" to metatheoretic existence of a structure which interprets it (David Hilbert once embraced such a "distributive" notion).

Returning to "perfection" of theories, one could then stipulate formally that

20 a theory U is "more perfect" than another theory V, if and only if one can *rendre raison* in U for V's existence (prove in U that V is consistent, or equivalently for interesting U's, that V has an interpretation).

This formal relation—definable in reasonable set-metatheories—is sometimes called proof- or consistency-theoretic strength, and informal antecedents of it have also played a role in assorted "proofs" of the "existence of god" (consider, for example, Descartes' implicitly metatheoretic hierarchy of "formal" and "objective" "realities" in Meditation III (ATVII, 40-42).

It might therefore have seemed quite reasonable for Leibniz to expect
21 that new adumbrations of "perfection" would emerge from them in his hoped-for *characteristica magna*; and
22 that these notions would augment, or at least be compatible with, the proof-theoretic elegance (epistemic "perfection") he sought to reconcile with them.

In the event, however, another metalogical "complementarity" principle applies to 11, 13 and 20. For it is not hard to prove in metatheories of the sort mentioned above—relatively weak subtheories of ZFC, for example—that
23 "epistemically perfect" theories in the sense of 11 or 13 cannot be "perfect" in the sense of 20, if such "perfection" is maximal with respect to proof theoretic strength.

Shifting again to perfection of interpretations, however, one can make another observation which applies to a wide variety of interesting theories T (Peano arithmetic and assorted set-theories among them): that
24 interpretations **w** of such T that are "ontologically perfect" in the sense of 14 (satisfy appropriate "saturation"-properties) are "more perfect" (in the sense of 20) than any of their "epistemically accessible" substructures or subthe ories (in particular, "more perfect" than any of their subtheories that are "recursively axiomatisable" in various ways that generalise the original sense mentioned above in 11).

To me at least, these lines of argument suggest that such "complementarity" properties lie along a persistent unsharp boundary between epistemology and metaphysics, and that the *Unschärfe* of this boundary has interesting implications for attempts to make rigorous certain forms of "the ontological argument", as well as assorted roughly cognate "arguments from (or to) design".

In the concluding section of this chapter, I will draw further on the evidence of the modal semantics developed in Boos (2003b) to suggest that the "confusion" Leibniz (unlike Descartes) gracefully attributed to all finitary monadic "knowledge" (including his own)—and tried also intermittently to put to fideist uses—may be systematically ineliminable.

More precisely, I will argue

25 that the structures **w** of 1.5, 3.15 and 3.17—as Solovay, Chaitin and other investigators have observed in many forms—will almost certainly turn out to be random elements of the Stone space St(T) of T—or random binary decimals (cf., once again, Leibniz' attempts to model "contingency" in (Couturat) 1–3, reviewed above in 1.15–1.18);

and finally

26 that recurrent randomness of this sort refracts (metalogical counterparts of) Leibnizian "perfection" and "plenitude" in potentially interesting ways. (*Vielleicht "würfelt" also "der Alte" doch?*)

5 Conclusion

In this concluding section, I will try to sketch some interrelations and analogies between
1 the epistemic "virtuality" of "Virtual Modality"'s modal semantics in ([Boos], 2003b);
2 the incompleteness, plurality of interpretation and "liminality" of discursive modalities; and
3 the diffusion, underdetermination and "confusion" of Leibniz' *promenade sur le rivage de la mer*".

In a period in which popularisers of "chaos theory" have tried to describe sensitivity to initial conditions in terms of "butterfly effects", many readers may also understand the sort of holism and semantic realism about "*Fait(s)*" that Leibniz tried to express in the passages that follow.

4 **Raison(s), (In)definability and analyses infinitorum**

("Monadology", Sections 33, 35 and 36, *G* VI 612–613, *AG* 217)

Il y a aussi deux sortes de Verités, celles de Raisonnement et celles de Fait. Les Verités de Raisonnement sont nécessaires et leur opposé est impossible, et celles de Fait sont contingentes et leur opposé est possible . . . .

Et il y a enfin des idées simples, dont on ne sauroit donner la définition . . . .

Mais la raison suffisante se doit aussi trouver dans les verités contingentes et de fait, c'est à dire dans la suite des choses répandues par l'univers des Créatures, où la Résolution en raisons particulières pourroit aller à un detail sans bornes, à cause de la variété immense des choses de la Nature et de la division des corps à l'infini. Il y a une infinité de figures et de mouvements présents et passés, qui entrent dans la cause efficiente de mon écriture présente, et il y a une infinité des petites inclinations et dispositions de mon âme présentes et passées, qui entrent dans la cause finale.

(But there must also be a sufficient reason in contingent truths, or truths of fact, that is, in the series of things distributed throughout the universe of creatures, where the resolution into particular reasons could proceed into unlimited detail because of the immense variety of things in nature and because of the division of bodies to infinity. There is an infinity of past and present shapes and motions that enter into the efficient cause of my present writing, and there is an infinity of small inclinations and dispositions of my soul, present and past, that enter into its final cause.)

In an attractive metaphor, mentioned earlier—an obvious prototype of Fourier-analysis—Leibniz also likened the "confusion" of "*detail sans bornes*" and "*la varieté immense des choses de la Nature et . . . la division des corps à l'infini*" to our pleasant but somewhat vague auditory perception(s) of "the waves at the seashore" in Section 13 of "Principles of Nature and Grace" (*G* VI 604, *AG* 211).

A less poetically evocative but comparably defensible likeness for such "confusion" (and for Nicholas of Cusa's universally distributed "*centre partout*", cited by Leibniz two sentences further on), might be found in the **C**-valued superpositions of this essay's semantics, its "mixtures" and its virtual "intensions".

Such superpositions can be "localised" and "sharpened", but only provisionally, in partial and processive ways, adequate to "decide" data that emerge in countable simulacra of a given set-theoretic structure **M**.

M cannot localise itself, however—cannot judge its own adequacy and "sharpness". It may only "think", for example, that it is uncountable, in ways that dissolve in the yet-"sharper" focal resolution of wider set-metatheoretic contexts.

As I argue in Boos (2003b), accordingly, the "actual worlds" **w** that determine Leibnizian models and countermodels for modal extensions of T are not models of T, but "virtual" models of T, or equivalently, topologically random variables of models of T. And interpretations of designation (naming) and predication are distributive and graduated along similarly random lines.

By this, for example, I mean that the interpretations of "Sophie Charlotte" and properties of "Sophie Charlotte" expressible in a theory T that talked about "Sophie Charlotte" would be "random", **C**-valued interpretations of these notions in the inherently "virtual" model **u**.

By results of Robert Solovay, these **C**-valued interpretations of T's constants and atomic predicates "distribute" uniformly over other **C**-valued models **w** of T, as definable borel images of their values in **u**.

Leibniz' search for higher-order "*principes*" and efforts to discern different levels of "necessity" might also bear witness to another early insight—into the recurrent but elusive distinction between theory and metatheory. Indeed, one might draw a thumbnail distinction between his metaphysics and Spinoza's by observing that Leibniz' "god" "chooses"—metatheoretically—among alternative interpretations of Spinoza's "*deus sive natura*".

Work of Solovay, Chaitin and others has also suggested that "randomness"—which I've assimilated in admittedly rather heuristic ways to Leibniz' "confusion"—is theory-relative, rather like other, more obviously relational semantic notions, such as "consistency" (Gödel), and "truth" (Tarski). In particular, randomness-in-a-theory might be assimilated to what is underdetermined or *imprécisable* in that theory (compare Leibniz' many allusions to what "only [a] god could know").

In "The World, the Flesh, and the Argument from Design" (Boos, 1994), I observed

5   that most models for interesting first-order theories T are "random", in a sense that can be made precise in topological as well as measure-theoretic ways; and
6   that attempts to define models in "inductive" ways, "from within", lead precisely to such randomness (whose topological variant is usually called "genericity") in the models they "induce" (cf. again 4.13 above, and the remarks that follow).

Let me once again call such modal, alethic and quasi-alethic notions—which lie at the boundary or epistemic horizon of theories for which they may be formulated—"liminal" (at least with respect to those theories to which they are "intended" to apply). The consistency of Peano number theory, for example, a theorem of ZF, is liminal for Peano number theory itself, as is the consistency of ZF in ZF. What is determinate or expressible in a given theory T, similarly, is liminal for T, but becomes object-theoretically determinate and expressible in a variety of metatheories for T—for which analogous questions can be posed in their turn.

Such considerations suggest that some outlines of a network of interrelations may be discerned between

7   "modality",
8   "counterfactual support",
9   "potentiality",
10 "dispositionality",
11 "probabilism",
12 "distributive" notions of designation and predication, and

13 *"randomness"* of epistemic and theoretical boundaries,

where all of these notions are relativised to particular theories T and expressed in metatheories U for T.

On my readings of such "liminality"—most of them prompted by one or another result in the metalogical textbook and monograph literature—one might also draw on such patterns to formulate a few preliminary conjectures.

One is that semantic "bivalence" and sharpness of measurement may be both
14 transcendent, in the sense that "the" distinction between what is determinate and indeterminate may be indeterminate, as well as undecidable "from within", and
15 transcendental illusions, in the sense that thoroughly indeterminate and undecidable interpretations may look incontrovertibly determinate, again "from within" ("the" "universal" structure **u**, for example—a **C**-valued "redundant truth-definition"—"looks" bivalent in the multivalent context of the random V(**C**)).

Another is that one should perhaps expect some sort of "fuzziness", "randomness", or semantic gradualism to emerge in studies of modality and counterfactuality—as updated and more complex forms, perhaps, of Leibniz' "continuity"-principles.

If what is "factual", for example, were as illimitably complex as Leibniz thought—not "chaotically" determinate, but recurrently random and indeterminate, with respect to "our" theories and metatheories as they evolve—might not "the" (modal) distinction(s) between factual and counterfactual be comparably graduated and nondeterministic as well?

Whatever the merits of such conjectures, these late-twentieth and twenty-first-century metamathematical miniatures of Leibniz' ideas (heard at the seashore, as it were) prompted me to propose another hypothesis (cf. Boos, 2003b), with which I will also conclude here.

The two anachronistic clauses of this conjecture are:
16 that Leibniz may have been "wrong" about the semantic adequacy for modal discourse of pairs of structures {**u**,**w**} of the sort mentioned above;

but
17 that Leibniz may have been "virtually" right about the semantic adequacy of such pairs, in the sense that
18 Leibniz was "almost surely" right about their semantic adequacy—with value 1 in T's Stone algebra **C**.

6 "One Who Works for the People"

Consider for a moment Gödel's "other" theorem—the completeness theorem—which flowed in part from
1. David Hilbert's enigmatic remark that consistency in some sense implies existence (the "difficult" half of Gödel's result, if one interprets "existence" as "existence of an interpretation or model for the theory");

and later eventuated in
2. Leon Henkin's discovery that a maximal (consistent) theory U, complete and "perfect" enough to decide all sentences in its language and provide witnesses for its existential claims, would secure its own "existence" (in the sense just introduced).

One might try to assimilate such a "perfect" U to Aristotle's vision of "divine" thought—*noesis tou theou*—as *noesis noeseos*—self-sustaining and self-sufficient thought which would—as an aspect of its "divinity"—decide "everything" about [itself].

Henkin did show (refining earlier ideas introduced by Thoralf Skolem) that there is a sense—expressible only in a metatheory for T and U—in which U might be said to "realise" itself. But he also showed that if the original theory T is "intelligible" (can "account for" its axioms in an orderly and reproducible way),
3. there will be many such interpretations U, and furthermore (in a sense which also can be made precise),
4. "almost all" such interpretations would be "unintelligible" to themselves, and therefore to any "scientists" "in" them who might wish to provide intelligible and communicable accounts of them.

(Spinoza lives . . . .)

Plato, Aristotle, Chrysippus, Anselm, Descartes and Spinoza would all have concurred (on the available textual evidence) in a monist view of "universal" notions such as "truth" (and a fortiori, perhaps "god"). What most of them—including Spinoza—failed to anticipate (unlike Leibniz) was that such "universality" and "maximality" might not be unique.

Leibniz, of course, was no pyrrhonist. But he did surmise—on proto-metalogical grounds, I suspect—
5. that nothing can "conceive" a unique metatheoretic interpretation in which all "intelligible conception" maximises [itself].

6. that there was something intrinsically incoherent about claims to define or conceive a unique "ultimate" "essence", "existence" or "limit of conception" from within; and
7. that attempts to remove that incoherence subtly vitiated the "firstness" of "first causes", the "primality" of "prime movers" and the "conceivability" of *"aliquid quo maius nil concipi potest"*; but also
8. that "our" inadequacy "from within" might be remedied by a kind of metatheoretic *"demiourgos"* ("one who works for the people"), into which he was free to project his deeply held hopes that an intelligible metatheory may be mindful of "us."

6 Berkeleyan Metalogical "Signs" and "Master Arguments"

1 Introduction

In this chapter, I will draw on analyses of "semantic paradoxes" and other twentieth-century metalogical motifs to sketch conjectural "analogies, harmonies and agreements" (cf. *Principles* 105) for George Berkeley's

1.1 proto-semeiotic reconstruction of "signs", "general ideas" and counterfactual concept formation (cf. *Principles* I 11-20, 66, 83, 108, *Alciphron* 4, 7, and 3.331–335);

1.2 the elusive dialectical *elenchos* he saw in (what has come to be called) the "master argument" (cf. *Principles* 23 and *Dialogues* II 200); and

1.3 the carefully argued "coherentist" analyses he outlined for scientific and mathematical inquiry (cf. *Principles* 31, 58–66, 105–108 and *Dialogues* III 243–246, 251–252).

(An account of the citations appears in footnote 2 at the end of this section.)

These analogies, in turn, will lead to an interpretation of Berkeley's "immaterialism" as a dogmatically inflected theory-relativism or metatheoretic contextualism about the entities he called "signs".[1]

In an effort to discern "dogmatic" from more skeptical or "undogmatic" versions of such contextualism (a problem considered at length in Section 3), one might consider the following paraphrases of Berkeleyan or quasi-Berkeleyan assertions or premises (1.4 through 1.8). All of them can be elicited from his principal metaphysical writings.

1.4 "General ideas" and other "notions"—Locke's "nominal essences", for example—are essentially semeiotic or syntactical in nature ("signs").

1.5 Something must therefore interpret such "signs" or "notions", if they are to be non-vacuous, and acquire the status of "ideas".

1.6 Certain agencies or active "powers" (which one might or might not call "minds" or "spirits") tacitly or explicitly "intend" such interpretation(s), and such intention is constitutive of ordinary existence-claims.

---

[1] Glosses of the "master argument" and some of its (putative) antecedents and implications may be found (e.g.) in (Bolton, 1987), (Grayling, 1986), (Tipton, 1987), (cf. pp. 113–117), (Muehlmann, 1992) (cf. pp. 20, 147ff. and 163), and (Winkler, 1994) (cf. pp. 187–190 and 290). Nontechnical studies of related issues also appear in (Brook, 1973), (Flage, 1987), (Jesseph, 1993) and (Thrane, 1982).

1.7 These "intentions" are stratified, and perhaps even hierarchically organised, in intrinsically asymmetric ways ("minds", for example, can "perceive" and "conceive" their "ideas", but not conversely).

1.8 Attempts to ignore this stratification, finally, or collapse the hierarchy or hierarchies to which it may give rise (construe "spirits", for example, in terms of any partial "ideas" they may have of themselves) are likely to lead either to paradoxes, or outright "absurdities" (one of which he believed he had isolated in the "master argument" of *Principles* 23).

Skeptical and latitudinarian interpretations of 1.4–1.8 compatible with Berkeley's avowed finitism might simply accept indefinite iterations and ramifications of the hierarchies sketched above. He himself seems to have been well aware that one could inflect the nominalist implications of 1.4 "skeptically" as well as "dogmatically" (cf., e.g., *Dialogues* III, 227–230). More pointedly: he was also aware, I think, that the view he called "immaterialism" might not in fact follow from 1.4–1.8 alone.

These quasi-Berkeleyan assertions do, however, seem to imply that there is something elusive, or at least extra-"ordinary", about existence-claims not based on 1.7's conceptual or perceptual intentions.

In particular, they seem to lead to a natural trichotomy.

Either one

1.9 postulates the existence of an "existence"-predicate (?) that eludes "all" of 1.7's intentions (cf. *Commentaries* 671, 725, *Principles* 17, 18, *Dialogues* II 222).

Or one

1.10 skeptically suspends judgment about the tenability of "existence"-assertions at the margins of "all" such "spirits'" intentions (cf., e.g., Philonous' struggles to deflect Hylas' most searching criticisms in *Dialogues* II 218, 223, 225, and the exchange about "fair dealing" in *D* III 231–233).

Or one

1.11 dogmatically denies the "possibility" of such (inconceivable and apparently inexpressible) assertions (the orthodox immaterialist view which Philonous ultimately induces Hylas to adopt).

Berkeley also knew that pyrrhonist and academic skeptics had deployed 1.7's hierarchies and 1.8's paradoxes as skillfully as had their Augustinian/Cartesian opponents—so much so that the Stoics, Augustine, Anselm, Descartes, Pascal, Spinoza and Leibniz (among others) all resorted to various shades of *petitio* (or postulation of *principes*) to part the ontological waters.

The "intentional" or "theory-relative" ontologies of 1.1–1.8 may, in other words, be compatible with any evidence we have (Berkeley, 1979, Hume, 1975), or could have (Kant, 1956). Indeed, I will argue in Section 5 that they may be regulative of such evidence. But they are also incomplete, in several complementary senses of the word. They would not, therefore, sustain the aspirations of Arthur Lovejoy's "monistic pathos" (though they may be compatible with its "organismic" counterpart; cf. (Lovejoy, 1965), 10–17, and the comments and quotations from P 105 and D I 211).

A fortiori, they would not provide the demonstrative arguments Berkeley sought. Nor do they provide definitive rationales for the underlying monist impulses that animate monotheists such as Berkeley and scientific as well as metaphysical realists alike.

What, after all—skeptics had asked—could re-activate or re-interpret the interpreters' disparate activities; reconcile conflicting (much less "incommensurable") interpretations; and ensure that such processes of re-conciliation, re-interpretation and re-activation do not eventuate in "vicious" hierarchies of infinite regress?

Rather curiously, perhaps, few western commentators construed such hierarchies as "ascents"—except, ironically, when they wished to impose theologically motivated linearity-, directedness- and maximality-conditions on them. For an especially well-known historical example of such an effort to secure a "final" place to stand, for example, consider Descartes' claims in Meditation III that "formal" and "objective realities" converge uniquely in "god".

Or consider the ostensibly quite different ontological claims of a later, more "secular" semeiotician drawn to premises very like 1.1–1.8. Charles Sanders Peirce vigorously embraced the infinite extendibility of "inquiry", in ways that essentially affirmed 1.7's hierarchies, as well as "tychism", "synechism" and much, much more. To secure a univocal outcome of such inquiry, however—a kind of Diodorean as well as "scholastic" "realism" (persuasively reviewed in (Fisch, 1986), 184–197)—he then begged (or at least presupposed) the unicity of its convergence to an all-encompassing limit.

Berkeley, by contrast—a committed christian monotheist as well as mathematical and physical finitist—responded to skeptical interpretations of 1.1–1.8 as dogmatically as did Peirce, albeit in somewhat different ways, and a moment's reflection suggests why. He was personally loath to "sit down a forlorn skeptic", whatever David Hume and other critics (cf. the beginning of Section 4) may have thought of his system's implications.

Prototypes of 1.4–1.8's iteration also figured prominently in assorted well-known ancient, medieval and early-modern skeptical *elenchoi* all the same. *A baculo rejections* of such "regresses", in turn, were essential to "foundational" responses to such arguments.

Such rejections took two potentially interrelated forms. One could accept the regresses' indefinite recurrence, but normalise them as Peirce did (and Leibniz effectively had done). One of the most straightforward ways to do this was to impose a Diodorean necessity- or linearity-constraint on them: presuppose that inquiry "must" temporally converge, or that such temporal convergence is constitutive of world-internal "necessity". Or one could cap them, as Aquinas and Descartes had done—impose a cutoff, in effect, on the number of such iterations or recurrences.

Berkeley's finitism, I believe, inclined him toward the second of these alternatives. He tacitly accepted a number of uniformity and convergence assumptions, of course, both

1.12   synchronically—in his assumption that different minds' perceptions are interrelated by "general ideas" which structure them in significant and communicable ways; and
1.13   diachronically—in his meliorist accounts of scientific progress.

He grounded these presuppositions, however, in the timeless, unitary and apparently inexpressible hermeneutic agency of an Entity he called "the Author of Nature". I will sometimes follow the example of (Popkin, 1979) and call Berkeley a "fideist", with this rather idiosyncratic blend of potentially skeptical premises and christian-deist conclusions in mind.

If one inquires, however, into the constitutive epistemic and metaphysical attributes of this "Authorship"—to the extent they might be discernible "from within"—they sometimes turn out to be rather minimal. The epistemic and metaphysical reduct of such "Authorship" essentially normalises the metatheoretic regresses just mentioned, and sustains the synchronic and diachronic uniformities postulated in 1.12 and 1.13.

Berkeley's "god" therefore looks rather like a necessary-precondition-for-the-possibility of something, and his appeals at several points to arguments to design in the *Principles* and *Dialogues* (cf. 3.20, D II 211) seem to bear this out. I will return to such characterisations in Section 4, but underscore here only that the underlying metaphysical structure of this night-watchman-god (if I may borrow a phrase) might be characterised in metalogical terms, as

1.14   a—dogmatically delimited—two-level type-theory, in which
1.15   finite, second-order, active "minds"—indefinitely many of them—intend and interpret first-order, passive "ideas"; and
1.16   "God"—an allegedly universal and self-interpreting second-order Mind—aligns, normalises and uniformises these lesser minds' semeiotic activities

(*tota simul*, to borrow a phrase from Boethius), and ensures that their intentions form a kind of directed hermeneutic system with "God" at its (unique) limit.

Berkeley's "God", in short, might be identified with
1.17 an "ultimate" metatheoretic limit-structure of first-order inquiry—an inexpressible limit, moreover—like the "occasions", "substrates" et al he so effectively mocks—in the sense that it is beyond (first-order) ideation, in Berkeley's sense of "idea", though we can have a (notoriously elusive) "notion" of It.

This was not the end of the matter, however. Such a "God" might be identified with the (putatively) benign genius of Descartes and Leibniz. But assent to any "notions" we might have of it, therefore (or of its explicitly amoral counterpart "*deus sive natura*"), would require fideist acquiescence in the simultaneous intelligibility and inscrutability of a great many quasi-scientific and quasi-theological ancillary "notions"—rather like the assent Descartes had demanded to a significant number of scholastic dicta, allegedly given by the "natural light" Hylas required for the "occasions" of corpuscular "matter", and Peirce later postulated for his inchoate limiting conceptions of convergent "inquiry".

Fideists, in short—as others have observed—might be viewed as skeptics who calibrated their skeptical glass, in "practical" or religious terms, so that it appeared half-full rather than half-empty. Berkeley evinced his very substantial skeptical and dialectical capacities at many points—in the *Analyst* and *De Motu*, in the eighteen (!) objections he canvassed in *Principles* 38–84 ("But, say you, ..."), and in Hylas' most cogent critiques in *Dialogues* II and III.

In keeping with this view of Berkeley as an episodic fideist-skeptic-*malgré lui*, therefore, the principal burden of this chapter will be to argue that Berkeley's restless attempts to probe the limits of his hypotheses, as well as "save" them, led him (as similar impulses later led Kant) into thought-experiments of unprecedented dialectical acuity.

He was well aware, for example, that problems of mutual interpretability are pervasively synchronic as well as diachronic (recall the mite's foot at *D* I 188–189)—a point that is often elided, and an observation he shared with his skeptical predecessors—and that these problems were inherent to abstraction as well as theory-formation. His insights into the implications of 1.7 and 1.8, moreover, more particularly his acknowledgment that the structural aggregate of experience might be inexpressible in experiential terms (cf. 1.17) foreshadowed the theory/metatheory distinction, a metalogical subtlety which eluded many early-twentieth-century pioneers in formal logic (cf., e.g., (Goldfarb, 1979)).

Finally, the elenctic arguments he developed to vindicate 1.8 anticipated the "semantic paradoxes", which recapitulated the insight of 1.17, and motivated the work of these pioneers' successors (could 1.13's "God" interpret *itself*?).

If I am right, at any rate, that Berkeley's thought experiments anticipated uses of semantic paradoxes and other central motives of twentieth-century metalogic, such paradoxes—and the metalogical theories devised to accommodate them—may themselves have hermeneutic value, as reflections of the classical metaphysical dilemmas in formal miniature. The work of Peirce, for example, mentioned above, can readily and usefully be analysed in such terms.

Less obviously, perhaps, similar cross-disciplinary analogies might also clarify some *aporiai* which emerged in the mathematical and philosophical writings of Gottlob Frege. For Frege defended—curiously enough—a "logicist" counterpart of the view that I have attributed to Berkeley: that metamathematical miniatures of "psychologism", relativism and underdetermination would disappear in a sufficiently complete and "absolute" second-order logic.

Unlike Berkeley, however—who acknowledged that we might have to content ourselves with a merely "notional" understanding of certain limiting concepts—Frege also hoped to formulate "all" aspects of this system's anticipated semantics from within. His fallacious *Grundgesetz V*, moreover—which essentially postulated that "all" conceptual constructs expressible in his system had unique referents—foundered precisely on the illusion Berkeley identified (*mutatis mutandis*) in *Principles* Intro 23:

> that the immediate signification of every general name [is] a *determinate*, abstract idea.

(Interestingly enough, the formal entities warranted within Frege's system by *Grundgesetz* V are also called "abstraction-terms".)

The interrelation just sketched is somewhat nonstandard, perhaps. If so, so much the better. It is my hope that a more general canvass of such underlying contrasts and affinities might lead to a kind of metalogical structuralism, or at least a heuristic analogical study of common patterns that cross conventional philosophical disciplines and identities in sometimes unexpected ways.

To this end, the three main sections of the present chapter will offer miniature studies of the three analogies sketched in 1.1–1.3—between metaphysical and metalogical "abstraction" (1.1), semantic-paradoxical readings of the "master argument" (1.2), and the *Glanz* and *Elend* of his metaphysical and scientific coherentism (1.3). More precisely, I will sketch metalogical reconstructions of "general ideas" and counterfactual concept formation in Section 2 (cf. 1.1); outline metamathematical analogues of the elusively dialectical *elenchos* Berkeley saw in the "master argument" in Section 3 (cf. 1.2); and critique some of the presuppositions of the carefully argued "coherentist" analyses he outlined for scientific and mathematical inquiry in Section 4 and the conclusion, Section 5 (cf. 1.3).[2]

2 Iterated Abstraction and Metatheoretic Ascent

In this section, I will sketch several "finitary" metamathematical analogies for Berkeley's elusive "general ideas" and his critique of counterfactual concept-formation. These reconstructions are openly anachronistic, and my outline of them here will be brief and allusive; a more thorough technical survey may be found in "Virtual Modality" ([Boos] 2003).

One can sketch these details in qualitatively accurate terms, however, in ways that seem useful and persuasive. It will also be helpful from time to time to formulate these analogies in straightforwardly prescriptive terms drawn from the philosophy of language and philosophy of science (two of the principal catch-basins of late-twentieth-century metaphysics).

The first steps in this formulation will be to assimilate

2.1 "mere signs" to "syntactical" or "object-theoretic" terms and formulae of first-order theories;

2.2 "ideas" to metatheoretically definable theoretical contexts and interpretation(s) for such terms and formulae (cf. 2.11), and definable elements and relations of such interpretations;

and, finally,

---

[2] Above and in what follows, I have cited passages from several well-known longer texts in Luce and Jessop's nine-volume edition of Berkeley's *Works* by paragraph or entry numbers, when the editors preserved or introduced such divisions ("*Principles* 23" or more briefly, "*P 23*", e.g., refers to paragraph 23 of *The Principles of Human Knowledge*, and *P* Intro 11 to paragraph 11 of the "Introduction" to the *Principles*).

One such longer text—*Three Dialogues between Hylas and Philonous*—lacks such an item or section enumeration, and I have therefore identified passages from it by page number in *Works*' second volume (e.g., "*Dialogues* II 200" or "*D* II 200"). All other passages from the *Works* are identified "lexicographically", by their volume and page number (e.g., "3.332" for a passage from the "Appendix" of *Alciphron* on page 332 of III).

2.3   1.4–1.8's "activities" of "perception" and "conception" to dynamic, metatheoretic representation(s) of such definable contexts and interpretations.

Within this semiformal framework, one can compare the (quasi)-Berkeleyan asseverations set out in 1.4–1.8 with (quasi)-metalogical counterparts of them, in 2.4–2.8.

2.4   Theoretical terms, mathematical "objects" and counterfactually-supported "modal" notions ("causality", for example) are essentially syntactical or "object-theoretic" in nature ("formal"; "mere signs").

2.5   Only in "metatheoretic" contexts, moreover, do such object-theoretic "notions" acquire interpretations, which render them non-"vacuous", fix their semantic referents, at least locally, and confer on them (for example) the provisional dignity of potentially communicable scientific data.

In this sense, theoretical notions of the sort mentioned become inherently relational and context-dependent. One might also call them liminal, since the threshholds (*limines*) of such metatheoretic contexts may be expected to shift and evolve in time.

2.6   Only "intentional" dynamical entities capable of partial (self)-representation can "observe" or "intend" such interpretations as in 2.3, "in real time". Only such entities, moreover—interpretive systems, let us call them, or systems for short—can iterate 2.5's postulation of new metatheoretic contexts; quantify over indefinite ranges of interpretations in them; or individuate or discern such contexts in local ways—all, again "in real time."

(It may therefore be no accident that the twentieth-century physicists Eugene Wigner, John Wheeler and others have made intermittent efforts to assimilate such systematic activities to more general counterparts of "consciousness").

2.7   Such discernments and individuations are open-ended, provisional and hierarchically organised in temporally and structurally asymmetric ways (interpretative systems may "perceive", "represent" or "measure" temporally prior "subsystems" in certain contexts, but such "subsystems" cannot "perceive", "represent" or "measure" them).

2.8   Implicit or explicit attempts to blur or efface 2.7's metatheoretic or metasystematic hierarchies—"define" "the" (?) limits of "definition", for example, or "conceive" "the" limits of "conception"—lead to paradoxes, one of which Berkeley adumbrated in the "master argument".

In Chapter 1, ([Boos], 2003) and elsewhere, I have called indefinite iterations of 2.4–2.6's type-hierarchies and theory-to-metatheory transitions (instances of) metatheoretic ascent. Such iterations and objections to them have a long history, as mentioned in the introduction to this chapter, and it may be useful to begin with a brief response to some of the objections.

Several pioneers of early twentieth-century logic and a good many late twentieth-century analytic philosophers have dismissed indefinitely extended metatheoretic ramifications of the sort sketched in 2.1–2.8. Russell as well as Frege, for example—like their stoic and cartesian predecessors—never really questioned the unicity of "(mathematical) truth", or subjected claims about it to the sorts of hermeneutic ambiguation considered in 2.7.

An impressive roster of Frege and Russell's spiritual descendents sought to "save" such hypotheses in suitably modulated forms, but their results seldom confirmed their ontological realism (or "monism"). The patterns of metatheoretic ascent sketched in 2.1–2.5 provide a fairly natural "phenomenal" gloss of the "internal" structure of the subject, as the innovators of its second and subsequent generations developed it. It has, indeed, become a metalogical datum—a straightforward consequence of analyses of semantic paradoxes and results of Skolem, Gödel, Tarski, and a long line of innovative successors—that "well-defined", ostensibly "bivalent" "truth"-predicates are plurally interpretable syntactical constructs, and "universally" "bivalent" "truth"-predicates are therefore ill-defined. In other chapters, I have likened this predicament to a noetic variant of Bohrian "complementarity"— in this case between completeness of a theory and expressibility of its axioms and consequence-relation.

Guided by such analogies, I have also argued that 2.1–2.8's patterns of indefinitely ramified metalogical ascent offer a generative framework for potential insights. We are all guided by our metaphysical temperaments, of course, and a number of historical ironies lurk in the analogies I have drawn between 1.4–1.8 and 2.1–2.5. But dismissal of the processive and indeterminate aspects of the theory-relativism I have sketched may be a mistake.

In any event, recurrent metatheoretic reconstructions of abstraction as theory-formation—or rather, in more orthodox quasi-Berkeleyan terms, of "general ideas" as syntactical entities in first-order theories—have certain virtues.

Consider, for example, Berkeley's formulation of the "one-over-many"-problem, exemplified by the notorious general idea of a triangle, which is, neither oblique, nor rectangle, equilateral, equicrural, nor scalenon, but all and none of these at once .... (P I 17)

For a straightforward metalogical counterpart, consider an incomplete first-order theory T, and a metatheoretically identifiable collection M of "empirically observed" interpretations of T (elements of T's *Stone space*). In this context, one might identify

2.9   the theory T(M) "abstracted from" M, or "the general idea" which holds "indifferently" of everything in M.

metatheoretically with

2.10   the set of sentences in the language L(T) of T which hold in every element of M (which logicians sometimes call, appropriately enough, the theory of M).

In this framework, a metalogical counterpart of Berkeley's polemic becomes a theorem. It would indeed be "absurd" to conflate T(M) with any particular "intended interpretation" of T (or T(M)) in M. For

2.11   T(M) itself—in an appropriate metatheory U—is the "general idea" (in a weaker sense Berkeley endorses), or the "sign", of all its extensions S in L(T).

This simple metatheoretic reconstruction also clarifies an equivocation in the specious view Berkeley derided: each proper extension S of a given (incomplete) theory T (of "trianglehood", say) "is a T"—in the weak sense that every interpretation of S satisfies T (a datum one can model as an intersection in T's Stone space). But no such S, of course, "is" T (is identical to T; proves the same theorems as T).

Both Berkeley's critique, in short, and his semeiotic analysis of more tenable forms of "abstraction" might be reconstructed as insights into informal completeness assumptions about sufficiently "general" incomplete theories T, insights which disambiguated in informal but thoroughly cogent ways "is"-as-"equal to", from "is"-as-"is-a-subtheory-of" (and *a fortiori*, therefore, from "is"-as-"is-an-interpretation-of").

Berkeley's polemic would have fallen away, of course, if one could identify significant empirical "general ideas" T(M) with unique "intended" extensions M of them in M—if such theories T(M), in other words, were complete. Here, however, Gödelian arguments become relevant: expressible theories—in plausible senses of the word "expressible" (cf. Boos, 1998)—are inherently incomplete. To the extent the generalisations we derive from our experience(s) are expressible or communicable, therefore, they will be incomplete as well.

Historically, Berkeley directed his critique of "abstract ideas" in his Introduction to the *Principles*—fairly or unfairly—against Locke. It may therefore be worth observing that one might assimilate certain metalogical "abstractions" T(**M**) to (formal counterparts of) Locke's "nominal essences" (Essay III.3.17 ff.), and contrast them with interpretations **M** of T and T(**M**), which would bear many of the marks of Lockean "real essences" (*Essay* III.6.6)—and, ironically, of Leibnizian "complete individual notions".

On one reading of Berkeley's critique, moreover, he directed it ultimately not at "abstraction"—for which he had an alternate semeiotic account—but at just such "real essences"/"complete individual notions"—which violated his avowed finitism. *A fortiori*, it would run counter to that finitism to assimilate Berkeleyan "ideas" to arbitrary metatheoretically complete extensions **M**/interpretations **M** of T(**M**).

What should one assimilate them to, then, on the analogical account I have sketched? Here is one—metatheoretically finitary—candidate, a more precise counterpart of 2.2.

2.12   An idea of T (or T(**M**)) is a metatheoretically definable extension of T (or T(**M**) in L(T)).

Notice first that finite extensions and metatheoretically definable interpretations **M** of T(**M**) would satisfy the requirements of 2.11. We will return to this point in the next section's study of metalogical "master arguments".

This reconstruction almost immediately suggests another.

2.13   A notion of T (or T(**M**)) is any metatheoretically postulated (but not necessarily metatheoretically definable) extension of T (or T(**M**) in L(T)).

This more latitudinarian interpretation will also reappear in the sequel.

A first observation to make in this context is that there will in general be many more notions than ideas, by straightforward metatheoretic counting-arguments. A second is that metalogical "ideas" and "notions" that are complete (that decide every sentence in the language of T) may be assimilated to the elusive notion of a theory T's (metatheoretically) "intended" interpretation.

David Hilbert once argued, in effect (cf., e.g. [Hilbert, 1900], 300–301, [Hilbert, 1904], 182–183 and the 1899–1900 exchange of letters with Frege in [Frege, 1976], 66–68), that a theory T "exists" if and only if it is interpretable. A quasi-Berkeleyan philosopher of mathematics, in turn, might maintain that interpretations of T can "exist" if and only if they are "intended" in one or another metatheory T_1. If Hilbert and the quasi-Berkeleyan are *both* right, then

2.14 contextual "existence" may alternate in ramified and indefinitely iterated hierarchies T, T_1, T_2, T_3, ...

Earlier I have called the pattern of such ramified hierarchies "metatheoretic ascent". Generation of such patterns conflicts with tacit presuppositions many analytic philosophers seem to make about "intended" interpretations of assorted T, when they formulate assertions about "the" "truth" of T. For the language of such assertions often suggests that these interpretations "exist" in some "ultimate" but elusively extratheoretic sense.

Indeed, it may support the fundamental analogy I am offering that some of the makers of these assertions have tried to identify such existence with the semantic existence of a theory-independent realm of structural invariants (in Chapter 2, for example, I assimilated such invariants to platonic *eide*). Unlike Berkeley (and perhaps unlike Kant), I would not undertake to "prove" (or "deduce")—say—that ramified hierarchies of the sort considered in 2.14 are all that exist—much less that some sort of metatheoreticians' "god" or "*transzendentale Einheit des Verstandes*" provides a bridge of intersubjective uniformity across the branches that appear in them. I will, however, argue that they may well be all that expressive and (intermittently) ratiocinative beings can find expressions for, and reason about.

Prompted by such reflections, I will also argue in the chapter's last two sections that the recurrence of such hierarchies also provides alternatives to Kant's notions of "*Erfahrung*" and "*Dinge an sich*".

3 Semantic Paradox(es) and "Master Arguments"

In the last section, I suggested that Berkeley's recourse to "signs" anticipated (the need for) metalogical completeness results in insightful ways. In this one, I will extend this metalogical analogy to the "*esse* is *percipi*" doctrine, and suggest that Berkeley's "master argument" might be taken as a prototype of the semantic paradoxes and **in**completeness arguments that generate the last section's patterns of metatheoretic ascent.

More precisely, I will sketch *two* metalogical analogues for the "master argument" (a phrase Berkeley himself never used), analogues which also turn out to be interrelated (cf. (Chaitin, 1975), (Boolos, 1979), (Vesley, 1992) and (Boos, 1998)). As the reader may already have inferred, I will construe these interrelations and their potential implications for Berkeley's metaphysics along lines that assimilate metatheoretic "definition(s)" to Berkeley's "perception(s)", and metatheoretic "interpretation(s)" to his "conception(s)".

In principle, as discussed in Chapter 1, Berkeley disjoined "perception" from "conception". In practice, however—as commentators are well aware—he tended to conflate "perception" and "conception" in tacit ways, especially when such assimilations served immediate methodological and elenctic aims (in Philonous' more elenctic and defensive responses of *Dialogues* Two and Three, for example); and disjoin them, often with an air of great rigor, when they did not (most conspicuously, perhaps, in the first *Dialogue*).

Perhaps all that one can reliably say about this shifting distinction is that Berkeley consistently considered "perception" a somewhat more "immediate", "extensional" and even *eidetic* activity of "spirits" (an activity which generates something roughly comparable to Hume's "impressions"), and "conception" a more "mediate", "intensional" and hermeneutic form of such ideation (which gives rise to something rather like Hume's "ideas"). Though he never quite said so explicitly, he may also have believed that one conceives something—and perhaps even that one has a "notion" of it (cf. 2.13)—when one perceives a *sign* for it.

Against the background of these historical and textual ambiguities, I will focus in this section on two metalogical reconstructions of Berkeley's master argument.

The first, more "conceptual" analogue, applied to terms t and assertions s, is

3.1   that every "conceivable" s and "existent" t must be interpretable.
For if not, one could conceive an uninterpretable t (an "absurdity", for "... do not you yourself [interpret it] all the while?").

The second, more "perceptual" is

3.2   that every "tenable" existence claim s must be witnessed by a definable t.
For if not, one could perceive or individuate an indefinable t (another "absurdity", for "... do not you yourself [define it] all the while?")

Consider, first, the initial claims in 3.1 and 3.2:

3.3   that every "conceivable" s and "existent" t must be interpretable, and

3.4   that every "tenable" existence claim s must be witnessed by a definable t.

If one identifies "conceivability" of a first-order theory T with consistency of T, first, and assimilates "consistency" to "existence" as Hilbert did (2.14), 3.3's metalogical counterpart of a Berkeleyan identification of *esse* with *concipi* (or *concipi posse*) would be correct, in any metatheory for T strong enough to prove the Henkin completeness theorem (cf., e.g., (Bell and Machover, 1977), 117–122 and 191–202, or 2.1–2.11 of (Boos, 1998)), some of whose implications are sketched in Section 2.

Complications arise, however, when such a theory T serves as a metatheory for *itself*. For Gödel's results, reviewed below, establish that it is metatheoretically conceivable that [*T* may be uninterpretable] in T, or equivalently, that [*T* may be inconceivable] in T—a metalogically coherent analogue of the predicament Berkeley thought patently absurd in *Principles* 23.

In fairness to Berkeley, however, many mathematical logicians—most, perhaps—find witnesses to the consistency of [*T* is inconceivable] with T "counterintuitive", even "pathological". I have already remarked that this response seems to me mistaken, and will return to it later in this section and in Section 5.

Secondly, one might consider 3.4's thought-experimental stipulation that every object-theoretical existence-claim be witnessed by something definable as a metalogical counterpart of the explicitly Berkeleyan dictum that *esse* is *percipi* (or once again, perhaps, *percipi posse*). Here too one can find a variety of metatheoretic settings (interpretations of T in metatheories U for T in which T's consistency is axiomatic) in which Berkeley's dictum is simply correct.

If s, for example, has the form [there exists an x such that p(x)], and T proves [there is a unique x such that p(x)] (cf. 3.19), it is well-known that one can define metatheoretically conservative extensions T(c) of T (extensions which prove no new theorems in the language L(T) of T), in which the existence-claim s is witnessed (somewhat uninformatively) by a "new" constant c.

In the case of more general object-theoretic existence-claims s, one can also employ subtler metatheoretic methods, originally devised by Thoralf Skolem, to define canonical extensions S of T and interpretations **M** of such S, in which object-theoretic existence-claims—as well as stronger assertions of "functionality" in the language L(S) of S—are provably witnessed by terms of L(S). There are even metatheoretic senses (cf., e.g., 3.36 and 7.9 of (Boos, 1994)) in which "most" interpretations **M** of S have such "constructive" or "substitutional" characteristics.

Here too, however, one can ask whether the original theories T (or U) must verify in some way Berkeley's dictum for themselves, and the essentially negative but complexly ramified answers to such questions reflect parallel subtleties in the implications of his master argument. Let us return then to the putative metalogical analogues sketched above in 3.1 and 3.2.

My gloss of the first analogue (3.1) will be Gödelian, though its use in this context is not: a comparison of conjunctions such as "s but s is inconceivable/uninterpretable" (where s might be an assertion that there are "trees, for instance, in a park, or books existing in a closet", in the language of some metatheoretically consistent background-theory T) with (also metatheoretically consistent) assertions of the form "s, but [s is inconsistent with T]"; or "s, but [s holds in no interpretation of T]".

I will interpret the second (3.2) as a prototype of the Berry/Richard paradox (cf. 3.15–3.17), applied to existence assertions s of the form [there exists an x such that p(x)], and assimilate conjunctions such as "s but s is imperceivable" to assertions of the form "s but [no t can be defined or individuated such that p(t)]". On this account, the existence of an imperceivable tree is as elusive—if not "absurd"—as the existence of a least integer which cannot be defined in less than (say) forty syllables....

The quasi-Berkeleyan "marks and signs" in such reconstructions will once again be syntactical or object-theoretic entities in (recursively axiomatisable) first-order theories T, which may, in turn, be metatheoretically interpreted in (also recursively axiomatisable) first-order metatheories U. The early twentieth-century logical pioneer Thoralf Skolem argued for a kind of theory-relativism about such interpretation(s), and I will argue below that Berkeleyan finitism and nominalism provide natural rationales for the first-order "*Semeiotik*" he developed.

I will also draw on on an extensive canon of metalogical results, discovered and refined by Skolem, Gödel, Tarski, Henkin, Robinson, Keisler, Barwise and others. In precise ways these results have clarified some of the implications of

3.5   U's capacity to "encode" syntactical entities of T (and S); and
3.6   U's capacity to talk about interpretations of these syntactical entities.

The former will permit us to apply Skolem's and Godel's relativisation and completeness results to T in U. The latter will allow us to adduce Gödel's incompleteness results, as well as the analogical arguments and "semantic paradoxes" which historically provided a heuristic for their discovery. It will also be relevant to observe that 3.1–3.2's reconstructions of the master argument are also interrelated. In Boos 1998, for example (summarized briefly in 3.24), I vary arguments of Chaitin and others to show that Gödel's First Incompleteness Theorem follows from an explicit formalisation of the Berry Paradox, in the same way that the Second Incompleteness Theorem follows from the First.

Let us begin, then, with the first, more "conceptual", Gödelian reconstruction in 3.1 of the thought experiment in *P* 23.

Consider a first-order theory T (consistent by the lights of S), in which one can do arithmetic, catalogue books, reason semantically and inductively, and carry out other cognitive tasks. For example, T might include a significant but finite fragment of U.

One can then

3.7 formulate consistent sentences s in T which assert (by T's lights) that [there exist "trees ... in a park, or books ... in a closet"], and encode such s as [s] in T.

In the theory T + s (which is metatheoretically consistent, by assumption, in U), one can then define

3.8 [conceivability of such [s]] in T

as (provable)

3.9 [[validity] of [s] in some interpretation of T]; or, equivalently in T, as [[consistency] of [s] with T].

In defense of the first assimilation in 3.9, one might adduce the plurally interpretable semeiotic qualities Berkeley attributed to concept-formation in *P* Intro 23, and the plausibility of an ontologically neutral first-order semantics as a metalogical carrier for those qualities.

In defence of the second, one might recall the force and tenacity with which Berkeley defended a maxim that everything that is consistent (not "repugnant" or "absurd") must be conceivable (and conversely). I will return to these points in Section 4.

In this framework, finally, one can assimilate

3.10 "s but [s] is inconceivable" to
3.11 $s^* = [s, \text{but } [s] \text{ holds in no interpretation of T}]$; or equivalently,
3.12 $s^* = [s, \text{but } [[s] \text{ is inconsistent with T}]]$.

On the analogical account I am offering, Berkeley considered informal assertions of the sort I have assimilated to s^* "absurd", and a reduction to such "absurdity" yields the *elenchos* of the "master argument".

Working with the assumptions in 3.5–3.6, however, Gödel showed that

3.13 consistency of the theory T + s^* is metatheoretically provable in U.

Indeed, it turns out in U (cf., e.g., (Boos, 1994), 7.9) that

3.14 "most" interpretations of T model infinitely many "pathological" assertions of the form

$s^* = [s, \text{but } [[s] \text{ holds in no interpretation of T}]]$.

Now let us turn briefly to the second, more "perceptual" reading of the "master argument", which assimilates it to the Berry/Richard paradox, first cited in (Russell, 1906, 645), according to which we seem to be able to "define" all too concisely such entities as

3.15 "the least natural number that is not definable by a property which can be expressed in fewer than one hundred English words".

As I suggested earlier, this will amount to assimilation of
3.16 "the nearest (or most readily conceived) unperceived or imperceivable tree" (say) to
3.17 "the simplest indefinable real number", or
"the least integer which cannot be defined in less than forty syllables"....

In exploring the implications of this assimilation, I will also draw on the usual gloss or provisional resolution of the paradox, according to which
3.18 (object-theoretic) "definability" with respect to a given theory T is not itself (object-theoretically) expressible by a predicate in the language of T.

When one tries to localise the problem in a particular formal theory T, rather than "English", the following definitions will also turn out to be helpful.
3.19 a closed term t in L(T) is nameable or definable in T if and only if for some property $s(x)$ in L(T) that does not occur in t, [there exists an x such that $s(x)$ and for all z $(s(z) \Rightarrow (z = t))$] is provable in T, and
3.20 an n-ary relation symbol p in L(T) for n ≥ 1 is expressible in T if and only if for some formula $q(x_1,...,x_n)$ of L(T) in which p does not occur, [for all $x_1,..., x_n$ $(p(x_1,...,x_n)$ if and only if $q(x_1,...,x_n)$] is provable in T.

Most analytically trained philosophers will see in *Principles* 31 a formal recapitulation of the underlying idea behind Bertrand Russell's classical study of "definite descriptions", "the present king of France", *et al*, and *Principles* 32 is sometimes called "explicit definability" in expositions of Beth's Theorem (cf., e. g., (Bell and Machover, 1977), 455–456).

As I suggested above, one way to clarify Berry's paradox is to interpret it as evidence that Principles 36 and 37 are both liminal notions, in the sense that they lie outside the "threshhold" or "horizon" of T—or perhaps contextual, in that their "sense" requires recurrent appeals to metatheoretic contexts that are underdetermined by T.

Formalisation of Berry's implicit argument, in any event, leads to the following ("folklore"-)
3.21 **Metatheorem (Berry et al.)**

Working in an appropriate metatheory M, let a consistent first-order theory T interpret enough formal arithmetic to encode its syntax and proof-procedures as in 3.1–3.6, and extend T consistently to a new and presumably consistent theory T* by adding

a new binary predicate d(u,v)

(whose "intended" interpretation is that "u codes a definite description of v"), and

a scheme of new axioms (one for each unary formula s(x) in L(T)),

(which asserts, *seriatim*, once for each predicate s in the language of T) that

[for all u and all v

(d(u,v) if and only if (u is a numerical code and v is a number), and

for all y, d([s(x)],y) if and only if (y is the unique x such that s(x))].

Then the binary predicate d in the language of T* is inexpressible by any formula in L(T).

A metatheoretic proof of this assertion can be sketched as follows. Working in the metatheory U, in which we assume T and T* are consistent, we observe that if there were a formula *def*(u,v) in the language L(T) of T which expressed d, one could import Berry's informal argument into T, and reconstruct Berry's paradox as an object-theoretic contradiction in T. Since T and T* are consistent, we conclude in U that no such binary formula *def*(u,v) expresses T*'s predicate-symbol d in L(T).

Might some sort of encodable variant of metatheoretic undefinability permit us to "encode" the Berry paradox, and obtain a specific counterpart of Gödel's undecidable sentence y equivalent to [the consistency assertion for T]?

The answer is "yes". Formalisation of the metatheoretic proof just sketched yields an explicit counterpart δ of y which is provably equivalent to [[its own] inconsistency] in T, and therefore to y.

Gödel could, in other words, have employed the Berry- rather than the liar-paradox, and achieved essentially the same result.

To see this, observe first

3.22 that the liar paradox ("[This sentence] is false"), by the diagonal lemma (cf., e.g., (Boos, 1998), 2.5.1–2, 61) would yield a contradiction in T (of the sort we have been considering), if "truth"-for-T were expressible in T; and

3.23 that Gödel obtained the sentence y which is equivalent to [T is consistent] by diagonalising [unprovability], which *is* expressible in T, rather than "truth", which is not.

Varying this plan slightly, one can argue as follows.

3.24 **Metatheorem** (cf. (Boos, 1998))

Let T be a first-order theory which encodes [its own syntax] in plausibly normalised ways (cf. (Smorynski, 1977), 827), and let δ in L(T) be an appropriate unabbreviation—(cf. 3.12 of (Boos, 1998)) of the following sentence:

3.24.1 [it is unprovable in T that [metatheoretic pointwise definability is expressible in T]].

Then

3.24.2 δ is equivalent to the Gödel-sentence [T is consistent] in T.

By Gödel's original argument, therefore, δ is undecidable in T.

Sketch of Proof.

Observe first that δ provably implies [T is consistent] in T (as does any sentence of the form [[a given sentence] is unprovable].

For the converse implication—that [δ is provable] ⇒ [Con(T) is provable] in T—one uses the antecedent to "internalise" the proof of 3.21 in T, as [the proof of 3.18 in T]. The code for [this argument] will then witness the [provability of [a contradiction] in T].

Readers familiar with the contents of (Smorynski, 1977) or (Boolos, 1989) (e.g.) will quickly recognise that the rather straightforward idea of this sketch is to obtain Gödel's First Incompleteness Theorem from a formalisation of 3.21, in much the same way as Gödel formalised the First Incompleteness Theorem to obtain the Second.

To see the potential relevance of these metalogical complexities for the quasi-Berkeleyan dictum that "everything must be definable", consider again one of Skolem's metatheories U in which T is (provably) consistent. In such metatheories, one can indeed define interpretations of T in which everything is definable, as I mentioned above. But neither T nor U can express Berkeley's dictum for itself, by 3.21.

Moreover, in a minimal finite extension U = T* of T in which one can express [definability in T], the [unprovability in *T*] of the claim that [[this notion of definability] is expressible in T] is equivalent to the assertion that [*T* is interpretable].

In plainer language: it follows metatheoretically from 3.21 that

3.25 T cannot express its own powers of individuation.

If one construes "unprovability" as a form of "underdetermination", moreover, it follows from 3.24 that

3.26 [Underdetermination of the range of [*T's powers of individuation*]] is equivalent in T to [T's best effort to formulate its own notion of "conceivability" (or interpretability, or "existence")],

and this in turn—by Gödel's theorem in its original formulation—to [underdetermination of [T's best effort to formulate its own notion of "conceivability" (or interpretability, or "existence")]].

More provocatively, 3.25 and 3.26 suggest that

3.27 Rational trust in *one's own* existence is equivalent to acknowledgment that one cannot define or determine *one's own* perceptual and conceptual limitations.

Most provocatively of all (by contraposition),

3.28 A kind of metalogical pyrrhonism would seem to follow from claims to define or determine *one's own* perceptual and conceptual limitations. For such claims would imply (and be implied by) the assertion(s) that [their premises make no sense].

These glosses, and the antecedent interrelations sketched above in 3.13 and 3.21,

3.29 the metatheoretic unprovability of object-theoretically encoded consistency and

3.30 the object-theoretic inexpressibility/indefinability of metatheoretic definition (of Russell's "definite descriptions"),

also seem to me to bear out two of Berkeley's fundamental insights:

3.31 that "existence" is a relational notion (relative, in his case, to the intentional activity of "spirits"); and

3.32 that "spirits" have notions but not ideas of *their own* "conception", "perception" and other (characteristic) modes of ideation.

Indeed, the former is prompted by a reading of 3.11–3.12, according to which one cannot "really" define metatheoretic consistency of T as [consistency of *T*] in T, since the two can come into conflict. And the latter seems to follow from 3.21 and 3.24.

For the semantic paradoxes Berkeley's thought experiment anticipated suggest that "existence" varies with metatheoretic contexts in ways one might call "intentional", or at least "observational", to the extent the notion of "existence" can be discursively clarified at all.

Detached from such local metatheoretic contexts, moreover, "existence"-assertions—like claims about "satisfiability" and "truth"—may not even be well-defined. The characters I have variously called Berkeley-the-intentionalist, Berkeley-the-dialectician and Berkeley-the-semeiotician seem therefore to have been well-served by the master argument.

Less so Berkeley-the-fideist—and more particularly, Berkeley-the-believer-in-god's-ultimately-unitary-design. For metalogical counterparts of Berkeley-the-dialecticians's arguments suggest that conceivability, interpretability, perceivability and what might be called *discernibilty* are also relational, in complexly interrelated ways.

The interrelations and thought-experiments Berkeley explored—and the semantic paradoxes that gave them metalogical form—do, after all, seem to point to recurrently liminal, contextual and underdetermined aspects of discursive attempts to express all these notions. They therefore suggest that expressibility, communicability and intersubjective uniformity of our "ideas" pose persistent underlying problems for the more dogmatic aims of Berkeley's "immaterialist" ontology.

Metatheoretic "existence" / conceivability / interpretability, for example, does not imply its object-theoretic counterpart(s), and similar observations may be made about "individuation" / "definability" / "discernibility". For metatheoretic interpretability (consistency) of the object-theoretic data {T, s and [s is uninterpretable in *T*]}—a metalogical reading of the "absurd" thought-experimental predicament of Berkeley's *reductio*—is not absurd. Indeed, Gödel showed that it is a theorem for theories T in which one can encode enough induction to write sentences such as [s is uninterpretable in T].

In a sense, therefore, the deeper tenets of Berkeley's intentionalism may have been correct, if one consolidates them in an agnostic portmanteau-acknowledgment that "everything" might "eventually" be (relatively)

3.33 conceivable (metatheoretically interpretable),
3.34 perceivable (metatheoretically definable), and/or
3.35 communicable (via "signs") (metatheoretically expressible).

But one could not appeal in any consequential way to 3.29–3.30's patterns of indefinitely ramified metatheoretic ascent (cf. 2.14) to impose a theist cap on them ("Is this fair dealing?"). For such finitism / definabilism / contextualism, on the analyses of 3.11, 3.21 and 3.24, would be as liminal as what it analysed, and could not univocally or unitarily express, much less interpret *itself*.

If these assimilations are tenable, then, Berkeley may have identified and begun to explore natural metaphysical archetypes of a deep and recurrent question in twentieth-century metamathematics: how does one interpret such paradoxical—or at least liminal—"definitions"?

If one construes "the inexpressibility of 'definability'", for example, as evidence for the theory-relativity of "definition" (cf. Section 2), might one not credit Berkeley (or at least the figure I have called "Berkeley-the-dialectician") with insight into the theory-relativity ("subjectivity"?) of perception?

Similarly, if one construes "the unprovability of consistency" as evidence for the theory-relativity of "interpretation" (cf. Section 3), might one not credit Berkeley with insight into the theory-relativity ("subjectivity"?) of conception?

Finally, might one not construe the recurrent metatheoretic consistency of extensions of the form $T^* = T + s^*$ (cf. 3.11) as a metalogical hermeneutic for "normal" (non-dogmatic-Berkeleyan) intuitions about "nonconstructive" existence-assertions, and use the arguments of Berkeley-the-semeiotic-dialectician to confront Berkeley-the-reductive-dogmatist?

Be that as it may, these metalogical assimilations do seem to me to suggest, at the very least,

3.36 that interpretations of "pathological" (?) theories of the form S + [S is inconsistent], for finite extensions S of T (which are modeled by "most" interpretations of T; cf. again (Boos, 1994), 7.9), may in fact provide natural metalogical counterparts for our "normal" (non-Berkeleyan) intuitions about "nonconstructive" existence assertions.

3.37 that Berkeley the stringent critic of mathematical infinity touched on deeply aporetic questions of metaphysics that reemerged in the "miniature" underdetermination and indeterminacy results of twentieth-century metamathematics.

Such anachronistic assimilations might also become more plausible if one recalls the force with which Berkeley defended his maxim (set forth in passages reviewed below) that everything that is consistent (not "repugnant" or "absurd") must not only be conceivable, but conceivable in feasibly finite time, and conversely. In the next section, I will argue that such maxims amounted to higher-order *petitiones* of the basic dogmatic-"immaterialist" tenets at issue in his work.

In any event, the metalogical concomitants of semantic paradoxes adduced in this chapter are theorems, once again, not "absurdities". They are also natural consequences of metalogical readings of Berkeley's intentionalism, theory-relativism and contextualism in hierarchies of metatheoretic ascent. But they do not yield the reductio-arguments Berkeley anticipated in Principles 23 and D 200, exactly because conceivability, interpretability and definability are stratified and intentional, as he himself had claimed.

On this account, then, Berkeley's "master argument" may have been just as deep as he thought it was, and its implications more complex and far-reaching than he anticipated.

4 Semeiotic "Coherentism"?

In what senses was Berkeley a scientific or metaphysical coherentist? Might he also have been a *sceptique empeché* (*ou du moins un fidéiste*), *malgré lui*? Should he have been?

His fellow empiricist David Hume seems to have thought so, if one may judge by his dismissive remark that

> Berkeley's arguments admit of no answer and produce no conviction. Their only effect is to cause that momentary amazement and irresolution and confusion which is the result of skepticism. (*E* I, 155)

Berkeley himself, by contrast, vigorously rejected "forlorn skepticism" as a form of epistemic despair:

> Prejudices and errors of sense do from all parts discover themselves to our view; and endeavouring to correct these by reason we are insensibly drawn into uncouth paradoxes, difficulties, and inconsistences, which multiply and grow upon us as we advance in speculation; till at length, having wander'd through many intricate mazes, we find ourselves just where we were, or, which is worse, sit down in a forlorn scepticism. (*Principles*, 1)

At times, however, Berkeley seemed to seek refuge from such despondency in dogmatic phenomenalism (a *fuite en avant* which arguably anticipated Hume's "skeptical solution of these doubts"; cf. *E* I, 44 ff).

In elenctic passages such as the following, for example, Berkeley vigorously claimed to embrace "particular" and "immediate" "phenomenal" entities-of-"sense" ("ideas"), and proscribe indefinite and mediate "noumenal" entities-of-thought ("abstract ideas", "ideas of reflexion"):

> ... assert the evidence of sense as high as you please, we are willing to do the same. That what I see, hear and feel doth exist, that is to say, is perceived by me, I no more doubt than I do of my own being. *But I do not see how the testimony of sense can be alleged as a proof for the existence of anything, which is not perceived by sense.* We are not for having any man turn skeptic, and disbelieve his senses; on the contrary, we give them all the stress and assurance imaginable; nor are there any principles more opposite to scepticism, than those we have laid down .... (*Principles*, 40 (emphasis mine)

As a refutation of skepticism, however, such "principles" might indeed be the *fuite en avant* mentioned above. For the italicized sentence essentially recapitulated a classical *pyrrhonist* claim—that one cannot infer "inevident", "noumenal" conclusions from "evident", "phenomenal" premises. Sextus Empiricus tried to interpret such assertions in a self-referentially "undogmatic" manner (*adoxástos*). But more doctrinaire "phenomenalists" (Hume, for example) did not, and Hume may have had such comparisons in mind.

In this section, I will argue that the metalogical analogies offered earlier help clarify which aspects of Berkeley's arguments supported Hume's dismissive remark (or oblique praise, or whatever it was), and which did not. This may have a certain historical or at least hermeneutic relevance, since Berkeley's keen elenctic abilities led him to develop dialectic arguments that implicitly clashed with his ultimate ontological aims.

To personify the point, I will try to employ these analogies to contrast the positions of two distinct characters, whose shadow outlines—anticipated earlier—I will call, respectively "Berkeley-the-(dialectical)-semeiotician" and "Berkeley-the-(dogmatic)-ontologist".

I have already argued that "Berkeley-the-semeiotician" employed prototypical insights into (analogues of) completeness-results and semantic paradoxes to argue, in effect,

4.1   that "existence" (of "signs", at least) is equivalent to metatheoretic determination of interpretations for those signs (or metatheoretic postulation of such interpretations, if the signs are mere "notions").

If Berkeley then defined

4.2   a "spirit" as a source of such metatheoretic agency,
      it would indeed follow that
4.3   "existence" is an inherently relational notion, and requires the metatheoretic activity of such "spirits" to sustain it.

It would seem to follow from this line of argument

4.4   that attempts by members of a given class of "spirits" to decide claims of "existence" witnessed by metatheoretic interpretations beyond their expressive power would be "... words to no manner of purpose, without any design or signification whatsoever" (cf. *Dialogues* II, 233).

It would also seem to follow, however, that

4.5   No members of a given class of "spirits" can conclusively deny that such claims may (or may not) be decided in indefinitely ramified hierarchies of such metatheoretic venues.

More than most metaphysicians—more than Hume, for example, except perhaps in the most searching passages of the *Dialogues Concerning Natural Religion*—Berkeley was well aware that there is a problem here. The point becomes conspicuous in the latter part of *Dialogue* II and beginning of *Dialogue* III, in which Hylas (literally, "matter-son") makes a number of well-argued points—more, arguably, than any other dialogical opponent in the history of European philosophy.

Coincidentally, Hylas' queries also appear in passages—which have no counterpart in the *Principles*—in which he derives skeptical conclusions from the apparent relativism of Philonous' prior arguments ("Our faculties are too narrow and too few", *D* III 227), and calls into question the methodological consistency of his author's conclusion.

Finally, several of these metacritical remarks would do honor, I think, to that *venerabilis cautor et interpres*, Umberto Eco's fictional "William of Baskerville". The best known are perhaps the following.

> [Hylas] We have therefore no idea of any spirit. You admit nevertheless that there is spiritual substance, though you have no idea of it; while you deny that there can be such a thing as material substance, because you have no notion or idea of it. Is this fair dealing? (*Dialogue* III, 232)

> [Hylas] Since therefore you have no idea of the mind of God, how can you conceive it possible that things should exist in his mind? Or, if you can conceive the mind of God without any idea of it, why may not I be allowed to conceive the existence of matter, notwithstanding that I have no idea of it? (*Dialogue* III, 231)

A few pages earlier, Philonous had provided an opening for these remarks at the end of *Dialogue* II, in passages which may bear on debates about the relevance of Gödelian arguments in the philosophy of mathematics, as well as the metalogical reconstructions introduced in earlier sections:

> [Philonous] ... can any more be required to prove the *absolute impossibility* of a thing, than the proving it *impossible in every particular sense, that either you or I understands it in*?

> [Hylas] But I am not so thoroughly satisfied that you have proved the impossiblity of matter in the ... most ... indefinite sense.

> [Philonous] When is a thing shewn to be *impossible*?

> [Hylas] When a repugnancy is demonstrated between the ideas comprehended in its definition.

> [Philonous] But *when there are no ideas, there no repugnancy can be demonstrated* between the ideas (*Dialogue* II 225) (italics mine)

In effect, Hylas ventured here to assert that consistency assertions that are nonconstructive may nevertheless be (metatheoretically) tenable. To this, Philonous then responded that Hylas'

4.6 failure to sustain any of the four or five interpretations of "matter" he has proposed (rendering it "... impossible in every particular sense, that you or anyone else understands it in...").

would suffice one to conclude that

4.7 "matter" is uninterpretable ("absolutely impossible"), and therefore inconsistent ("repugnant") (cf. *Dialogue* II 225).

The formal fallacy of such an "inference", on reflection, is rather obvious (only "four or five interpretations"). But the contextual power of such arguments is pragmatically undeniable. In practice, of course, one often adduces (dubious) metatheoretic arguments about "burdens of proof" in an effort to settle the disputes that emerge. But such appeals notoriously lead to little genuine conciliation of views.

Indeed, there may be an "inductive", even "soritical" undertow here. It would be fatuous, for example, to claim that I have no conceptual limitations. But it would be equally hubristic to claim that I can "determine"—or even conceive in precise terms—what they are. Might it not be comparably impossible for a collegium of observers ("... you or anyone else ....") to conceive precisely or determine what their (and presumably our) conceptual limitations are?

On the one hand, such reflections tend to confirm—once again—Berkeley's insight that "spirits" cannot have ideas of themselves or other spirits (though he makes oddly Cartesian claims of "immediate" or "intuitive" notions of oneself at D III 231). On the other, they also suggest that ineliminable penumbrae of nonconstructive concept formation may attend any serious epistemological or metaphysical inquiry, and that Hylas' "fair dealing"-objection is indeed methodologically (and metalogically) apt.

Consider, for example, the following brief remark—thoroughly plausible on pragmatic grounds, or as a warrant for something like the "certainty" Wittgenstein later tried again and again to characterise in *Über Gewissheit* (cf., e.g., 115, 125, 163, 337, 341, 546, 638).

> It is to me a sufficient reason not to believe the existence of anything, if I see no reason for believing it. (*Dialogue* II 218)

Perhaps. But notice what happens when one introduces italics into this sentence ("It is *to me* ..., if *I* see ...."). A number of mildly notorious remarks in *Über Gewissheit* and the *Philosophischen Untersuchungen* turn on such shifts. They arguably mark shifts to metatheoretic points of view—provisional or hypothetical, in my case—of *me*. And they strongly suggest (at least to *me*)

4.8   that metatheoretically constructive arguments point to the inadequacy of object-theoretically constructive arguments to decide certain implicitly self-referential issues.

And this, in turn—another echo of pyrrhonist "lapse into infinite(s)" (*ekptosis eis apeiron*)—suggests a role for skeptical *epoche* (suspension of judgment) after all. Or rather it suggests

4.9   that "I" (or "we") should suspend judgment about (metatheoretic) demands for (object-theoretic) constructivity. For cognate demands for "philosophical rigor" may cut both ways (cf. also Wittgenstein's wryly "polite" gloss in

(ÜG 467), that *"Dieser Mensch ist nicht verrückt. Wir philosophieren nur."* By way of side-comment, a rather interesting and very conspicuous Wittgensteinian "master argument" also appears at (ÜG 595)).

More particularly, "dialectical Berkeley" might have had sufficient reason to suspend judgment about the "existence" of either "the" structure of "the" physical universe or "god", on rather similar grounds. One might gloss the dichotomy between "dialectical" and "dogmatic" Berkeley, therefore, along the following rather straightforward metalogical lines.

"Dialectical", "skeptical" or "elenctic" Berkeley would claim only to demonstrate (or have demonstrated) that "immaterialism" itself is a coherent and defensible world-view—that

4.10 "immaterialism" is consistent (not "absurd", "repugnant", *et al.*).

"Dogmatic" or "deist" Berkeley, by contrast, would claim to demonstrate (or have demonstrated) that "materialism" is, roughly speaking, an incoherent and *in*defensible world-view—that

4.11 "materialism" is inconsistent ("absurd", "repugnant", et al.).

Even that *subtilis doctor* James Boswell seems to have recognised that "dialectical Berkeley" may indeed be irrefutable (Boswell, I, 292). Most readers of the *Principles* and *Dialogues*, on the other hand, believe that "dogmatic Berkeley" failed to prove his case.

Why? Well, Hylas' "fair dealing"-objection tracks the response of many cogent readers. How does Philonous respond to it? More particularly, what is the "case" Berkeley undertook to prove?

Attempts to answer these questions may clarify some recurrent sources of strain in Berkeley's system. For the character I have called "Berkeley-the-(dogmatic)-ontologist"—who most definitely did not aim to "sit down a forlorn skeptic"—had (at least) three additional ontological aims.

The first was

4.12 to leave no penumbra of dialectically underdetermined metatheoretic reinterpretations at the margin of his (implicitly fixed) class of "spirits" and their "perceptions"—no counterparts, in particular, of 2.1–2.5's metatheoretic ascent.

The second was

4.13 to provide a deist rationale for a divinely "perceived" system of determinate interrelations between lesser spirits' perceptions.

And the third,

4.14 to integrate this structure of systematic interrelations—comparable in many respects to Leibniz' system of *phaenomena bene fundata* (cf. Leibniz's thoughtful remarks in the flyleaf of his copy of the *Principles*, reproduced in (MacIntosh, 1971), translated in (Leibniz, 1989), 307, and glossed rather dismissively in (Wilson, 1987)—into an ontology adequate to any hypothetical reconstructions his corpuscularian opponents might devise.

The first desideratum above—4.10—was especially difficult to satisfy, since arguments to perceptual and (by implication at least) conceptual relativity formed such an integral part of his dialectical deconstruction, in *Dialogue* I and the *Introduction* and early sections of the *Principles*, of the corpuscularian scientific orthodoxy of his time. The thought-experimental counterparts of semantic paradoxes I have sketched aided him in this deconstructive project, but they also tended to eventuate, as remarked earlier, in something very like informal skeptical counterparts of Section 2's "regressive" patterns of metatheoretical ascent. I also remarked that Berkeley's thought experiments suggest we are endowed with (and "condemned" to employ) a capacity for nonconstructive concept-formation. Hylas, ironically, comes closest to the essential point in dispute when he formulates the following objection at the end of *Dialogue* II:

> [Hylas] Be that as it will, yet I still insist upon it, that our not being able to conceive a thing, is no argument against its existence.

Berkeley was capable of great rhetorical concision, but Philonous' response below to Hylas' query is uncharacteristically diffuse and (literally) "long-winded"—a claim the reader might verify by reading the italicised sentence aloud. Readers who already agree with Berkeley might find it a rhetorically effective peroration all the same. Those who do not might be forgiven if they see in it a "refutation" *a baculo*.

> [Philonous] That from a cause, effect, operation, sign or other circumstance, there may reasonably be inferred the existence of a thing not immediately perceived, and that it were absurd for any man to argue against the existence of that thing, from his having any direct or positive notion of it, I freely own. *But where there is nothing of all this; where neither reason nor revelation induce us to believe the existence of a thing; where we have not even a relative notion of it; where an abstraction is made from perceiving and being perceived, from spirit and idea: lastly, where there is not so much the most inadequate or faint idea pretended to: I will not indeed thence conclude against the reality of any notion or existence of a thing: but my inference shall be, that you mean nothing at all: that you employ words to no manner of purpose, without any design or signification whatsoever.* (*Dialogue* II 223)

If Berkeley is indeed pounding the table here to a degree that was unusual for him (a somewhat less attractive way in which a philosopher-rhetorician might "think with the learned and speak with the vulgar"), the passage may bear closer scrutiny. In aid of Hylas, as it were, I will argue

4.15 that Berkeley's remarks begged, in effect, a metatheoretic / higher-order counterpart of the basic "immaterialist" / "intentionalist" tenet at issue in his work; further,

4.16 that a cognate *petitio* underlay his presupposition that the hypothetical conclusion(s) of his master argument(s) were absurd, rather than paradoxical-but-not-inconsistent manifestations of semeiotic contextuality and underdermination (patterns he himself had observed in his *Introduction* to the *Principles*); and finally,

4.17 that the metalogical analogies I have outlined may indeed clarify the structure of this *petitio* in philosophically relevant ways (relevant, in particular, to 4.12–4.14).

The "higher-order counterpart of the basic 'immaterialist'… tenet at issue", once again, is simply the claim Hylas queries—that nonconstructive existence-assertions are inherently illegitimate.

Perhaps the culmination of Berkeley's rhetorical excursus in D II 233 should not, after all, be that we "employ words to no manner of purpose, without any design or signification whatsoever". Perhaps it should simply be that we employ them nonconstructively, in ways that we acknowledge have more than one potential signification—mindful, perhaps, of Berkeley's warning, in the *Introduction* to the *Principles* (P I 23), that we should not expect "the immediate signification of every general name" to be "a *determinate … idea*"—"abstract" or otherwise (the italics are Berkeley's). In doing so, we might also acknowledge—plausibly enough, given that we are finite sentient beings—that "our" conceptual framework is (inevitably) incomplete, no matter who "we" are (cf., again, 4.8–4.9)—an insight Berkeley's own dialectical and thought-experimental explorations of the implications of perceptual and conceptual relativity anticipated.

It has often been suggested that Kant's transcendental idealism eventuated in a "two truths"-doctrine—one "speculative", and the other "practical". In roughly comparable ways, one might argue that Berkeley's metaphysics and philosophy of science come in two variations—one more "skeptical" (or "dialectical", or "semeiotic") and the other "dogmatic" (or "analytic", or perhaps even proto-"transcendental").

Texts in which Berkeley rather clearly seemed to express more skeptical or semeiotic views include the following:

> The connexion of ideas does not imply the relation of *cause* and *effect*, but only of a mark or *sign* with the thing *signified*. (*P* 65)

> Hence it is evident, that those things which under the notion of a cause co-operating or concurring to the production of effects, are altogether inexplicable ... have a proper and obvious use, ... when they are considered only as marks or signs for our information. (*P* 66) (cf. also *Dialogue* III, 241–243)

> Those men who frame general rules from the *phenomena* and afterwards derive the *phenomena* from those rules, seem to consider signs rather than causes (*P* 106).

In a later passage in the *Principles*, he even suggested, in pointedly "Humean" terms, that such scientific "marks" and "signs" may have to be merely "notional".

> It is also to be remarked, that all relations including an act of the mind, *we cannot so properly be said to have an idea, but rather a notion of the relations and habilitudes between things*. (*P* 142) (italics mine)

In still other passages, however, Berkeley warmly endorsed—in rather conspicuously un-"phenomenalist" and un-"operational" terms—the counterfactual-supporting attractions of scientific "notions" and "general ideas", and he often shifted his focus in these contexts to the (alleged) "uniformness" of signs' "analogies, harmonies and agreements".

> If therefore we consider the difference that exists between natural philosophers and other men, with regard to their knowledge of the phenomena, we shall find it consists, not in an exacter knowledge of the efficient cause that produces them, for that can be no other than the will of a spirit, but only in a greater largeness of comprehension, whereby analogies, harmonies and agreements are discovered in the works of Nature, and the particular effects explained, that is reduced to general rules ... which rules are grounded in the analogy, and uniformness observed in the production of natural effects, are most agreeable, and sought after by the mind; for that they *extend our prospect beyond what is present, and near to us, and enable us to make very probable conjectures touching things that may have happened at very great distances of time and place, as well as predict things to come; which sort of endeavour towards omniscience is much affected by the mind*. (*P* 105) (italics mine)

> There are certain general laws ... learned by the study and observation of Nature, and ... applied ... to the explaining the various *phenomena* ... which explication consists only in discovering the *uniformity* there is in the production of natural effects. (*P* 62)

> ... the ideas of sense ... have a steadiness, order, and coherence, and are not excited at random ... but in a regular train or series, the admirable connexion whereof sufficiently testifies the wisdom and benevolence of its Author. Now the set rules or established methods, wherein the mind we depend on excites in us the ideas of sense, are called the Laws of Nature .... (*P* 30)

> This gives us a sort of foresight .... And without this we should be eternally at a loss .... all this we know, *not by discovering any necessary connexion between our ideas*, but only by observation of the settled laws of Nature .... (*Principles* 31; italics mine)
>
> The ideas imprinted by the Author of Nature are called *real things*. (P 33)

Perhaps Berkeley's coherentism contributed an oblique influence to Hume's basic insights into the conceptual underdetermination of "necessary connexion". If so, it might have been natural for an agnostic with a grudging and ambivalent view of "animal *nisus*" (cf. the footnote to *E* I51) to posit a new source of intersubjective uniformity as a replacement for the "volition(s)" of Berkeley's great spirit.

Like the Humean observation in *P* 142, cited above, however, the italicised disclaimer of insight into "necessary connexions" flowed in Berkeley's case from his quasi-animist (or perhaps quasi-monadic) view that genuinely "causal" forces must emanate from the "wills" of "spirits". In this connection, it may also be relevant to recall that a "rationalist" counterpart of Berkeley's prescient observation had already appeared in Leibniz' preface to the *Nouveaux Essais sur l'Entendement Humaine*, which also provided a plausible exemplar for Hume's best-known "matter of fact" (cf. *E* I21).

> Or tous les exemples qui confirment une verité generale de quelque nombre qu'ils soyent, ne suffisent pas pour établir la necessité universelle de cette même verité, car il ne suit point, que ce qui est arrivé, arrivera de même. Par exemple ... tous les ... peuples de la terre connue aux anciens, ont tousjours remarqué qu'avantlte decours de 24 heures, le jour se change en nuit, et la nuit en jour. Mais ... celuy ... se tromperoit encor qui croiroit, que ... c'est une verité necessaire et eternelle qui durera tousjours, puisqu'on doit juger que la terre et le soleil même n'existent pas necessairement, et qu'il y aura peutestre un temps où ce bel astre ne sera plus, au moins dans la présente forme, ny tout son systeme.

> But all the examples which confirm a general truth, however many they may be, do not suffice to establish the universal necessity of that same truth, for it does not follow, that what has happened will happen again. For example .... all the ... peoples of the world known to the ancients always observed that before the passage of twenty-four hours, day changes to night and night to day. But ... anyone would still be deceived who believed that ... this is a necessary and eternal truth which will last forever, since one must judge that the earth and even the sun do not exist necessarily, and that there will perhaps be a time when this beautiful star will no longer be, at least in its present form, or any of its system.
>
> (Leibniz, G V, 43)

These reminders and historical cross-references suggest rather strongly that all three of these authors were well aware of a common *Problematik*—one which persists in late twentieth and twenty-first century philosophy of science.

A "Berkeleyan" / "Leibnizian" formulation of this common problem—arguably the basic begged question of Berkeley's "system" (cf. 4.14)—might be: "How 'uniform' and '*bene fundata*' are the 'phenomena' of *P* 62 and *P* 106, quoted above?"

In particular, an assumption that

4.18   what is nonconstructive to one is nonconstructive to all,

seems essential to the claims of intersubjective uniformity Berkeley wished in the end to uphold. How could it possibly be justified "from within"?

In his "dogmatic phenomenalist" mode, Berkeley "proved" 4.18 in practice from a tacit premise that

4.19   an ultimate *discriminatio* discerns "actual" fact, from "possible" counterfact, in the (ostensibly plural and emergent) *medias res* of "spiritual" activities.

In his system, in turn, 4.19 is equivalent to an assertion that

4.20   an "ultimate" *discriminatio* discerns factual from counterfactual ideation and concept formation.

Only "god", of course, in Berkeley's system—operating in effect as a kind of final metatheory or Form of the Good—could distinguish a unique "factual" path through potential patterns of alternative interpretation and/or metatheoretic ascent, and ensure intersubjective uniformity of the mental activities of disparate "spirits". Accordingly, Berkeley's attempts to describe the workings of this divine *ratio* (from "*reri*", which meant to "judge", "deem" or "think"—an ontology Berkeley would surely have endorsed) also suggested (with Leibniz) that the principal role of 1.14–1.17's "night-watchman-god" would be to sort factual from counterfactual, in ways our "signs" might retrace and "support"—ensure, in short, that "the tree" of Russell's parody "will continue to be".

He acknowledged that such a normalising or uniformising *discrimen* would have to lie outside the realm(s) of determinate ideation, but hoped it might lie within the discursive structure(s) of more vaguely hermeneutic "notions". Appropriately enough, he also resorted to subjunctive conditionals of an elusively "notional" sort when he needed to relax his near-operationalist demands for immediate perception, and settle for counterfactual thought experiments:

> The question whether the earth moves or no, amounts in reality to no more than this, to wit, whether we have reason to conclude from what hath been observed by astronomers, that *if we were* placed in such and such circumstances, and such and such a position or distance, both from the earth and sun, *we would perceive* the former to move among the choir of the planets, ... *and this, by the established rules of Nature, which we have no reason to mistrust, is reasonably collected from the phenomena*. (*P* 58; italics mine)

In more obviously liminal contexts, such thought-experimental accounts segued directly into certain sorts of "occasions"—in this case, for the normalising powers of "god"'s observation.

> ... I imagine that *if I had been present* at the Creation, I should have seen things produced into being, that is, become perceptible, in the order described ....
>
> But, Philonous, you do not seem to be aware, that you allow created things in the beginning, only a relative, and consequently hypothetical being; that is to say, upon supposition there were men to perceive them, ...
>
> In answer to that, I say, first, created beings might begin to exist in the mind of other created intelligences, besides men .... I say farther, in case we conceive the Creation ... as ... a parcel ... produced by an invisible power ... where nobody was present: that this way of conceiving accords with my principles .... (*D* III 251–252); italics mine)

(It seems an interesting thought experiment, by the way, to try to discern a difference between attempts to conceive inconceivable objects and Berkeley"s "imagination" of a subjunctive conditional whose premise is that he might have "been present at the creation.")

Since Berkeley had very little tolerance for Hylas' appeals to "occasions" in other, less explicitly empyrean contexts, it might be worth a brief pause to reconsider the word's Latin origins. Long before medieval scholastics and Cartesians began to employ the word, an "*occasio*" in Latin was something dispositional or emergent—something that might "fall to" one ("*occidere*")—and its usage often approximated our use of the word "opportunity" (which derives from Latin words for something that might be "brought to" one).

Berkeley notoriously derided Hylas' resort to such an elusive and perhaps indefinable notion, and his dialectical analysis of such "signs" independently suggested that semeiotic concept formation which supported such "counterfactuals" might have to be "relative, and consequently hypothetical" indeed. Yet he warmly praised, above and in other passages (cf. *D* II 189) the ineffable *ratio* of a spirit which so disposed and aligned lesser spirits' potential ideas that they "fell"—in effect—into place.

Consider, for example, quasi-hieratic passages such as the following, which cited prototypical examples of Kantian *Erhabenheit* and *Teleologie* to invoke a personal counterpart of *KdprV*'s *bestirntem Himmel uber* [uns], *und ... moralischen Gesetz in* [uns].

> At the prospect of the wide and deep ocean, or of some huge mountain whose top is lost in the clouds, are not our minds filled with a pleasing horror? ... How aptly are the elements disposed? ... How exquisitely are all things suited, as well to their particular ends, as to

constitute apposite parts of the whole? ... The motion and situation of the planets, are they not admirable for use and order? ... Do they not measure areas around the sun ever proportional to the times? So fixed, so immutable are the laws by which the unseen Author of nature actuates the universe .... The feeble narrow sense cannot descry innumerable worlds revolving around the central fires; and in those worlds the energy of an all-perfect mind displayed in endless worlds .... Though the labouring mind exert and strain each power to its utmost reach, there still stands out a *surplusage immeasurable* .... Is not the whole system immense, beautiful, glorious *beyond expression and beyond thought*! what treatment then do those philosophers deserve, who would deprive these noble and delightful scenes of all reality? ... can you expect this scepticism of yours will not be thought extravagantly absurd by all men of sense? (*D* II 211; italics mine)

(Interestingly enough, Hylas' immediate response to this rhapsodic passage focuses unerringly on its essentially fideist ground-base: "My comfort is, you are as much a skeptic as I am.")

Disputes about alleged phenomenal "uniformness" beyond semantic and conceptual horizons had always been integral, of course, to theodicies, but also to theological and emerging scientific and methodological debates about "design"—later reviewed and refined by "Philo" and "Cleanthes" in Hume's great set-piece, *Dialogues Concerning Natural Religion*.

There is, also, a sense in which the tension between the figures I have called Berkeley-the-skeptical-dialectician and Berkeley-the-dogmatic-fideist may have anticipated cognate tensions between Hume-as-Cleanthes, the proponent of "design", and Hume-as-Philo, the quasi-pyrrhonist ("you are as much a skeptic as I am").

On the one hand, Hume-the-skeptical-pluralist acknowledged that

Nothing, at first view, may seem more unbounded than the thought of man, which not only escapes all human power and authority, but is not even restrained within the limits of nature and reality .... while the body is confined to one planet, along which it creeps with pain and difficulty; the thought can in an instant transport us into the most distant regions of the universe; or even beyond the universe, into the unbounded chaos, where nature is supposed to lie in total confusion. What never was seen, or heard of, may yet be conceived; nor is any thing beyond the power of thought, except what implies an absolute contradiction. (*E* I13)

On the other, Hume-the-"Newton of the Moral Sciences" aspired to encompass "all" (?) this liminal concept formation in a simplistic corpuscularian psychology no more empirically adequate than Berkeley's theological coherentism, and conflated indicative with counterfactual usages in mildly notorious passages such as the following:

.... we may define a cause to be *an object followed by another, and where all the objects similar to the first are followed by objects similar to the second*. Or in other words, *where, if the first object never had been, the second never had existed*. (P I60; italics Hume's)

Such ambiguities, in turn, raise other historical and quasi-historical questions.

Was Berkeley, for example, a "speculative" semeiotician (to borrow a Kantian usage), whose ability to call a "best explanation" *deus* rather than *natura* saved a fideist hypothesis dear to his ethical and pastoral ("practical") heart?

Was Hume, in turn, Philo-the-skeptic, whose ability to call such a place-holder explanation "Custom" "saved" Cleanthes' design-hypothesis—and his own "practical" equilibrium—when he exited the study to practice "the easy philosophy"?

In his role as "Philo", for example, Hume's answer in the *Dialogues* to most of these questions was clear. As "Cleanthes", however, he intoned passages quite comparable to Berkeley's hymn-to-design in *D* II 211.

Finally, ought not an honest (and perhaps therefore cautious and processive) "phenomenalist" simply suspend judgment about intersubjective uniformity and "ultimate" coherence? Might it not be as uniform as "we" (or some of us) think (or hope) it is, or anticipate little that is "ultimate" if it is? (Methodological uniformities, after all, might not "rise tomorrow" either—"*au moins dans la forme présente*")?

If "the whole system", moreover, is "beyond thought", might attempts to see its "*surplusage*" of *D* II 211 above not be—on Berkeley's own account—"to no manner of purpose, without any design or signification whatsoever"? Could we even claim to "conceive" it as "the whole system", if we do not beg Berkeley's deism, or at least his apparent presupposition that some emergent "occasion" aligns spirits' "perceptions"?

Such questions suggest that one might characterise Berkeley's particular form of this predicament, perhaps (with apologies to Descartes and Arnauld) as

4.21 "*Berkeley's Circle*": an infinitary "inference" that

[belief in [methodological "uniformness" of judgments and "very probable conjectures" about phenomena]
would enable us to discern, in
[belief in [methodological "uniformness" of judgments and "very probable conjectures" about phenomena]
would enable us to discern, in .....
[belief in ....
["the" intelligible signature of "god"] ....]]]

Similarly imbricated paradoxes would also seem to attend ontological, cosmological, "transcendental" and "inference-to-the best-explanation"-arguments, in which one substituted

"*die transzendentale Einheit der möglichken Erfahrung*", or
"the veridicality of the best explanation" for
"the intelligible signature of God",

and such parallels suggest—to me, at least—that scientific "inferences to the best explanation" are nominally "secular" *residua* of late-seventeenth and early-eighteenth-century design-arguments.

All these iterated quasi-dialectical "inferences", in particular, might work, if "god"'s (or "experience"'s or "nature"'s) metatheoretic structure of "explanation" continued to align with "ours". But is this not what we set out to "infer"?

In a sense, the most important aspect of the *petitio* I am trying to characterise might lie in question-begging uses of the definite article "the". None of the arguments I have offered could presume to decide whether some form(s) of metatheoretic uniformity might constrain human inquiry, in ways inaccessible to those who engage in it. They do suggest, however, that such uniformities would not be expressible or characterisable—much less deducible—from "our" premises and arguments from within.

And this, finally, may point to an underlying reason why ontological, grandeur-of-"god"- or grandeur-of-the-universe-arguments (cf. again *D* II 211 above) cannot do the job, as the "fool" Gaunilo and "skeptic" Hume, among others, have observed. For no "grandeur" expressible in "our" (or [our]) terms could even discern what might be definable in "our" (or [our]) terms, much less account for more transcendent forms of "grandeur" in terms we would be likely to find both discursive and intelligible. And we, in turn, can hardly prove the consistency—much less unique interpretability—of what we cannot express.

If no discursive individuation of this grandeur were communicable, moreover, impeccably constructive arguments—brilliantly anticipated by "Berkeley-the-(dialectical)-semeiotician"—would suggest it may not matter what we call an ineffable "grandeur" that might lie beyond our evolving conceptions of what is conceivable—the *rerum natura*; "the Universe"; "what all call god"; "*deus sive natura*"; "*le meilleur des mondes possibles*"; a "final" (inexpressible) theory (or quasi-mystical Aristotelian *theoria*); "*das moralische Gesetz in uns*"; or perhaps, simply (in another *Flucht nach vorne*), "*das Unbedingte*".

Leibniz could therefore pray to *le bon dieu*; Berkeley to the "Author of Nature"; Johann Bachofen to the *Göttinnen des Mutterrechts*; and Einstein to "*dem Alten, der nicht würfelt*"; ....

I will return to the problems of Berkeley's ambivalence about alternate forms of underdetermination and inexpressibility in the chapter's conclusion, and close this section with some (very) brief remarks about Kant and interrelations between Berkeley's idealism and "*transzendentalem Idealismus*."

At least one variant of Kant's "transzendental deduction" rests, I believe, like the "master argument" and a natural inferential reading of the "*cogito*", on an informal semantic paradox (cf. (Boos, 1983b) and (Boos, 1998), 2.23 ff.). Like the "master argument" and the "cogito", also, it can be construed as an attempt to secure consistency and (unique) interpretability from within.

As I suggested in (Boos, 1995), 3.4–3.6 and (Boos, 1998), 5.11 and 5.52–5.57, for example, one might try to (re)construct a metalogical "transcendental deduction" as a dialectical *reductio*-argument that

4.22   one cannot coherently question the coherence of a cognitive structure from within.

For—one might argue—

4.23   one cannot coherently conceive that that *one's own conception* has no coherent interpretation.

Arguments analogous to the Gödelian gloss of the "master argument", however, suggest that such conception may metatheoretically be quite coherent. Indirectly, Gödelian diagonal arguments might also suggest that any "transcendental deduction" of the coherence of a given cognitive structure would have to appeal to wider semeiotic frameworks, of the sort that would seem to permit Berkeley's "greater largeness of comprehension".

In these wider contexts, however,

4.24   nontrivial interpretations that witnessed such coherence would provably be underdetermined from within, and therefore non-unique; and

4.25   the theoretical sources of this plurality would seem to lie in indeterminate properties of number—the very "form of inner sense", associated with time, that Kant considered most constructive, as well as most constitutive of "synthetic *a priori*" intuition" and "the transcendental unity of the understanding".

Perhaps an awareness that certain liminal problems of this sort might in some appropriate sense be undecidable hovered over Kant's formulations of the *Antinomien* (in which he took special—and well-deserved—pride). He did not interpret Berkeley's *dogmatischen* (*KrV* B274) or *mystischen und schwärmerischen Idealismus* (*Pr* 293), however, as the *Satz* of an *Antinomie*.

Indeed, Kant remarked with a certain unsubtlety in the *Prolegomena*—well after he had published his laborious "*Widerlegung des* (["*problematischen*"]) *Idealismus*" in the first edition of the first *Kritik* (B274 ff.)—that the position attributed above to "dialectical Berkeley" "never crossed my mind":

> Denn dieser von mir genannte Idealismus betraf nicht die Existenz der Sachen (die Bezweiflung derselben aber macht eigentlich den Idealismus in rezipiérter Bedeutung aus), *denn die zu bezweifeln, ist mir nie in den Sinn gekommen* ....
>
> For this idealism as I have called it did not bear on the existence of things (the doubting of which constitutes idealism in its received meaning), *for to doubt this never crossed my mind* ....[Pr 293] (Italics mine)

A pity. One can question Kant's reading of Berkeley, of course, or quibble about the distinction(s) between *Sachen* and *Dinge an sich*. But serious engagement with Berkeley's "dialectical" and "dogmatic" positions—with the latter, in particular, as a potentially "antinomial" *Idee der Vernunft*—might have offered Kant opportunities to clarify his own analyses of such dialectical oppositions, and revise some of his own sharpness-claims for the "*Grenzbestimmungen*" to which he remained so unalterably committed.

Indeed, I have argued that such "boundary-determinations" typically amount to (movable) boundary conditions for semantic paradoxes such as those mentioned above, and that the most salient lingering problem for "dogmatic" Berkeley remained the "intersubjective validity" he tacitly invoked to ground the mutual intelligibility of our separate acts of ideation (the "*phenomena*" of P 62), the proto-*Semeiotik* of his "general ideas", and the "coherentism" of his philosophy of science.

This problem with its air of *petitio* also remained for Kant, of course, in his appeals to the "*notwendiger Allgemeingültigkeit*" "*für uns jederzeit ... und ebenso für jedermann*" ("necessary general validity" "for us at every time ... and also for everyone") which constituted the alleged "objectivity" of *Erfahrungsurteile* in [Pr 298]; and in his invocations of the "*Regeln, welche den Zusammenhang der Vorstellungen in dem Begriff eines Objekts bestimmen*" ("rules which determine the connection of representations in the concept of an object") which distinguished "truth" from "dreams" in [Pr 290–291].

Kant claimed, of course, that he had "deduced" these constitutive uniformities in the first Kritik, in some appropriately qualified sense of the word "deduce". But it may indeed be the case that it never really "crossed his mind" to query certain "transcendental unities" of the sort Berkeley had projected into his benevolent Friend, "the Author of Nature".

5 Conclusion(s)

In the previous sections, I have tried to identify a recurrent tension between the positions of an elenctic figure I have called "Berkeley-the-dialectician", and those of a more dogmatic counterpart who might be called "Berkeley-the-design-theorist", or "Berkeley-the-inferrer-to-the-best-(theological)-explanation".

I have also tried to argue more briefly that similar tensions might also be traced in the critical writings of Kant, and suggest that this strain provides a measure of support for the ancient charge that Kant's enormous architectonic eventuated, in the end, in *"two truths"* ("the" speculative and "the" practical). If one credits certain elusive remarks in the third *Kritik*), there might even be a third ("the" teleological), and two more vaguely alethic modalities which inflected all three ("the" constitutive and "the" regulative).

I have also suggested that most of these tensions emerged from a common dilemma both authors tried to analyse and characterise, in ostensibly very different ways—Berkeley in his well-known reflections on "abstract" concept-formation and concessive comments about "notions", and Kant in his even more famous remarks in the first paragraph of the first *Kritik* about the *Schicksal der Vernunft* (see Chapter 8).

In both cases, moreover, one might elicit from these more conspicuously dialectical observations a common insight, or at least conjecture,

5.1 that nonconstructive concept formation, however deeply and recurrently aporetic it may be, is essential to thought, sentience, consciousness and other forms of reflective grace.

This conjecture about the experiences of somewhat more complex Leibnizian "souls", in turn, prompts some more, in the spirit of the figures I have called Berkeley-the-semeiotician and Kant-the-dialectician, though not of their more "dogmatic" counterparts, Berkeley-the-theologian and Kant-the determiner-of-"transcendental"-boundaries:

5.2 that *"Erfahrungen"* might consist of expressible, communicable and (conjecturally) consistent data in potential branches through 2.14's hierarchies, whose "contrar[ies]" (like Hume's "matters of fact") are "still possible" (cf. *E* I21).

5.3 that one might assimilate putative entities not expressible or communicable in a given range of theoretical frameworks or contexts to (metatheoretic) *"Dinge an sich"* with respect to those frameworks or contexts.
Certain subsidiary conjectures would then follow.

5.4 Expressibility and communicability in patterns of metatheoretic ascent—not fixation *via* an allegedly unique structure of "perceptual" *Anschauung (en)*—might be (regulative of the structure(s) of) "experience".

More precisely,

5.5 Metatheoretic richness and variety of such expression and communication—which need not be "verbal", but should be susceptible of some sort of information-theoretic analysis—might distinguish such "*Erfahrungen*" from "mere" "*Wahrnehmung(en)*", but only in provisional and graduated ways.

In effect, I have replaced the unitary second-order design of Berkeley's cognitive ontology structure—an avowedly christian (or at least theist) variant of Spinoza's "*deus sive natura*"—with indefinitely ramified higher-order designs of metatheoretic inquiry (a skeptical counterpart perhaps, of *natura naturans*). I have then used this thought-experiment notion of pluralistic "design" to reinterpret and relativise Kant's notions of "*Erfahrungen*" and "*Dinge an sich*".

In Kant's architectonic, secular intuitions' (alleged) indefeasibility and intersubjective uniformity—underwritten by "god" in Berkeley's *Principles* and *Dialogues*—distinguish determinate *Verstandesbegriffe* from indeterminate *Vernunftideen*.

The account I have sketched, by contrast, assimilates *Verständlichkeit* (not a Kantian term of art, interestingly enough) to expressibility and communicability. So conceived, "understanding" does not lend itself to unique interpretation from within, and demarcations between "understanding" and "reason"—like the distinctions between "*Erfahrungen*" and "*Dinge an sich*"—are local and graduated, not global and sharp.

Are there any "true" *Dinge an sich*, as in 5.3—beyond "every" "conceivable" conceptual horizon, as it were? Perhaps. But then "we" presumably cannot discern them or talk about them in definitive ways (*per definitionem*).

More to the point, it would appear that we could not even decide whether we might eventually conceive of agents or metatheoretic intelligences capable of "greater largeness of comprehension" that could "discern them or talk about them in definitive ways".

At this point I may have proposed a thought experiment that parallels Berkeley's own "master argument" in significant ways, and the arguments sketched in Section 4 recommend suspension of judgment about the "existence" of such "*Dinge an sich*". All one can assert with some confidence is that such entities—if they "existed"—could not individuate this "existence" in discursively intelligible terms.

In part at least, I have developed this quasi-Berkeleyan, quasi-Kantian line of argument in an effort to provide evidence for a twofold metaphilosophical conjecture, also set out in (Boos, 1998) and elsewhere:

5.6 that awareness of semantic paradoxes and other "pathological" metamathematical observations may clarify the aporetic complexity of their classical archetypes; and conversely,

5.7 that the depth and elegance of these metalogical analogues come more sharply into relief when one sees them as formal ectypes of informal metaphysical archetypes.

Assorted questions arise immediately. What dilemmas? Which "ectypes"? What "archetypes"?

I have tried above to provide a number of partial answers to the first question. A more responsive effort would suggest that most such *aporiai* seem to arise from—thoroughly natural—"first-philosophical" and "foundational" desires to "conceive" "the" limits of "conception", or "experience" "the" limits of "experience".

The "ectypes"—patterns or templates in their own right—are admittedly relatively few. But they apply to any first-order theory T that "can count" and formulate "inductive" arguments about [its own] syntax, and they have a number of potential—historical and other—interpretations. They include

5.8 multiple interpretability;
5.9 underdetermination of semantics by syntax;
5.10 inexpressibility of ("the" limits of) "expressibility";
5.11 relativity of the range(s) from which the multiple interpretations in 5.8 may be chosen.

Hovering over these four patterns, I believe (and argued at greater length in (Boos, 1998)), is a dynamic (cf. 2.3 and 2.6) of

5.12 metatheoretic ascent, or recourse to wider metatheoretic frames in which one hopes to bring these patterns into provisional focus.

This "dynamic", in turn, is a natural consequence and concomitant of the recurrent contextuality of any metatheoretic venue(s) in which one might attempt to express, subsume, suspend or delimit the "phenomena" of 5.8–5.12, and resolve apparent contradiction into "mere" paradox. They seem, in short, to furnish "constructive" conceptual analyses of "nonconstructive" concept-formation. Seen in this light, they might seem to provide the very models of "transcendental arguments", but they also suggest, once again, that any boundaries one may hope to draw between "constructive" and "nonconstructive" modes of concept formation will have to be deeply and recurrently "nonconstructive".

These analyses also provide a basis for the allusion above to the patterns of metatheoretic ascent in 5.12 as a "dynamic".

Russellian diagonalisation, for example (and therefore Russell's "paradox", as a particular special case) might be regarded as a kind of infinitesimal generator for an indefinitely ramified dynamic of class-formation (cf. (Boos, 1987a), 3–8).

Gödelian diagonalisation, of the sort sketched in Section 3, may also be regarded as such a generator, in this case for a roughly parallel dynamic of linguistically expressible concept-formation, in 5.12's patterns of indefinitely extendible metatheoretic ascent.

Mathematical logicians typically invoke "intended interpretations" to "cap" the ascent(s) mentioned in 5.12, at least provisionally. Such invocations normalise the problems raised by the "phenomena" of 5.8–5.12, and provide local contexts for their resolution. More or less equivalently, such invocations also allow them to say—in indefinitely revisable ways—what they "mean" by certain especially problematic assertions about intensionally infinite entities in number theory, analysis, set-theory and model theory.

A partial canvass of such "intensions"—in roughly ascending order of Kantian *Transzendenz* and Cantorian *Inkonsistenz* (cf., e.g, (Cantor, 1966), 443–444; a gloss of Cantor's observation appears in (Boos, 1987a), Section 1), but also of the "generative dynamic" just mentioned—would include nominally "extensional" notions such as

5.13 "the" hereditarily finite sets HF, and their numerical counterparts;

5.14 "the" set N of "all" natural numbers; and

5.15 "the" continuum, which may be construed alternately as "the" collection of "all" sequences of N consecutive coin tosses; or "the" collection of "all" countable interpretations of a consistent, recursively axiomatisable first-order theory T.

More extravagantly, they might even encompass such "multiplicities" (to borrow Cantor's usage) as

5.16 "the" class of "all" sets ("the" set-theoretic universe),

5.17 "the" (hyper)class of "all" classes, and even

5.18 Cantor's original example of extravagant "intentional" *Inkonsistenz* (in a letter to David Hilbert),

5.19 "the" "*Inbegriff 'alles' Denkbaren*" ("aggregate of everything thinkable").

The last "aggregate" would seem to be a particularly expansive example of Aristotle's *noesis noeseos* (self-referential "thought about thought", attainable only by "a god"), applied to Anselm's *ens quo nihil majus cogitari potest* (cf. 2.9)—the ineffable in pursuit of the unthinkable.

Cantor, however, a devout man, seems to have thought of this "aggregate" in more or less Kantian terms, as a kind of *Grenzidee* (cf., e.g., KrdrV, B 310–311)—a notion Hilbert himself later invoked, in a reflective passage of his essay "Über das Unendliche" (Hilbert, 1903, 288); compare also a related sense of "*ideale*" *Objekte*, discussed earlier in the same essay. Such allusions seem clear invocations of the common underlying *aporiai* mentioned above. In any case, as we have seen, one can nominally or provisionally secure the semantic "existence" and "extensionality" of metalogical notions such as these in a wide assortment of intensionally stronger finite or recursively axiomatisable (meta)theories.

Among the many semantic notions and intensional constructs that can provisionally be "secured" in this way, one might cite the following—in roughly ascending order of their likely appearance in efforts to ground (render provable) the "extensional" unicity of the notions listed in 5.13–5.17:

5.20 "consistency";
5.21 "interpretability";
5.22 "existence";
5.23 "definability";
5.24 "truth" ("satisfaction"); and
5.25 "standardness" ("well-foundedness"; "intendedness").

These are—almost transparently, I think—metaphysical as well as metatheoretic notions. They are also interwoven with the (meta)metatheoretical venues that "ground" them in indefinitely complex ways. A kind of conceptual interaction between the "extensional" (set-theoretic) notions of 5.13–5.17 and the "intensional" (semantic) ones of 5.20–5.25 generates the dynamic of 5.12's metatheoretic ascent (in our heads, if nowhere else).

The metaphysical "archetypes" of these metalogical "ectypes" are—ostensibly, at least—more varied, and have much greater historical resonance—as indeed they should.

In the history of metaphysics, for example, the "'first-philosophical' and 'foundational' desires to 'conceive' 'the' limits of 'conception', or 'experience' 'the' limits of 'experience'", mentioned above, have drawn people to (attempt to)

5.26 "define" what it means to "define" something;
5.27 engage in contemplative "thought about thought";
5.28 understand "what is in our power" (and accept what is not);

5.29 conceive, revere (and "prove" the existence of) "that than which nothing greater can be conceived";
5.30 transcend "every" "conceivable" doubt;
5.31 see ourselves (and our capacities for "sight") *sub specie aeternitatis*;
5.32 discern why we cannot discern that god is just (without equivocation about "discernment", or alienation of "god" or "justice" beyond recognition);
5.33 accept "empirical" judgments grounded in "real essences" beyond our comprehension;
5.34 critique "abstract ideas" in language which (allegedly) does not employ them;
5.35 understand "the" distinction between "intuition" and "understanding" (and "determine" from it "the" boundary between "understanding" and "reason");
5.36 know (or be "certain") whether "we" can "know" something (or only be "certain" about it);
5.37 understand the ethical implications of our lack of understanding of "the" boundary-conditions of our actions;
....

On my interpretation of these analogies, they suggest that any *Grenzbestimmungen* one might hope to draw between "phenomenal" and "noumenal" are "noumenal"—or at least "liminal", in the sense sketched above in Section 2. Similarly, any putative demarcation between "immanence" and "transcendence" will be "transcendent"—or at least deeply and ineluctably "nonconstructive", and expressible only in provisional and theory-relative ways, somewhere in 5.12's dynamic of metatheoretic ascent.

One might attempt to formulate this claim—itself elusively "liminal"—as an implication.

5.38 Every candidate for such an Archimedean pou sto or point-of-ultimate-discernment requires another, to "grasp", "discern", or "ratify" it as such.
5.39 Therefore, one cannot express a stable sense in which such a point-of-discernnment might be said to "exist", much less characterise the alleged vantage-point "itself".

The ancient stoic logicians—Gödel's ancestors, perhaps—called such states-of-ultimate-discernment *katalepsis* (comprehension). On my reading of the analogies sketched above, the "theory-relativism" implicit in 5.38–5.39 provides a reasonable skeptical response to claims that such states "exist" (or "must" exist).

Indeed, the premise 5.38 formally recapitulates a skeptical "mode" (*tropos*) of "lapse into unlimited" (indefinite regress, *ekptosis eis apeiron*). The middle-academic skeptics introduced this mode to confute the stoics' invocations of putative *katalepsis* as a "criterion of truth". One might even compare this "mode" with Ludwig Wittgenstein's conundrum about "*Sehen*" and "*Sehen als*". One cannot "see" the (object-theoretic) notions without seeing them "as" (metatheoretically) "intended" intepretations—whether one can "see" the "intensions" themselves or not.

In any event, my efforts to explore these analogical interrelations between metalogic and metaphysics and apply them to the reflections of one kindly iconoclast (whose religious beliefs were thoroughly orthodox—another fideist paradox, perhaps) seem to me to confirm a number of tentative conclusions and working hypotheses.

One is that counterparts and analogues of semantic paradoxes and other metalogical dilemmas do indeed pervade metaphysics and epistemology of every variety—including reductive attempts to dismiss or overcome or deconstruct metaphysical speculation, in the name of pragmatism, positivism, or other varieties of "empiricist" metaphysics (which are as speculative as the claims they dialectically negate).

Another is that these patterns arise from a common metalogical and metaphysical need to "conceive" "the" limits of "conception" (including moral or practical conception). Studies of metalogical miniatures of this need have prompted researchers to formulate and study theories T that are characterisable by their potential usefulness as tacit metatheories—not only for other theories, but ultimately—and most intriguingly—for [themselves].

A third is that studies of these metalogical analogues and ectypes—this one, for example—are unlikely to solve problems associated with their metaphysical or metaethical counterparts. They may, however, clarify the recurrence and apparent intractability of their metaphysical and epistemological archetypes, about which we seem drawn to speculate again and again—ineluctably, perhaps, as Kant suggested in his well-known comments about the *Schicksal der Vernunft* (fate of reason) (cf. Chapter 1, Section 1.22).

A fourth is that "experience" cannot—indeed should not—be construed as a well-defined structure or Kantian "*Inbegriff*" (aggregate), as most ancient, early modern and twentieth- and twenty-first-century European and North-American metaphysicians seem tacitly or explicitly to have assumed, but should rather be understood as a processive, underdetermined and "problematic" (B 310) notion, along the lines of a Kantian idea-of-reason.

A fifth is that nonconstructive concept-formation may be regulative of (every discernible form of) self-awareness and sentient inquiry, and may emerge in the activities of Leibizian monadic "souls" at "earlier" and more modest levels than one might think. So too may capacities—themselves dispositional—to actuate or realise more than "we" can express, however underdetermined to "us" its potential interpretations may be (*Vielleicht "glaubt" die Katze doch, dass es eine Maus gibt* ... ?; cf. Wittgenstein *ÜG* 478).

A sixth is that graduated abilities to consider, examine, and in some sense realise such theories T—in which theoretical [self]-reference and conceptual thought experiments about "inconceivabililty" are possible—might therefore be regulative of the sorts of [self]-consciousness Aristotle attributed to *theoria*, as well as Section 2's "organismic pathos" (a nice phrase of Lovejoy's) that resonated in Berkeley's invocations, also quoted above, of the

> ... greater largeness of comprehension, whereby analogies, harmonies and agreements are discovered ..., and ... reduced to general rules ... which ... are most agreeable, and sought after by the mind; for that they extend our prospect beyond what is present, and near to us, and enable us to make very probable conjectures touching things that may have happened at very great distances of time and place, as well as predict things to come; *which sort of endeavour towards omniscience is much affected by the mind.*
>
> (*P* 105; the italics are mine)

Correlatively—if paradoxically—they might, finally, provide a measure of reconciliation between "sympathy"-based ethics, and more complex pluralistic forms of "categorial imperative", and an alternative interpretation of Kant's oddly moving remark, that

> Der Begriff einer Verstandeswelt ist also *nur ein Standpunkt*, den die Vernunft sich genötigt sieht, außer den Erscheinungen zu nehmen, *um sich selbst als praktisch zu denken.*
>
> The concept of a world of understanding/intelligible world is *only a standpoint*, which reason sees itself constrained to take outside the appearances, *in order to think* [*of*] *itself as practical*. ([*Grd* 457]; the second set of italics are Kant's own)

If so, indefinitely ramified and graduated degrees of such "theoretical" and "practical" "understanding" and [self]-awareness might be realised in varying degrees of Leibnizian monadic "confusion" in all sentient beings.

Be that as it may, the more dialectical thought experiments George Berkeley developed to vindicate his "immaterialist" worldview, particularly in passages such as the "master argument", quoted above from *P* 23, enjoy a certain philosophical notoriety. The subtlety of the views and responses sketched in his semeiotic reflections and the dialectical tensions that emerged in his scientific "coherentism" seem to me equally interesting, but less well-known.

The metalogical analogies I have adduced also bear witness to the complex interrelations between the various aspects of Berkeley's thought, and in the process to the paradoxical subtlety and remarkable acuity of Berkeley's efforts to "think with the learned and speak with the vulgar". However wrong his particular readings of these arguments may have been, or begged his denials of some of their more skeptical or pluralist implications, his "greater largeness of comprehension" helped rouse Kant from his dogmatic slumbers, and provided a theistic prototype for the latter's secular-but-"transcendental" coherentism.

More particularly, I have argued

5.40 that Berkeley developed in his doctrine of "signs" an informal semeiotic to avert the antinomial implications of such paradoxes;

5.41 that his attempts to explore the implications of this semeiotic led him to consider forms of "immaterialism" that effectively paralleled (and may therefore have anticipated) finitist, nominalist and theory-relative interpretations of formal logic in interesting and potentially informative ways;

5.42 that his attempts to resolve the antinomial implications of such paradoxes and provide a warrant for the mutual interpretability and intersubjective uniformity of our disparate perceptions and conceptions led him to postulate (and attempt to deduce) a theological prototype of Kant's "*notwendige Allgemeingültigkeit*";

5.43 that his theological idealism therefore anticipated "transcendental" variants thereof, in its attempts to vindicate the "antithesis" of what might be described as an *Urantinomie der reinen Vernunft*; and finally, by way of anachronistic appreciation;

5.44 that Berkeley's thought-experimental "master argument" may indeed have been as significant as he thought it was. On my account, it anticipated significant aspects of Gödel's diagonal lemma and semantic-paradox analysis by more than 200 years.

7 The Second-order Idealism of David Hume

1 Introduction

David Hume's "consequent skepticism" still elicits analogues of the "Answers by Reverends and Right Reverends," which he wryly recalled at the end of his life as evidence that

> my former publications (all but the unfortunate Treatise) were beginning to be the subject of conversation. "My Own Life," [*Essays*], 610

Such "Answers" now come more than "two or three in a year," of course; and their eristic motivations are presumably metaphysical rather than religious. Nevertheless, their very persistence may recall Hume's dismissive remark that George Berkeley's arguments

> admit of no answer and produce no conviction. Their only effect is to cause that momentary amazement and irresolution and confusion which is the result of scepticism. [*E*] I, 155

Berkeley, of course, saw the matter differently, though he defined the problem in similar terms:

> Prejudices and errors of sense do from all parts discover themselves to our view; and endeavouring to correct these by reason we are insensibly drawn into uncouth paradoxes, difficulties, and inconsistencies, which multiply and grow upon us as we advance in speculation; till at length, having wander'd through many intricate mazes, we find ourselves just where we were, or, which is worse, sit down in a forlorn scepticism.
> [*Works of George Berkeley*] I, 25

In the event, however, Berkeley's "solution" was a brilliant *fuite en avant* whose ambiguities resonate in Hume's ironic remark. For he embraced enthusiastically all and only all those allegedly phenomenal entities-of-"sense" ("ideas") which are (again, allegedly) "particular" and "immediate;" and proscribed rigorously all "noumenal" entities-of-thought ("abstract ideas," "ideas of reflexion") which are indefinite and mediate. In his own words,

> ... assert the evidence of sense as high as you please, we are willing to do the same. That what I see, hear and feel doth exist, that is to say, is perceived by me, I no more doubt than I do of my own being. *But I do not see how the testimony of sense can be alleged as a proof for the existence of anything, which is not perceived by sense. We are not for having any man turn sceptic, and disbelieve his senses*; on the contrary, we give them all the stress and assurance

imaginable; nor are there any principles more opposite to scepticism, than those we have laid down .... [WB] I, 57 (emphasis mine)

The first italicized sentence, ironically, may be regarded as a near-verbatim restatement of a classical skeptical—in fact pyrrhonist—claim: that one cannot infer "inevident," "noumenal" conclusions from "evident," "phenomenal" premises. Sextus Empiricus took pains to interpret this (itself rather suspiciously noumenal) claim "in an undogmatic way" ("ἀδοξάστως"); but it remained one of the two or three major points of doctrinal contention between the academic and pyrrhonist skeptics and their stoic opponents. Hume presumably had something rather like this in mind when he made his remark.

Nevertheless, Berkeley did not take Sextus' pains. He did not make significant attempts to provide a metacritique of his own arguments, even a *pro forma* one. And he was right, therefore (or so I think), to insist that he was no "forlorn sceptic." Read the other way around: critics who opt for a rather doctrinaire, "phenomenalist" reading of pyrrhonism may owe their readers an explanation why the position they read into Sextus did not eventuate in something like Berkeley's ....

At any rate, Berkeley is usually called an "idealist," of course, not a skeptic; by contrast with Hume, who is usually considered, with varying degrees of attention to the classical antecedents, England's great skeptic—"*der scharfsinnige Hume*" ("the sharpwitted Hume"), as Kant called him.

Was Hume a skeptic? I will aver that he was not, for reasons parallel to the ones sketched above for Berkeley. For example, I will argue that, except for very few passages (most prominently, his famous disclaimer of a warrant for personal identity at [*Treatise*] 633–636), he never undertook a serious variant of the pyrrhonist metacritique referred to above.

Instead, I will suggest, Hume was what might be called a "second-order idealist": a reductive metaphysician, very like "that very ingenious author" (Hume's description of Berkeley at [E] I 155; compare the phrase of Kant just quoted) who chose to "prescind", or at least reject, not first-order non-immediate objects (which he seems to have regarded with a slightly uneasy sort of benign neglect), but rather second-order non-immediate relations ("necessity," "causation") between such objects. In the sense in which Berkeley called himself an "immaterialist," therefore, one might appropriately call Hume an "irrelationist."

Any such programmatic denial of second- (or higher-) order "I know not what"'s is likely to encounter sooner or later some variant of an essentially metalogical problem, which confronted Berkeley, and whose "irrelationist" variant may be formulated as follows. Hume argues brilliantly that we may consistently deny the existence of relations whose alleged "universality" we cannot secure; but this does not (on his own principles, or anyone else's) prove that we "must" deny the consistency of their (hypothetical) existence. To vary his own insight, such an implication might be a "matter of (agreed-upon, metatheoretic) fact;" but it is not a provable "relation of (metalogical) ideas."

Berkeley's vulnerability to this objection is of course well-known. Hylas struggles to formulate a prototype of it in *Dialogue* II, and Boswell's prosaic observation of it in [*Life of Johnson*] I, 292 is what precipitates Johnson's attack on the unoffending stone. For some reason, the corresponding lacuna does not seem to console as many nonskeptical opponents of the less obviously dogmatic Hume, or suggest to them that the great English skeptic, were he such, might have done well to embrace this dilemma as an emergent occasion for skeptical ἐποχή, or at least for a Scots judicial verdict of "not proven."

2 Some Dogmas of Empiricism

Traditionally, one invokes metaphysical principles to secure the uncertain boundary conditions of philosophical "proof." They are the methodological premises which one exempts, at least provisionally, but more often dogmatically, from further scrutiny, and which one uses, openly or "subreptively" (in Kant's language) to bridge lacunae of the sort I tried to identify above. Hume formulates several rather high-metaphysical and "first-philosophical" "principles," remarkably many for a skeptic, especially when one considers their concentration in the more youthful, but also more passionately skeptical *Treatise*. In this and subsequent sections, I will mainly concentrate on four or five of these "principles" (his own word), the first three of which are

2.1 a clarity-and-distinctness principle for consistency rather than "truth": that whatever is clearly and distinctly conceived or perceived is at least consistent ("implies no contradiction");
2.2 the converse of 2.1: that any notion which is inconceivable is impossible (Berkeley's "mere nothing"); and
2.3 a principle of the identity of indiscernibles for "impressions" and their "corresponding" "simple ideas."

By way of a first gloss on these principles (whose actual formulations in the *Treatise* and first *Enquiry* will be discussed later in more detail), notice that 2.3 seems to elude scrutiny on Hume's own terms. Indiscernible when? By whom? For how long? As long as the "impressions" and "corresponding" ideas remain "simple?" As long as the sun rises tomorrow?

Neither 2.1 nor 2.2, moreover, seems itself particularly "clear and distinct" (a phrase whose various modulations both Berkeley and Hume employ rather often, in just such contexts). And finally, the "forward" implication 2.1 ((c & d)-conceivable implies consistent) is weaker than Descartes'; but its converse 2.2 (consistent implies (c & d)-conceivable) is stronger. Each of these points will be taken up in subsequent sections.

All three of these "maxims" (also Hume's usage) are in fact variants of tenets which one may also find in the work of Berkeley. In keeping with this chapter's underlying argument, I will want to assert that Hume characteristically applies them one or more theoretical or methodological "type-levels" higher.

Several passages of [*T*] and [*E*] I further suggest that Hume considered 2.1–2.3 jointly equivalent to

2.4   a principle of the well-foundedness of "sensation" and "conception" (often called "Hume's microscope").

This principle, of course, is a central dogma of Hume's empiricism—his cardinal tenet that first-order "impressions" and "corresponding" (also first-order?) "simple ideas" not only "exist," but exist as virtual atoms of perception and intention, which can be examined to "any" desired degree of precision under the "microscope." Hume's own metaphor for this "microscope or new species of optics" ([*E*] I 62) also recalls both the metaphysical pathos of Pascal's humbling shifts of scale, and Berkeley's meditations on the noninvariance under different perceptual resolutions of a mite's foot ([*WB*] I, 60–61; cf. collateral reflections at [*T*] 189–190).

Indeed, the "microscope"'s near-axiomatic assertion of conceptual as well as perceptual well-foundedness may be an almost paradigmatic antiskeptical attempt to block potential instances of infinite or (or at least indefinite) regress. Recall, for example, Hume's characterization of the process by which we "carry up our inference from one testimony to another," at [*E*] I 46:

> In a word, if we proceed not upon some fact, present to the memory or senses, our reasonings would be merely hypothetical; and however the particular links might be connected with each other, the whole chain of inferences would have nothing to support it, nor could we ever, by its means, arrive at the knowledge of any real existence. If I ask why you believe any particular matter of fact, which you relate, you must tell me some reason; and this reason will be some other fact, connected with it. But as you cannot proceed after this manner, *in infinitum*, you must at last terminate in some fact, which is present to your memory or senses; or must allow that your belief is entirely without foundation.

When one reads such claims, it may be well to keep in mind that various dilemmas of infinite regress ("lapse into the infinite;" cf. [*PH*] I, 166) provided the pyrrhonists with many of their most telling counterarguments to stoic dogma; and that a parallel impulse to escape the implications of what he called "*analysis infinitorum*" ironically prompted Leibniz to look outside the physical for his perceptual "atoms" (cf., e.g., [*OF*], 17–18). Denials of such regress of course have a venerable history too—Aristotelian *archai* ("beginnings") were essentially postulated to instantiate such denials. My point here is once again that such denials—including 2.4—are paradigmatic antiskeptical thumps on the dialectical table, and should be considered as such in "radical empiricist" contexts as well.

Closely related to 2.1–2.4, and inherent to the doctrines I have called "first-" and "second-order" idealism, is the essentially dogmatic claim, alluded to earlier, that one must in certain cases deny what one can (at least consistently) deny—in-principle-unperceivable objects, for example; or underdetermined "forces." For the cases at hand, we may formulate this claim—whose metalogical counterpart is quite unsecured, and whose tacit metatheoretic transition from "may" to "must" might be regarded as one alethic counterpart of the deontic one Hume made famous, from "is" to "ought"—as a supplementary enthymematic "principle:"

2.5 that any relational notion or concept which is not uniformly definable in terms of perceptual "objects" (e.g.: "matter;" or "force;" or "necessary connexion;" or "cause") is "inconceivable" and therefore (by 2.2) "impossible."

Later, I will argue that all these principles are closely related to Hume's slide from the indicative / "factual" to the subjunctive / "counterfactual," in his mildly notorious definition of "cause" at [*E*] I, 76:

> we may define a cause to be *an object, followed by another, and where all the objects similar to the first are followed by objects similar to the second. Or in other words where, if the first object had not been the second never had existed.*

Hume's apparently unconscious ambiguity about "the" boundary conditions of "causation" seems to reflect both

(a) unwarranted confidence that one can distinguish unequivocally between experiential παινόμενα (the pyrrhonists' "appearances;" compare Hume's "impressions," and "corresponding simple ideas"), which he trusted; and "inevident" *noumena* (literally "things thought;" compare again the empiricists' "ideas of reflexion"), which he did not; and

(b) unawareness such confidence is difficult to reconcile with equivocation about the metatheoretic boundaries between factual and counterfactual implication(s).

The former is essentially the prototype of W. v. O. Quine's first "dogma of empiricism." The latter is an issue which famously arises both in Quine's own work, and in critical evaluations of it.

To the extent one interprets the slide of *E* I 76 as an unwarranted assimilation of "is" to "might be," it also bears an ironic resemblance to the unexamined assimilations of "is" to "ought" which Hume rightly queried. Efforts to clarify or eliminate modal-alethic ambiguities such as this one, or the one mentioned several paragraphs above from *E* 1 46 (roughly speaking, the two are dual to each other), tend to take one of two basic directions.

Attempts to reduce "might be" to "is," first, would seem to yield a genuinely "dogmatic" empiricism, which may be polemically effective but condemned to observe Wittgenstein's silence about [itself]. Attempts to reduce (or at least assimilate) "is" to "might be", on the other hand, may yield a tenable sort of metatheoretic skepticism. But they would also render Hume's "irrelationism" (and Berkeley's "immaterialism," for that matter) little more than intriguing hypotheses—internally consistent, perhaps, if carefully stated; but hardly capable of "proof."

"Reverends and Right Reverends" have (sometimes) been less scandalized by "mere" hypotheses—the pope enjoined such a formula on Galileo, after all, and variants of it probably saved the lives and livelihoods of a good many seventeenth-century "fideists" and *"libertins érudits"*—so a more skeptical Hume might ironically have seemed less provocative, in some contemporary circles at least. Whether or not alternative *Scharfsinn* of this sort would also seem less provocative now, or would have aroused an alternative Kant so vigorously from his dogmatic slumbers, I will leave the reader to decide (or leave in appropriate skeptical suspension).

It is also just here, by the way, that the precedents of classical pyrrhonism become most equivocal and problematic. On the one hand, both Sextus Empiricus and Cicero have a tendency to lapse into a near-dogmatic "phenomenalism," and refer all pragmatic decisions to a suspiciously uniform and single-sorted class of φαινόμενα ("appearances"), which have an alleged quasi-phenomenal (and thus quasi-acceptable) superstructure of ἔθη ("customs"), συνήθεια ("habits"), and even φύσις ("nature"), which in turn give rise, in some "apparently" coherent way, to "undogmatically" followed πείσεις ("experiences/ persuasions") and ἀνθρώπεια πάθη ("human feelings").

A good pyrrhonist (or academic skeptic; there is no great division here) is even advised to "follow the local customs" ("undogmatically," of course); the better, presumably, to hold at bay the ravages of the Achaean reverends and right reverends (or so at least Sextus claimed: recall the particulars of Socrates' arraignment).

On the other hand, as I observed earlier, some of classical skepticism's subtlest elenctic arguments exploit the indefinitely regressive relativity and contextuality of just such demarcations between "evident" φαινόμενα and "inevident" νούμενα. The skeptical "modes" (τρόποι) of infinite regress, "hypothesis," and criterial ascent—the need to postulate ever-wider criteria of metatheoretical "truth;" (cf. e.g., [PH] I, 20 and 164–169)—all suggest that reexamination of one's (meta)theoretical presuppositions may be inherent to any metaphysical "thought about thought." And one such presupposition might well be the empiricist (and, on some accounts, pyrrhonist) dogma that there is some sort of uniquely determined transition from "the" "phenomenal" (factual, objectual, verifiable) to "the" "noumenal" (hypothetical, intentional, unverifiable).

In effect, I wish to argue that Hume—like Berkeley, hardly a partisan of the infinite, even in its more indefinite and skeptical guises—largely ignored these "infinitary" and metacritical aspects of the skeptical tradition (whose great advocate in early modern philosophy was probably Pascal); and tended naturally toward the first, more reductive, "phenomenalist" (and dogmatic) view of received skeptical argumentation.

The strains of this reductive view have arguably (re)emerged much later, of course, in the ruins of positivism: as indeterminate boundaries between "observation" and "theory," and "synthetic" and "analytic;" in the methodological ambiguities of doctrines such as Quinean "physicalism" and "naturalized epistemology;" and in all manner of Wittgensteinian conundrums.

Most of these dilemmas were already prefigured, I wish to suggest, by the strains of efforts such as Sextus', to avoid "dogmatic" phenomenalism, yet do things like "prove" that there is no "proof" (cf. [PH] II, 97–192); and programs such as Hume's, to "run over libraries" and "commit volumes" cheerfully "to the flames," yet retain dogmatic "maxims" and "principles," whose underlying metatheoretic ambiguities are marked by slides such as the one between factual and counterfactual, remarked above, at [F] I 76.

Confronted perhaps by such implicit tensions, Hume undertook what might be called his own *fuite en avant*, which led to a conclusion I will paraphrase—in non-Humean language—as a last, summary (and highly speculative) "principle" of Humean metaphysics:

2.6 that one not only can, but "must" ground (what amounts to) a second-order instrumentalism about physical relations, in (what amounts again to) a third-order essentialism about
(a) mental relations ("force and vivacity"); and
(b) "moral" relations ("custom" and "human nature").

The burden of this chapter, in effect, is to provide some warrant for the strong language in which I have phrased 2.6. The claims Hume makes for the uniformity of these higher-order mental and moral entities are, I would argue, deeply and thoroughly "dogmatic," and there is some evidence that he was aware of this. On the one hand, he expressed his well-known afterthoughts and reservations about the self-referential limit of the mental at [T] 633–636. On the other, he offered wryly high-metaphysical praise for the wondrous uniformity he had come to attribute to "Custom" at [E] I 54–55:

> Here, then, is a kind of *preestablished harmony between the course of nature and the succession of our ideas, and though the powers and forces by which the former is governed be wholly unknown to us; yet our thoughts and conceptions have still, we find, gone on in the same train with the other works of nature. Custom is that principle, by which this correspondence has been effected*; so necessary to the subsistence of our species, and the regulation of our conduct, in every circumstance and occurrence of human life. Had not the presence of an object instantly excited the idea of those objects, commonly conjoined with it, all our knowledge must have been limited to the narrow sphere of our memory and senses; and we should never have been able to adjust means to ends, or employ our natural powers, either to the producing of good, or avoiding of evil. *Those, who delight in the discovery and contemplation of final causes, have here ample subject to employ their wonder and admiration.* (except for the phrase "final causes," the emphases are mine)

Hume never really explained why we should not query the metatheoretic "force" and "uniformity" of "Custom," in the same terms as the ones he invoked to question both "necessity" and "causality." Indeed, one of the principal motives of my arguments in the sections that follow is a—skeptical—suspicion: that Hume's anacoluthic decision to hold harmless the "moral" from his judgment of *non liquet* against the physical, effectively makes his "custom" as immune to criticism as Berkeley's "God."

3 "An Establish'd Maxim in Metaphysics"

At [T] 233, Hume writes that

> Whatever is clearly conceiv'd may exist; and whatever is clearly conceiv'd, after any manner, may exist after the same manner. *This is one principle*, which has been already acknowledg'd. Again, every thing, which is different, is distinguishable, and every thing which is distinguishable, is separable by the imagination. *This is another principle*. (emphases mine)

These "principles" provide textual prototypes for 2.1–2.3 of the previous section; 2.3 is also mentioned at [T] 24 as

> ... the principle above explained, that all ideas which are different, are separable.

There is no explicit earlier reference to this claim as a "principle," but it is essentially stated at [T] 10. Hume presumably considers it explained by the perceptual and conceptual atomism of "impressions" and "correspondent" "simple ideas" he has developed in [*T*]'s opening pages—and this, in turn, is essentially the "foundationalist" doctrine I have tried to paraphrase above in 2.4. This already suggests that Hume may have considered the "microscope"'s claim of experiential well-foundedness in 2.4 and the metaphysical "principles" in 2.1–2.3 to be almost interdeducible—or at least different aspects of a single axiomatic (and in this sense "dogmatic") assertion or scheme of assertions, which he later "derives" from them in the text(s).

In any case, 2.3 is concisely recapitulated in the "Appendix," along with several of its not-necessarily-equivalent nuances, as follows:

> Whatever is distinct, is distinguishable; and whatever is distinguishable, is separable by the thought or imagination. All perceptions are distinct. They are, therefore, distinguishable, and separable, and may be conceiv'd as separately existent, and may exist separately, without any contradiction or absurdity. [*T*] 634

The sheer density of all these pronouncements seems to me once again to merit some comment. Vast brooding issues of high metaphysics hover over such gnomic assertions, and some of them would not seem out of place in a newly discovered autograph letter of Leibniz or Descartes. It may be appropriate, then, that Hume states 2.1 at [*T*] 32 as

> ... *an establish'd maxim in metaphysics, That whatever the mind clearly conceives includes the idea of possible existence*, or in other words, that nothing we imagine is absolutely impossible. We can form the idea of a golden mountain, and from thence conclude that such a mountain may actually exist. We can form no idea of a mountain without a valley, and therefore regard it as impossible. (emphasis mine)

Notice also the last sentence's appeal to the example Descartes uses to illustrate the ontological argument, and the ([*E*] I 76)-like slide, from "may exist," in [*T*] 233, to the doubly- (and perhaps even trebly-) intentional "includes the idea of possible existence," here.

Hume also phrases 2.1 in quasi-Cartesian language as a principle that whatever is "distinctly conceived" is indeed consistent ("implies no contradiction"):

> ... whatever is intelligible, and can be distinctly conceived, implies no contradiction, and can never be proved false by any demonstrative argument or abstract reasoning *à priori*. [*E*] I 35

Hume's elaborations of the principle 2.1 and 2.2, including his famous definition at [*E*] I 25 of "matter of fact" as a thing whose

> ... contrary ... is still possible, because it can never imply a contradiction,

make it rather clear that he also considered "consistency" (noncontradiction) equivalent, in turn, to (possible) existence. The latter equivalence reemerged in Rudolf Carnap's attempt to reconstruct "the" analytic-synthetic distinction; and may plausibly be construed as an anticipation (shared by Leibniz, and implicitly affirmed by Kant) of a metalogical result now known as "the completeness theorem." As we have seen, Hilbert first adumbrated this result in the opening years of the twentieth century, and Gödel provided it with a rigorous formulation and proof in 1930. According to this now well-known metatheorem, the consistency of (e.g.) one of Hume's "relations of impressions" *is* equivalent to the "existence" of some "possible" (though not necessarily "factual"?) interpretation for it.

On the same account, Hume's "relations of ideas" (compare again Sextus' φαινόμενα, and Kant's *Noumena*) would correspond to "propositions" provable without appeal to "nonlogical" axioms, which may hold in some interpretations but fail in others. In Hume's words,

> Propositions of this kind are discoverable by the mere operation of thought, without dependence on what is anywhere existent in the universe. [*E*] I 25

It's not really clear, of course, on Hume's own terms, how or in what venue "the mere operation of thought, without dependence on what is anywhere existent in the universe" could actually occur—a latent difficulty which may well underlie a long and rather quizzical footnote which appears at [*E*] I, 43–45. We will take up this question in more detail in Section 6 below.

In any case, Hume's endorsement of the problematic transition from consistency (... "can imply no contradiction") to "conceivability" (whatever that turns out to be) is further evident in the full text of his characterization of "matter of fact" at [*E*] I 25, just quoted:

> The contrary of every matter of fact is still possible, because it can never imply a contradiction, *and is conceived by the mind with the same facility and distinctness, as if ever so conformable to reality.* [*E*] I 25 (emphasis mine)

Hume's lapidary assertions in the first part of this sentence have been somewhat clarified, I believe, by the rather natural metalogical analogy I've introduced (with due apologies for anachronism). But the extralogical nuances of the italicized remarks in the second half of the sentence open a whole horizon of further questions, all of them metaphysical, and most of them vexed.

Is "facility and distinctness," for example, another variant of the "clarity and distinctness" of 2.2 above? Predictably, perhaps, I suspect that it is. Why the same "facility and distinctness," moreover, when this might seem to strain the careful gradations of "force and vivacity" he will later associate with degrees of "belief?"

First, I would argue, because Hume subscribed to the Berkeleyan particularity-doctrine about "ideas," reconstructed above in 2.3, even "counterfactual" ideas (which he sometimes simply conflated with ideas "ever so conformable to reality;" cf. [E] I 76 once again, and the question raised above about the very legitimacy of "relations of ideas"). And second, because he disassociated (as did Berkeley and Descartes) "phenomenal" presentation(s) of ideas from the affective and volitional confirmation(s) of belief (itself perhaps a somewhat curious position for a skeptic to take).

Is the metatheoretic transition represented by Hume's "as if " (in "... as if ever so conformable to reality") itself a "relation of ideas?" Or a "matter of fact?" (Another way, in effect, to pose the problem created by [E] I 76).

If the latter, from what "impression" is it "deriv'd," and may we not then consistently (and perhaps even "conceivably") deny any particular instance or construal of it? If the former, have we not clarified *obscura per obscuriora*? If it is neither, do we "commit it to the flames?"

Hume's slide from "conceivable" to "consistent" (and "possibly existent"), in other words, may itself be "consistent," in some still wider (meta)theory; but it does not seem on Hume's own terms to be a "relation of ideas" (a dubious class, in any case, whose nominal extension sometimes seems to include little more than throwaway propositions about triangles—and perhaps not even them); and it may or may not be "conformable to reality" either. Once again, we will return to these problems at greater length in Section 6.

The "principle" 2.1 also appears in an interesting contrapositive form, which involves a kind of metatheoretic redoubling of its object-theoretic modality, at [T] 43:

> 'Tis in vain to search for a contradiction in any thing that is distinctly conceiv'd by the mind. Did it imply any contradiction, 'tis impossible it cou'd ever be conceiv'd.

In effect, Hume came very close to entertaining here a "conception" which would assert [its own] "impossibility," a thought-experiment which almost cries out (with hindsight) for interpretation in terms of Gödel's diagonal lemma (according to which every formula $\psi(r)$ in sufficiently strong formal theories T, gives rise to a "diagonal" sentence $\psi = dia(\varphi)$ with the property that $T \vdash \psi \leftrightarrow \varphi[\psi]$; cf. [Smorynski, 1977], 827).

In this context, it may also be relevant to observe that Hume's strongest rhetorical formulation of the "principle" 2.1 appears in one of the few contexts in which he explicitly considers—in order to dismiss—a skepticism (his term) which does reflect upon itself: the mathematical skepticism of infinite regresses, ascents and descents, discussed earlier; a skepticism which also poses, for example, the possibility of

> [a] real quantity, infinitely less than any finite quantity, containing quantities infinitely less than itself, and so on *in infinitum*; this is an edifice so bold and prodigious, that it is too weighty for any pretended demonstration or support, because it shocks the clearest and most natural principles of human reason ....
>
> An infinite number of real parts of time, passing in succession, and exhausted one after another, appears so evident a contradiction, that no man, one should think, whose judgement is not corrupted, instead of being improved, by the sciences, would ever be able to admit of it.
>
> *How any clear, distinct idea can contain circumstances, contradictory to itself, or to any other clear, distinct idea, is absolutely incomprehensible*; and is, perhaps, as absurd as any proposition, which can be formed. *So that nothing can be more sceptical, or more full of doubt and hesitation, than this scepticism itself, which arises from some of the paradoxical conclusions of geometry or the science of quantity.* ([E] I 156 and 157–158, emphases mine)

Perhaps the most eloquent (and familiar) statement of the converse principle I have called 2.2 (that whatever is consistent is conceivable) appears at [E] I 18:

> ... thought can in an instant transport us into the most distant regions of the universe; or even beyond the universe, into the unbounded chaos, where nature is supposed to lie in total confusion. What never was seen, or heard of, may yet be conceiv'd; nor is any thing beyond the power of thought, except what implies an absolute contradiction.

I have already suggested that there may be clear tensions between the genuine metaphysical pathos of such passages, and the reductive implications of Hume's own methodological "principles," whose finitism and perceptual atomism would seem to eventuate in a more narrow view of concept formation, of the sort he endorses (say) in [E] I, 43–45.

As an example of the rather complex propositional interrelations between 2.1 and 2.3 and the reductive, even foundational doctrine I've called Hume's perceptual "atomism," consider the following argument from [T] 36:

> In order to know [whether] any objects, which are join'd in impression, be separable in idea, we need only consider, if they be different from each other; in which case, 'tis plain they may be conceiv'd apart. *Every thing, that is different, is distinguishable; and every thing, that is distinguishable, may be separated, according to the maxims above-explain'd. If on the contrary they be not different, they are not distinguishable; and if they be not distinguishable, they cannot be separated.* ([T] 36, emphases mine)

Hume also characterizes this amplification of 2.3 explicitly as a biconditional assertion about (imaginable) "objects", at [T] 18:

> We have observ'd, that whatever objects are different are distinguishable, and that whatever objects are distinguishable are separable by the thought and imagination. *And we may here add, that these propositions are equally true in the inverse, and that whatever objects are separable are also distinguishable, and that whatever objects are distinguishable are also different.*

At this point, it may be worth observing that it is not at all clear what has happened here to the "unbounded chaos" and "total confusion" which are supposed to lie within our "power of thought" at [*E*] I 18. All *principia individuationis* may be equally indispensable to "thought and imagination;" but some are apparently more indispensable than others.

On my construal, the [*T*] 18 passage just quoted does clarify an obvious sense in which 2.3 may be derived as a special case of the conjunction of 2.1 and 2.2: that is, if "every" (possible?) state of affairs can be clearly and distinctly conceived (perceived), then "every" instance of (potential? hypothetical?) inequality must be discernible. The fact that the parenthetical modifiers in the last sentence's conclusion would greatly strengthen an already strong Leibnizian claim is, I would argue, still another variant of the problem which Hume unconsciously blurs at [*E*] I 76.

In effect, the claims of 2.1–2.2 also amount to a rather striking assimilation of alethic to epistemic modality, as I suggested above: "(im)possibility" is equivalent (in some appropriate sense) to "(in)conceivability." The degree to which one may be willing to countenance this extraordinarily strong biconditional would seem to vary with one's acceptance or nonacceptance of

3.1 a parallel assimilation of the hypothetical to the "factual" (with a tacit bias toward the latter); and

3.2 a presupposition that "our" perspective of what is, or may be (what is conceivable to "us") is "the" only (potentially) relevant or correct one; its findings therefore determine what "implies a contradiction," and what does not.

Acceptance of 3.1, once again, would undercut rather severely the expansiveness of [*E*] I 18 ("thought can in an instant transport us ..."). But such acceptance would accord rather well with the elenctic and reductive effects of Hume's "principles," and make more plausible the slide of [*E*] I 76.

The "principle" 3.2, similarly, would already seem to establish that Hume's "custom," or something functionally akin to it (the pyrrhonists' φαινόμενα? Kant's "*Form(en) der Anschauung*"?) is—for "us," at least—a virtually indefeasible criterion of (metalogical) truth.

To the extent one really accepted 3.2, in fact, the famous assertion at [*E*] I 44–45 that

> Custom, then, is the great guide of human life .... Without the influence of custom, we should be entirely ignorant of every matter of fact beyond what is immediately present to the memory and senses,

would itself become a "relation of ideas" (another way, perhaps, of saying that 3.1 and 3.2 essentially beg—or, less provocatively, express—the most fundamental "empiricist" points which are here at issue).

Rather mild malign-genius arguments usually suffice to call 3.1 severely into question, of course. And even milder benign-genius arguments undermine 3.2.

Who are "we," after all, to assert that "our" conception exhausts what there is? One does not need to be a platonist to "conceive" that one may emerge from whole hierarchies of "caves," as from infancy (or reflect that "When I was a child, I thought as a child ...."); though it might be difficult, even perhaps impossible to "comprehend" in any one of the "caves" what such emergence(s) might mean.

Thoroughly Humean "observation and experience," in fact, would seem to confirm Paul's comment, just quoted, as well as Simone Weil's more pointed remark (which the thoughtful Hume of "My Own Life" would readily have endorsed, I think), that intellectuals who are proud of their intellects are like condemned prisoners who are proud of their large cells ....

In any case, the strong biconditional claim of 2.1 ∧ 2.2—that things conceivable are possible, and (may) exist, and things inconceivable are impossible, and cannot exist—would seem at the very least to embody a form of doctrinaire verificationism, to which no skeptic in traditional senses of the word need—or should—subscribe.

I have already argued that implicit appeals to this equivalence also underlie both Berkeley's (dogmatic) first-order rejection of non-"perceptual" objects, and Hume's (equally dogmatic) second-order rejection of non-"perceptual" relations between "perceptual" objects (the "second-order idealism" which I referred to earlier). In the remainder of this section, I will try to trace the more immediate implications of these "established maxims" for a slightly different cardinal tenet of Hume's system, which is closely related to 2.4 and 2.5 above: his insistence that "impressions" must not only be 2.4's metaphysical atoms (even "substances," of a sort, as Hume wryly remarks; see below); but thoroughgoing epistemic "particulars" as well (Compare Kant's insistence that—e.g.—intuitions and their "*Formen*" be "*durchgängig bestimmt*").

In the passage from the abstract at [*T*] 634 quoted earlier, Hume simply assumes his implicit atomism (the principle that "All perceptions are distinct"), as a premise, from which he derives the conclusion that all perceptions may (hypothetically) be disjoined for purposes of mental examination—the only classical characterization of "substance" which he believes has any substance. Compare the following remarks from [*T*] 222 and [*T*] 244:

> Every quality being a distinct thing from another, may be conceiv'd to exist apart, and may exist apart, not only from every other quality, but from that unintelligible chimera of a substance. [*T*] 222

> I have already prov'd, that we have no perfect idea of substance; but that taking it for *something, that can exist by itself*, 'tis evident every perception is a substance, and every distinct part of a perception a distinct substance .... [*T*] 244

In essentially the same terms, he also confirms his doctrine that perceptual and conceptual discernment provide an adequate principle of individuation, in the sentences which immediately follow [*T*] 233's "two principles"-passage, quoted earlier:

> My conclusion from both (principles) is, that since all our perceptions are different from each other, and from every thing else in the universe, they are also distinct and separable, and may be consider'd as separately existent, and may exist separately, and have no need of any thing else to support their existence. *They are, therefore, substances, as far as this definition explains a substance.* [*T*] 233 (emphasis mine)

This particularity-doctrine is of course part of what grinds the fine structure of the "microscope"'s lens, through which Hume enjoins us to view all doctrine(s)—including, presumably, the doctrine that all perceptions can be individuated as the particularity-doctrine would require. It has an immediate precedent in the eidetic precision of Berkeley's "ideas," and a more remote one in another "established metaphysical principle" about "ideas," which Descartes mentions in his responses to Pierre Gassendi as

> ... id quod vulgo ajunt Philosophi, essentias rerum esse indivisibiles; Idea enim repraesentat rei essentiam, cui si quid addatur, aut detrahatur, protinus fit alterius rei Idea.

> ... what the Philosophers commonly say, that essences of things are indivisible; for an Idea represents the essence of a thing, such that if something is added to it or removed from it, it immediately becomes the Idea of another thing .... [*AT*] VII 371

One usually assumes Hume's and Descartes' "ideas" are too disparate to credit such juxtapositions. Here again, I am not so sure. Perhaps, at any rate, Hume would have been warier of his perceptual *principium individuationis*—which is also a rather clear ancestor, I would argue, of Schlick's "*Fundament*," and Carnap's *Protokolle*—if he had taken more seriously the extent to which intentional atoms could indeed be aligned, all too well, with "that unintelligible chimera of a substance."

In any case, Hume does also argue that the particularity-doctrine is essentially equivalent to the "indiscernibility" principles, discussed earlier. At [T] 18, he specifically invokes his claim that "different" implies "distinguishable" to "prove"

> that the mind cannot form any notion of quantity or quality without forming a precise notion of the degrees of each .... [The emphasis here is Hume's.]

The more Berkeleyan qualities of his argument also emerge quite strongly in such passages as the following:

> We know nothing but particular qualities and perceptions. As our idea of any body, a peach, for instance, is only that of a particular taste, colour, figure, size, consistence, etc. [T] 658

> That our senses offer not their impressions as the images of something *distinct*, or *independent*, and *external*, is evident; because they convey to us nothing but a single perception, and never give us the least intimation of any thing beyond .... [T] 189

The doctrine I have called Hume's "finitism" or "perceptual atomism" is essentially derived from this particularity-assumption, in turn, at [T] 228:

> ... 'tis impossible to conceive extension, but as compos'd of parts, endow'd with colour or solidity. The idea of extension is a compound idea; but as it is not compounded of an infinite number of parts or inferior ideas, it must at last resolve itself into such as are perfectly simple and indivisible.

Curiously, Hume also mixes this claim to extensional particularity in the Abstract with a counterpart of [E] I 76's "smooth and easy transition" from the factual to the counterfactual:

> When we simply conceive an object, we conceive it in all its parts. *We conceive it as it might exist, tho' we do not believe it to exist.* [T] 653, emphasis mine

Once again, I wish to suggest, Hume's fluent arguments for such conception "in all its parts" confer a specious completeness and specificity on our conceptions — a *durchgängige Bestimmung*, in Kantian language—a specificity which empirical psychologists, Hume's intellectual descendants, long ago observed few of us ever have. Every trial lawyer, for example, knows that witnesses refabricate details they could not credibly have perceived. Alexander Luria's psychophysical studies of people who do perceive and remember in overwhelmingly precise (often synaesthetic) detail, moreover, strongly suggest that such remarkable capacities may be an arcane burden for the rare individuals (autistic savants such as "Rain Man," for example) who seem to possess them.[1]

The claim that counterfactual conception occurs in this way would seem to be especially—well—counterfactual. Let us attempt a short thought-experiment. In the opening moments of the film-version of *Amadeus*, Salieri's alarmed servants break into his study, discover their maestro has slashed his throat, and transport him to the hospital through the snow-covered streets of Vienna (actually, Prague), as the opening bars of Mozart's "little" G-minor symphony reverberate through the soundtrack, with sinister and strikingly brilliant emotional force. Do you hear that opening G-minor chord (actually an open octave), harshly attacked by Neville Marriner's oboes and strings—"in all its parts?"

Good. Life has favored you with something of Salieri's remarkable gift—which Peter Shaffer's character extravagantly contemns, because "god" has not made him a "genius" (whatever that is .... The capacity to "hear" scores with the clarity the film confers on Salieri is a rare ability; I would give all of *Amadeus*' ticket-window receipts to have it).

Now "conceive," further, that Milos Forman and Neville Marriner had chosen instead to accompany Salieri's grim journey through the frozen streets with another orchestral work,[2] also in a minor key; but recently discovered, say, wrapped around a bundle of psalters in a monastery in the Steiermark; and that this work also begins with a tonic chord, in the minor key. Can you hear that chord too? "In all its parts?" (How was it scored? ...)

We will consider related problems of bogus specificity in greater detail below in Section 5.

Other statements of Hume's particularity-doctrine parallel also rather closely Berkeley's rejection of "abstract ideas." One also confirms a reductive tenet, which is interesting in its own right, and which might be called Hume's redundancy theory of "existence":

> *We have no abstract idea of existence, distinguishable and separable from the idea of particular objects.* 'Tis impossible, therefore, that this idea of existence can be annex'd to the idea of any object, or form the difference betwixt a simple conception and belief. [*T*] 623

Since this comes from the "Abstract," which Kant may have known in German translation, it is interesting to speculate for a moment whether Hume's assertion may have confirmed Kant in his better-known view that

---

1 Alexander Luria, *The Mind of a Mnemonist*, trans. Lynn Solotaroff, New York: Basic Books, 1968.
2 Director and musical conductor respectively of the 1984 *Amadeus*.

> Sein ist offenbar kein reales Prädikat, d. i. ein Begriff von irgend etwas, was zu dem Begriffe eines Dinges hinzukommen könne. [Kr I] A 598 / B 626
>
> Existence is clearly not a real predicate, that is, a concept of something or other which could be attached to the concept of a thing.

All of these interrelated arguments and "principles," in any case, do seem to support Hume's famous claim, concisely stated in the "Appendix" at [T] 658, that "our idea of any mind is only that of particular perceptions" (compare Berkeley's familiar insistence that we have no "ideas" of other "spirits," only "notions"). The last clause in the passage from [T] 623 just quoted also underwrites Hume's equally well-known tenet that "belief is nothing but a peculiar feeling" (of "force and vivacity"; cf. [T] 624).

Indeed, Hume's disclaimer about "existence," just quoted, suggests that any entity which relates "ideas"—not just the author's personal identity, but his philosophical argumentation as well—may on his "principles" be underdetermined by them—another "I know not what," which the microscope can never bring into focus. This might provide a provisional basis for a rather drastic version of the metacritique mentioned earlier, but it is presumably not one which Hume himself—especially the Hume of [E] I—would have been likely to accept. More about such questions of methodological coherence later.

One of the more interesting concomitants of the doctrine of particularity is Hume's apparent phenomenalist belief in the incorrigibility of "impressions" and their "corresponding" "simple ideas"—another "dogma of empiricism" which is oddly reminiscent of earlier claims made by that distinguished empiricist, René Descartes (at [AT] VII 37, quoted below). Hume applies it with an almost Cartesian zeal at [T] 189–190 to "the actions and sensations of the mind":

> Upon this head we may observe that all sensations are felt by the mind, such as they really are .... [T] 189
>
> Add to this, that every impression, external and internal, passions, affections, sensations, pains and pleasures, are originally on the same footing; and that whatever other differences we may observe among them, they appear, all of them, in their true colours, as impressions or perceptions. And, indeed, if we consider the matter aright, 'tis scarce possible it shou'd be otherwise, nor is it conceivable that our senses should be more capable of deceiving us in the situation and relations, than in the nature of our impressions. [T] 190

Compare Descartes'

> Jam quod ad ideas attinet, si solae in se spectentur, nec ad aliud quid illas referam, falsae proprie esse non possunt; nam sive capram, sive chimaeram imaginer, non minus verum est me unam imaginari quam alteram. Nulla etiam in ipsa voluntate, vel affectibus, falsitas

est timenda; nam, quamvis prava, quamvis etiam ea quae nusquam sunt, possim optare, non tamen ideo non verum est illa me optare.

As for ideas, if they are viewed solely in themselves, and not to something else to which I refer them, they cannot, properly speaking, be false; for if I should imagine a goat or a chimera, it would be no less true *that* I imagine the one than the other. Nor is any falsity to be feared in the will itself, or the affections, for though I may [be able to] choose things which are wrong, and even things which never are, it is not thereby untrue *that* I choose them. [*AT*] VII 37 (emphases mine)

Parallels such as this, reinforced by Berkeley's and Hume's free use of phrases such as "clear and distinct," "clear and determinate," et al., suggest to me a mildly heretical (re)definition of their common project: to restrict the "proper" objects of philosophical and scientific discourse to a (somewhat ill-defined) subclass ("simple ideas," "impressions") of Cartesian "clear and distinct ideas".... "ill-defined," for example, in one rather straightforward methodological sense: that the clarity and distinctness of (our ideas of) the "compounding" of these "simple ideas" to form other, "deriv'd" ideas remain themselves elusively obscure and confused.

Finally, Hume's dilemma about personal identity may also sharpen Berkeley's problem of the indefinability of "spirit." Directly after the passage just quoted, Hume deepens his dogmatic assertion that "every impression, external and internal" appears in its "true colours," and extends it explicitly to the mental:

For since all actions and sensations of the mind are known to us by consciousness, they must necessarily appear in every particular what they are, and be what they appear. Every thing that enters the mind, being in *reality* a perception, 'tis impossible any thing shou'd to *feeling* appear different. This were to suppose, that even where we are most intimately conscious, we might be mistaken. [*T*] 190

The elusive indefinition and underdetermination of the "self" which so troubles Hume at [*T*] 633–636 seems to me relatively straightforward, at least by comparison with the implications of the "common-sensical" passage just quoted.

It is the quasi-Berkeleyan expectations expressed in that passage, by contrast—that one "must" have minutely detailed phenomenal knowledge, not only of (e.g.) the veins in the leaves on the (bump on the branch on the) tree in the quad; but also, by dubious analogical inference from this dubious premise, of [one's own] mental qualities as well, and "in every particular,"—which seem to me distinctly bizarre.

Perhaps what has happened is that Hume has conjoined here Berkeley's particularity assumptions with a residual variant of Descartes' "conclusion" from *Meditation* II's thought-experiment with the wax: that what (we hope) perceives most clearly and distinctly is (or at least "shou'd" be) what is most clearly and distinctly perceived. If so, it may have been this needlessly rigoristic construction of the self which made the underdetermination of that self which he proposed in the "Appendix" so troubling for him—not only for the methodological foundationalism of his arguments (which it did undermine, as he scrupulously acknowledged); but for his metaphysical peace of mind (which it might otherwise have left more or less untouched).

In the end, of course, this section's tesselation of glosses and quotations can only begin to canvass Hume's use of a variety of metaphysical "principles" which might (I believe) have seemed highly "inevident" and abstruse to a classical skeptic. In subsequent sections, I will

(a) extend the parallels I've sketched between Hume's arguments and those of Berkeley;
(b) explore in greater detail these arguments' equivocal distribution over "the" counterfactual;
(c) offer further remarks and conjectures about the ambiguous methodological limits of Hume's skepticism;

and finally,

(d) examine several senses in which Hume's non- and anti-skeptical uses of "Custom" may have provided a persuasive prototype for Kant's notion of the synthetic *a priori*.

4 Immaterialism and "Irrelationism"

In this section, I will try to clarify several ways in which Hume's "principles" and "established maxim(s) of metaphysics" may have eventuated in the dogmatic position I have called "irrelationism" or "second-order idealism."

A preliminary observation is that Hume does not—on pain of inconsistency?—apply his principle of the equivalence of conceivability and consistency to Berkeley's assimilation of "object" to "intentional" object ("idea"): does not, in other words, endorse a dogmatic denial of the independent existence of external objects. Hume does endorse, in effect, the epistemic cogency of Berkeley's analysis; but he seems to leave its metaphysical "truth" in partial suspension, somewhat ambiguously, and in ways his own principles should almost certainly forbid.

Consider, for example, the next passage's wryly dialectical (or simply equivocal?) conclusion to an otherwise model Berkeleyan argument.

> Bereave matter of all its intelligible qualities, both primary and secondary, you in a manner annihilate it, and leave only a certain unknown, inexplicable *something*, as the cause of our perceptions; *a notion so imperfect, that no sceptic will think it worth while to contend against it*. [*E*] I 155 (second emphasis mine)

This final epigrammatic turn is the source of my earlier characterization of Hume's attitude as "benign neglect." Why we cannot extend the same offhand acceptance to the second-order "inexplicable something" of causality he never takes the trouble to explain.

In the following passages from [*T*] 241–242, Hume clearly argues that "whatever difference we may suppose [betwixt ... a perception. And ... an object or external existence] 'tis still incomprehensible to us." Yet he does so in a context which seems to presuppose that a comparative study of the "connexions and repugnances betwixt" hypothetically distinct (though indiscernible) "objects" and "impressions" is—quite contrary to his own "principles"—a meaningful thing to undertake:

> ... 'tis impossible our idea of a *perception*, and that of an *object or external existence* can ever represent what are specifically different from each other [Compare the "principle" 2.3 above]. *Whatever difference we may suppose betwixt them, 'tis still incomprehensible to us; and we are oblig'd either to conceive an external object merely as a relation without a relative, or to make it the very same with a perception or impression* .... [*T*] 241

> I say then, that since *we may suppose, but never can conceive a specific difference betwixt an object and impression*; any conclusion we form concerning the connexion and repugnance of impressions, will not be known certainly to be applicable to objects; but that on the other hand, whatever conclusions of this kind we form concerning objects, will most certainly be applicable to impressions. The reason is not difficult. As an object is supposed to be different from an impression, we cannot be sure, that the circumstance, upon which we found our reasoning, is common to both, supposing we form the reasoning upon the impression. [*T*] 241–242

> Thus we may establish it as a certain maxim, that we can never, by any principle, ... *discover a connexion or repugnance betwixt objects, which extends not to impressions*; tho' the inverse proposition may not be equally true, that all the discoverable relations of impressions are common to objects. [*T*] 242 [emphases mine]

These are not, of course, the only passages in which Hume seems to endorse the "supposition" but not the "conception" of external objects, and thus declines to accept an apparent inference from his own principles to Berkeley's notoriously awkward "dogmatic" conclusion: that such objects do not "exist." In the next section's comments on Hume's attitude toward "hypothesis," I will consider briefly his alternative derivation of this "opinion of the double existence of perceptions and objects," from a "common hypothesis of the identity and continuance of our interrupted perceptions" (cf. [T] 211).

For now, let us consider the following "evident" "corrollary," which Hume appends to his analyses of "cause" and "necessary connexion" at [T] 172:

> I shall add as a fourth corrollary, that we can never have reason to believe that any object exists, of which we cannot form an idea. For as all our reasonings concerning existence are deriv'd from causation, and as all our reasonings concerning causation are deriv'd from the experienc'd conjunction of objects, not from any reasoning or reflexion, the same experience must give us a notion of these objects, and must remove all mystery from our conclusions. [T] 172

For me, at least, the "mystery" of this passage and the ones quoted earlier remains Hume's reluctance to derive from them Berkeley's conclusion, or at least some proto-Kantian variant thereof. The "corrollary" itself might be regarded as a weaker, more plausible and more skeptical variant of 2.2—not that what is inconceivable is inconsistent; but simply that we "never have reason to believe" it is consistent.

But it is well known that Hume also considers certain *relations*—"connexion(s) or repugnance(s) betwixt objects" "incomprehensible" and "undiscoverable"—"cause," for example, and "necessity," when the latter means something other than "constant conjunction." And it is equally well-known, as I have already remarked, that he does not consider these second-order notions "so imperfect, that no sceptic will think it worth while to contend against" them.

In the rest of this section, I will review several aspects of Hume's famous critique of mind-independent "connexions and repugnances," first in quotation and then in paraphrase; and ask again why we should *not* query 2.1 ∧ 2.2, and replicate the suspension Hume applies to Berkeley's arguments one level higher: consider, less dogmatically, Hume's second-order hypothesis: that there are no relations between "objects" and/or "impressions" except those which are introduced ("determin'd") by the mind; and say not that this hypothesis "must" be correct, but only that we cannot "know" it is wrong.

First the review. At [E] I 31, Hume avers that:

> When we reason *a priori*, and consider merely any object or cause, as it appears to the mind, independent of all observation, it never could suggest to us the notion of any distinct object, such as its effect; much less show us the inseparable and inviolate connexion between them.

At [E] I 63:

> When we look about us towards external objects, and consider the operation of causes, we are never able, in a single instance, to discover any power or necessary connexion; any quality which binds the effect to the cause, and renders the one an infallible consequence of the other. [E] I 63

Directly after his remark, quoted earlier from [T] 189, that "all our sensations are felt by the mind, such as they really are," he continues:

> ... when we doubt, whether they present themselves as distinct objects, or as mere impressions, the difficulty is not concerning their nature, but concerning their relations and situation. [T] 189 (emphasis mine)

At [T] 223, he argues that "a just inference" from the observation

> that 'tis not from a view of the nature and qualities of objects that we infer one from the other, but only when in several instances we observe them to have been constantly conjoined ....

would be that

> we have no idea of power or agency, separate from the mind, and belonging to causes. [T] 223 (emphasis mine)

So far, the argument remains compatible with a second-order version of the first-order agnosticism Hume decided to apply to Berkeley's external objects (indeed, it closely parallels it; cf. the "fourth corrollary" at [T] 172, quoted above). Such objects may exist (though we can presumably have no "microscopic" analysis of "existence"—a restatement of one of Hume's recurrent methodological problems); but they are "incomprehensible" to us: quasi-noumenal entities which "we may suppose, but never can conceive." For all we may "know," there may be then no "real" "connexions or repugnances" which "give rise to" our "propensities;" just as there may be no "real" "objects" which "give rise to" our "impressions."

But note the subtle shifts of tone in the following passages, from Hume's central critique of "the idea of necessary connexion," in [T] 165–169:

> The several instances of resembling conjunctions lead us into the notions of power and necessity. *These instances are in themselves totally distinct from each other,* ...
>
> [Why? .... Compare the stringent criteria of Hume's *first*-order "principle(s)" of "perceptual atomism," discussed earlier ....]

> ... and have no union but in the mind, which observes them, and collects their ideas.
>
> *Necessity, then, is the* **effect** *of this observation, and is nothing but an internal impression of the mind, or a determination to carry our thoughts from one object to another* ....
>
> *Upon the whole, necessity is something that exists in the mind, not in objects*; nor is it possible for us ever to form the most distinct idea of it, considered as a quality in bodies. [*T*] 165 (emphasis mine)
>
> ... the necessity or power, which unites causes and effects, lies in the determination of the mind to pass from one to the other. *The efficacy or energy of causes* is neither plac'd in the causes themselves, nor in the deity, nor in the concurrence of these two principles; but *belongs entirely to the soul*, which considers the union of two or more objects in all past instances. 'Tis here that the real power of causes is plac'd, along with their connexion and necessity.
>
> .... power and necessity ... are consequently qualities of perceptions, not of objects, and are internally felt by the soul, and not perceiv'd externally in bodies. [*T*] 166 (emphases mine)

In effect, these passages modulate almost imperceptibly from the more skeptical premise, that there may be no mind-independent necessity (because we cannot cite any observation of it under the microscope, to "prove" that there is), to the doubly dogmatic conclusion,
(a) that there "is" ("can be?") no such necessity "in objects," and
(b) that "[t]he efficacy or energy of causes" does nevertheless "exist," and in fact "belongs entirely to the soul."

How—on Hume's own terms—can this conclusion even be formulated, much less "verified" or "proved?" I will try to retrace some of Hume's ancillary arguments and auxiliary assumptions in the next section, but would first like to point out a curious *prima facie* resemblance between this transition (from inability to locate "efficacy" in the physical, to insistence that it must be "plac'd" in the mental), and another well-known argument from the teak chest of the philosophical opposition: Descartes' "inference," at [*AT*] VII 31, from the (plausible) premise that the "comprehension" of a given piece of "beeswax"'s "waxhood" is not "accomplished by the faculty of imagining [alone] (*"nec igitur comprehensio haec ab imaginandi facultate perficitur"*), to the (implausible, or at least unwarranted) conclusion(s) that it is accomplished "by the mind alone" (*"sola mente;"* whatever this means); and that "perception" (*"perceptio"*) of the wax is entirely "an inspection by the mind alone" (*"solius mentis inspectio"*).

The wax-sequence is usually considered one of the *Meditations*' less successful arguments. For one thing, it looks suspiciously like a straightforward quantifier-mistake (denial of "inspection by the imagination alone" need not imply "inspection by the mind alone"). Could one suggest that something like that is going on in Hume's argument?

In one sense at least, yes. A metatheoretical quantifier-shift *is* clearly relevant to Berkeley's and Hume's arguments: consistency of the (object-theoretic) negation of a with a given theory T does *not* imply (metatheoretic) negation of the consistency of a with T. Essentially I have tried to argue that the "very ingenious author" and his successor *der scharfsinnige Hume* both recognized this lacuna, at least implicitly; and that both tried to cross it, over a *pons philosophorum* of "principles" of the sort summarized above.

Berkeley actually acknowledged the "bridge" his counterpart of the "principle" 2.2 must cross quite explicitly at one point. In one of Hylas' fitful awakenings from his dogmatic slumbers, in the second of the Three *Dialogues*, he remarks to Philonous that he is willing to grant in effect that "matter" may be indefinable, and in this sense "inconceivable;" but he remains unwilling to agree that what is inconceivable is impossible.

> Be that as it will, yet I still insist upon it, that our not being able to conceive a thing, is no argument against its existence. [*WB*] I 223

Philonous' reply is perhaps less interesting for its *a baculo* dismissal of the argument than for a nuance in its first sentence.

> That from a cause, effect, operation, sign or other circumstance, there may reasonably be inferred the existence of a thing not immediately perceived, and that it were absurd for any man to argue against the existence of that thing, from his having no direct and positive notion of it, I freely own. But where there is nothing of all this ... where an abstraction is made from perceiving and being perceived ... I will not indeed thence conclude against the reality of any notion or existence of any thing: but my inference shall be, that you mean nothing at all: that you employ words to no manner of purpose, without any design or signification whatsoever.

It is a curious and rather interesting fact, I think, that Berkeley, who denies the "existence" of unperceived first-order objects, is here willing to grant, by way of what "may reasonably be inferred," just the sort of relata of a "cause, effect, operation, sign or other circumstance" which Hume's second-order idealism would relegate to the "custom" or "habit" of the beholder.

The passages from Hume and Berkeley we have adduced so far recapitulate two of the most famous lines of arguments in the history of western metaphysics, and I have tried to array them in a way which highlights the remarkably close parallels which can be traced between them. In the same spirit, I will devote the remainder of this section to a concise paraphrase of Hume's "Berkeleyan" derivation of a rejection of second-order "objects" ("relations" between first-order "objects") from his "principles," reviewed above in Section 2. In the next section, I will then offer further metalogical and metaphysical arguments which bear on that derivation's implicit premises and ultimate tenability.

Inevitably, both the paraphrase and its subsequent examination will also involve more talk of "theories," "metatheories" and the like. Lest this seem gratuitous, it may be helpful to recall briefly how many of Hume's arguments and conclusions he expressed in the language of (admittedly informal) "demonstration," "inference," and "proof." Consider, for example, the following passages from [T] 650–651, which restate the "principle" 2.1, as well as the completeness argument outlined earlier:

> It is not any thing that reason sees in the cause, which makes us infer the effect. *Such an inference, were it possible, would amount to a demonstration, as being founded merely on the comparison of ideas. But no inference from cause to effect amounts to a demonstration. Of which there is this evident proof.* The mind can always conceive any effect to follow from any cause, and indeed any event to follow upon another: whatever we conceive is possible, at least in a metaphysical sense: but *wherever a demonstration takes place, the contrary is impossible, and implies a contradiction. There is no demonstration, therefore, for any conjunction of cause and effect.* [T] 650 (emphases mine)

> 'Tis evident, that *Adam*, with all his science, would never have been able to *demonstrate*, that the course of nature must continue uniformly the same, and that the future must be conformable to the past .... This conformity is a *matter of fact*, and if it must be proved, will admit of no proof but from experience. [T] 651

Hume's language is emphatically not that of twentieth-century logic or metamathematics, of course. But it often seems to express insights which do have rather straightforward metalogical parallels.

Two such observations are provided by [E] I 25–26's informal anticipation of the completeness theorem, and concomitant recognition that one should not expect to derive significant "nonlogical" results from purely logical axioms, both cited earlier. An even more straightforward parallel appears in [T] 650's remark, italicized above, that "when a demonstration takes place, the contrary is impossible, and implies a contradiction." It would seem perverse not to interpret this as an informal expression of the "easy half" of the completeness theorem (sometimes called the "soundness lemma").

To the extent such parallels and assimilations are appropriate, I believe, study of them may help pose questions and clarify dilemmas which confront almost any serious attempt to characterize "experience" in terms which involve (talk about) "demonstration," "inference" and "proof."

For example: are Hume's rather narrowly stratified characterizations of "experience" themselves "experiential" (defeasible "matters of fact," rather than definitional "relations of ideas)?" Is the answer to *this question*? These questions and conundrums themselves have metalogical analogies, and I will attempt to consider these in more detail in the chapter's final sections.

On my reconstruction, in any case, Hume's "irrelationism" begins with a hypothetical acceptance of external objects which "give rise" to "correspondent impressions," however "incomprehensible" these objects or process(es) of "giving rise to" may be. I have already argued that such acceptance is not really compatible with Hume's "principles" 2.1 and 2.2, but will not press the point further here. Hume has already scandalized enough philosophical Right Reverends, perhaps, without imposing on him the burden of Berkeley's immaterialism. Perhaps ironically, his initial tendency to suspend judgment about external objects would also be more compatible than the reductive "principles" 2.1 and 2.2 with the skeptical "intense view" he expresses in the last section of [T] (cf. [T] 269), or the more urbane "consequent skepticism" he endorses in [E] I XII.

Though Hume apparently suspends application of 2.1 and 2.2 to "objects" in the end, it may also be interesting to note that he continues to hold in this context to its heuristic corollaries 2.3–2.6. He does, for example, clearly consider "all" first-order "ideas" and "impressions" capable of complete individuation, at least "in principle" (a problematic qualification in itself, by his own "principles;" more also about this later). Such completely individuated "ideas" and "impressions" effectively become (would become?) the intentional "atoms" discussed above in Section 3.

But here the individuation stops. Suppose one has some perceptual objects under the "microscope"'s lens. Can these objects "determine" the "connexions and repugnances" between them? No, of course not. No more than, say, the base-set of a model or relational structure can "determine" its relations. If this were the only point Hume wished to press, there would be little to quarrel about. It is his evaluation of this predicament—his insistent tendency, in effect, to overdetermine first-order "perception," but underdetermine second-order "hypothesis"—which generates the controversy (a line of thought which suggests to me that the "irrelationist" Hume may also have been the first "logical atomist").

Can we ever discern such relations, at any rate, through the same microscopic lens Hume has given us to focus on the atoms? On his account, no. We would have to set the focus to infinity, and our finite processes of "compounding" and "deriving" would be too limited and coarse-grained to canvass the relations' entire extensions.

One can make this especially clear for relational "matters of fact" which may extend over "the whole" course of future inquiry; but less obvious cases may also follow from the intentional and objectual finitism that are embodied in Hume's principles. (On rather similar grounds, in fact, one could also argue that we cannot "see" Hume's perceptual "atoms" as atoms. Something like the antithesis of Kant's Second Antinomy might follow from this line of argument, however. And it is obviously not one Hume would have been likely to accept.)

Suppose we grant, then, that we can neither individuate these relations under the microscope, nor "compound" them in an exhaustive and finitely surveyable way. Do they therefore elude conception altogether?

By the auxiliary principle of Hume's "fork," yes; for strait are the epistemic gates through which "all" experience and "reasoning from experience" must pass:

> *All* reasonings may be divided into two kinds, namely demonstrative reasoning, or that concerning relations of ideas, and moral reasoning, or that concerning matters of fact and existence. [*E*] I 35 (emphasis mine)

Does this mean that these "incomprehensible" relations—"cause," for example—do not "exist?" Yes, it does, to the extent that the "principles" 2.1 and 2.2—at the relational but not the object-level—are given their full reductive force.

But does this not "bereave" experience of its relational coherence? Fortunately, no. For "Custom" and an "inextricable contrivance of Nature" (Cf. [*E*] I 55, 57) provide us (whoever "we" are) with "propensities" and "determination(s) of the mind", which regulate an otherwise unordered, nonrational experience. This "preestablished harmony" (this functional equivalent of "forms of intuition" and "categories of the understanding?") also ensures that we may not only "suppose" that "our" experience is so regulated; but 'must' assent to the fact ("matter of fact?" "relation of ideas?") that it is entirely "determin'd" thereby; and that such intentional relations—such "propensities" and "determinations"—operate "by an absolute and incontroulable necessity" (cf. [*T*] 183), and "belong entirely to the soul."

The foregoing is only a paraphrase, of course, supported by a few quotations. Moreover, it is a paraphrase of Hume's more explicitly 'radical' response—exemplified by the passages from [*T*] 166, quoted above—to the last two paragraphs' opening questions. Other passages—the two paragraphs which straddle [*T*] 168–169, for example—are phrased with qualifications (that there is no "real intelligible connexion" between external objects; that a "power or necessary connexion" between such objects "is what we can never observe in them") which do not explicitly invoke 2.2.

Nevertheless, I believe the paraphrase preserves a degree of fidelity to the spirit and letter of Hume's principal argument and dominant conclusion. It may serve as a brief summary of the complex of arguments and presuppositions I have in mind, when I attempt to characterize Hume's metaphysical position as a variety of "second-order idealism".

5 "Whatever Is Clearly Conceived ... .

In the following sections, I will subject the foregoing reconstruction of Hume's principal argument to a range of "sceptical doubts" about its methodological tenability, and sometimes its metatheoretic consistency. The first of these doubts arises from a sense that there is something which verges on begging the question in

(a) Hume's (and Berkeley's) defense(s)—implicit and explicit—of the "principles" 2.1 and 2.2;
(b) both authors' persistent and characteristic use of such "principles" in metatheoretic "refutations" of (what amount to) their object-theoretic negations.

Consider these "principles" once again. Suppose I (with the hapless Hylas) query the "empiricist" warrant for them, especially 2.2. In response, my "sharpwitted" and "very ingenious" interlocutors offer to show (in effect)

5.1 that I cannot conceive what a distinction between conceivability and possibility might be, for
5.2 any attempt I may make to discern the two is either a mere conception or supposition that such a distinction exists (compare Hume's "I may suppose, but never can conceive ...."); or an equivalently vacuous second-intentional conception-of-conception, which would not apply in a uniform way to any particular instance of "my" conception (compare "my" attempts to conceive of an unperceived / unconceived tree, or, *a fortiori*, of [myself]); and is, therefore, microscopically "incomprehensible."

In either case, therefore,

5.3   the alleged distinction does not itself "exist" (is a "mere nothing").

I'll later offer some partial attempts to disambiguate the elusive notions "conception what" and "conception (or "supposition") that"—which might be called *de re* and *de dicto* conception. For now, I would like to make two points which I hope the foregoing arguments will have made more plausible.

The first is that underlying many of the Berkeleyan and Humean arguments we have considered is a strong dialectical and rhetorical tendency to reject more or less out of hand not only
(a)   all "nonconstructive existence claims" (a phrase from the philosophy of mathematics); but sometimes, at least (in certain set-piece arguments, and in the more sweeping formulations of the elenctic "principles" which drive them),
(b)   all nonconstructive concept formation whatsoever.

The second is that the transition from 5.2 to 5.3 *does*, in fact, invoke a methodological or metatheoretic counterpart of the very principle which Hylas (and I) have questioned.

From the (*de re*) "inconceivability" for "me" of "my" notion of "(in)conceivability," namely, it may follow that such second-intentional conception is in some sense theory-relative, contextual and nonconstructive (of which more later); or that it is merely *de dicto*, in some sense; or all of the above. But not that "(in)conceivability" is equivalent to "(in)consistency" or "(non)existence."

Indeed, this alleged equivalence itself may give rise to the thesis of an antinomy. For the same argument which establishes that (in some sense) I don't know what I'm talking about when I try to discern (in)conceivability from (in)consistency, would seem to establish (in the same indefinite sense) that I don't know what I'm talking about when I try to identify them either.

And this dialectical response, in turn, is simply one among many lines of argument which suggest that outright refusals to countenance nonconstructive claims of (non)existence are methodologically untenable.

And if this is so, then perhaps Hume and Berkeley have after all been "insensibly drawn into" one of Berkeley's charmingly described

> uncouth paradoxes, difficulties and inconsistencies, which multiply and grow upon us as we advance in speculation; till at last, having wander'd through many intricate mazes, we find ourselves just where we were ....

(I refrained for a moment from (re)completing Berkeley's sentence with its continuation, "... or, which is worse, sit down in a forlorn skepticism," because the outcome seems to me neither worse nor forlorn.)

The "principles" 2.1 and 2.2, at any rate, do also resemble metalogical principles which might be proposed, and whose formulations straddle theory and metatheory in ways which narrowly avoid semantic paradox. Not surprisingly, perhaps, the more well-defined versions of these metalogical principles are metatheoretic "matters of fact"—a philosopher of mathematics acquainted with the notions involved may (as far as we can tell) consistently accept or decline them. Their very formulations are local and theory-relative, moreover, as are the notions of "consistency" and "conceivability" on which they are based.

Consistency, at least, would seem to be a rather straightforward metalogical notion: one simply formalises the classical "principle of noncontradiction." But the well-known second incompleteness theorem of Gödel suggests that this definition is both recurrently metatheoretic and theory-relative.

Metatheoretically consistent (conceivable?) theories of the form T + [T is inconsistent] (T + [T is inconceivable]?), for example, strongly suggest that attempts to provide unambiguous definitions of "consistency" may never quite mean what we "intend" them (quite nonconstructively) to mean. Even if we relativize our metatheoretic intentions to assorted background meta(meta)theories, the meta(...)theoretic "consistency" can be queried in turn, in latter-day variants of the pyrrhonists' dilemma of criterial ascent.

"Consistency," in other words, might be only a bit less obviously (meta)theory-relative than "truth."

Conceivability, of course, is *not* a standard metalogical notion. That one should not expect to find a single, uniform theoretical $\varphi(v)$ for "v is conceivable" already follows from Gödel's diagonal lemma, which generates from φ a sentence $\Psi \leftrightarrow \neg\varphi([\Psi])$, which provably asserts [its own] inconceivability.

If we accept a background metatheory T in which a completeness-theorem holds (as in Hume's passage from [E] I 25–26, quoted above), then "consistency of an assertion σ with a theory t" within the metatheory is provably equivalent to "existence of an interpretation or model of t in which σ holds". Let us nominate this "internal" notion—call it [consistency]—as a candidate for a minimal notion of conceivability *de dicto*, for assertions or states of affairs that can be coded or expressed in T. (This definition itself is then expressed in a metametatheory for the metatheory T).

Thus we have, in the framework just sketched,

5.4.1 A theory t about a relation r ("cause," say) is conceivable (*de dicto*) in a metatheory T if and only if T ⊢ "t has a model **m**".

A simple counterpart to (part) of the stronger, "conceptual atomist" notion Hume often seems to have in mind, might be the following:

5.4.2 A theory t about a relation r ("cause," say) is conceivable *(de re)* in a metatheory T if and only if
(a) T proves that "t has a pointwise-definable model **m**"; i. e.,
(b) there is a Russellian definite description ϕ in the language of T such that T proves that "the unique x such that $\phi(x)$ is a model of t".

I would like to emphasize immediately that 5.4 is not offered as a reconstruction of "my" notion of conceivability, either *de dicto* ("conceivability that"), or *de re* ("conceivability what") (I'm not sure whether I have one). Rather, 5.4.1 and 5.4.2 are intended as formal analogues for notions one might associate with (different calibrations of) Hume's demands for conceptual clarity and "microscopic" particularity.

One could also introduce a concomitant notion of conceivability *de re* for arbitrary objects **m** in T (**m** is conceivable *de re* if and only if **m** is pointwise definable). Neither this notion nor 5.4.2, however, strong as they are, would quite correspond to the full strength of Berkeley's and Hume's "finitist" demands.

If one required, for example—with appropriate metatheoretic care—that the model or arbitrary object **m** be "hereditarily" pointwise definable (roughly, that **m** and all elements of elements ... of **m** be pointwise definable; consider again Berkeley's thought experiment with the mite's foot), then some some rather everyday definable "objects" of a set-metatheory T—"the real numbers," for example—would become "inconceivable" *de re* in T. Which would have been fine with Berkeley and Hume.

Some infinite objects in the sense of certain T's would still be conceivable *de re*, however. The least infinite ordinal, for example, would be conceivable *de re* in quite reasonable set-theories T.

If one further demanded that every relation on **m** be pointwise definable, one would have something like the full force of the view I have called Berkeleyan and Humean "finitism:" only theoretical objects which are provably finite in T would be "conceivable" de re in T.

Another example of nonconstructive concept formation may help clarify a sense in which even the relatively "weak" 5.4.2 is counterintuitively strong. Since I have resorted to Russell's "definite descriptions," let me summon up that philosophical Ghost of Hamlet's Father, "the present king of France."

If T is some theory about as-yet unspecified intensional objects, and t some appropriate theory about "France", "royalty," 'the present," etc., then 5.4.2 would require that there be some uniquely definable thought-object **m**, which provably satisfies t in T. If I search [my] field of mental vision in good Humean fashion, it seems to me that I find no such object. It would then follow that "I" have (or T has) no conception (*de re*) of "the present king of France;" and consequently (if T or "I" accept 2.2 for this form of "conception") that "the present king of France" is not only nonexistent, as Russell observed; but incoherent ("impossible;" "inconceivable" in T) as well.

What is wrong here? Essentially, I would argue, an equivocation between what I have called conception *de dicto* and conception *de re*, in the formulation and justification of 2.2. It may be helpful here to recall a related medieval distinction: that between *conceptio quod* (conception that—essentially what I have been calling "conception *de dicto*"), and *conceptio quid* (conception what—what I have been calling "conception *de re*").

As I indicated briefly above, I have offered 5.4.2 as a rough metalogical reconstruction of conception *what*, not conception *that*. I have also argued that the former seems to me the minimal sort of "conception" one would need to satisfy Hume's (and Berkeley's) demands for "microscopic" precision and particularity, and reflect his view of more obviously underdetermined and nonconstructive notions such as "necessity" and "cause."

The notion which may be more plausibly assimilated to nonconstructive "conceivability," of course, is the weaker one in 5.4.1, of conception *de dicto*, or conception *that*. This follows from a straightforward application of the completeness theorem, which we have assumed provable in T, which asserts that object-theoretic consistency of t is equivalent in T to object-theoretic "possibility," construed as interpretation in some (not necessarily definable) structure, model or "possible world" **m**.

Returning to my example, it seems to me that I can quite straightforwardly conceive that there is a present king of France: say, by "supposing," in Hume's words, that some interpretation of this phrase might exist; and it further seems to me that I may do this quite "nonconstructively," without conceiving what (or in this case who) such a personage might be.

This analysis of "conception" (which closely parallels arguments about *de re* and *de dicto* modality in philosophical logic, and analyses of quantifier-"scope" and "exportation" in the philosophy of language) also suggests that not only "cause" and "necessity" are shaved by Hume's razor, but almost *any* (nonconstructive) instance of *conceptio quod* which is not extensionally canvassed in "all" its particulars by an instance of *conceptio quid*: not only our old friend "the present king …," but "the waltz-king" (which one was it, anyway?), the "king of the road," "*ein König von Thule*," "*le roi de Prusse*," "the King of Glory," and "king's x" …. Wittgenstein would have a field day …. (In fact, he did).

Not to mention the essentially metatheoretic notion of *conceptio quid* / conceivability *de re* itself. Several early-twentieth-century logicians (Russell, Grelling, Berry, Richard) discovered in effect the following phenomenon, later refined by Gödel and others:

5.5   the notion of "pointwise-definability in the language of T" (definability by one of T's "definite descriptions") is not itself definable by a predicate in T.

The definition 5.4.2a, for example, is not itself expressible in T. The implicit existential quantifier (over "definite descriptions") in its paraphrase 5.4.2b is metatheoretic with respect to T (an instance of metalogical quantifier "exportation").

This result seems sufficiently suggestive to warrant comparison with the self-referential elusiveness of "my" notion of (in)conceivability (*de re*) noted earlier, and perhaps with the more notorious self-referential indefinability (for "me") of "my" identity as well. Hume's original formulation of this aporia is *de re* (in effect), but Gödel's results also permit a plausible *de dicto* version of this as well (cf., e.g., [Boos], 1983).

In a metametatheory for T, one can say of certain interpretations **M** for T that every theory t which is "possible" or consistent in **M** (i.e., is such that **M** $\models$ "t has a model") is "conceivable *de re*" in **M** (that there is a pointwise-definable element **m** of **M** such that **M** $\models$ "**m** is a model of t").

But the very fact that one must have recourse to interpretations **M** of T in metametatheory $\mathcal{T}$ means that the associated reconstruction of 2.1 and 2.2 is a "matter of (metametatheoretic) fact," and not even expressible in the language of T's "relations of ideas."

In the metametatheory $\mathcal{T}$, one can even consider models **M** of T which have the property that every element of **M** *is* pointwise-definable in **M**: such structures can be obtained whenever one can introduce sufficiently many "Skolem functions" for T in the metametatheory (a requirement which depends in general on various forms of the axiom of choice in $\mathcal{T}$), and they are sometimes called "Skolem models" or "Skolem hulls." What one can*not* do, however, is say in T that 2.1 and 2.2 are the case.

Finally, notice that the appeal to a metametatheoretic $\mathcal{T}$ begins still another ascent into successive meta(meta)theories, of the sort which would seem to undermine in itself any truly Humean project of trimming one's metaphysics to fit one's epistemology (Where could a "final" calibration of the closeness of trim take place?).

Other instances of recourse to metatheoretic ascent arise from related points of dialectical stress. Consider, for example, the strain between 2.2 and Hume's professed finitism about the "compounding," "mixing" and "deriving" of "ideas" and "impressions." To apply 2.2 in its contrapositive form ("inconceivable *de re* → impossible") to a given theory t or relation r in T, as in 5.4, Hume would need to show that "all" attempts to define an interpretation m for t in T fail.

But his own methodological finitism should presumably forbid him to undertake such a project, much less make such a claim; for it would presumably involve checking each of an infinite scheme of potential definitions of such an **m** for verification of "all" the axioms of t in T (On similarly "Richardian" grounds, it is hard to see how we could determine that "'everything' is finite," unless "we"—or something to which we somehow had epistemic access—were not part of "everything"). In effect, any attempt(s) to amalgamate such infinitary processes of verification into something finitely surveyable would also seem to require consideration of higher-order meta(meta)theories .... And these would all, presumably, be even more conspicuously at the wrong end of the microscope.

At one point in [E] I, Hume seems to sense that his argument of inductive incompleteness might be metatheoretically applied to his claim ("relation?") of indemonstrability for at least one assertion: "the supposition that the future will be conformable to the past:"

> I must confess that a man [sic] is guilty of unpardonable arrogance who concludes, because an argument has escaped his own investigation, that therefore it does not really exist ....
>
> Even though we examine all the sources of our knowledge, and conclude them unfit for such a subject, *there may still remain a suspicion, that the enumeration is not complete, or the examination not accurate.* [E] I 38–39 (emphasis mine)

His response to this brief examination of methodological conscience promises

> considerations which seem to remove all this accusation of arrogance or suspicion of mistake ....

Unfortunately, these turn out to be an urbane review of Humean commonplace,

> It is certain that the most ignorant and stupid peasants—nay infants, nay, even brute beasts —improve by experience ....

followed by a reassignment of the burden of proof:

> If you assert, therefore, that the understanding of the child [who is "careful not to put his hand near any candle" after having been once burned] is led into this conclusion by any process of argument or ratiocination, I may justly require you to produce that argument; nor have you any pretence to refuse so equitable a demand. [E] I 39

Many things might be said about this short passage, in the context of the arguments we have considered so far. We consider three.

The first and simplest is that it simply fails to redeem the expectations prompted by the remark which introduces it, quoted above ("... considerations which seem to remove all this accusation of arrogance or suspicion of mistake ....").

The second is that this passage essentially "refutes" a charge of *ignoratio elenchi* with an *ignoratio elenchi*. Hume seems to suggest, in effect, that any contemporary readers who failed to provide precise alternative explanations for this particular instance of (what we would now call) conditioned response, would be "justly" constrained to assent that no "process of argument or ratiocination" (or even of information processing?) could play a significant role in any such "supposition that the future will be conformable to the past." Among other things, this also looks perilously like another question-begging (as well as "enumerative") attempt to appeal to a metatheoretic version of 2.1 $\wedge$ 2.2 (If we've looked and looked for a conceptual analysis of x, and found none, then there must be no conceptual analysis of x) to vindicate 2.1 $\wedge$ 2.2.

The third point which seems to me to bear emphasis is that the passage does also appeal quite clearly to an induction, at the level of Hume's metatheoretic (?) observations about "stupid peasants ... infants ... [and] brute beasts ...;" and that this inductive argument, in turn, *is* a (rather weak) subsidiary line of inference in a larger enumerative "ratiocination," as Hume for a moment acknowledges. But this "ratiocination" is not one from which he will permit us (on pain of "unjust" obduracy) to withhold our assent.

In the chapter's last section, I will reconsider the problem of self-application of Hume's elenctic "principles" at some length. The argument we have just been considering, for example, might be paraphrased as an imploding assertion, that

5.6   Only "Custom" could provide enough persuasive force for the implicitly inductive arguments Hume would need to convince us that [only "Custom" could provide enough persuasive force for the implicitly inductive arguments Hume would need to convince us that only ....]

Further examination of the indefinite but suggestive distinction between *conceptio quod* / conception *de dicto* and *conceptio quid* / conception *de re* may also help clarify some other, interrelated problems of Hume's metaphysics and epistemology. Consider once again, for example, the King of Glory. Ironically, an assertion that we might be able to conceive that such an Entity might exist, without any conception what "He" might be, would seem to parallel rather closely Pierre Gassendi's proto-empiricist analysis of nonconstructive concept formation about the christian god, in his comments on Descartes' *Meditations* (cf. [*AT*] VII 287–288, 293, and 294–295).

"Ironically," because Descartes also makes strong use of a distinction which closely resembles the one between *conceptio quid* and *conceptio quod*: the distinction between "*imaginari*" ("imagine") / "*comprehendere*" ("comprehend"), and "*intelligere*" ("understand") / "*concipere*" ("conceive"); (cf. [*AT*] VII 46, 52, 65-65, and 72-73; one "comprehends," e.g., an equilateral triangle, but "conceives" a myriagon, or god).

Hume worked with rather similar notions, I believe, in transposed language; his "supposition" retains the more nonconstructive aspects of Descartes' "conception;" and his "conception" resembles Descartes' "comprehension" (cf. again [*T*] 241's "I may suppose, but never can conceive ...."). Descartes attached "clarity and distinctness" primarily to "conception" (not "comprehension"), and associated it with "truth." Hume, on the other hand, attached "clarity and distinctness" primarily to his stronger notion of "conception" (which I have assimilated to Descartes' "comprehension"), and associated it with "possibility" and consistency.

In many ways, the Cartesian "clear and distinct conception" (or "idea") resembles the stoics' paradigmatically "dogmatic" notion of the καταληπτική φαντασία ("apprehensive presentation"), a major object of pyrrhonist scorn. Hume, on the account I have offered, so strengthened the latently dogmatic elements of pyrrhonist "phenomenalism" that the criterial and dogmatic aspects of stoic κατάληψις migrated into "phenomena," his "ideas" and "impressions," and the "conceptions" (finitely) "deriv'd" from them. Thus my characterization of him as a "dogmatic" phenomenalist and "second order idealist."

And thus it may also be appropriate that a skeptical critique of Hume's transposition of κατάληψις into "the phenomena" concentrate on the dogmatic "principles" which effect this shift, and on several aspects of the "principles" which remain recurrently marginal to theoretical and speculative attempts to formulate, much less decide them. I believe, for example, that I can conceive of unconceived relations, possibilities and "*possibilia*," in much the same way I conceive of unperceived trees; but such "supposition" is precisely the sort of conception whose relevance to such assertions both Berkeley and Hume would strenuously deny. I am also willing to admit that I cannot conceive what "the" distinction between conceivability and possibility might be; but would argue, once again, that this does not imply that I cannot conceive that such a distinction may exist, unless my interlocutor begs a metatheoretic counterpart of the original question at issue.

Formal reductions of this sort of dilemma are also familiar to philosophical logicians who work in the margins of "epistemic", and "doxastic" as well as "alethic" modality. Metalogical counterparts also include questions about the degree to which a given theory T's quantification is "substitutional" (instantiatable by terms of the theory; this counterpart of 2.2 for objects in T may be ensured metatheoretically—for example, in one of the "Skolem models" M, mentioned above); as well as the metaphysically more suggestive problem of the theory-relativity and indefinability of "the definable."

Roughly speaking, the more "ideal" one's observer, the more that observer's epistemic, doxastic and alethic modalities will converge; and the more "ideally" comprehensive one's theory (or restricted the metatheoretic range of its "intended" interpretations), the more "substitutional" will be its quantification (as interpreted in this restricted class of "intended" models). But no theory which encodes enough inductive structure to make questions of internal definability or "conceivability" interesting can define or conceive (*de dicto* **or** *de re*) what "its own" notions of "definability" or "conceivability *de re*" should be.

It may also be worthwhile to (re)emphasize that on either construal of "conceivability"—*de dicto* **or** *de re*—one can prove the existence of countermodels **M** of T for the natural analogues of "Humean" as well as "Berkeleyan" claims, in quite reasonable metametatheories $\mathcal{T}$ for T.

More precisely, one can prove the following assertions in such a $\mathcal{T}$.

5.7   There are models **M** of T, finite subtheories t of T, and assertions α in the language of t such that

1   **M** $\models$ "t $\cup$ {α} is 'possible' (consistent)"; but no interpretation **m** of t which is conceivable *de re* (definable) in **M** is such that **M** $\models$ "**m** is a model of α".

2   **M** $\models$ "t $\cup$ {α} is '*im*possible' (inconsistent)"; but **M** *also* satisfies (the conjunction of all the sentences in) t $\cup$ {α}.

Informally, 5.7.1 says that there are assertions α which are both true and inconceivable *de re* in **M**, and 5.7.2 that there are even α's which are true and inconceivable *de dicto* in **M**. The former assertion follows from work of Richard et al. and Skolem; the second is essentially a consequence of Gödel's theorem.

Either or both of the theories T and the t in 5.7.1 and 5.7.2, moreover, can be higher-order: an instance of either assertion for a first-order t and α would provide a countermodel for (one of two obvious formal analogues of) Berkeleyan "immaterialism;" corresponding instances of 5.7.1 or 5.7.2 for (suitably chosen) second-order t and α would provide countermodels for the view I have called (Humean) "irrelationism."

I will close this section with a brief but serious attempt to convince the reader —if the point is not already clear—that the arguments of the last few paragraphs (and of this chapter more generally) are not formal efforts to "deconstruct" (Berkeley and) Hume in some way. I hope the reader will regard them as genuine—if somewhat wry and indirect—tributes to "the very ingenious author" and *der Scharfsinnige Hume*. For both the (genuinely) "ingenious authors" got straight to the dialectical point, as it were. On a positive interpretation of the arguments I have tried to present, Hume and Berkeley both innovated informal metaphysical counterparts of limitative theorems and semantic paradoxes which surprised, even shocked the brilliant mathematicians who first encountered them two centuries later.

On my view, the Richard paradox, Skolem's theorem and Gödel's diagonalisation lemmas provide persuasive evidence for the theory-relativity (subjectivity?) of "definition," formal "universality," and "consistency" (among other things). If the analogies of this chapter are correct, the brilliantly original arguments of Berkeley and Hume provided equally persuasive evidence for the theory-relativity (subjectivity?) of "perception," informal "universality," and "conceivability." ... No mean accomplishment, that.

We will attempt to consider more aspects of this "evidence" in the next section.

6 "[Metatheoretic] Matters of [Object-theoretic] Fact"

Familiar metalogical arguments strongly suggest that another Humean notion may subtly elude the "microscope"'s focal field: his well-known distinction between "relations of ideas" and "matters of fact."

Consider again the completeness-theoretic characterization at [*E*] I 25 of these latter "objects of human reason":

> The contrary of every matter of fact is still possible; because it can never imply a contradiction, and is conceived by the mind with the same facility and distinctness, as if ever so conformable to reality. *That the sun will not rise tomorrow* is no less intelligible a proposition, and implies no more contradiction, than the affirmation, *That it will rise*.

"Contradiction," one may ask, with what? With (or in) what theory? "With an empty theory" is (I believe) the implicit answer; or at least with some theory about "the sun," etc. whose nonlogical axioms (as they would now be called) do not decide (prove or refute) the assertion that "the sun will rise tomorrow." Should one propose axioms for some physical theory, T, which does decide the issue (given appropriate boundary conditions, say, which may be built into T),

Hume need only point out the "factuality" (and defeasibility) of these axioms in order to reassert his point.

In any case, this line of argument suggests to me that one (re)consider the following straightforward reconstruction, which parallels Rudolf Carnap's classical attempts to formalise an (he would have said "the") analytic / synthetic distinction, in more contemporary metamathematical language.

6.1.1 A sentence σ in the language of a (metatheoretically) consistent T is a matter of fact for T if and only if T does not decide σ: that is, both σ and $\neg\sigma$ are consistent with T (cf. "The contrary of every matter of fact is still possible ....");

6.1.2 ρ is a relation of ideas for T if and only if ρ is not a matter of fact for T: that is, ρ is provable or refutable in T.

Quine's "holistic" objections to such reconstructions are of course well-known. Here, I wish to make a different point. By standard Gödelian arguments,

6.1.3 a sufficiently comprehensive T encodes [its own] metatheoretic provability of σ from T and consistency of σ with T, as object-theoretic predicates [[σ] is provable] and [[σ] is consistent], respectively (We will omit mention of T when no confusion seems likely to arise).

Using these predicates, one can also

6.1.4 encode the metatheoretic notions "ρ is a matter of fact" ("relation of ideas") for T, as object-theoretic predicates MF([ρ])(RI([ρ])).

Notice that one can at least write these notions in a natural way in T, unlike "conceivability *de re*" for T; and the [consistency]-predicate is identical with the one I called "conceivability *de dicto*" for T in the last section, by the very completeness result (in T) which Hume's "relation of ideas" / "matter of fact" distinction prefigured.

Nevertheless, metalogical problems once again persist. Like consistency itself, the MF / RI-distinction may be formally expressible in T, but any attempt to ensure that it "mean" what one "intends" it to mean is once again a matter of metatheoretic fact, and this is so even for theories T which are metatheoretically "sound" as well as consistent (i. e., such that $T \vdash \sigma$ if and only if $T \vdash \text{Prf}_\tau[\sigma]$).

For example, by Gödel's arguments, there are metatheoretic matters of fact (undecidable sentences) φ for such T, such that it is also a metatheoretical matter of fact (for T) whether [φ] is an object-theoretic "matter of fact" (for T). That is, there are metatheoretic interpretations for T in which MF([φ]) holds, and others in which it fails.

The principal (even canonical) example for this is Gödel's original sentence, $dia(T) = \varphi$, which is provably equivalent to the sentence Con(T) = [[0 = 1] is not provable] which encodes the putative metatheoretic consistency of T (and in a sense, therefore, the assertion that T itself is "conceivable *de dicto*" "in" T).

Furthermore, it is even a matter of metatheoretic fact whether φ itself is object-theoretically *"conceivable (de dicto)."* For there are also interpretations of T (indeed, of T $\cup \{\varphi\}$), in which [[φ] is consistent]) ([[Con(T)] is consistent]) holds, and others in which it fails.

All this suggests, to me at least, that Hume ought to find it extraordinarily hard, on his own principles, to bring "the" "relation of ideas" / "matter of fact"-distinction—a cardinal tenet of his philosophy, after all—into clear resolution under his own "microscope." And this is indeed the case.

For Hume bases the notion of "relation of ideas" rather explicitly on one of metaphysical necessity: a "relation of ideas" is precisely something of which it is not the case that "the contrary" is "possible" (cf. again [E] I 25). And I have construed this implicit necessity as metatheoretic validity in all interpretations, or equivalently (by completeness), to metatheoretic provability.

But Hume's rejection of physical relations such as "cause" and "necessary connexion" is based precisely on his refusal to countenance such a metaphysical necessity, as something which cannot be microscopically "deriv'd," or even "conceiv'd."

On this construal, Hume's "refutation" of the existence of "necessary connexion" also bears a curious resemblance to the pyrrhonist project to "prove" in some sense that "there is no proof," mentioned in an earlier section (cf., once again, [PH] II, 97–192).

More tellingly, perhaps, I would also argue that a sustained Humean attempt to "conceive" or "derive" a determinate "relation of ideas" / "matter of fact distinction" would have to face a fork of its own. It would either beg for itself a metatheoretic exemption from Hume's own "empirical" or "experiential" criteria; or it would formulate an "internal" distinction, which (like MF and RI above) will be subject to the self-referential dilemmas of semantic paradox sketched above.

And in either case, the troublesome margins of metatheoretic questions would persist. In the first, they drive the argument into what I have called "criterial ascent." In the second, they burden it with the opaque complexities of what might be called "[intentional descent]" (informal formulation and consideration of things like MF([MF([ρ])]), ...); and with (related) semantic paradoxes of any attempts to decide within T what is decidable in T.

Finally, both these lines of argument—metatheoretic and theory-internal—are likely to involve explicit consideration of metaphysical claims which are no less underdetermined by "every" finite stage of inquiry into their theoretical tenability than the physical relations of "cause" and "necessary connexion."

Lest such allusions to "criterial ascent" and informal counterparts of self-referential dilemmas seem pointless and avoidable, recall also that there is an obvious sense in which the whole debate between Carnap and Quine can be construed as an an attempt to adjudicate whether "the" distinction between "relation of ideas" and "matter of fact" is a "relation of ideas" or a "matter of fact."

Nor does it end there: plausible assertions, that such attempts at adjudication are "in fact" [?] impossible, are essentially metametatheoretical, "matter-of-factual" claims that the object theoretic distinction between "relation of ideas" and "matter of fact" is a metatheoretic "matter of fact" ....

"*O Altitudo*," Thomas Browne might have said (Others may find other forms of exclamation or apostrophe more appropriate).

In any case, these metalogical analogues are in some sense expressly non-Humean. But they do, I think, nevertheless suggest several concomitant lines of strain in Hume's own arguments. And this strain is also reflected in interesting shadows of doubt and reconsideration which do recur from time to time, not only in the Appendix to [*T*], but occasionally in [*E*] I as well.

Consider, for example, Hume's own, rather striking deprecation of the fork itself, alluded to earlier, in the immensely long footnote which spans [*E*] I 43 to 45:

The same distinction between reason and experience is maintained in all our deliberations concerning the conduct of life .... Though it be allowed, that reason may form very plausible conjectures with regard to the consequences of such a particular conduct in such particular circumstances; it is still supposed imperfect, without the assistance of experience, which is alone able to give stability and certainty to the maxims, derived from study and reflection.

But notwithstanding that this distinction be thus universally received, both in the active and speculative scenes of life, I shall not scruple to pronounce, that it is, at bottom, erroneous, at least, superficial.

If we examine those arguments, which, in any of the sciences above mentioned, are supposed to be the mere effects of reasoning and reflection, *they will be found to terminate, at last, in some general principle or conclusion*, for which we can assign no reason but observation and experience. *The only difference between them and those maxims, which are vulgarly esteemed the result of pure experience, is, that the former cannot be established without some process of thought, and some reflection on what we have observed, in order to distinguish its circumstances, and trace its consequences ....* [emphases mine]

The tentative conclusion of this rather remarkable argument—with which one might compare Quine's speculations about "the" "empirical" nature of logic—seems to be at least twofold: that there may be no sharp, "fork"-like demarcation between "reason" and "experience;" and that such distinction(s) as may exist may be relative to some process of thought, and some reflection on what we have observed, in order to distinguish its circumstances and trace its consequences.

This seems to me quite compatible with my—metalogically straightforward—observations that the demarcation of such "internal" reconstructions as MF and RI is a matter of metatheoretic revision, … And so on, *ad indefinitum*.

Had Hume explored this footnote's insights somewhat further, he might have seemed a vaguer and more latitudinarian "empiricist" to his contemporaries and intellectual descendants. But he would also have been (in my view, at least) a better and more consequent skeptic.

7 Hypotheses Non Fingimus?

In the last two sections, I have suggested several times that Hume's own "establish'd maxims" and philosophical arguments often lack—and perhaps in some sense "must" lack—the clear microscopic resolution he demands of his philosophical and scientific opponents. The arguments of preceding sections have also suggested that Hume-the-metaphysician may be constrained to engage in exactly the sort of indeterminate speculation which he so vigorously denies his scientific opponents. Such (second-)intentional metatheoretic notions not only elude his own criteria of clarity, distinctness, and extensional particularity; they also conflict with the tacit claim of universality which he makes for his metaphysical arguments:

> *All* the objects of human reason or inquiry may naturally be divided into two kinds, to wit, Relations of Ideas, and Matters of Fact. [*E*] I 25

> We have said that *all* arguments concerning existence are founded on the relation of cause and effect; that our knowledge of that relation is derived entirely from experience; and that all our experimental conclusions proceed upon the supposition that the future will be conformable to the past. [*E*] I 35 [emphases mine]

("All?" Have we enumerated them? Extrapolated from our meager finite evidence about them—and thus reached "universal" conclusions about "Custom?"—with the aid of "Custom?")

For the most part, Hume seems quite understandably to have assumed that one could simply merge the methodological / "metatheoretic" conceptual apparatus of his "principles" with his applications of them, on the one hand (relatively more "object-theoretic" arguments), and his own and others' analyses of them (metametatheoretic, or perhaps metaphilosophical, considerations of their consequences), on the other. Rather naturally, perhaps, he seems also to have assumed most of his most important methodogical presuppositions would simply be granted—as hypotheses, if necessary (or "principles," or "establish'd maxims")—by his readers.

In this section, I will explore further some reflections of these assumptions and meta-assumptions in his persistently ambiguous attitude toward hypothetical and counterfactual aspects of "the universe of the imagination" (cf. [T] 68).

It may be useful to recall, first, the extent to which it is also a characteristic mark of the problematical notions Hume wishes to query, that they involve "smooth and easy" concept formation about entities which are "counterfactual," or at least (hypothetically?) marginal to his (allegedly) well-defined first-order sensory data ("impressions" and "ideas"). Mathematical equality, of course, is an almost archetypical example (recall the argument of the *Phaedo*).

> This appears very conspicuously with regard to time; where ... the various corrections of our measures, and their different degrees of exactness, have given us an obscure and implicit notion of a perfect and entire equality. [T] 48

The paradigmatic physical example is of course "causation."

> ... of those three relations, which depend not upon the mere ideas, the only one, that can be trac'd beyond our senses, and informs us of existences and objects, which we do not see or feel, is *causation*. This relation, therefore, we shall endeavour to explain fully before we leave the subject of the understanding. [T] 74

The existence of external objects is cited at [T] 211:

> There are no principles either of the understanding or fancy, which lead us directly to embrace this opinion of the double existence of perceptions and objects, nor can we arrive at it but by passing thro' the common hypothesis of the identity and continuance of our interrupted perceptions.

In the end, I have argued, Hume's decision to grant provisional and somewhat equivocal assent to the existence of Berkeley's "external objects" virtually requires that he consider such broad and vague "hypotheses," which are of a sort he elsewhere seems quite unequivocally to condemn (more of his more general view of "hypothesis" later).

"Continu'd existence" of objects in general comes under scrutiny at [T] 197:

> ... 'tis evident, that whenever we infer the continu'd existence of the objects of sense from their coherence, and the frequency of their union, 'tis in order to bestow on the objects a greater regularity than what is observ'd in our mere perceptions.

The closely-related property which might be called "continu'd identity" (still of objects) is subjected to his reservations about the *"principium individuationis* or principle of identity" (with interesting implications for the "principle" 2.3), at [*T*] 201:

> This fiction of the imagination almost universally takes place; and 'tis by means of it, that a single object, plac'd before us, and survey'd for any time without our discovering in it any interruption or variation, is able to give us a notion of identity .... Thus the principle of individuation is nothing but the *invariableness* and *uninterruptedness* of any object, thro' a suppos'd variation of time ....

Personal identity, similarly, suffers a miniature version of its familiar dissolution at [*T*] 254:

> ... we often feign some new and unintelligible principle, that connects the objects together, and prevents their interruption or variation. Thus we feign the continu'd existence of the perceptions of our senses to remove the interruption; and run into the notion of a *soul*, and *self*, and *substance*, to disguise the variation.

But does not Hume also embrace the "counterfactual," in all its resplendently hypothetical glory, in the famous passage, for example, quoted earlier, from [*E*] I 18?

> [T]hought can in an instant transport us into the most distant regions of the universe; or even beyond the universe, into the unbounded chaos, where nature is supposed to lie in total confusion. What never was seen, or heard of, may yet be conceived; nor is any thing beyond the power of thought, except what implies an absolute contradiction.

Yes and no. Two potentially conflicting motives seem to me worth comment here. The first is that there is a certain slide from [*E*] I 18's "thoughts" which "can in an instant transport us into the most distant regions of the universe; or even beyond the universe ...," to [*E*] I 25's

> [p]ropositions of this kind ["relations of ideas"] [which] are discoverable by the mere operation of thought, without any dependence on what is anywhere existent in the universe. Though there never were a circle or triangle in nature, the truths demonstrated by Euclid would for ever retain their certainty and evidence.

Perhaps this is one reason why Hume called them "relations of ideas"—"thought-objects," in effect, rather than "merely logical" relations (compare again the literal meaning of the pyrrhonists' παινόμενα). An assimilation of "relations of ideas" to "thought[s] ... beyond the universe" might also resonate with Hume's tendency to dismiss mere "supposition" altogether, and clarify somewhat his virtual abandonment of "relations of ideas" and "the distinction between reason and experience" in the lengthy footnote at [E] I 43–45, quoted earlier.

A similarly suspicious tendency to conflate what is "independent" of (metatheoretic) "fact" with what is "beyond" all fact also (re)appears, I believe, in Kant's ambiguous characterization of "the noumena." It seems to me at least a collateral ancestor of the early twentieth century cliché that "logic" is "tautological," and thus "vacuous."

In any case, a second point I wish to make about Hume's equivocal views of hypothesis and "the" counterfactual involves another, quite different assimilative slide: from (what I have called above) *conceptio quod* or "conception *de dicto*" to (what I have called) *conceptio quid*, or "conception *de re*."

Immediately after the passage from [E] I 18 quoted above, Hume continues as follows:

> But though our thought seems to possess this unbounded liberty, we shall find, upon a nearer examination, that it is really confined within very narrow limits, and that all this creative power of the mind amounts to no more than the faculty of compounding, transposing, augmenting, or diminishing the materials afforded us by the senses and experience. [E] I 19

(His subsequent example of such "compounding"—the "golden mountain"—is another one of those great philosophical commonplaces; it presumably lies in the demesne of the present king of France).

In an earlier section, I have argued that such "compounding, transposing, augmenting or diminishing" is inadequate to account for conceptual thumbnail sketches one might propose (e.g.) for "the present" and other "kings". One might also pose a naturally related, almost soritical problem: how much can one "diminish" or "augment" a clear and distinct "impression" before it loses its clarity and distinctness? An orthodox "microscopic" answer would seem to be conceivable, but this would indeed require something like the dogmatic "finitism" of Hume's and Berkeley's minimum visibles, tangibles, etc., discussed earlier as particularly doctrinaire (and methodologically strained) realizations of a hypothetical demand that "everything" be "conceivable *de re*."

Notice, moreover, that a parallel move to save-the-"impressions" for "augmentation" would seem to require that there be maximum visibles, tangibles, and the like—a considerably less intuitive solution to an inversion of the same metaphysical (metamathematical?) problem.

Why, then, does Hume's "compounding" metaphor (which I would defy him to "trace up" to well-defined "impressions," along relevantly determinable paths) seem to him (and perhaps to us) as plausible as it does—in the case of the "golden mountain," say? Perhaps because this "idea" involves a relatively simple propositional conjunction, or perhaps a comparably simple instance of (relatively) "atomic" predication. Whether I predicate "goldenness" of some "mountain," or conjoin "goldenness" and "mountainhood," I may slide the gold-filter, as it were, over "mountainhood," or "in front" of a (slightly fuzzy) "general idea" of a mountain, in quasi-Berkeleyan fashion.

In the case of "the present king," however, quantification is involved, not only in the obvious way Russell envisioned (over "people"), but perhaps in other, implicit ways as well; and it seems to be "nonconstructive" quantification over indefinite ranges of "intended" interpretation. The need to accommodate such quantification, in fact, lies at the heart of the "Richard paradox"—the indefinability of definability, mentioned above.

Does "... of France," for example, mean over "all" "the French people?" Or only those who do not accept another pretender (if there are any)? Do "I" have (as Kant presumably wondered) a Humean "impression," suitably "compounded," of "the present"—as opposed to "the former"—much less of "the present" in general?

Simple arguments of this sort suggest to me that a tacit willingness to engage in "nonconstructive" quantification over indefinite, "intended" domains—what might be called "domains of reflexion"—is pervasively involved in this as well as other common instances of "hypothetical" concept formation which fail to satisfy Hume's precisian demands. Indeed, I would argue that our background-readiness to engage in "nonconstructive" quantification of this sort may be one of the reasons why paradoxes such as Richard's bring us up so short, and have the purchase they do: we tend to think we "ought" to be able to "fix" conundrums of this sort, since we do "define" (or at least refer to) things like the first integer not definable in less than forty syllables "all the time," do we not? ... Lest this seem strained, consider again Berkeley's "counterintuitive" claim that we can*not* "conceive" / "define" the nearest "inconceivable" / "indefinable" tree).

Metalogical counterparts of such "supposition" (as opposed to Hume's "conception," or Descartes' "comprehension") and "intention" were also central, of course, to Section 4's reconstruction of a distinction between *conceptio quod* and *conceptio quid*. Hume's apparent willingness to preach the latter and engage in the former, is not, on my account, a deliberate consequence of some sort of intellectual hypocrisy, as I tried to make clear at the end of the last section. It is, rather, a near-inevitable concomitant of his (ironically hypothetical) foundationalism: his "suppositions" of conceptual particularity, and related (and equally hypothetical) assumptions that quantification "must" be "substitutional," over well-defined "internal" domains of "ideas" and "impressions," (always instantiated by a particular one of these "ideas" and "impressions," and preferably one of the latter).

Some aspects of this methodological (dialectical? hermeneutic? vicious?) circle emerge in the following well-known passages from [T] 168–169, just prior to [T]'s counterparts of [E] I 76's modally ambiguous definition of "cause." Hume has just reiterated the more cautious version of his "irrelationalist" metaphysics, which even begins with a (prudential?) concession.

> As to what may be said, that the operations of nature are independent of our thought and reasoning, I allow it .... But if we go any farther, and ascribe a power or necessary connexion to these objects; this is what we can never observe in them, but must draw the idea of it from what we feel internally in contemplating them.

He follows this, however, with a rare and interestingly explicit excursion into methodological self-reference.

> And this I carry so far, that I am ready to convert my present reasoning into an instance of it, by a subtility, which it will not be difficult to comprehend .... when we change the point of view, from ... objects to ... perceptions; ... the impression is to be considered as the cause, and the lively idea as the effect; and their necessary connexion is that new determination, which we feel to pass from the idea of the one to that of the other. The uniting principle among our internal perceptions is as unintelligible as that among external objects, and is not known to us any other way than by experience. Now the nature and effects of experience have been already sufficiently examin'd and explain'd. It never gives us any insight into the internal structure or operating principle of objects, but only accustoms the mind to pass from one to another. [T] 169

There are several questions one might ask about this "subtility." The first involves the strength of the "must" in "must draw from them ...." One can credit, once again, the transparent sincerity of Hume's deprecation of a "dogmatical spirit" at the close of Book I ([T] 274), and still question such usages, on his own terms.

Even more pointedly, one could also query the closely related "determination of the mind" on [T] 169. Not only is "the mind" itself not unitary, as Hume observes in the Appendix. The "uniting principle" of this "determination" by "the mind" is also, Hume admits, "unintelligible"—which I take to mean that no microscopic analysis for it exists.

But then this would seem to be a straightforward instance of what is not "known to us ... by experience," according to Hume. Not only does such speculation suggest that "accustoming" and "determination" may after all be "secret powers" (compare the ambiguous outcome of Hume's struggle to distinguish such "force" or "determination" from "animal *nisus*," at [*E*] I 67, discussed in the next section). It might also serve just as well as a genuine *reductio* for his claim that "the nature and effects of experience have already been sufficiently examin'd and explain'd."

A basic methodological problem hovers over this issue, as over its counterparts in twentieth and twenty-first century philosophy of science. It goes beyond the underdetermination of "personal indentity" which Hume scrupulously acknowledged, and it is as troublesome (I believe) for Kant's "solution" to it as it is for Hume's: how can one characterize "experience" in a way which does not

> draw any inference concerning any object beyond those of which we *have had experience*. [*T*] 139

(The emphasis is mine; the entire passage is italicized in [*T*], as part of two "principles" of which we should be "fully persuaded".)

In Kantian terms, for example, one might ask why Kant's own characterization of "*Erfahrung*" is not itself an "*Idee der Vernunft*." On Hume's *own* "principles," of course, we cannot "generalize" from "have had experience of" to "may have experience of," any more than Adam can; so we cannot slide from indicative to subjunctive, or from past experience to possible experience (compare Kant's often-repeated assimilation of *Erfahrung* to "*mögliche Erfahrung*"). In a similar sense, it might seem that Hume ought to hold, by the same criterion he applies to physical generalization and scientific concept formation, that we (literally) don't know what we are talking about, when we generalize metaphysically about "experience."

If we cannot "conceive" (or "derive," or "compound") metaphysical necessity, in short, neither can we (presumably) "conceive" metaphysical possibility. And if what is conceivable is what is possible, according to 2.1 and 2.2, neither can we conceive what (possible) conception may be, as I have argued on somewhat different grounds in Section 5 above. And if "matters of fact" are what is possible but not necessary, how can we "conceive" "the" scope of experiential "fact?"

To me, these considerations also bring into different relief such confident remarks as the following, from [*T*] 83:

> 'Tis impossible for us to carry on our inferences *in infinitum*; and the only thing, that can stop them, is an impression of the memory or senses, beyond which there is no room for doubt or inquiry.

Why? By whose criterion is there "no room?" That of "Custom," of course. But whose "Custom?" And what secures its alleged (intersubjective) uniformity? We will return to this "unintelligible" question in the next section.

Further down the page, Hume instantiates his claim that we need not concern ourselves with the classical pyrrhonist trope of "lapse into infinites" with an examination of our warrant(s) for a belief that "Caesar was kill'd in the senate-house on the *ides of March*:"

> Tis obvious all this chain of argument or connexion of causes and effects, is at first founded on those characters or letters, which are seen or remember'd, and that without the authority either of the memory or senses our whole reasoning wou'd be chimerical and without foundation. Every link of the chain wou'd in that case hang upon another; but there wou'd not be any thing fix'd to one end of it, capable of sustaining the whole; and consequently there wou'd be no belief nor evidence. And this actually is the case with all *hypothetical* arguments, or reasonings upon a supposition; there being in them, neither any present impression, nor belief of a real existence.

Yes? ... This would be a natural skeptical "supposition" about the tenuous nature of historical evidence. Hume (the skeptic?) essentially begs here the well-foundedness of a particular "inference" from "experience," a reasonable enough response, perhaps, since we seem in this case to have no plausible alternative.

But a stronger denial of Leibnizian *analyses infinitorum* also seems to underlie the dissatisfaction Hume expresses in the quotation's last sentence—a denial which finds positive expression in his "principles," I have argued, and which posits a quantifier-shift that is thoroughly unwarranted on Hume's own terms:
from

7.1   Every "chain of argument" has a beginning (which we may denominate its "experiential" "evidence");

to

7.2   There exists a single uniform beginning ("Experience," "the" *source of* "Evidence") for every "chain of argument."

The first of these assertions may well be plausible (*pace* Leibniz). But the second, however plausible it may sometimes look in Hume's (and Berkeley's) fluent prose, does not follow from it, unless one invokes auxiliary—and essentially question-begging—assumptions, which would ensure "the" alleged uniformity—intersubjective and somehow structural—of ("all?") human experience. I have already suggested that Hume—that wry believer in "preestablished harmony"—was surprisingly willing to accept, even posit such auxiliary hypotheses, whenever he could express them primarily in assertions about "Custom" and "the mind," rather than claims about comparably uniform relational structures, in an (essentially featureless) physical external world. And this way, I will also wish to argue, lies Kant.

I will close the section with a few remarks about some implications of this apparent quantifier-shift (an instance of something logicians sometimes call a "uniformization argument") for a putative Humean counterpart of Newton's famous dictum, that *Hypotheses non fingo* ("I do not frame hypotheses").

On the interpretation I have offered, Hume's apparent uniformization argument does seem to postulate at least tacitly ("subreptively," in Kant's language) a uniform and essentially nonempirical ("absolute?") reference-frame—"Experience"—for "moral" (matter-of-factual), rather than "physical" (mind-independent) phenomena. Instead of "absolute space," in other words, the Newton of "Moral" Science (cf. [*E*] I 14's

> there is no reason to despair of equal success in our enquiries concerning the mental powers and economy, if prosecuted with equal capacity and caution ....)

may have "framed," in effect, a thoroughly hypothetical and equally non-experiential entity, which I have called "Experience" (and which is structured uniquely by "Custom"); but which one might also call, in honor of Hume's great predecessor, "absolute experience."

And if so, it may simply have remained for Kant to formulate an allegedly "necessary," "synthetic a priori" postulate (hypothesis?) that such postulation (with names duly changed) is "necessary" and "synthetic a priori." Tainted rationalist phrases like "absolute experience" need only be replaced by other, more modest (if essentially untranslatable) usages like "*Erfahrung überhaupt.*" ...

If Hume was Newton, perhaps Kant was not Copernicus, but Laplace?

8 "A Kind of Pre-established Harmony?"

In one of the *Pensée*'s aperçus, Pascal remarks that

> J'ai grand peur que cette nature ne soit elle même qu'une première coutume, comme la coutume est une seconde nature.
>
> I'm much afraid that this nature may be only a first custom, as custom is a second nature. [*Pensées*] 126/93

Consider, for comparison, the utterly "inconceivable" operation of Hume's "Nature," which by an absolute and incontroulable necessity has determin'd us to judge as well as to breathe and feel .... [*T*] 183

Hume's "Custom" and (Human) "Nature," in fact, are almost synonymous. He defines the former as the "principle" by which

> a person ... immediately infers the existence of one object from the appearance of another .... [*E*] I 42
>
> This principle is Custom or Habit. For wherever the repetition of any particular act or operation produces a propensity to renew the same act or operation, without being impelled by any reasoning or process of the understanding, we always say, that this propensity is the effect of *Custom*. By employing that word, we pretend not to have given the ultimate reason of such a propensity .... [*E*] I 43
>
> ... reason is nothing but a wonderful and unintelligible instinct in our souls .... This instinct,'tis true, arises from past observation and experience; but can any one give the ultimate reason, why past experience and observation produces such an effect any more than why nature alone shou'd produce it? Nature may certainly produce whatever can arise from habit: Nay, habit is nothing but one of the principles of nature, and derives all its force from that origin. [*T*] 179
>
> Perhaps we can push our enquiries no farther, or pretend to give the cause of this cause; but must rest contented with it as the ultimate *principle*, which we can assign, of all our conclusions from experience. It is sufficient satisfaction, that we can go so far ....
>
> This hypothesis seems even the only one which explains the difficulty, why we draw from a thousand instances, an inference which we are not able to draw from one instance, that is, in no respect, different from them ....
>
> no man, having seen only one body move after being impelled by another, could infer that every other body will move after a like impulse. All inferences from experience, therefore, are effects of custom, not of reasoning. [*E*] I 43

At times such passages almost seem to brush against an inchoate counterpart of the notion of a physiological *a priori*. Such a construal of Hume's "moral" arguments might render them oddly acceptable to twenty-first century psychologists, social psychologists, students of artificial intelligence and philosophers of mind. But (at least) two major difficulties would stand in the way of such a construal.

The first, and more straightforward, is that Hume explicitly rejected such physiological accounts as were available to him—most conspicuously, in his lengthy repudiation (set forth in three paragraphs and two extended footnotes of [*E*] I; cf. 64–69 and 77–78) of any hypothesis that our notion of "power" might reflect an experience of "animal *nisus*," for example; or be "copied from any sentiment or consciousness of power within ourselves" ([*E*] I 67).

The "animal spirits" posited in such accounts were simplistic and implausible, no doubt. But so (for Hume) were his arguments for their rejection (though he does grudgingly acknowledge that a "vulgar, inaccurate idea" may be formed of such evidence):

> It may be pretended, that the resistance which we meet with in bodies, obliging us frequently to exert our force, and call up our power, this gives the idea of force and power. It is this *nisus*, or strong endeavour, of which we are conscious, that is the original impression from which this idea is copied. But, first, we attribute power .... to the Supreme Being, who never meets with any resistance [*This* argument from Hume?] .... *Secondly*, This sentiment of an endeavour to overcome resistance has no known connexion with any event: What follows it, we know by experience; but could not know *à priori*. It must, however, be confessed, that the animal *nisus*, which we experience, though it can afford no accurate precise idea of power, enters very much into that vulgar, inaccurate idea, which is formed of it. [*E*] I 67

Indeed. Given the historically anthropomorphic origins of notions of "power" and "force" in early science, such a "vulgar inaccuracy" might seem to be a model case of the "founding" of an idea which Hume himself elsewhere calls "unintelligible" upon an obvious "antecedent impression." Why this is so much more "vulgar" and "inaccurate" than other first approximations and impressions, he does not explain.

The second difficulty with an attempt to represent Hume as a believer in some kind of natural or experiential *a priori* can perhaps be expressed in quasi-Kantian terms: for Hume, I will argue, "Custom" and "Habit" are essentially constitutive of causal analyses of the "nature" which he refers to as their "origin," in one of the passages quoted above.

The existence at least of a close interrelation between Hume's "nature" and "custom" (and between both and "causality") is witnessed again and again in the passages we have cited thus far, and further confirmed by the following remarks, which immediately follow [*T*]'s "definition" of "CAUSE," as (hypothetically?) "constant conjunction":

> when I consider the influence of this constant conjunction, I perceive, that such a relation can never be an object of reasoning, and can never operate upon the mind, but by means of custom, which determines the imagination to make a transition from the idea of one object to that of its usual attendant, and from the impression of one to a more lively idea of the other. [*T*] 170

German equivalents of the word "custom" ("*Brauch(tum)*," sometimes "*Sitte*") play no significant role whatsoever in Immanuel Kant's metaphysical writings. Furthermore, Kant explicitly identifies Hume's "habit" ("*Gewohnheit*") with (merely) "*subjektive Notwendigkeit*" (subjective necessity") at [*Pr*] 258.

Nevertheless, I will argue in this section that Hume effectively ascribed both to "custom" and to "habit" (cf., again, [*E*] I 43's "This principle is Custom or Habit") most of the attributes of *objektive Notwendigkeit* Kant later associated with the synthetic *a priori*.

In this connection, first, it may be of some interest that one notion which does bear a significant functional resemblance to Hume's "*Custom*" is Kant's "*Natur*," which the latter explicitly characterized at B 164 of the first *Critique* as "*das verknüpfende Vermögen*" ("the connecting faculty"). At [*Pr*] 319 Kant similarly refers to "*Natur überhaupt*" ("nature in general") as "*die Gesetzmässigkeit in Verknüpfung der Erscheinungen*" ("lawlikeness in the connection of appearances;" compare again "*Custom*"'s role as the "faculty" which produces "necessary connexions").

Also reminiscent of "*Custom*"'s capacity to order "possible experience" is [*Pr*] 36's claim that "*Natur und mögliche Erfahrung*" ("nature and possible experience") are "*ganz und gar einerlei*" ("utterly the same [thing]"), and Kant's characterizations at [*Pr*] 318 of *natura formaliter* (as opposed to *materialiter spectata*, as

> *der Inbegriff der Regeln, unter denen alle Erscheinungen stehen müssen, wenn sie in einer Erfahrung als verknüpft gedacht werden sollen.*
>
> the aggregate of rules [cf. the "general rule" at the bottom of [*T*] 141], *under which all appearances must stand, if [or when] they are to be thought of as connected in an experience.*
> (emphases mine)

More even than Kant dutifully acknowledged at [*Pr*] 257–262, then, I wish to suggest that Hume's arguments represented as "skeptically" half-empty a glass that Kant made "transcendentally" half-full. Past commentators have already pointed out several senses in which the *kopernikanische Wende* may in part have been a (brilliant) shift of metaphysical *Gestalt* (cf, for example, the articles by Beck, Dauer and Malherbe cited in [Capaldi] 171), in terms which would seem to be compatible at least with my account of Hume as a "second-order idealist." More recently, John Wright has analyzed Hume as a "skeptical realist," in terms which do *not* focus on the metatheoretical arguments of serious academic and pyrrhonist skepticism, as I have attempted to do here (cf. [Wright], 27–32, and notes 25, 31 and 34; from such a "metatheoretical" viewpoint, the notion of a "skeptical realist" would be a contradiction in terms); but which do seem to support a measure of assimilation between Humean "principles" and *Kantische Grundsätze des transzendentalen Idealismus*.

There are, at any rate, several fairly clear senses in which a "second-order" idealism about the relational structure(s) of experience is an essential part of Kantian *transzendentaler Idealismus*. Consider, for example, the following rather straightforward juxtaposition. First Hume:

> What! the efficacy of causes lie in the determination of the mind! As if causes did not operate entirely independent of the mind, and wou'd not continue their operation, even tho' there was no mind existent to contemplate them, or reason concerning them. Thought may well depend on causes for its operation, but not causes on thought. This is to reverse the order of nature .... [*T*] 167

Then Kant:

> ... so klingt es zwar anfangs befremdlich, ist aber nichtsdestoweniger gewiss, wenn ich in Ansehung der letzteren [der "ursprünglichen Gesetze des Verstandes"] sage: *der Verstand schöpft seine Gesetze (a priori) nicht aus der Natur, sondern schreibt sie dieser vor.*

> ... thus it sounds indeed disconcerting at first, but is certain nonetheless, when I say in regard to the latter [the "original laws of the understanding"]: *the understanding does not draw its (a priori) laws from nature, but prescribes them to her.* [*Pr*] 320 (emphasis Kant's)

Similar instances of such "prescription" may be found in the first *Critique* at (e.g.) B 163–165; cf. especially the characterization of "*Kategorien*" at B 163.

Differences remain, of course, between the "idealism(s)" of Hume and Kant; and one of them has to do with their attitude toward Berkeley's "I-know-not-what"'s. In an earlier section, I argued that Hume's attitude toward unperceivable "external objects" wavers somewhat between amused suspension-of-judgment and prudential acceptance (both of which conflict with his own "principle" 2.2), that such objects do, of course, in some sense at least, "exist." (Perhaps he has exited his study, or "left the shade?" cf. [*E*] I 159).

Kant's position in this matter is notoriously complicated, among other things by the many, rather scholastic distinctions which he tried to make between "*transzendentale Gegenstände*," "*Noumena*" and "*Dinge an sich.*" But the latter, at least, do definitely "exist", as (e.g.) *Korrelat(e) der Erscheinung* ([*KrV*] B 45; cf. also [*Pr*] 351 ff).

More revealingly, perhaps, Kant's initially spirited attempt in the *Prolegomena* to distinguish his position from Berkeley's (and presumably from what I have called Hume's "second-order" counterpart of it as well), ultimately collapses into the remarkably flat and even rather deadpan claim at [*Pr*] 293, that

> ... dieser von mir sogenannte Idealismus betraf nicht die Existenz der Sachen (die Bezweiflung derselben aber macht eigentlich den Idealismus in rezipierter Bedeutung aus), *denn die zu bezweifeln, ist mir nie in den Sinn gekommen ....*

> ... this idealism, as I have called it, did not affect the existence of things (doubt of which really constitutes idealism, in its received meaning), *for to doubt this has never crossed my mind* .... [Pr] 293 [emphasis mine]

(Perhaps one doesn't need to exit the study after all?)

Other textual parallels also seem to me of interest. Compare, for example, Kant's analyses of *"Vernunftideen"* ("ideas of reason," which characteristically involve unending *Reihen der Bedingungen*) with Hume's "skeptical" comments on a primary *Vernunftidee*, "this theory of the universal energy and operation of the Supreme Being:"

> Though the chain of arguments which conduct to it were ever so logical, there must arise a strong suspicion, if not an absolute assurance, that it has carried us quite beyond the reach of our faculties, when it leads to conclusions so extraordinary, and so remote from common life and experience. We are got into fairy land, long ere we have reached the last steps of our theory; and *there* we have no reason to trust our common methods of argument, or think that our usual analogies and probabilities have any authority. Our line is too short to fathom such immense abysses. And however we may flatter ourselves that we are guided, in every step which we take, by a kind of verisimilitude and experience, we may be assured that this fancied experience has no authority when we thus apply it to subjects that lie entirely out of the sphere of experience. [E] I 72

Or the "dialectic"'s antinomial aspirations to *"Vollständigkeit der Bedingungen"* ("completeness of the conditions;" cf. again Hume's "chains of argument" in the last quotation) with the following passages from [T] 266–267:

> Nothing is more curiously enquir'd after by the mind of man, than the causes of every phaenomenon; nor are we content with knowing the immediate causes, but push on our enquiries, till we arrive at the original and ultimate principle.

> This is our aim in all our studies and reflections: And how must we be disappointed, when we learn, that this connexion, tie or energy lies merely in ourselves, and is nothing but that determination of the mind, which is acquir'd by custom ....

> ... when we say we desire to know the ultimate and operating principle, as something, which resides in the external object, we either contradict ourselves, or talk without a meaning ....

> I have already shewn, that the understanding, when it acts alone, and according to its most general principles, entirely subverts itself, and leaves not the lowest degree of evidence in any proposition ....

(A minor stylistic parallel in passing: Kant also likes to use metaphors involving "*Schwingen*" or "*Flügel*" to characterize Icarian "flights" of reason. Further up on [T] 267, Hume compares "Men of bright fancies ... to those angels, whom the scripture represents as covering their eyes with their wings").

Or compare Kant's analyses of the "mathematical" antinomies with the following rhetorical question from the end of [E] I Section 130, whose answer follows in Section 131:

> ... can we ever satisfy ourselves concerning any determination, which we may form, with regard to the origin of worlds, and the situation of nature, from, and to eternity? [E] I 162

> It seems to me, that the only objects of the abstract sciences or of demonstration are quantity and number, and that all attempts to extend this more perfect species of knowledge beyond these bounds are mere sophistry and illusion. [E] I 163

Hume almost never uses the phrase "(the) appearances" in [T] or [E] I, either in the sense of Sextus' φαινόμενα, or that of Kant's *Erscheinungen*. So it seems to me mildly interesting that a rare exception, toward the end of [T]'s Appendix might almost be read as a rendering into eighteenth-century English of some previously undiscovered remarks by Kant:

> As long as we confine our speculations to *the appearances* of objects to our senses, without entering into disquisitions concerning their real nature and operations, we are safe from all difficulties, and can never be embarrass'd by any question.

> The appearances of objects to our senses are all consistent; and no difficulties can ever arise, but from the obscurity of the terms we make use of.

> If we carry our enquiry beyond the appearances of objects to the senses, I am afraid, that most of our conclusions will be full of scepticism and uncertainty. Thus if it be ask'd, whether or not the invisible and intangible distance be always full of *body*, or of something that by an improvement of our organs might become visible or tangible, I must acknowledge that I find no very decisive arguments on either side .... [T] 638–639

Also striking is a near-verbatim parallel between the concluding words of the *Treatise*'s last paragraph,

> Nothing is more suitable to that philosophy, than a modest scepticism to a certain degree, and *a fair confession of ignorance in subjects, that exceed all human capacity*.
> [T] 639 (emphasis mine)

and the famous concluding words of the first Critique's first paragraph, in its original version (A VII), where "questions" posed by "the nature of reason itself"

> ... übersteigen alles Vermögen der menschlichen Vernunft ....
> ... exceed all capacity of human reason .... (emphases mine)

Passing textual parallels aside, however, my principal conjecture in this section is that Hume's "Custom" and "(Human) Nature," essentially equivalent in the relevant passages of [T] and [E] I, as I have remarked, also uniformize "the" structure of Hume's "experience," in ways which strongly anticipate the *durchgängige Bestimmung* (literally, "thoroughgoing determination" / "definition") of Kant's "Nature," "forms of intuition" and "categories of the understanding."

In several senses, I have already argued, Hume's "Custom" and "(Human) Nature" in [T] and [E] I do seem to organize his "experience" in patterns which normalize rather rigidly the (allegedly) undogmatic ways in which the pyrrhonist's ἔθη ("customs"), συνέθεια ("habits") and φύσις ("nature") organize their φαινόμενα ("phenonema"). A further contrast with his academic and pyrrhonist predecessors arises from the fact that Hume also claims to have in hand a "principled" and essentially dogmatic analysis of what "the phenomena" "are"—namely, that they are "ideas, " which are "ultimately" well-founded in (clear and distinct) "impressions."

In earlier sections, I have also compared the "microscopic" precision of Hume's analysis with a "phenomenal" version of stoic κατάλεψις; suggested that this analysis eventuates in his postulation of a kind of counterfactually (?) universal reference-frame for "all" Experience; and compared this frame with Newton's "absolute space." Here, I will argue that a transition from the pyrrhonists' custom(s) to Hume's "Custom" closely parallels the one just sketched, from φαινόμενα and πάθη—appearances and human experience(s)—to Hume's Experience.

It is "Custom," after all, not "custom," which is the majestic entity of the passages from [E] I 43, quoted earlier, and of [E] I 44–45:

> Custom, then, is the great guide of human life. It is that principle alone which renders our experience useful to us, and makes us expect, for the future, a similar train of events with those which have appeared in the past. Without the influence of custom, we should be entirely ignorant of every matter of fact beyond what is immediately present to the memory and senses. We should never know how to adjust means to ends, or to employ our natural powers in the production of any effect.

Even [T] 102's more pedestrian remark that

> we call every thing CUSTOM, which proceeds from a past repetition, without any new reasoning or conclusion.

leads again to the ringing conclusion that

we may establish it as a certain truth, that all the belief, which follows upon any present impression, is deriv'd solely from that origin.

Hume's many variations on the argument that "all" concept-formation is "deriv'd solely from custom" also seem to me oddly reminiscent, once again, of Descartes' (anacoluthic) transition in his wax-argument, from "not recognized solely by means of (isolated applications of) the senses," to "recognized 'solely by inspection of the mind'."

And perhaps with good reason. For Descartes' argument is also an attempt to explain (an allegedly representative instance of) concept formation, as a process which interrelates otherwise disjoint (repetitions of) sensory data about the various physical states of his piece of beeswax, "fresh from the hive."

Hume's "custom," similarly, is an "invisible intelligent principle" ([E] I 69), which interrelates ("conjoins") otherwise disjoint (repetitions of) sensory events, into coherent conceptual patterns. Hume almost never explicitly uses the phrase "*a priori*", except in association with "relations of ideas," but there are tantalizing partial exceptions, such as the oddly Kantian sentence at the end of the dismissal of "animal *nisus*," quoted above, in which he observes that

> What follows it [the sentiment of an endeavour] we know by experience, but could not know it *à priori*. [*E*] 1 67

Aside from such exceptions, his virtual dismissal of such "relations," in passages such as the long footnote from [*E*] I 43–45, cited earlier, may obliquely suggest that some sort of *a priori* might have to remain, somehow, when the "relations of ideas" with which it is customarily associated have "in fact" vanished; rather like the Cheshire Cat's smile ....

Furthermore, if "custom" is what structures "all" experience and makes it coherent, as Hume claims, it would almost seem to become, in some (admittedly leading) sense, "constitutive" of experience, as I have already argued. This conclusion might also be suggested by the implicit regress, which would seem to be its most obvious (if awkward) alternative.

This arises as follows. For Hume, "All our reasonings concerning fact are of the same nature," namely that of the reasoning of

> [a] man finding a watch or any other machine in a desert island, [who] would conclude that there had once been men in that island. [*E*] I 26

Other questions aside, it seems to me striking that Hume here chooses—of all potential illustrative examples for his point—the set-piece scenario usually employed in pocket versions of the argument from design. If this scenario could really be taken as an a genuine example of "inference" both from and to "Custom"—which operates in a manner Hume wryly characterizes with allusions to "preestablished harmony"—it would become noticeably less clear why we could not take its notorious theological counterpart in the *Dialogues* as an "inference" both from and to the "existence of the Supreme Being."

Such "reasoning," at any rate, is not an *a priori* "relation of ideas," as Hume makes clear. But a dispositional ability to engage in it "must" in some sense be prior to one's encounter with the "fact"(s) it "concerns." Otherwise we would presumably remain in the state of Hume's "Adam," or of his "man [sic] with no tincture of natural philosophy," confronted by "two smooth pieces of marble" at [*E*] I 28. Or, even more explicitly, of the person at [*E*] I 42:

> Suppose a person, though endowed with the strongest faculties of reason and reflection, to be brought on a sudden into this world; he would, indeed, immediately observe a continual succession of objects, and one event following another; but he would not be able to discover anything farther.

In this context, Hume's familiar analyses in these passages suggest several obvious questions which lead in the direction of the "regress" just mentioned.

How, for example, does a "Custom" come to be? Or undergo a "modification"? And how could one interpret an initial "matter of fact?" (What did "Adam" do?) If custom provides, as it were, the initial conditions for ("reasoning" about) "ideas" and "impressions," what (if anything) provides initial conditions for "Custom?"

Hume's own tendency to make everything a "matter of fact" in the footnote to [*E*] I 43 might suggest that the answer to the last question ought to be ("must" be?) "experience" (or perhaps "Experience"). But this would rather obviously give rise to an indefinitely alternating regress of "custom" and "experience," each in some sense "prior" to, or a prerequisite (or "presupposition") for the other. Hume of course has little patience for any sorts of infinities, much less for "infinite regress."

Such a dilemma would therefore be deeply problematic for Hume, but I believe he ultimately concluded, on balance, that it *is* "Custom" which provides initial conditions for "Experience," rather than (or at least more than) the other way around. In the very paragraph he footnotes, on [*E*] I 43, for example (quoted earlier), he remarks as follows:

> Perhaps we can push our enquiries no farther, or pretend to give the cause of this cause; but must rest contented with it as the ultimate principle, which we can assign, of all our conclusions from experience. It is sufficient satisfaction, that we can go so far....

Perhaps the footnote's inordinate length also reflects a certain sense of unease at the new and problematic direction in which such arguments might lead.

In any case, if one rejects the regress, and rules out "experience(s)" as the foundation of "Adam"'s "experience," one might well find oneself drawn to an alternative self-referential conclusion, which may be no less awkward: that it is, indeed, "Custom" which somehow "constitutes" or "determines" [itself] as it exemplifies "a kind of pre-established harmony" (cf. once again, [E] I 55). Recall also that this "harmony" provides

> Those, who delight in the discovery and contemplation of *final causes* ... ample subject to employ their wonder and admiration. [E] I 55

These high-metaphysical properties which Hume explicitly attributes to "Custom" also contributed to my earlier extravagant comparison of Hume's "Custom" with Berkeley's "God." In the present context, it may be useful to recall that one of the chief pragmatic functions of the "god in the quad" for Berkeley was to confer an intersubjective validity and uniformity on our disparate "ideas." And that for Kant,

> objektive Gültigkeit und notwendige Allgemeingültigkeit (für jedermann) sind Wechselbegriffe ....
>
> objective validity and necessary general-validity (for everyone) are interchangeable concepts .... [Pr] 298

Perhaps, then, "Custom" does do more than mediate between us Berkeleyan "spirits." Perhaps it embodies some of the properties which Kant built into his more elaborate, synthetic *a priori* architectonic, of "forms of intuition" and "categories of the understanding."

Hume's "Custom" *is*, after all, completely prerequisite in Hume's view to any "discovery" or "enlargement" of our ideas (cf. e.g., the language of [T] 163). Compare Kant's characterization of the "synthetic" as "*erweiternd*" ("extending," sometimes translated pedantically as "ampliative"), rather than (merely) "*erläuternd*" ("elucidating"), at [KrV] B 11 and [Pr] 266.

"Custom" may thus be "*a priori*," in the sense of "priority" sketched in the regress-argument above—if it is, indeed, "the" self-referential "principle" which well-founds and "determines" initial experience, but is not "in" it.

I have already suggested that it may have to serve as a rather incongruously Aristotelian "first cause" of itself, as well as the "final cause" which Hume acknowledges with some bemusement at [E] I 55. It is, moreover, so pervasively constitutive of "all" experience that it even "conceals itself:"

> Such is the influence of custom, that, where it is strongest, it not only covers our natural ignorance, but even conceals itself, and seems not to take place, merely because it is found in the highest degree. [*E*] I 28–29

This preestablished harmony thus shares its elusive "subtility" with the music of the spheres ....

"Custom" is, moreover, relevant to several *other* contexts in which Kant would invoke synthetic *a priori* "principles," "categories" and "forms." For example, Hume conjoins the notion with "general rules" at [*T*] 141, and remarks many times that custom in some sense "produces" generality, for example at [*T*] 24:

> If ideas be particular in their nature, and at the same time finite in their number, 'tis only by custom they can become general in their representation, and contain an infinite number of other ideas under them.

Recall further that on Hume's account, both physical counterfactuality and mathematical infinities are not only intensional, in some sense. They are guided by intensions which are themselves prescribed by "custom." The former claim has been well-supported by twentieth-century mathematical and metamathematical arguments, I believe (an assertion which many philosophers of mathematics would reject). The latter, however, poses a genuine *dilemma*.

Either "custom" itself is plural and in flux (which Hume does not seem to think is the case); in which case we would lose the invariance and unicity most of Hume's readers and followers have continued to associate with (certain) mathematical notions, however residual their conception(s) of "relations of ideas" might otherwise be. Or "custom," which allegedly regulates our understanding of uniformity, regularity and invariance, is itself essentially uniform, regular and invariant, under "conceivable" perturbations of "fact."

Aside from the methodological problem of figuring out what such (higher-order?) self-referential uniformity might "mean," a critical reader of Hume's texts (Kant, for example) might observe that these properties put unbearable weight on the ordinary, "customary" senses of the word "custom." What are some of the principle attributes of this latter, "foundational," "constitutive" notion of "custom," which such a critical but essentially nonskeptical reader might wish to retain, after discarding the word itself as inadequate, and in several senses even misleading?

Well, ... this renamed "final cause" of Hume's would somehow have to be both "necessary," and constitutive of "necessity," first; and similarly with "invariance," "uniformity" and the rest. Above all, it would have to be structurally determining of "Experience," in some of the senses considered earlier, but it would not itself be part of "experience."

Such a critical commentator might further conclude—especially if he or she were a genuine admirer of Hume's *Scharfsinn* in other respects—that the "dilemma" proposed above really is one, and that we must posit at least a "weak," "regulative" version of Hume's "custom" (or whatever we wish to call it), if we wish to solve the problems Hume has justly posed. And, finally, that the only alternative to an unacceptably "weak," "skeptical," "pluralist," and "merely regulative" version of the guiding principle Hume called "Custom" might have to justify as well as "necessitate" [itself].

This line of development, at any rate, is the one I had in mind above, when I suggested that "that way lies Kant." Whether or not Kant actually took this path, any serious attempt to preserve "realist" assumptions of semantic unicity in a metaphysics based on "custom" (as in Wright's "skeptical realism"), would have to normalize that "custom" in the direction of something like the *Formen* and *Kategorien* of Kant's "Transcendental Analytic."

One might even call the resulting position "empirical realism and transcendental idealism," as Kant did, if one (re)attached the word *"transzendental"* to the puzzling sort of [self]-"constitution" which this renamed strong version of "custom," sketched above, seems to require. And one might also wish to retain, in a "Transcendental Dialectic," something of the elenctic spirit of several of Hume's more genuinely skeptical critiques: of extravagantly underdetermined notions, for example—such as attributes of "god"—which not even this strong "transcendental" notion of "custom" could plausibly decide.

And this last methodological decision, finally, might further lead us to propose that we stratify philosophical notions and assertions: roughly speaking into "custom"-determined "concepts, which we can "understand" (*"Verstandesbegriffe"*); and undecidable "ideas of pure reason" (*"Vernunftideen"*), which we can only "speculate" about, when we remain (as we should, of course, whenever we approach such matters as careful "empirical realists") within the (allegedly) secure bounds of (strong) "custom."

My own "speculation(s)," that these dialectical tensions are already present quite noticeably in Hume's own texts, and that Hume himself did resolve that tension, however awkwardly, in the direction of a "strong" construal of his own notion of "Custom," would obviously benefit from further evidence that Hume himself did give this notion most of the attributes I have ascribed to it: uniformity, in particular; regularity; semantic unicity; and some functional equivalent ("the absolute and incontroulable necessity" of [*T*] 183 and 179, perhaps) of the "necessary connexion(s)" whose place it effectively takes.

In the rest of this section I will try to provide more textual evidence along these lines, and draw some tentative final conclusions from it in the section which follows.

Hume insists on the physical relevance of "Custom" quite explicitly and directly, in his quaintly thought-experimental accounts of laws of impact for "Billard-balls," for example; and more obliquely in his valiant attempts, in sections 65–70 of [E] I, to assimilate the degrees of uniformity in the physical and "moral" science(s). He considers "the constant and universal principles of human nature" ([E] I 83) and "regular springs of human action and behavior" (ibid) neither more nor less uniform than phenomena of Newtonian physics, and (also at [E] I 83) asserts quite flatly that

> ... human nature remains still the same, in its principles and operations. The same motives always produce the same actions. The same events follow from the same causes.

It might appear for a moment that he wishes to qualify such assertions with the following remark at [E] I 85:

> We must not, however, expect that this uniformity of human actions should be carried to such a length as that all men, in the same circumstances, will always act precisely in the same manner, without making any allowance for the diversity of characters, prejudices, and opinions.

But he immediately trims physical uniformity to match:

> *Such a uniformity in every particular, is found in no part of nature.* On the contrary, from observing the variety of conduct in different men, we are enabled to form a greater variety of maxims, which still suppose a degree of uniformity and regularity. (emphases mine)

The next two sentences, moreover, provide a small and rather charming masterpiece of casuistry and strained rationalization:

> *Are the manners of men different in different ages and countries?* We learn thence the great force of custom and education .... [E] I 85–86 (emphases mine)

His "examples" for this assimilation are of necessity rather selective, but some are chosen with his usual wit and forensic skill. Consider, for example, the following miniature exchange, about an "honest, opulent friend" who has suddenly stabbed him, and stolen his "silver standish:"

> I no more suspect this event than the falling of the house itself, which is new, and solidly built and founded.—*But he may have been seized with a sudden and unknown frenzy.*—So may a sudden earthquake arise, and shake and tumble my house about my ears. [*E*] I 91

Other alleged "examples" are unfortunately little more than bland expressions of social prejudice and economic false consciousness, which undermine rather than confirm his argument:

> Is the behaviour and conduct of the one sex very unlike that of the other? It is thence we become acquainted with the different characters which nature has impressed upon the sexes, and which she preserves with constancy and regularity. [*E*] I 86

> The poorest artificer, who labours alone, expects at least the protection of the magistrate, to ensure him the enjoyment of the fruits of his labour. He also expects that, when he carries his goods to market, and offers them at a reasonable price, ... he shall be able, by the money he acquires, to engage others to supply him with those commodities which are requisite for his subsistence. [*E*] I 89

At one point, his polemic makes a claim that was, I believe, demonstrable nonsense—historical as well as scientific nonsense—in 1748:

> Thus it appears, not only that the conjunction between motives and voluntary actions is as regular and uniform as that between the cause and effect in any part of nature; but also that this regular conjunction has been universally acknowledged among mankind, and has never been the subject of dispute, either in philosophy or common life. [*E*] I 88

At another, he appeals to physical "secret powers," in a rather bizarre *L'Homme Machine*-like passage (especially bizarre for Hume, that steadfast opponent of physiological cant, "secret powers," and dormitive virtues of all sorts), at [*E*] I 87:

> They know that a human body is a mighty complicated machine: That many secret powers lurk in it, which are altogether beyond our comprehension: That to us it must often appear very uncertain in its operations: And that therefore the irregular events, which outwardly discover themselves, can be no proof that the laws of nature are not observed with the greatest regularity in its internal operations and government.

Notice, finally, that these "laws of nature" are not only "inferred" here, under the familiar "determining" guidance of "Custom." They are actually "observed."

As my glosses of these quotations probably make clear, Hume's arguments for the uniformity and universality of "Custom" seem to me both explicit and remarkably weak—weakest of all when he actually attempted (unlike Kant) to "derive" social and ethical conclusions from such claims. To some extent, I have suggested, it is simply not possible to avoid accusing the great critic of facile generalization of facile generalization, and the great critic of "induction" of providing a failed "inductive" "proof" that "custom" is the driving force of "induction."

Nor do I believe that the strains between Hume-the-dialectician and Hume-the-enunciator-of-"principles" can be entirely alleviated by this section's assimilation of "Custom" to (near-)synthetic *a priori* principles of "empirical" concept formation. They reemerge in Kant's metaphysics, for example, in the thoroughly *unscharfe Grenze(n)* (in my view, at least) between "*Verstand* and "*Vernunft*." We might paraphrase earlier questions, for example, and ask once again whether establishment of this "boundary" is itself a *Geschäft des Vestandes oder der Vernunft*.

It was perhaps inevitable, moreover, that Hume's "strong" notion of "Custom" reflected a number of unexamined presuppositions of the contemporary physical science it was supposed to interpret and extend—as, of course, did Kant's *kritische Grundsätze* as well.

One such presupposition, for example, expressed a somewhat facile physical-scientific tendency to claim a universality for its conclusions which they simply did not have (very few realistic physical systems, for instance—even thoroughly "classical" ones—are exactly solvable in Newtonian terms). Hume's attempt simultaneously to uniformize and universalize "Custom"'s range of application was very much in this tradition. Somewhat more prudently, Kant relativized the scope of the forms of intuition and categories of the understanding to the "speculative," first; and within that, further, to the "phenomenal."

Another, related tendency has always tempted scientists to raise "principles" to the level of their incompetence, and apply them to limiting cases of dubious experimental and experiential relevance (Consider again, for example, the case of *L'Homme Machine*). Counterparts of this, I have suggested, might be sought in the imprecise demands for perceptual and intellectual precision represented by Hume's "atomism," and Kant's kindred insistence, referred to several times already, that phenomena and their analyses be *durchgängig bestimmt*.

Other examples might include the very demands for uniformity, invariance and homogeneity which underlie the semantic monism and metaphysical-realist impulses I have discussed. Kant expressly built "transcendentally" rationalized versions of such demands into his architectonic. Hume lavished them more informally on the "moral," as if to compensate for the deficiencies he had localized in the physical.

Kant, of course, explicitly repudiated metaphysical skepticism, most conspicuously in the relevant (and relatively little-read) passages of the "Transcendental Methodology" (B 784–B 797). Hume, by contrast, continued to embrace a "consequent skepticism" in [*E*] I, however much he relativized its "proper sphere" to "the shade" and "the schools" ([*E*] I 159).

My own tentative conclusion to this section—indeed, to the chapter as a whole—is that Hume may have been too quick to abandon that "shade." In his zeal to formulate his new "science" of "Human Nature," and make good the implausibly strong claims this construct imposed on his notion(s) of "Custom" and "Habit," Hume the antecedent sociologist may once again have ignored too readily the scruples of Hume the consequent skeptic.

9 Conclusion

Throughout the foregoing sections, I have suggested that Hume's arguments are rich in textures of highest metaphysics, and merit both appreciation and criticism in terms this characterization may make appropriate. In the process, I have also retraced several ways in which "dogmatic" elements of this metaphysics predominate over the early, near-pyrrhonist views he expressed (e.g.,) in [T] 263–274 (whose tone, moreover, is often closer to fear-and-trembling than to ἀταραξία), as well as the later, more mellow, "consequent" skepticism of [E] I.

The more dogmatic and high-metaphysical aspects of his views also emerge more clearly, I have argued, when one takes seriously his own "principles" and "establish'd maxims of metaphysics," especially his overriding argument for the alleged intersubjective uniformity of "Custom" and "(Human) Nature"—two cardinal tenets of his metaphysics which may seem no less "hidden," "secret" and "occult" than the physical "forces," "causes" and "powers" they replace.

Early in [E] I, Hume expressly places his new "Moral Science"—the science of "*mos*," or "custom"—beyond "suspicion," and enumerates some of its "principles," as near-Cartesian (or perhaps Leibnizian?) "powers and faculties," from which he even claims to be able to derive a—strikingly dogmatic—"criterion of truth:"

> Nor can there remain any suspicion, that this science is uncertain and chimerical; unless we should entertain such a skepticism as is entirely subversive of all speculation, and even action. *It cannot be doubted, that the mind is endowed with several powers and faculties, that these powers are distinct from each other, that what is really distinct to the immediate perception may be distinguished by reflexion; and consequently, that there is a truth and falsehood in all propositions on this subject, and a truth and falsehood, which lie not beyond the compass of human understanding.* [E] I 13–14 (emphases mine)

Basta .... Compare this with Descartes' preliminary but confident canvass of "mental" certitudes at AT VII 28–29, which follows the assertion that he is a *res cogitans* at AT VII 28. Notice also that Hume's claim in the italicized passage looks remarkably like a full-blown clarity-and-distinctness criterion, not only for (mental) "consistency," as remarked above in Section 3; but for (mental) "truth."

The implicit analogy Hume draws between "moral" and Newtonian science appears quite clearly at [E] I 14:

> ... shall we esteem it worth the labour of a philosopher to give us the true system of the planets, and adjust the position and order of those remote bodies; while we affect to overlook those, who, with so much success, delineate the parts of the mind, in which we are so intimately concerned?

Once again, such passages strongly suggest that Hume's desire to normalize and uniformize the undefined notions of his metaphysics, in his "unitary principles" of "Custom" and "Human Nature," impelled him to accept, at least tacitly, "rationalist," even "transcendental" arguments, as constitutive metatheoretical "principles" for his allegedly empirical object-theories.

One of these "principles"—the distinction between "relation of ideas" and "matter of fact"—implicitly begs a metametatheoretic claim that "empirical" object- and "metaphysical" metalevels can be so disjoined; such complexities only deepen the problems of interpretation (let alone "justification") which Hume's "consequent" "skepticism" and "empiricist" metaphysics seem to me to pose.

As I've retraced Hume's "principles," I've also tried to gloss them in terms of two recurrent metalogical phenomena, which might be outlined roughly as follows:

9.1   patterns of criterial or metatheoretical ascent, which permit escape into metatheoretic contexts to resolve object-theoretic problems (for example, to decide what is "matter-of-factual"—theoretically undecidable—and what is not; or to provide a metatheoretic criterion of object-theoretic "truth");

and

9.2   patterns of object-theoretic self-reference (best known by their traditional associations with various forms of semantic paradox discussed earlier); these permit (among other things) formulation and consideration of things like $([\psi])$, $([\text{Prf}([])$, ..., *ad indefinitum*.

These patterns seem to be interrelated in various ways that are still not well understood. Systematic metatheoretic separation of theory and metatheory in mathematical logic (as in 9.1) arose historically from a need to (re)interpret and explain away formal counterparts of the semantic paradoxes of 9.2. In effect, meta- or type-theoretic hierarchies processively beg questions which we cannot answer, in indefinitely many ascending stages. The stages, in turn, are carefully calibrated (we hope) so that

(a) we have some understanding—we think—of the questions we have prorogued at the various stages, and

(b) no *primum mobile* of "real" question-begging, or (The Absolute forbid) "real" inconsistency, will "ever" "actually" be reached. "All" "regresses" may be pernicious; but some, we hope, are less pernicious than others.

I have also remarked on the resemblance between such recourse to successive metatheories and its dialectical response, the pyrrhonist critique of "criterial ascent." The very theory-marginality of notions of the sort we (with Hume) may wish to employ—"definability," "consistency," "conceivability" (both "-what," and "that"), even "undecidability" (the mark of Hume's "matters of fact")—suggests that these notions' place(s) in such hierarchies may be both indeterminate and in flux (an indefinitely theory-marginal complex of metatheoretic "matters of fact"); and that they may be subject to an indefinite variety of meta-(...)theoretic "context-principles."

Intelligible uses of such notions—"context-principles" included—would also seem to be more local than global, however, and more indefinite than particular. Their metatheoretic boundary conditions await further clarification; and we may not in many cases be able to claim much ability to "trace them up" to their "antecedent impressions."

More "global" uses of such notions—and of pronouncements about their use(s), and pronouncements about *their* use(s) ... (*pace* Wittgenstein)—by contrast, either seem to beg significant questions about [their own] scope ("all" instances of [themselves], for example); or perhaps worse, to eventuate in genuinely self-referential problems of semantic paradox and outright methodological inconsistency. Indeed, there is a sense in which "semantic paradoxes," of the sort Gödel exploited in his diagonal lemmas, are exactly what results when one systematically begs the question of the (in)admissibility of systematic question-begging (cf. [Boos], 1987a and 1987b).

Many steps, in any case, of the metatheoretic hierarchies discussed above, may be "locally" effaced, in various provisional, hypothetical, or theory-relative ways. But alleged attempts to efface them "all" "at once" do seem to collapse hierarchies of processively begged questions into the outright semantic paradoxes they provisionally resolve (or at least prorogue) ....

I will close with some final attempts to relate these remarks to the patterns I have traced in the metaphysics of Berkeley, Hume and (in the following chapter) Kant.

Resort to metatheoretic hierarchies may provide a kind of conceptual graduation and relativization of assorted "dogmatic" philosophical doctrines—aspects of Cartesian or stoic metaphysics, perhaps, or even "platonist" assumptions in the philosophy of mathematics—so long as one is willing to admit (*sotto voce*, perhaps) new metatheoretic decisions about "the" boundaries of "experience" (Hume's "matters of fact"); admit, in effect, that we may not have emerged from our "last" platonic cave (not such a depressing mental reservation, after all—recall Weil's remark about intellectuals who are proud of their intellect(s)).

Unfortunately, such resort seems peculiarly ill-suited to the empiricist "irrelationism" and nonlogical-positivism I've attributed to Hume. For on the reconstruction I have offered, Hume posits a (putatively) definite—though not, on pain of genuine paradox, definable—class of (allegedy) determinate ground-level "experience(s)" (his "ideas and impressions"), whose (supposed) global structure, in effect, constitutes an invariant reference frame for Section 6's "absolute Experience." Anything not (finitely) "mixed," "compounded" or "deriv'd" from such atomic "experience(s)" is "inconceivable;" and therefore, by 2.2, inexistent (a "chain of argument" which also provided the original context for my allusion to the remark of Weil).

If one takes such claims of experiential closure and uniformity seriously, however, they would seem to proscribe exactly the sorts of deferred clarification and conceptual relaxation-time which Hume's own arguments tacitly require; and which these metatheoretic hierarchies do provisionally seem to provide.

This implicit strain between Hume's metaphysical "principles" and the framework he employed to formulate them also gives resonance, I have argued, to his well-known slide from indicative to subjunctive at [*E*] I 76. It even ambiguates (or at least graduates) the purchase of the "fork" itself—the dichotomy between "relations of ideas" and "matters of fact" (Some final remarks about this shortly).

Finally, this strain also suggests once again that the methodological procedure(s) of the "microscope" itself may not be well-founded. What counts as an "impression" may, after all, be contextual or thought-experimental; "observation" may even have been "theory-laden" in eighteenth-century "moral science" as well; and what seemed at first to be perceptually "simple" and first-level "atomic" phenomena may be indefinitely (re)resolvable: as zeroth- (minus-first-, minus-second-, ...) level "ideas of reflexion," whose "intended interpretations" involve the [νοούμενα *PH* 1 96] or "thought-objects" of Sextus Empiricus, as well as the *Vernunftideen* (if not "*Noumena*") of Kant.

In this context, it may also be historically appropriate that such emergence(s) of "the" "noumenal" in "the" "phenomenal" later divided two brilliant logical positivists, both of whom were sincere admirers of David Hume. The Schlick-Neurath dispute ([Schlick], 1932–33, 1943, and [Neurath] 1932–33, 1934), and its reflections in subsequent debates over "observation" and "theory," only clarified and made explicit, I believe, the deep and long-standing strains referred to at the end of the last section: between Hume the antecedent sociologist, and Hume the consequent skeptic.

Dilemmas which arise from the possibility just discussed—that apparently "empirical," object-level entities might have to be reconstrued as "conceptual"—may also be mirrored with the aid of the "coding"- or quoting-procedures ("[...]") we introduced and employed in earlier sections. On one construal, for example, self-referential *aporiae* about such ("empirical?" "conceptual?") entities as "conceivability," and the like arise with special force when one attempts to make Humean metaphysics precise within one, allegedly global theory—some putatively "last" metatheory, say.

We close with a recapitulation of several of these queries and dilemmas, mentioned earlier, with a few final supplements and marginal nuances. Is [Hume's rigorous empiricism], once again, rigorously empirical? Is [the distinction between "relations of ideas" and "matters of fact"] a "relation of ideas" or "matter of fact?" Is there a clear resolution under the "microscope" of ["the microscope"]? (Compare Hume's own passing reference at [*T*] 637 to the "very abstract term simple idea"). From what "impressions" are [our "ideas" of "compounding," "impressing and "deriving"] "compounded" or deriv'd?" Is it "conceivable" (*de re*), or only "possible," that [conceivability] is (in)equivalent to "possibility"]? ...

[What the hammer?] .... [What the chain?] ....[3]

Such dilemmas and begged questions, in short, seem to reflect complexities which arise quite naturally in mirror-inversions of metatheoretic ascent, which one might call [intentional descent]. "Custom" gives rise to especially suggestive instances of these *aporiae* of [intentional descent], mentioned earlier. One of these may be characterized in somewhat greater detail as follows.

The great critic of induction, who (inductively) inferred that "All our reasonings concerning fact are of the same nature" ([*E*] I 26), and "All inferences from experience are effects of custom, not of reasoning," (*E*] I 43), skillfully refashioned an implicit regress, described above (of ... experience which presupposes custom, which presupposes experience, which ....), as a self-referential (self-confirming?) assertion, which one might call

9.3   *Hume's Circle:* Only "Custom" can "determine" that [Only "Custom" can "determine ....]

(An epicycle results if one substitutes "what" for "that.")

---

[3] William Blake, "The Tyger," *Songs of Experience*, 1794.

Perhaps Kant saw the circularity (or at least [circularity]) of such a claim, as I tried to suggest in the last section. If so, he may have engaged in his own *fuite en avant* (*Flucht nach vorne*, in this case): to reinterpret that very circularity as a "transcendental" necessity .... (What "we" cannot avoid—epistemically, "transcendentally", or perhaps metalogically—is in some sense "necessary," is it not? Or is it? Compare the discussion above of the "principle" 2.2 ....)

If one carefully analyzes "Custom," in manner of the last section, at any rate, one might well hope to rewrite the Circle 9.3, in transcendental bold face as it were: as a "transcendentally necessary" assertion that

9.4.1 "we" "must" encode "the" structures of ("our") "Nature" and "Custom," as "forms of intuition" and "categories of the understanding;"

and, therefore (—if "we" can "understand" what "we" are talking about; compare earlier sections' arguments about the theory-relativity and -marginality of demarcations between *Verstand* and *Vernunft*—and for that matter, between "*transzendental*" and "*transzendent*"),

9.4.2 that "we" "must" "transcendentally" assert ["the" "necessity" of ["transcendentally necessary" assertion]].

One analogical problem with such a heroic *petitio* of [the Circle]'s (and, allegedly, [its own]) "necessity" is that the most natural theory-"internal" reconstruction of "necessity"—encoded [provability]—need not decide anything, beyond what it "internally" "necessitates."

A suggestive proof-theoretic illustration of this predicament may perhaps be found in the following fundamental results of Gödel and Löb:

9.5 Suppose a theory T can process enough "arithmetic" information (roughly speaking) to encode its own syntax, and therefore to talk about [itself]. Then

1   T need not prove ($[\varphi]$ is provable) $\rightarrow \varphi$.

In fact,

2   T proves ($[\varphi]$ is provable) $\rightarrow \varphi$ if and only if T proves φ.

(Formal background and proofs for these results can be found in the very well-written and accessible survey article ([Smorynski], 1977); Löb's theorem appears on p. 845).

Perhaps these formal considerations are reflected in the fact that the *petitio* of "Hume's Circle" seems to filter through the barriers of Kantian "transcendental necessity" and "deduction" rather well; and even reemerge anew, in the earlier paragraph's "transcendental" "bold face," as

9.6 Kant's Circle:
Only synthetic a priori argument can provide apodictic "proof" that [only synthetic a priori argument can provide apodictic "proof" that [only synthetic a priori argument can provide apodictic "proof" that [....]]] .

I hope to explore some implications of these lines of argument for Kant's "critical" method with greater care in Chapter 8. Here, they simply seem to me to suggest, once again, that Hume would have given more coherent and persuasive arguments for the mitigated skepticism he sincerely professed, had he focused somewhat more of the redoubtable *Scharfsinn*—with which Kant justly credited him—on the methodological presuppositions of [his own arguments] (the "principles" and "establish'd maxims in metaphysics" with which he began, of course; but also some of the less "establish'd" ones he derived from them, in his role as the "Newton of the Moral Sciences"); if he had questioned, for example (among other potentially "inconceivable" relations), the "microscope's own resolution (perhaps it is more like Heisenberg's thought-experimental "microscope" than Leeuwenhoek's physical one?); and granted a certain correlative *Unschärfe* to the "fork"'s dichotomy between "relations of ideas" and "matters of fact".

As a theory-relative notion, Hume's dichotomy has fundamental importance, and it seems likely to retain that importance as long people engage in any sort of serious theoretical talk about what they experience (including [themselves]). But it also seems to be an excellent example of what Kant later called a "regulative" (as opposed to "constitutive") principle, which guides but does not "determine."

And if this is so, finally, it may only be Hume's (and Kant's) gratuitous further assumption(s)—that this and several other "principles" and "establish'd maxims of metaphysics" "must" be made uniform as well as universal—that we would be well-advised to "commit ... to the flames."

8 Kantian Ethics and "the Fate of Reason"

What particular privilege has this little Agitation of the Brain which we call Thought, that we must make it the Model of the whole Universe?

(Hume, 1976, 168)

******

... at once it struck me, what quality went to form a Man (sic) of Achievement especially in Literature and which Shakespeare possessed so enormously—I mean *Negative Capability*, that is when someone is capable of being in uncertainties, Mysteries, doubts without any irritable reaching after fact and reason.

(Keats, 1959, 261)

******

Die menschliche Vernunft hat das besondere *Schicksal* in ihrer Gattung ihrer Erkenntnisse: dass sie durch Fragen belästigt wird, die sie nicht abweisen kann; denn sie sind ihr durch die Natur der Vernunft selbst aufgegeben, die sie aber auch nicht beantworten kann; denn sie übersteigen alles Vermögen der menschlichen Vernunft.

(Opening lines of the *Kritik der reinen Vernunft*, A VII; the italics in the text are mine)

Human reason has the particular *fate*[1] in one branch of its investigations: that it is harassed by questions which it cannot dismiss out of hand, for they are posed to it by the nature of reason itself; but which it also cannot answer, for they exceed all capacity of human reason.

******

Der Verstand macht für die Vernunft ebenso einen Gegenstand aus, als die Sinnlichkeit für den Verstand. Die *Einheit* aller möglichen empirischen Verstandes-handlungen *systematisch* zu machen, ist ein Geschäft der Vernunft, sowie der Verstand das Mannigfaltige der Erscheinungen durch Begriffe verknüpft und unter empirische Gesetze bringt. Die *Verstandeshandlungen* aber, ohne Schemata der Sinnlichkeit, sind *unbestimmt*; ebenso ist die *Vernunfteinheit* auch in Ansehung der Bedingungen, unter denen und des Grades, wie weit, der Verstand seine Begriff *systematisch* verbinden soll, an sich selbst *unbestimmt*. Allein, obgleich für die durchgängige *systematische Einheit* aller *Verstandesbegriffe* kein Schema in der Anschauung ausfindig gegeben werden kann, so kann und muß doch ein *Analogon* eines solchen Schema gefunden werden, welches die *Idee des Maximum* der Abteilung und der Vereinigung der Verstandeserkenntnis *in einem Prinzip* ist.

(*Kritik der reinen Vernunft*, B 692; the italics are again mine)

Understanding constitutes an object for reason just as the sensory [does] for understanding. To render the *unity* of all possible *operations of the understanding systematic* is an affair of reason, [just] as understanding structures and brings under empirical laws the manifold of appearances. The operations of the *understanding*, however, without schemata of what is sensory, are *undetermined*; just as the *unity of reason* in itself is *undetermined* with

---

[1] It may be metaphysically relevant in this well-known context that *"fatum"*—before the Romans hijacked the expression for the decrees of their gods—was simply "what was spoken", and in that sense "deemed", or "judged" (from *"fari"*, *to speak*, akin to *phemi* and *phasko* in Greek).

respect to the conditions under which and degree to which understanding is supposed to structure and interrelate its concepts. But though *no comprehensive schema of systematic unity* of all concepts of the understanding can be found in intuition, *an analogon* of such a schema, namely the *idea of the maximum* of (sub)division and unification of conceptual knowledge *in a principle*, can and must be found.

Preface

In this chapter, I
1 conjecture that Galileo's "book of philosophy" (not "nature") may not be written in Galileo's mathematics, but in Hilbert's and Gödel's metamathematics;
2 outline a common metalogical analysis of four "cogito"-like arguments: Descartes' original, Berkeley's "master-argument", and Kant's "metaphysical" and "transcendental" "deductions";
3 advocate a "skeptical"[2] form of *"transcendental idealism"*, which "locally" distinguishes "experiential" assertions from metatheoretic "preconditions" for their "intended" interpretation;
4 and argue

(i) that "experience" is a rational but inherently "problematic" regulative ideal (cf. KdrV, B 100);
(ii) that the integrity of "essentially incomplete" but "intelligible" inquiry is the only "unity" such "experience" could have; and
(iii) that "almost all" "maximal limits" to which such "experiential" inquiry might converge would be intrinsically "unintelligible". (*"Es ist das Mystische ...."*)

1 Introduction

"Neo-Kantians" such as Hermann Cohen and Ernst Cassirer "relativised" Kant's attempts to ground *"Erfahrung"* (experience) in an *"Architektonik"* derived from Newtonian spacetime, in an effort to refine his "critical" presuppositions, and accommodate more recent mathematical and scientific inquiries into the hermeneutics of differential geometry and mathematical physics undertaken by Gauss,

---

2 As discussed in chapters 2 and 3, the "skeptics" were "reflective" (*"skeptikoi"*) "seekers" (*"zetetikoi"*), who "sought" to "see (more)" (*"skeptesthai"*). What they "sought" was not 'the' "truth"—ironically, derived in English from a Germanic word for "trust" or "belief"—but 'locally' adequate "vantage-points" (*"skopoi"*, *"skopia"*) of "insight" (*"skemma"*) and conceptual 'tranquility' (*"ataraxia"*).

Lobachevsky, Poincaré, Lorentz, Hilbert, Einstein and others. In this chapter I "relativise" Kant's attempts to explore ascending levels of "reflective inquiry" and "regulative principles" in his three *Critiques*, in an effort to refine his "dialectical", "practical" and "teleological" extensions of the first's "theoretical" *Architektonik* of (Newtonian scientific) "experience", and accommodate more recent metalogical and metamathematical inquiries into the hermeneutics of concept formation undertaken by Hilbert, Skolem, Gödel, Tarski, Keisler, Scott, Solovay and others.

In potted-historical terms, a preliminary rationale for such a relativisation, and for the "skeptical" as well as "locally transcendental idealism" I will derive from it, might be sketched as follows.

There is a sense—to which people like Ernest Cassirer were particularly attentive —in which Kant merely emulated Leibniz' and Leonhard Euler's impassioned attempts to respond to the metaphysical resonance of Galileo's (often-misquoted) suggestion in the *Assayer* (*Opere* 6:23:2): that the "vast book" of "philosophy" would be "written in the language of mathematics". More than a century later, working scientists and historians of science among Cassirer's predecessors and contemporaries such as Hertz, Helmholtz, Duhem and others also attempted to respond to Galileo's dictum in new ways, and Peano, Frege and Russell more or less simultaneously formalised the "neo-Leibnizian" recursive framework of a quantificational logic, which gave rise rather quickly to intriguing "semantic-paradoxical" refinements of Kant's "*mathematische und dynamische Antinomien*".

During the twentieth century, logicists and logical positivists—most of them convinced that such paradoxes would be swept aside by a new "empiricist" synthesis—drew on work of Pierre Duhem and others to propose "reductive" translations of Galileo's "vast book" into rudimentary forms of first-order logic, Russellian type-theory and Zermelian set theory, and sidestep skeptical implications of these antinomies implicit in the work of Skolem and Gödel.

Against the grain of this (admittedly partial as well as summary) historical sketch, I wish to suggest that

1. efforts on the part of the logical positivists' more nuanced "analytical" successors to titrate conventional empiricism with a bit of "logic" have *not* yielded a stable response to these and other antinomies (and never will),

and propose a conjectural "*metalogische Wende*", or (in more pretentious emulation of Kant's preface to the first *Critique* (cf. B XVI ff.)), "Revolution": that

2. future "books" which record inquiries into boundaries of (inquiries into (boundaries of (.... "philosophy"))) will be "written in the languages of metamathematics";

and that

3   the "texts" of such "books" will be "intelligible" (recursively axiomatisable, essentially incomplete) theories, representable as "closed" subsets of a "universal" topological and measure-space adumbrated in the "dyadic" writings of Leibniz, and reintroduced in nineteenth- and twentieth-century guises by Georg Cantor and Marshall Stone.

In opposition to Kant and his empiricist successors' efforts to "define" or "demarcate" a "canonical" or "surveyable" "*Bereich der Erfahrung*" ("range of experience") —*via* "critical methods", "transcendental" analyses or more alembicated *ad hoc* methodological titrations of "empiricism" of the sort just mentioned—I will also
4   outline "locally transcendental" rationales for the epistemic and metaphysical adequacy of essentially incomplete (first-order) theories;
5   interpret Kant's "*kritische Methode*" and "constitutive" / "regulative" distinction as partial anticipations of metalogicians' "theory" / "metatheory" distinctions;
6   invoke well-studied arguments to argue that such "local" distinctions and methodological principles generate heuristic iterative hierarchies of concept formation; and
7   characterise (but not "define") "experience" as a heuristic, distributive, "problematic" (cf. *KdrV*, B 100) and "merely regulative" temporal process of ramified "ascents" within such hierarchies.

A few preliminary remarks may anticipate immediate and obvious objections to these proposals, and my eclectic and impressionist use in them of metalogical terms-of-art such as "object-" and "metatheory". Philosophers often find such technical terms-of-art reductive; and logicians, for their part, find philosophical uses of them jejune. Perhaps eclectic uses of them fall into the sort of conceptual *intermundium* C. P. Snow wrote about many years ago.

Be that as it may, both the views just cited seem to me prejudices, as the proposals in 2 and 3 make clear. If these proposals have heuristic value, the methodological partitions and preconceptions just cited will erode as needs to
8   provide coherent accounts of "artificial" as well as "natural" sentience, and
9   develop natural-scientific analyses of notions such as "existence", "complexity", "randomness" and "experimental isolation"

become more apparent.

Kant, in any case, was no more acquainted with "first-order metatheoretical hierarchies" than he was with "time-orientable locally Lorentzian manifolds". He did, however, make a number of suggestive attempts to distinguish "regulative" from "constitutive" applications ("*Gebräuche*") of broadly problematic "higher-order" "principles" and metaphysical assertions (cf. the passage *KdrV*, B 692, quoted at length above).

Such distinctions—or superpositions of distinctions—first appeared in B 536–611 of the *Transcendental Dialectic*. Kant then took them more or less for granted thereafter, and applied them *passim* to more abstruse and underdetermined ranges of practical (ethical) as well as teleological assertions and principles in *Dialectic* as well as the second and third *Critiques*.

At a kind of asymptotic limit of such "principles", he even appealed at various points in all three *Critiques* and their ancillary texts to the notion of a regulative Ideal—an interpretative *Grenzidee* which seems to me as "reflexive", suggestive and conceptually underdetermined as Aristotle's *theoria*, and shadowy "aggregates" ("*Vielheiten*") of the sort Georg Cantor aptly called "*inkonsistent*" ("the" paradoxical "aggregate" of "all" consistent recursively axiomatisable first-order theories, for example; cf. Cantor, 1966, 443–444, reprinted in van Heijenoort, 1967, 114).

In Section 2 below, I will consider certain attempts to coopt and normalise ancient skeptical notions of "*ta phainomena*" in ways which might
10 ensure that a fixed "maximal" realm or "aggregate" of such *phainomena* "must" exist, and
11 furnish "design"-arguments for the "existence" and "unicity" of "experience" (rather than "*god*").

In Sections 3 and 4, I will
12 examine "pre-theoretical" or "pre-metalogical" characterisations of "'cogito'-like" arguments;
13 argue that such arguments can be construed as informal prototypes of "fixed-point arguments"; and
14 "test" these anachronistic characterisations against four historical examples:
(i) Descartes' "cogito"-argument, and its stoic and Augustinian antecedents;
(ii) Berkeley's more original but strangely ill-regarded "master-argument"; and in somewhat greater detail in the next section,
(iii) Kant's "transcendental" and "metaphysical deductions" of the "necessity of the possibility" of "synthetic *a priori*[3] judgment(s)", and

---

3 I will attach little or no epistemic or metaphysical significance in the sequel to 'adverbial' and 'adjectival' readings of the ablative phrase "*a priori*".
  I do this in part because
  (i) there is little or no syntactical distinction between such usages in German; and in part because

(iv) "the" allegedly unique structures of "experience" which such judgments "determine" when they are "constitutive", and "underdetermine" when they are "("merely") regulative").

In Sections 5–8, I will

15   focus on metalogically syntactical and semantic as well as informal interpretations for the "unicity" of "rule"-bound theoretical "experience" Kant hoped to derive from his *"transzendentale Deduktion"*,

16   argue for metamathematical interpretations of such "rules", and offer "skeptically transcendental" rationales for the recurrent and interpretive uses of first-order theories in the metatheoretical hierarchies mentioned in 6 and 7 above, and

17   adduce a modest result of metalogical "folklore": that no "faithful" syntactical interpretation of an "experiential" "object-theory" can be defined in any metatheory of "concepts of the understanding" (*"Verstandesbegriffe"*) which proves that such an object-theory is consistent.

On the one hand, the metamathematical observation in 15 suggests that

18   no metalogically tenable "transcendental deduction" of the "existence" of a "faithful" interpretation of a provably consistent "experiential" object-theory can be formulated.

19   On the other, it also suggests (to me at least) that

(i) indeterminate as well as underdetermined recourses to "higher"-order metatheories, which semantically interpret ("lower" levels or stages of) "experience"

introduce ever-higher-order mathematical as well as metamathematical analogues of

(ii) Kant's "merely regulative" *"intelligibele Ursachen"* ("intelligible causes") and *"Vernunftbegriffe"* ("concepts of reason")

into the object-theoretical processes they interpret, and that

(iii) the resulting "eternal return"—or more precisely, "eternal recurrence" of alternating "object"- and "metatheoretic" ("epistemic"and "metaphysical") "boundary conditions"—is indeed the "fate of reason". But may also be—in several senses—its "grace"....

(ii) there is little or no semantic distinction, at least in Latin, English, French or German, between

(iii) 'adjectival' phrases which qualify 'processive' nouns ("judgment" (or "judgment") *a priori*; "*ex cathedra* decree"; "off the cuff estimate"; "*prima facie* plausibility"); and

(iv) adverbial phrases which qualify 'processive' verbs and adjectives ("judge" (or *"juger"*) "*a priori*"; "decree *ex cathedra*"; "estimate off the cuff"; "*prima facie* plausible").

2 [Reflective Inquiry into "Ultimate" Limits of [Reflective inquiry into "Ultimate" Limits of [Reflective inquiry into "Ultimate" Limits of [...]]]]

Pyrrhonist skeptics often claimed to follow "undogmatically" what they called *ta phainomena*—a Greek neuter plural participle which is usually "objectified" in (oddly quasi-Kantian) English as "the appearances". (It might be more accurately rendered as "whatever things seem (to be)".)

Whatever the translation, relational and intentional implications of such usages—"seem" or "appear", under what circumstances, and to what?—have always fostered "idealist" interpretations of "*ta phainomena*" as ("mere") "intentional objects". These interpretations, in turn, crossed metaphysical boundaries in interesting ways: "dogmatic" proto-"cogito"-arguments might be found in them, as well as rationales for (otherwise rather puzzling) eighteenth-century claims that George Berkeley was a "skeptic".

This ambiguity itself has also been historically generative, in ways genuine "suspenders-of-judgment" and "skeptics" ("observers", "searchers") might applaud. Much of the history of early modern philosophy as well as the science which crystalised out of it might be read as a series of brilliantly informative attempts to coopt skeptics' "phenomenalism" for dogmatic purposes, with the aid of arguments which "normalised" or "uniformised" their "*phainomena*". Galileo, for example, sought such uniformities in the syntax and semantics of mathematics, in the passage from the *Assayer* cited earlier. And Descartes, Leibniz, Newton and Berkeley sought them in the hermeneutic (and mathematical) capacities of that great "analyst", "god".

Hume—even less plausibly, in my view—sought them in the persistence of human "Custom or Habit". Kant, finally, sought to "deduce" them, in a great secular Newtonian-scientific synthesis whose "transcendental" *adaequatio* would be "complete" enough to determine (or dispense with) its limiting *res* (or "*Dinge an sich*").

Intricately interrelated refinements in the languages of Galileo's mathematical "book", in short, have provided, and will continue to provide, heuristic ways to explore endless ramifications of "mere" phenomena. But these ramifications also bring into focus two deeper (and closely interrelated) skeptical reservations, which philosophical efforts to coopt skeptical *phainomena* have left unanswered (and, in my opinion at least, almost untouched).

The first is

1   that postulation of unique metaphysical limits of these practices (whatever they may be called) amount to secular—mathematical and "physical"—

refinements of "arguments from design"—not for the existence of "god", but for the existence of "the" *phainomena* (a slightly ironic echo, perhaps, of "*deus sive natura*").

Notice once again in this context the extent to which
2   the "structural" and "constitutive" uniformity and intertranslatability Kant attributed to his "phenomena" ("*Erscheinungen*")

may be assimilated to
3   the "structural" and "constitutive" uniformity and intertranslatability Berkeley attributed to his all-seeing "God" or "The Author of Nature".

A second and deeper reservation derives from one of the ancient skeptics' more acute insights and anti-stoic "*tropoi*" ("turns"):
4   that the very notion of "completeness" may therefore be conceptually "liminal" and "theory-relative", and in that sense incomplete; and in particular,
5   that conceptual as well as phenomenal "relationality" and "intentionality" may be iterable "*eis apeiron*"—"into (what is) infinite", and "into (what is) untried, or unexperienced" (both readings are etymologically defensible).

There are essentially two ways to respond to this extended skeptical "regress" (or "ascent"), and block "iteration", "ramification" and "indefinition" of the hierarchies to which it gives rise:
6   to "bound" such iterations in a unique "limit"; and
7   to "diagonalise" over them, in allegedly "canonical" ways which appeal to informal counterparts of metalogical "fixed-point"-arguments.

I will devote most of the rest of this section to the first, and the next to the second. "Semantic monist" commitments of any sort to the effect that
8   there must exist one and exactly one "universal" interpretation of a sufficiently "comprehensive" (as well as consistent) theory T of "experiential" phenomena (which must, then, be the "intended" interpretation of T, if there is one,

have traditionally had (at least) three venerable but fallacious arguments or argument-forms at their disposal, in my view:
9   **"existence-of-maximality"** arguments, often analysable as instances of object-theoretic quantifier error;

10 **"existence-of-determination"** arguments, which I will interpret as subtler and more plausible variants of metatheoretic quantifier error; and finally
11 **"uniqueness-of-maximality-and -determination"** arguments, which may be analysed as misreadings of informal patterns of metatheoretic Stone-"duality."

In **"existence of maximality"** arguments, for example, one "derives" the "existence" of a unique "maximal" bearer of a property ("maximal" in the sense of some "practical" or "metaphysical" ordering) from assumptions (often begged) about the structure of an ordering or hierarchy.

One might, for example, "infer"

12 "(necessary) existence" of a "maximal" relational entity ("cause"; "explanation"; "perfection"; ….), from
13 corresponding "medial" premises (every "effect" has a "cause"; every "phenomenon" has an "explanation"; every "quality" has less "perfection" than some other; …).

I characterised certain instances of "existence of maximality" arguments in 9 and 10 above as "quantifier errors". Kant had no symbolic treatments of such formal error at his disposal, in part because precise formal studies of indefinite quantifier alternation came very late to syllogistic logic. But his analyses of "transcendent" "*Regressus der Bedingungen*" in the Antinomies of the first *Critique* made it clear that he saw through several classical "mathematical" and "dynamical" "existence-of-maximality" arguments (at least in "constitutive" contexts: cf., e.g., *KdrV*, B 692).

Subtler and more metatheoretic **"existence-of-determination"**-arguments have been adduced since antiquity to justify, in some elusively absolute theory-marginal sense, the "existence" of a definitive "criterion of truth", ultimate "intended interpretation" or other extratheoretic form of semantic closure. Roughly speaking, such "existence-of-determination" arguments derive protometatheoretic claims about "final causality" (for example) from tacit or explicit counterparts of the quantifier errors "existence-of-maximality" arguments applied to their "formal" or "efficient" counterparts.

I will call this argument-type in both cases the **metatheoretic EA-argument**, from a standard logical usage for sentences with existential-universal (**"exists**-for **all"**) quantifier structure: sentences, in other words, of the form 'there **exists** an x such that for **all** y ($s(x, y)$). (*AE*- or universal-existential sentences, by contrast, have the form 'for **all** x there **exists** a y such that ($s(x, y)$).

Here, for example, are two thumbnail templates of such arguments, formulated in terms of "reasons" rather than "causes" ("*aitia*" meant both in Aristotle's Greek):

14 if for everything a contextual reason may be found, a noncontextual reason "must" (allegedly) be found for Everything;
15 if for everything a "better" reason may be found, a "best" reason "must" (allegedly) be found for Everything.

More often than not, such *EA*-arguments were (and still are) formulated in "elenctic", contrapositive forms. Such an argument played a prominent role, for example, in one of Socrates' "refutations" of the "vitiated" semantic relativity Plato attributed to Protagoras in the *Theaetetus*, where Socrates fallaciously claimed (in effect) that anyone who rejects the *EA*-claim,
16 that there "must" **e**xist a single "best" objective or intersubjective standard for **a**ll human judgments,

"must" also reject a cognate, but quantifier-reversed and weaker, *AE*-claim,
17 that for **a**ll human judgments, "better" contextual standards **e**xist for evaluation of those judgments' truth.

More subtly fallacious counterparts of such *EA*-arguments—to the effect that "not (there exists an x such that for all y s(x,y))" would "imply" "not (for all x there exists a y such that s(x,y))"—seem to me to enable "cogito-like"-arguments which are also implicit in Kant's "transcendental deduction", a view I will elaborate in the next section.

In a wide range of classical and contemporary metaphysical contexts (justifications of mathematical "realism", for example), the specious attractiveness of semantic EA-arguments is roughly equivalent to the specious attractiveness of "semantic monist" views they support.

When such "existence"-assertions are formulated in formally or informally "stronger" theoretical contexts **U**, however, (as they are in the next section's "cogito"-like arguments), parity of reasoning raises analogous criterial questions about the semantic credentials of **U**. And iterations of such questions and responses generate "natural" informal counterparts of the introduction's "metatheoretic ascents".

Such "regressive" hierarchies have typically been blocked by (fallacious as well as semantically equivocal) "modal" EA-assertions that
18 there "must" "exist" a "unique" last metatheory **U** for **T** which uniquely interprets [itself], and therefore "proves" or "grounds" [its own existence]

which "modally" beg the semantic-monist *desiderata* in question (with or without the Gödelian coding brackets "[…]", which "reduce" semantic "use" to syntactical "mention").

As the parenthetical remarks in 16–18 suggest, their existence claims have also been informally begged on the basis of the very "comprehensiveness" of the theories in question (an attractive but thoroughly dubious enthymeme, as Cantor saw when he called class-theoretic counterparts of such comprehension "*inkonsistent*").

It is correct, for example, that a theory in a given fixed language has fewer interpretations the more "complete" it is (the more assertions it decides). This is the fundamental underlying datum of a metalogical pattern called "Stone duality" (cf., e.g., Bell and Machover, 1977, 141–149).

19 But the same dual analysis also establishes that
(i) theories in fixed languages have more interpretations the fewer assertions they "decide" (the more "abstract" they are);

and as a corollary of Craig's Theorem (cf., e.g., Chang and Keisler, 1972, 84–87), that
(ii) theories in "augmented" languages have more interpretations when the terms in which their additional axioms are formulated are "new".

Sometimes, however, cognate "monist" arguments have been begged on the basis of a subtler, tacitly metatheoretic argument which anticipated "cogito"-like arguments (including Kant's "*transzendentale Deduktionen*") considered in greater detail in the next section:

20 that the position I called semantic monism above "must" be accepted;

for if it is not, then
21 incoherence would be the consequence, and we would not even be able then to query the claim.

(Cf. the miniature reconstruction of the *Theaetetus* argument in 16 and 17 above, and related claims that semantic-hierarchical relativism is "self-refuting", and therefore incoherent).

22 Dialectically attractive though such arguments may seem, they were really little more than "transcendental" second-order *petitiones* and quasi-metatheoretic reassertions of the supposition (closely related to 18 above):
(i) that any "sufficiently" "deep" and "comprehensive" as well as consistent theory must have a unique interpretation;

or its contraposition,
(ii) that any "sufficiently" "deep" and "comprehensive" theory without a unique interpretation would be inconsistent.

For not only is the inference just sketched refutable, for a wide range of philosophically relevant theories **T** (cf. 28–31 below), in any theoretical venue in which it can be expressed. It is essentially an amorphously "modal" type-raised counterpart of what is being "deduced".

Subtler but analogously dubious "existence of determination" arguments also appeared at several points in Kant's "speculative", "practical" and "teleological" "*Deduktionen*", usually in contrapositive forms.

Compare, for example,

23 "Experience" "must" have a unique "transcendental" structure. For many instances of it would otherwise be "merely empirical".
24 "Practical" judgment "must" have a unique "transcendental" structure (at least 'regulatively'). For otherwise, "our" actions would be governed by "mere interest".

Notice that
25 the first is a "critical" paraphrase of the simple but dubious argument that "experience" "must" be unique, and therefore "determinate" (for otherwise particular instances of it would be "mere chance");
26 the second is the "merely" regulative (but comparably anacoluthic) argument that "'the' moral order" "must" be unique, and therefore "determinate" (for otherwise particular instances of it would be "mere casuistry").

(In each of the formulations of 23–26, it may also be worth pointing out that "*bestimmen*" in ordinary German means "define" as well as "determine").

Notice also that these arguments have (or at least strongly suggest) a tacit quantifier structure—in the cases of 23 and 25, for example, the "inference" is that

27 if every particular experience has a metatheoretic "determination", a master-metatheoretic "determination" of "all" **"experience"** must exist;

and in 24, that

28 if every particular action has a metatheoretically defined moral "valuation", a master-metatheoretic moral **"valuation"** of "all" actions must exist.

(The **bold face** usages in 27 and 28 and the sequel are intended to reflect **"hypostatic"** shifts in each case—from particular object-theoretic "experiences", for example, to an "ultimate" self-referential *Grenzidee* called **"experience"**.)

Observe, finally, that the quantifier structures I have imputed to these "transcendental" arguments closely resemble that of traditional theological "arguments from design" (or more accurately, **design**):

29 if every particular phenomenon has a metatheoretically defined "design", a (numinous) master-metatheoretic **"design"** of "all" phenomena must "exist".

In none of these limit-arguments (or **"limit"**-arguments) could one seriously argue that the inference in question is (or "must" be) metalogically valid.

Whether any of them is a tenable "deduction" in some broader forensic sense of the word is open to question. But any hypostatic "limit" of such senses would be deeply "problematic", in Kant's usage.

Consider, for example, the analogies in 23–28. If they are ("forensically") "deducible" in some limiting hypostatic sense, why not their "physico-theological" counterparts in 29 as well?

Finally, one might perhaps hope to recover the validity of such "deductions" with the aid of **"uniqueness-of-maximality-and-determination"** arguments (cf. 11 above).

If, for example, semantic interpretations of theoretical designs (of whatever sort) formed (what is called) a "direct system", mathematical arguments would indeed yield a unique "direct limit" of such a system.

A good historical case can be made that Platonic, Stoic, Neoplatonic and other "dogmatic" metaphysicians believed that theoretical "cognition" "must" be directed with respect to mutual consistency. But if such "cognition" is **"intelligible"**, in a sense outlined below, it is not. To see this, call a "philosophically relevant" (first-order) theory **T** "intelligible" if and only if it is

30 *"parsable"* (its language is countable, and the metatheoretically defined set of Gödel-codes of its axioms is recursive; roughly speaking, it has an arithmetically decidable "axiomatisation"); and
31 *"autological"*: (it syntactically interprets a theory of finite sets or theory of arithmetic "strong" enough to include a mathematical-induction scheme which permits "encoding" of its proofs; roughly speaking, it has an arithmetically enumerable "proof structure").

The "philosophical relevance" of
32 **"parsability"** is that no inscrutable metatheoretic "oracle" (a word actually used in recursion theory) would be needed to recognise the theory's premises;

and of
33 *"autologicality"* that the theory can trace out, step by step, its consequence-relations, and pose (but not, Gödel discovered, answer) "internally" coded counterparts of semantic questions about "itself" ("Am 'I'

'consistent'"? "Do 'I' 'exist'"?) (cf., once again the first paragraph of the *KdrV*, cited above).

Together, these two conditions—which are most readily and neutrally formulated in first-order set- or class-metatheories which permit semantic interpretations of **T**—imply that Gödel's incompleteness-theorems apply to **T**.

For from Gödel's insights, it follows then that if one
34   partially orders a "system" $\mathfrak{S}_T$ of (metatheoretically) consistent "intelligible" extensions of a given "intelligible" theory T by setting $T' \prec T^*$ if and only if T' is a subtheory of T^*,

Then
35   **"inverse"** (but not **"direct"**)" limits or maximal "threads" of the system $\mathfrak{S}_T$ exist (they are the "models" or semantic interpretations of T) but
36   such "limits"—or "worlds" in which T "holds"—are **never unique**. For any such T, in fact, there will be as many such semantic interpretations of T as there are real numbers in the (tacit metatheory's) set-theoretic continuum.

Admittedly, the conditions in 32 and 33 are
37   intrinsically metatheoretic with respect to **T**.

But they also
38   apply equally well to the set- or class-metatheories just mentioned, as well as a wide variety of other theories which are not first-order.

Efforts to make sense of first- and non-first-order theories **T** would give rise quite naturally, therefore, to
39   "interleaved" *or* "interpolated" first-order semantic hierarchies, in which every consistent theory—first-order or not—would "eventually" find an interpretation.

Such hierarchies offer a first rationale for enlargement of Galileo's "*book*" to the metatheoretic compendia of interleaved commentaries proposed above, in which paths of inquiry would *not* be directed, but ramify *ad indefinitum*.

Would inquiry in such hierarchies be dialectically clever, but "practically", "speculatively" and "teleologically" jejune, as Socrates suggested Protagoras' sophistry would have to be?

In the following sections, I will continue to argue that it would not.

3 Reflective Inquiry and "Transcendent" "[Self]-Validation"

As I mentioned in Section 1, one way to respond to dispel skeptical hierarchies is to "bound" them in (allegedly) unique and determinate "limits", and another to "diagonalise" over them in (allegedly) stable and determinate "fixed points".

To clarify the latter, I will begin with a sense in which Kant (I believe) proposed

1. a "cogito"-like fixed-point argument (or "necessity-of-the possibility of" argument) to "refute" skepticism about the "necessity" *of a* "uniquely" "presupposed" structure of a generic "self," whose "intentions" he identified with "*Erkenntnis überhaupt*" (the "**metaphysical deduction**", usually associated with Kant's claims for the completeness of his "logical" categories); and
2. a second ("cogito"-like) fixed-point argument (or "necessity-of-the possibility of" argument) to "prove" that a uniquely "schematised" cognitive structure "must" "determine" isomorphic patterns in the "appearances" it "must" "experience" (cf. the "**transcendental deduction**", usually associated with wider templates of "intuition" and "the understanding" which also "determine" "the" structures of space and time).

Put somewhat differently, Kant

3. "deduced" from the first fixed-point argument above the "necessity of the possibility" of a uniform structure for "the" mind's (and "the minds'") "cognition" (the "existence" of which Berkeley had attributed to "God", and the agnostic Hume later acknowledged in the appendix to the *Treatise* he had essentially begged); and
4. "deduced" from the second fixed-point argument the "necessity of the possibility" of Hume's *petitio*, as a "unique" "precondition" or "presupposition" for the "apodeictic" "necessity" of four-dimensional Newtonian spacetime (which Kant—I will later argue—tacitly acknowledged in the third *Critique* he had essentially begged).

In effect, I am arguing that Kant believed he had devised more innovative "dialectical" Augustinian / Cartesian fixed-point-arguments, which would modulate

5. assertions a "Newton of the Moral Sciences" had made about highly normalised forms of "Custom or Habit" into

6   more rigorous assertions an "architect" of a "Copernican / Newtonian Turn" hoped to make about an even more highly normalised "*Architektonik*" "*der*" *Erfahrung*.

After I develop this interpretation (which can be considered and accepted or rejected in informal, non-"metalogical" terms), I will
7   consider straightforward but non-Kantian metamathematical glosses of "deduction"

proposed in 1, and
8   prove a straighforward metalogical proposition which bears on the tenability of metalogical analogues of the results Kant thought he had achieved, as well as the "transcendental methods" he devised to "deduce" them.

I will begin with an attempt to explain what I meant above by a "'cogito'-like" argument", and outline the four historical examples mentioned in the essay's introduction.

I will call an (formal or informal) argument "'cogito'-like" if it satisfies the following conditions (cf. 10 through 14 below).

The argument employs
9   notions (tacitly or explicitly) posited in a theoretical framework **T** and "wider" theoretical framework **U**; and
10 assertions about "existence" and "interpretation" of **T** and notions in **T**, which may be formulated in **T** as well as the "wider" framework **U**.

Moreover, the argument's
11 "derivations" and "demonstrations" of assertions *in* **T** (from "obvious" or "universally granted" "first principles" of **T**) also hold in **U**;

and its
12 "epistemic" and "ontolological" assertions about what is "conceivable" in **T** or "exists" in **T** (e.g.) can be formulated in **T** as well as **U**, and sometimes "proved" or "refuted" (from equally "obvious" "first principles") in **U**;

Finally, the argument (ostensibly)
13 "derives" the "epistemic" and / or "ontological" assertion φ about **T** from "first principles" of **U** (or "confutes" the "skeptical" negation Σ of φ *in* **U**);

and

14 "concludes" from this that the "epistemic" and / or "ontological" assertion ϕ is also "derivable" from "first principles" of **T** (so Σ "must" also be "refutable" in **T**) .

Admittedly, this definition is long and convoluted, but so, I would argue, were its historical prototypes. Part of my larger argument in the sequel will be that conceptual (or "critical") "theory / metatheory"-distinctions offer the only tenable remedies for such convolutions and ambiguities.

The best informal rationale or plausibility argument for the characterisation may be the reconstructions it provides for study of the three or four well-known "arguments" from the history of early modern philosophy, mentioned earlier: Descartes' "cogito"-argument, Berkeley's "master argument," and the two best-known Kantian "transcendental arguments" (or "deductions") in the *Analytic* of the first *Critique*.

Before offering these reconstructions, I would like to apologise in advance for my particularly extravagant use in them of ("scare")-quotation-marks. They are not, I believe, an affectation. For my underlying point in 15 through 26 below will be that such "scare-quoted" nouns, pronouns, adjectives and adverbs may be subjected to conceptually relevant forms of metalogical relativisation and metatheoretic ascent.

In the case of Descartes' (not entirely) original **"cogito"**,

15 **T** is "my" theory of "myself";
16 **U** is "god"'s counterpart of "this" theory; and
17 ϕ is an assertion of the "ontological" "existence" and unique interpretability of "my" theory of "myself" (which "I" had epistemically "doubted").

In the case of Berkeley's (more original) **"master argument"**,

18 **T** is "the" (theoretical) "idea" of a given "external object";
19 **U** is "the" theoretical framework in which "god" or another "spirit" confers "existence" on the "object" of **T** by "perceiving" (or "conceiving", or "interpreting") it;

and

20 ϕ asserts the "existence" or "interpretability" of the "object" of this well-defined "idea" **T** (the complex "object" determined by a "mite"'s idea of its "foot", for example, in one of his more whimsical examples).

In the case of Kant's (implicitly "Cartesian") **"metaphysical deduction"**,

3 Reflective Inquiry and "Transcendent" "[Self]-Validation" —— 323

21 **T** is "the" theoretical framework of "forms of intuition" and "categories of the understanding" which "constitute" *"Erkenntnis"* ("knowledge");
22 **U** is "the" (*"bloß regulative"*) *"systematische Einheit aller möglichen empirischen Verstandeshandlungen"* (("merely" regulative) "systematic unity of all possible actions of the understanding") which Kant characterised as an "affair of reason" *"Geschäft der Vernunft"* at *KdrV*, B 692, cited above); and
23 φ asserts "the" "unity", "completeness" and consequent "existence" of 21's "forms of intuition" and "categories of the understanding" **T** which "determine" or "constitute" corresponding "interpretations" of "experience".

In the case of Kant's (implicitly "Newtonian") **"transcendental deduction"**,
24 **T** is "the" theoretical framework or conceptual structure of "the" "necessary" Newtonian-scientific "preconditions" which "constitute" *"Erkenntnis"*;
25 **U** (once again) is "the" (*"bloß regulative"*) *"systematische Einheit aller möglichen empirischen Verstandeshandlungen"*; and
26 φ asserts "the" "existence", "completeness" and *"transzendentale Einheit"* of 24's Newtonian-scientific "preconditions" for the framework **T**, which "determine" or "constitute" unique "interpretations" of this framework in "experience".

In each of the cases just canvassed, I will argue, the assertion φ's initial plausibility derives from a dual premise:
27 that such an implicitly "metatheoretic" **U** "exists"; and
28 that **U**'s "attributes" ensure that all the relevant assertions of "existence", "unicity" and "interpretability" of T encoded in φ are "provable" *or* "derivable" or "demonstrable" in **U**.

In Descartes' case, these premises followed (via the "diagonal" *pons asinorum* of the "*circle*") from the veracity and "perfection" of (Augustine's, Anselm's and Aquinas') "god."

In Berkeley's, their counterparts followed from the kindness and hermeneutic agility of a somewhat more personal Anglican "god", whose "existence" he inferred from a "design" argument.

In Kant's case,
29 the ("merely" regulative) "existence" of the theoretical framework **U** (a secular *Vernunftidee*) is effectively secured by a "critical" "design" argument, and
30 the "existence", ("transcendental") "interpretability" and ("constitutive") "unity" of "experience" φ attributes to **T** is secured in **U** by the

"unicity", "completeness" and "universality" Kant hoped he had deduced from

31 the "unicity", "completeness" and "universality" of **T**'s "forms" and "categories" (in the "metaphysical" deduction), and
32 the "unicity", "completeness" and "universality" of the theoretical framework **T** of Newtonian science (in the "transcendental" deduction).

Whatever the merits of such 'cogito'-like arguments' assertions that the "existence"- and "unicity"-claims φ are "demonstrable" in **U**, I will also argue that

33 the "probative" force and philosophical purchase of these arguments derive from tacit metatheoretic assumptions in each case that [φ's demonstrability in **U**] implies [its demonstrability in **T**] (for metalogical counterparts of these assertions, cf. 4.6 and 4.32 below).

Shamans and high priests have told us since time immemorial, after all (but without philosophical "proof"), that "gods" and other higher "unities" **U** assure us "we" exist, and that our "experiential" theories **T** "make sense" in such **U**. The "philosophical" (as opposed to confessional) "purchase" of "cogito"-like arguments, by contrast, is to formulate subtler and more "secular" "experiential" conceptual apparatus **T** which "provably" ensure us that "we" exist, and that **T** "makes sense" in **T**.

The underlying project of "cogito"-like arguments, in other words, is to broaden conceptual and sensory "experience" so that it "validates" and semantically "interprets" "itself".

The relevance of the tacit premise 33 in the case of Descartes' "cogito" seems to me supported, or at least illustrated, by

34 the centrality of Descartes' otherwise abstruse (and classically quite "dogmatic") attempts to secure a "criterion of truth", where "truth" is identified *de facto* with "demonstrability in **U**");
35 the tenacity of his efforts to demonstrate that the epistemically and ontologically more "perfect" "god" (or more "divine" "theory" **U**) does not "conceal" such "truths" from its less comprehensive "image and likeness" **T**;

and

36 the swift emergence and enduring prominence of debates about the rhetorically begged "Cartesian circle", construed in the present context as a kind of meta-verification (whose formal counterpart would have to be formulated in

a metatheory for **U**) that [demonstrability of φ in **U**] implies [demonstrability of φ in **T**].

(In the case of 35, for example, the "Cartesian circle" may be paraphrased as a "fixed-point"-assertion that "'god' (or **U**) does not deceive us that ['god' (or **U**) does not deceive us]")

In Berkeley's "master argument", the tacit premise sketched in 18–20 above is a tacit conflation of

37 what "you" or "I" or **T** (may hypothetically be posited or "thought" to) "perceive" or "conceive" in **T** ("[that] a tree ["exists"] with "nobody by to perceive it");

and

38 what "god" (or **U**) (presumably) "perceives" or "conceives" about this thought-experimental framework **T** (that "you yourself [did] perceive it all the while").

To see that this tacit premise is problematic in Berkeley's system, recall that a finite "spirit" can only have a "notion" of (but cannot "perceive" or "conceive") another "spirit"'s "perception" or "conception."

The "master argument"'s triumphantly adduced auxiliary assumption, therefore—that "you yourself did perceive it all the while"—may or may not be "demonstrable" in **U**. But it is not even "expressible", much less "demonstrable" in **T**, on Berkeley's own principles. (In quasi-Berkeleyan terms, **T** might at most have a semantically underdetermined "notional" "sign" for this "second-order" self-referential assertion.)

The relevant φ in the "master argument"—which seems to have little intrinsically to do with "trees"—might also be paraphrased as an assertion that "I" (or "you" or "god") "perceive" or "conceive (something)" (and thereby confer "existence" on it). If so, then ("idealist") assertions that such "intentional" acts confer "existence" might be interpreted as natural generalisations of ("rationalist") "cogito-arguments".

Berkeley, finally—who acutely critiqued the contradictions implicit in semi-formal attempts to "define" Leibniz' "infinitesimals" and Newton's "fluxions"—also

39 offered interesting ("sign"-theoretic) "coherentist" reasons to accept the framework of Newtonian science, in the form of

40 a grand scheme of intersubjective interpretations between "all" "our" (otherwise disparate) "signs", "perceptions" and "conceptions" coordinated by his kindly Anglican god (the "design" argument mentioned earlier).

To the extent Kant acknowledged that (what he called) "*Erkenntnis*"—which is not quite the same as "knowledge" ("*Wissen*") in German—might "presuppose" secular forms of intersubjective interpretation between cognitive agents' "*Begriffe*" (and did not simply beg their conceptual or intersubjective "*Einheit*"),

41 there may therefore have been something "Berkeleyan" about the "design"-argument I attributed to Kant in 29 and 30 above; and

42 Berkeley might offer another conceptual bridge from Descartes' and Leibniz' "*rationalisme*" to Kant's "*Kritizismus*", despite the latter's "refutation" of his "*schwärmender Idealismus*".

In the next section, I will consider in some detail the senses in which Kant's "deductions" seem to me to fit "cogito"-like templates of the sort sketched above.

Here I will close with a suggestion that his "transzendentale Methode" in general (and the "deductions" in particular) fit another, complementary "cogito"-like template, or prescription: that of an "Archimedean"- or "fixed-point" confutation of "regressive" skeptical "doubt."

More precisely, I will attempt to interpret this "*Methode*" as a "dialectical"

43 "proof" (in some sense) that (something called) "synthetic a priori cognition" is a "proof" (or "sufficient reason", or "transcendental" "Archimedean point") for [itself];

or, in more or less equivalent adverbial terms, as a collection of roughly cognate "proofs" that

44 cognitive agents synthetically a priori "prove" or "determine" that

[cognitive agents synthetically a priori "prove" or "determine"]
(cf. the gloss of the "Cartesian circle" following 36 above).

For if not,

45 [cognitive agents fail synthetically a priori to "prove" or "determine" that

[cognitive agents fail synthetically a priori to "prove" or "determine" that
[cognitive agents fail synthetically a priori to "prove" or "determine" that
[cognitive agents fail synthetically a priori to "prove" or "determine" that
[...]]]].

But such a "Regressus" (a "descending" variant of ancient academic skeptics' "ascending" "problem of the criterion") would (Kant believed)

46 render "synthetic *a priori*" "proof" or "determination" "impossible" (i.e. counterfactual, Kant's standard "critical" gloss of "impossibility" in such contexts; cf., e.g., B 101, 137 and 628).

But

47 this—Kant argued—would be "absurd". For "synthetic *a priori*" "determination" is not only regulative (and in this sense "possible"). It is also "*konstitutiv*" (and in this sense "necessary"). For its "constitutivity" is instantiated by the ["synthetic *a priori*" capacity (*Vermögen*) to formulate this argument] ....

In effect, I believe, such a "fixed-point"-argument may be construed as an attempt to reconfigure 44's non-well-founded "descent" into a kind of "transcendental circle", in which the **T** of 33 above "validates" **T**'s "experience" in T. (More of this later.)

In metalogical opposition to such metaphysical "circles" (attempts to collapse "regresses" such as the one which would otherwise emerge in 36), I will argue in the next section that

48 notions of "doubt", "certainty", "perception", "conception", "knowledge", "syntheticity", "apriority", "cognition" and "determination", for example, cited earlier—are "liminal", in the sense that they straddle the "*limines*" or "thresholds" of whatever they "intend" or "qualify".

In quasi-Kantian terms, one might call them "locally" or ("situationally") "*transzendent*". Alternatively, in language Georg Cantor introduced in a letter to David Hilbert more than a century later, one might call them "locally" *or* "situationally" "*inkonsistent*".

To the extent one can make metalogical counterparts of such notions precise (more possible than many readers might think), [self-]referential attributions of such properties "generate" semantic paradoxes and corresponding patterns of iterated metatheoretic "ascent" and (what might be called) intentional "descent".

In Kant's case—and in particular in the allusion to "[this very argument]" just above—such notions will not only be "counterfactual" (or non-"experiential"). They will be ("locally") *transzendente Vernunftideen*, or perhaps more accurately "*Vernunftprädikate*" (predicates of "pure reason").

In plain German: "'*die*' *Rahmenbedingungen der Kantischen Argumente über* '*die*' *Bedingungen* '*der*' '*Möglichkeit*' '*der*' *Erfahrung sind lauter Vernunftideen* ...." (In even plainer English: "'the' boundary conditions of Kantian arguments about 'the' conditions of 'the' 'possibility' of experience are a bunch of ideas-of-reason ....")

"Worse" (or in my view, "better"): metatheoretically defined counterparts of such "frame"- or "boundary conditions" will always be "locally transzendent", in relative terms, with respect to any given "level" of what they (e.g.) "doubt", "conceive", "cognise" or "determine" to "be the case".

In particular, if one tried to "halt", with Kant and his neo-Kantian successors, the "object-theoretic descents" in 44 and 45—or the "metatheoretic ascents" they mirrored—in an effort to "define" "experience",

49 the terms in which one claimed to do so would become (locally) "transcendent" with respect to any fixed "neo-Kantian" object-theoretic notion of "experience" one might propose,

in the sense that

50 the terms in which one proposed to "define" a given "critically" demarcated notion of "experience" would not only fail to be "experiential"; they would be "experientially" inexpressible, and "paradoxically" as well as "conceptually" underdetermined by what they defined.

4 Reflective Inquiry and "*das Schicksal der Vernunft*"

In this section, I will assimilate the "locality" and essential incompleteness of "critical" inquiry to Kant's "fate of reason", poignantly characterised above in the opening lines of the *Kritik der reinen Vernunft*.

I will attempt in this section to

1 interpret more carefully the begged implication in Section 3 for the "'cogito'-like" reconstructions of *Kant's* "transcendental" method and "deductions";
2 characterise the views and analyses attributed above to Descartes, Berkeley and Kant as "idealisms" with successively broader and more "deductive ranges" of what is "conceived" in "experiential" **T** and "higher" **U**;
3 offer "skeptical" as well as "transcendental" arguments for assimilation of such analyses' "assertions" and "demonstrations" in **T** and **U** to counterparts of them in first-order metatheoretical hierarchies;
4 review well-known proofs that metalogical counterparts of the "dialectical" inferences and implications in 3.14 and 3.33 are either "locally" refutable or "locally" inexpressible in such hierarchies' theoretical frameworks **T**; and
5 adduce (and sketch the proof of) a simple but relevant bit of metalogical "folklore": that no "faithful" interpretation of a "weaker" theory **T** can be

found in a "higher" theory **U** which is "strong" enough to prove the consistency of **T**.

The breadth of the *"Rechtmäßigkeit"* (cf. B 116) Kant associated with *"Deduktion"* is well-known, as is the notion's association with "mediated" forms of reasoning (cf. B 761), and the role of *"transzendentaler Deduktion"* in the (curiously "subsumptive" or "reflective") *"Erklärung der Art, wie sich Begriffe a priori auf Gegenstände beziehen können"* (B 117).

In what follows, I will draw in fact on the latter quotation to argue

6   that (ostensibly) more latitudinarian notions of Kantian "deduction" do, after all, tend to close down on straightforward first-order consequence-relations in metatheoretical hierarchies which interpret such "predication" and "subsumption";

and that a good way to sustain Kantian "critical" insights (between the "**T**"s and "**U**"s of Section 3, for example) might be to

7   "localise" and iterate them, in the form of "locally critical" theory / metatheory distinctions;
8   exploit them as generators of "heuristic" notions of object- and concept-formation in the metatheoretic hierarchies they generate; and
9   interpret their unending extensions and ramifications in such hierarchies as "regulative" traces and counterparts of Kant's "fate of reason" (*"Schicksal der Vernunft"*).

I suggested earlier that Kantian or quasi-Kantian reconstructions of the generic "cogito"-like assertions φ in Section 2 might be

10 ("metaphysically" "deductive") claims that "the" "completeness" and "unity" of "forms of intuition" and "categories of the understanding" "determine" (or "constitute") **T** and its "unique" isomorphic counterparts in "experience"; and
11 roughly cognate "transcendentally" "deductive" claims that "the" "completeness" and *"transzendentale Einheit"* of "intuition" and "understanding" "determine" (or "constitute") **T** and *its* "unique" isomorphic counterparts in "experience".

I also suggested that all of the four "cogito-like" arguments outlined in Section 2 decompose into

12 (relatively uninteresting) "demonstrations" that interpretations of "thought", "perception" and "conception" in **T** (may) "exist" in "higher-order" frameworks **U**; and

13 (interesting but begged) "demonstrations" that corresponding interpretations of "thought", "perception" and "conception" in **T** ("must") "exist" in the "experiential" framework **T**.

In his "metaphysical" and "transcendental" variants of these arguments in the *Analytik*, Kant sought to work in a resolutely secular **U**, and dispense with conceptual "mathematical" and "dynamical" "physicotheological" arguments in his "constitutive" rationales for the "transcendental unity" of "*Erkenntnis*" to "*Erfahrung*".

Yet he insisted over and over again in the *Dialektik* of the first *Critique* that the "systematic unity"of "Reason" at KdrV, B 692—the theoretical framework I have called **U**—"must" remain a "merely regulative" but conceptually significant "*Aufgabe*", an *Ideal* which *zieht uns hinan* (so to speak) at the marginal horizons of such (scientific) "experience".

Logicians familiar with the semantic purchase of (relatively) "higher-order metatheories" such as **U** know that part of this "purchase" is their ability to "define" "internally canonical" interpretations of "weaker" subtheories such as **T**, and it is in that sense that such **U** may be "locally" "strong" enough to account for "scientific" "boundary conditions" of "experience" and "experiment(ation)".

But they also know, on the evidence of Gödel's results, if nothing else (cf. the proof of 38 below) that

14 the "strength" of such theories is "theory-relative" (or at least "theory-mediated")—if one stipulates that **U** *is* "(intrinsically) stronger" than T if and only if **U** proves the consistency of **T**, the (nonlinear) ordering of such "strength" is neither reflexive nor symmetric.

In particular, the [self]-sufficiency (and implicit [self]-validation) of Pierre-Simon Laplace's famous remark (quoted third-hand in Victor Hugo's autobiography) that "[s]*ire, je n'avait pas besoin de cette hypothèse(-la)*" might therefore have to be more carefully calibrated than Laplace knew.

If all Laplace meant was that he had no need of (secular *or* numinous) versions of (what Kant called) "the unconditioned" ("das Unbedingte"), he was quite right.

But if he (or Kant) meant

15 that his own or anyone else's intelligible formalisations of mathematics or mathematical physics could uniquely "determine" their own (mathematical or conceptual) boundary conditions, much less their own "intended" semantics,

he was subtly but deeply wrong (and Descartes, Berkeley and Kant with him). In the sequel, therefore, I will also continue to interpret

16   Kant's "transcendental deductions" as failed attempts to find such an "Archimedean point" of [self]-validation on the razor's edge between Newtonian science and its elusive "hypotheses" (which Newton had tried to disavow); and
17   the second and third *Critiques*' qualifications and relativisations of such claims as tacit admissions that his efforts to locate this "transcendental" limit had turned out to be more "problematic" (a Kantian term of art) than he had hoped.

Whatever the merits of Section 2's "'cogito'-like arguments", therefore, the tacit *petitio* of their claims that

18   [demonstrability of semantic assertions φ about **T** *in* **U**] "followed" from [demonstrability of such semantic assertions φ *about* **T** in **T**]

remained present in Kant's *Architektonik*.

In the remarks following 13 in this section, I appealed to analogies with metalogical hierarchies which have run in the background throughout the chapter's first two sections, and it is past time to offer the rationale for these analogies.

It would follow from the heuristic principles I have sketched that no one—on pain of circularity or semantic paradox—could claim to offer a "hypothesis-free" "deduction" that first-order "deductions" (much less metatheoretical hierarchies which employ it) are "canonical". One can, however, offer "skeptical" variants of Kant's "critical" or "transcendental analysis" which suggest that "eventual" recourse to first-order theories in "ramified" metatheoretical hierarchies may be ineluctable.

The "analysis" I have in mind is

19   **"skeptical"** in that it "suspends judgment" between (alternative interpretations of) "abstract logics" and their "consequence relations";
20   **"critical"** in that it postulates (on pain of semantic paradox) theory / metatheory distinctions to "make (syntactical) sense" of such interpretations;
21   **"transcendental"** in that recurrent theory / metatheory distinctions of this sort "regulate" but do not "determine" (semantic) efforts to assert such alternative interpretations' existence;
22   **"eventual"** in that such interpretive predications of semantic "existence" may "eventually" and "economically" be adjudicated in first-order metatheoretic set- or class-theories;

23 **"ramified"** in that "partial" interpretations of any "intelligible" theories which "encode" their own syntax and "consequence-relations" in "inductive" ways branch without limit;

and finally

24 **"ineluctable"** in that "all" sufficiently "finitary" theories introduced to examine and interpret other theories in such hierarchies are semantically indistinguishable from first-order counterparts, by a remarkable theorem of Per Lindström ([Lindström] 1969).

In somewhat plainer English:
25 the simplest and most "natural" way to understand abstruse theoretical frameworks in metatheoretical hierarchies is to interpolate set- or class-theories which interpret them;
26 "intelligible" set- or class-theories which satisfy "finitarity"-conditions are semantically indiscernible (via Lindström's theorem) from their first-order counterparts;
27 hierarchies of recursively axiomatisable first-order languages offer flexible, communicable and ontologically neutral "common languages" and frameworks of interpretation for "rational" discourse;

and finally,
28 such languages' "ontological neutrality"(cf. 27) and "skeptical" "suspension of judgment" cf. 19) offer "natural" conceptual interpretations of Kantian "Reinheit" and "apriority", as well as physicists' (equally elusive) ideals of "controlled experimentation" and "experimental isolation".

There are many reasons, of course, to dismiss such analogical conjectures and "skeptical transcendental arguments" out of hand. The first is that "ordinary" language is obviously not first-order at any given level of (potential) syntactic and semantic interpretation. So claims that it is are reductive.

My conjectures in 19–24 above—which I clearly cannot "prove" ("Chinese rooms" are lurking about here somewhere)—are
29 that the limitless syntactical and semantic complexities of "ordinary linguistic" usages may be interpretable in comparably limitless syntactical and semantic complexities of first-order hierarchies; ("hand over hand", so to speak); and

30   that sufficiently skilled speakers of "ordinary languages" often improvise locally adequate finitary substructures of such hierarchies in "real time" (otherwise—as non-native speakers well know—their interlocutors will "give up" on them).

A second objection is that one has seen unconvincing versions of such views before—in the writings of logical positivists, for example, or in "Quinean" assertions that "everything" "can be expressed" in "set theory".

On the account I wish to defend, "experience"—including linguistic experience—is certainly much broader than anything I am aware of in the writings of logical positivists. It is also broader than "set theory", or at least any particular set theory. For "set theories" (in my view at least) are just ontologically neutral way stations for attempts to talk about "predication"—what Kant would have called "*Subsumtion*", or "*das Besondere als enthalten unter dem Allgemeine zu denken*" in B 26 and B 32 of the third *Critique*.

And "critical" theory / metatheory distinctions, I would argue, are inevitable concomitants of attempts to iterate such "predication", and heuristic metatheories can be interpolated in such iterations almost at will. Such iterations occur "naturally" in hierarchies of "stronger" (possibly non-first-order) metatheories, with the proviso that assertions about such metatheories' consistency are "liminal" and remain open to further ("skeptical") inquiry.

The most "natural" way to clarify such consistency questions, in turn, is to seek semantic interpretations for them in yet-"stronger" theories of "predication". And the simplest and most "natural" venues for such searches are "stronger" as well as linguistically augmented first order set theories. Russell's early type-hierarchies were "linear" and begged questions about "the" limits of their "types" in ways which were anticipated by his own paradox and Cantor's observations about "*Inkonsistenz*", and were clarified later by metalogical observations of Skolem, Gödel and Henkin.

Wondrously subtle refinements of such hierarchies have been and will continue to be devised. But the more open and flexible they were, the more "natural" their interpretations in the aforementioned linguistically enhanced set theories became. Such remarks are obviously "heuristic" and subject to correction and revision. But they may make a provisional case for a conjecture outlined earlier: that "everything" we talk about and endeavor to interpret "is" (or "may as well be") "eventually" first-order.

In defense of this heuristic conjecture, finally, I would

31   appeal again to the openness, generality and intensional interpolability of metatheoretic hierarchies, given that extensional canvasses of potentially infinite evidence are beyond finitary "seekers'" ken; and

32 appeal finally to their pragmatic utility, in "ordinary" as well as metaphysical "discourse", as naturally "interleaved" venues for critique and interpretation of everyday "intentionality."

(My favorite thumbnail-critique of this sort is Wittgenstein's uncharacteristically disarming remark in *Über Gewißheit* (ÜG 12), that "[m]an vergißt eben immer den Ausdruck "*Ich glaubte, ich wüßte es*'" ("*one always forgets the expression 'I thought I knew'*"....)

The "subreptively" self-referential problem Kant sought to solve may be characterised as follows.

A genuinely *a priori* "*Deduktion*" would have to be

33 a conceptually idealised or "isolated" (thought)-"experiment", independent of "all" "experience" and "all" "external" ("merely empirical") verification;

34 a "deductive" act (in accordance with some sort of generalised "consequence"-relation) invariant under "all" changes of metatheoretic boundary conditions;

and most problematically (once again)

35 a "fixed point", "Archimedean point" or "self-validating" "demonstration" (in the sense once again of some sort of generalised "consequence"- relation), with respect to "all" attempts to "refute" it or "call it into question".

(It might also be assumed to "prove" its own ["provability"]—a curiously unproblematic assumption in metalogical contexts, as Löb observed (cf. Smorynski, 845), but one which also turns out to have little "transcendental" force or prooftheoretic purchase).

What it could not "prove", or even "express", remained, as before, its own "existence", "interpretability" or "consistency", much less the "necessity" of the "possibility" of its own "existence", "interpretability" or "consistency" .... (Kant's cogent observations about the ontological ambiguity of "existence" might be more relevant to "transcendental" analyses than he realised).

To the extent such "proofs" were truly "isolated", for example, "we" could not even adduce stable thought-experimental evidence to "prove" that they would not also be nonstandard (in quasi-Kantian jargon, that non-well-founded "*intensive Größen*"—metalogical counterparts of Hume's "problem of induction"—might occur in their formulation).

I will conclude this section with a brief semiformal sketch of the "folklore" result promised above in 1.17 and 3.5, which may clarify the nature of the barriers devisers of philosophical "cogito"-like arguments have faced. I've elided inductive definitions and blurred notational distinctions to simplify the exposition, but any competent logician can bring them into sharper focus.

The definitions first. A "syntactical interpretation" **E** of a (not necessarily first-order) theory **T** in another theory **U** is a systematic translation of formulae φ of **T** into counterparts φ^E *in* **U** such that

36   [**T** proves φ] (metatheoretically) implies that [**U** proves φ^E] in T^E, where T^E is the collection of such translates φ^E for φ in **T**.

Such a syntactical interpretation **E** is faithful if and only if
37   [**T** proves φ] if and only if [**U** proves φ^E] for all φ in **T**.

Since the "begged" metatheoretical assumptions of Section 2's "cogito"-like arguments asserted that the identity translation faithfully interprets **T** *in* **U**, the following may be relevant.

38   **Proposition.**

Suppose **T** is a consistent first-order theory which syntactically interprets Peano's arithmetic, and **U** is a (metatheoretically) "stronger" consistent theory which proves that [**T** is consistent]. Then no syntactical interpretation **E** of **T** in **U** faithfully interprets the assertion that [**T** is consistent] in **U**.

Proof (Sketch).

Suppose (for contradiction) that a given syntactical interpretation **E** did faithfully interpret the assertion that [**T** is consistent] in **U**. Then [**T** is consistent] would be provable in **U**, by hypothesis, and the biconditionals [[**T** is consistent] if and only if [T^E is consistent] if and only if [**T** is consistent]^E] would also be provable in **U**, by a series of inductive arguments.

So
39   [**T** is consistent]^E would be provable in **U**.

But it is a consequence of Gödel's original incompleteness results (cf. Bell and Machover, 1977 or Smorynski, 1977, 821–865), that [**T** is consistent] is not provable in the theory **T**.

The assumption that **E** is faithful and the remarks in the first paragraph would therefore yield that

40   [**T** is consistent]^E is not provable in **U**,

so we have derived a (metatheoretic) contradiction from the assumption that such a faithful interpretation **E** of **T** into **U** exists.

The informal originals of the assertions φ in the proposition made informal semantic claims—in "ordinary languages" (whatever they are)—about "existence", "interpretation" and "conceivability" of "experiential" frameworks, and I have assimilated such frameworks to formal theories **T**.

"Analytic realists" who dismiss such assimilations would presumably argue that such assimilations are inapt ("category mistakes", so to speak)—that "analogy" and "assimilation", for example, are not "identity". (The ancient skeptics' response to this particular argument was that "we" might not be able to discern the difference—arguably an anticipation of "equivalence-class" arguments employed in the proof of Leon Henkin's completeness theorem (Chang and Keisler, 1972, 61–67)).

I would respond that the intense eristic attention devoted to Descartes' "cogito", Berkeley's "master argument" and Kant's "deduction(s)" tacitly suggests that they were

41   the most vulnerable as well as "ambitious" thought experiments in their authors' work,

to which I have added a sustained argument that

42   the premises "begged" in their "proofs" and "deductions" become refutable in reasonable metalogical counterparts.

In the case at hand, moreover:

43   the notions of syntactical interpretation I have introduced are rather broad (they apply to theories which are *not* necessarily first-order, for example);
44   there are good (metalogical) reasons to assimilate the semantic claims of the informal assertions φ to consistency-assertions about first-order counterparts of such **T**; and
45   "skeptically transcendental" arguments for the recurrence of first-order interpretation in ramified metatheoretical hierarchies offer correlative "good reasons" to consider the methodological relevance of such constructions.

I will finish in the same spirit with a final heuristic argument (which may be all an honest skeptic should try to offer).

As I've suggested earlier, two interrelated hypotheses have animated this chapter: that

46 "precise" "transcendental" assertions of [self]-validation are "transcendental illusions";

and that

47 semantic paradoxes and metalogical analyses devised to accommodate them have hermeneutic value as thought-experimental refinements of their metaphysical and epistemological ancestors.

Such hypotheses—along with their companion-of-the-route, the "metalogical turn" postulated above in 1.1–1.3—cannot be "proved". Measured against [their own] criteria, they may be partially and provisionally sustained, but they will never be "conclusively" (much less "ultimately") "secured".

But these reflections suggest two more hypotheses. The next is that

48 critical awareness itself (and in particular critical [self]-awareness) may be a ("merely regulative") value, and Keats' "negative capability" (Keats, 277) one of its marks.

Such an awareness, for example, might (partially) reconcile

49 the apparent incompleteness of Kant's "practical reason" with the underdetermination of its "speculative" counterpart;
50 the *regulative Ideen* of Kant's "moral law within us" with our "desires" to "understand" the "starry heaven above us"; and
51 the *Würde* and *Schicksal* of Kant's "reasonable beings" (*"vernünftige Wesen"*), with the "dignity" and "fate" of Pascal's *roseaux pensants* ("thinking reeds") in a *"univers"* which *"n'en sait rien"*.

The last hypothesis is that

52 ideals of "heuristic" inquiry and "intelligible" verification and concept formation may be preconditions for "freedom", "autonomy" and mutual "respect";

and therefore that

53 recourses to such "merely" regulative ideals may be "reasonable beings'" cradle gifts as well as their "fate(s)."

5 Reflective Inquiry and "Das" Reich "der" Zwecke

Consider the following (fairly representative) passage, taken from a section in the second *Critique* entitled:

> **Wie eine Erweiterung der reinen Vernunft in praktischer Absicht, ohne damit ihre Erkenntnis als spekulativ zugleich zu erweitern, zu denken möglich sei** ....
>
> Hier werden sie [praktische Ideen, wie Freiheit, Unsterblichkeit, Gott und das höchste Gut] immanent und konstitutiv, indem sie Gründe der Möglichkeit sind, das notwendige Objekt der reinen Praktischen Vernunft (das höchste Gut) *wirklich* zu machen, da sie ohne dies *transzendent* und bloß *regulative* Prinzipien der spekulativen Vernunft sind, die ihr nicht ein neues Objekt über die Erfahrung hinaus anzunehmen, sondern nur ihren Gebrauch in der Erfahrung *der Vollständigkeit zu nähern* auferlegen.
>
> KdpV, B 241, 244 (last emphasis mine)

> **How an Extension of Pure Reason in Practical Intention is Possible to Be Thought, without at the Same Time extending its Cognition as Speculative** ....
>
> Here they [practical ideas, such as freedom, immortality, god and the highest good] become immanent and constitutive, in that they are grounds of the possibility for making the necessary object of pure practical reason (the highest good) *real*, since without this they are *transcendent* and merely *regulative* principles of speculative reason, which impose on it not the obligation to accept a new object beyond experience, but only to bring its use in experience *closer to completeness*.

Such appeals to "transcendent and merely regulative principles" provided later commentators and metaphysicians with a textual basis for assorted "two-truths"-interpretations of Kantian metaphysics, and the late-nineteenth century Kant scholar Hans Vaihinger, in particular, with one of the prototypes for his "*Philosophie des Als Ob*" (the most striking basis for such readings may perhaps be found in *KdU* §76, B 339–B 344).

In the first *Critique*, Kant employed such "merely regulative principles" as ladders to "higher" forms of "truth" (cf., e.g., B 83–87, 185, 269 and 670). But he never explicitly acknowledged—except for the *Dialectic* passages devoted to "intelligible causes" (cf. KdrV, B 566 ff.) ("critical" counterparts, so to speak, of epicurean "swerves")—that

1 "practical" recourse to regulative / constitutive distinctions might shift the ground of "speculative" (scientific) "experience",

or (in the jargon of the last section) that

2   "scientific" experience might not be faithfully interpreted in wider regulative frames for its "conscious" "practical" extensions ("… without at the same time extending its cognition as speculative ….").

In the opening section's potted history, I also
3   canvassed straightforward implications of the neo-Kantian observation that many "new objects" (and "new relations") have become part of "experience" in mathematics and mathematical physics, and
4   suggested that metalogical margins of "experience" and its thought-experimental counterparts might someday—"should" someday, if our descendants have the wisdom for it—encompass patterns of recurrently metatheoretical "object"- and concept formation undreamt of in Kant's philosophy (or mine).

Such arguments and the semantic pluralism which underlie them suggest alternative readings for certain remarks in the passage quoted above, and skeptical interpretations, in particular, for "practical" aspects of Kant's monist assumptions about "immanent" / "transcendent"- and "constitutive" / "regulative"- dichotomy. Suppose one accepted, for example, that "local" claims of "speculative", "practical" or "teleological" "completeness" (Kant's "*Vollständigkeit*") are only "locally" expressible, in essentially incomplete and therefore plurally interpretable metatheories.
    Then
5   the "ladders" mentioned earlier—an ancient skeptical image which appeared in the pyrrhonist writings of Sextus Empiricus (cf. M VIII, 481), and reappeared in Wittgenstein's "dogmatic" *Tractatus*—might branch and diverge "forever" in plurally interpretable metatheoretic patterns; and
6   the "practical" remarks in *KdpV*, B 241 and 244 and "teleological" counterparts of them in *KdU*, 301, 342 and 344 ff.) might begin to resemble ancient skeptical claims to engage in "dialectical" (and plurally interpretable) speculation "*adoxastōs*"—"undogmatically", "undoctrinally", and "not in the manner of the dogmatists" (cf., e. g., Sextus Empiricus, *PH* I, 24/16).

In what follows, I will argue
7   that the principal bar to such "skeptical" readings of Kant's non-skeptical texts is his assumption that the "hierarchies" of his "constitutive" / "regulative" and "immanent" / "transcendent" distinctions have exactly two levels;

8   that Kant worked with "speculative", "practical" and "teleological" counterparts of this linear two-stage ascent as a ground-bass in an extraordinary variety of assertions about *"die" Grenzbestimmung zwischen Immanenz und Transzendenz* (cf., e.g., *Prolegomena* 350–362 and *KdrV*, B 786); and
9   that this "capped", two-stage linearity of Kant's had an obvious historical precedent: the "cave parable", the most influential account of "practical" and "epistemic" perspective-shift ever written, in which Plato simply assumed that emergence from "the" cave would take place only once, in one "direction", and would never be iterated.

In Kant's case, as well as Plato's, why?

Why did Plato and Kant, both of whom seemed to countenance other sorts of *"infinitum"* and *"indefinitum"* readily enough, reject stages of semantic (re)interpretation and hermeneutic reflection out of hand? Why did both assume there was only one "line", or "ladder", or "form of the good", or "realm of ends", when partial realisations of noetic enlightenment would have accorded equally well with the "conversational" and self-"critical" aspects of Socratic *dialektikē* and Kant's *"Dialektik"*?

Why, above all, did both construe epistemic, metaphysical and practical "truth" and "goodness" as second-order templates, rather than hierarchies of relational interpretations elucidated in "heuristic" metatheories (or "caves" illuminated by "heuristic" "suns")?

In the case of Kant's "critical" work in general, and his "practical" philosophy in particular, the answer, I believe, had little to do with his alleged temperamental preoccupations with "duty" and dour fascination with unitary "law". He was a kindly teacher and brilliant conversationalist, who enjoyed musical entertainment and held down a place at his local *Stammtisch* (when he could afford it). If Kant felt a deeper sense of philosophical "duty" or "commitment", it was to "hieratic" philosophical traditions which enjoin the sorts of unicity and canonicity of interpretation I have assimilated to nominally secular "design arguments".

I have already argued that certain scientific ("speculative") counterparts of such arguments are metalogically untenable efforts to finesse the "fate" of "ontological" and epistemological "reason". Here I would suggest that such arguments are indeed more untenable—

10 not because ethical and "practical" "experience(s)" and (thought-)experiments are less complex than their scientific counterparts, but because they are vastly more so.

5 Reflective Inquiry and "Das" Reich "der" Zwecke — 341

(Consider, for example, that
11 ethicists are expected to acknowledge the potential relevance of illimitable complex "initial" and "boundary" conditions to moral "experience",

whereas
12 scientists are expected to require that experiments (*"expériences"* in French) be as "simple" (or at least "controlled", "isolated" and "replicable") as possible.)

In both cases, therefore (but *a fortiori* in the "realm" of "the practical"),
13 "intelligibility" and heuristic [self]-reference (however incomplete) are "regulative" "preconditions" of forms of communication and [self]-understanding worthy of the names.

The "regulative" force of such ideals of processive inquiry—without an awareness that such inquiry must be ramified if it is to be "intelligible"—clearly animated Descartes', Berkeley's and Kant's hopes that "design"-based "uniqueness-of-maximality-and-determination" arguments might be "provable";

But it could also provide a framework—mindful of the greater "depth" evoked in 10 above—for
14 acceptance of localised forms of Kant's *"Primat des Praktischen"*, but rejection of "the" inexpressible "unicity" of his *"Reich der Zwecke"* ("realm of ends").

The rest of the section will be given over to an attempt to outline such a "regulative" and "experiential" interpretation of Kant's *Primat*, in which significant notions of *"Zweck(e)"* and *"Autonomie"* will never "converge" to or eventuate in a unique moral or teleological *"Reich"*.

On my account, the "primacy of the practical" reflects
15 a "natural" desire to interpret one's experience in something "deeper"; more "conscious"; more "contemplative" (cf. Aristotle's *"theoretikon"*); and more "comprehensive" (and *compréhensive*); and
16 an "autonomously" self-imposed obligation to respect that desire in other "reasonable" (or at least "sentient") beings which "experience" it, in the sense that they can propose and test counterfactual "purposes" (*"Zwecke"*) they hope might "realise" (or "fulfill") it.
17 An "end" or "purpose", in this "practically" shifted context, is simply a (syntactically or semantically) interpretable object-theoretic assertion (the German *"Sachverhalt"*).

18 A "maxim" ("*Maxime*") is simply an object-theoretic assertion that certain object-theoretic "means" or actions imply or "bring about" a particular end.
19 A (moral) "law" ("*Gesetz*"), finally, is a modal as well as an (intelligible) metatheoretic assertion that certain "means" or actions "should" imply (or "morally entail") a particular object-theoretic "end".

In "Virtual Modality" (2003), I defined an adequate "virtual" semantics for alethic modal assertions in a given theory **T**, in which "worlds" $\mathbf{W_T}$ are boolean-valued extensions of an "initial" placeholder-"world" "generated" by a "virtual" structure which interprets **T**.

For what it is worth, one could define deontological refinements of such an alethic semantics which would single out

20 "practical" subclasses $\mathbf{W_T}$ of such "worlds" or "realm of ends" $\mathbf{W_T}$,

characterised by "reasonable" constraints on the "permissibility" or "proportionality" of means and "legitimacy" recognised in them, and in which "reality" "would" be as it "ought" to be (by the lights of $\mathbf{W_T}$).

The ranges of each such class $\mathbf{W_T}$ of counterfactual "deontological" interpretations would be

21 metatheoretic with respect to the first-order set-theory in which the "virtual" modal **T** mentioned above is interpreted;

and

22 metametatheoretic with respect to the original theory **T**, in which quasi-Rawlsian "ends" or "purposes" might "initially" be identified and debated,

but the semantic intricacies of such a deontological "stratification" might reflect the illimitable complexities associated with the "primacy of the practical" above.

But they are conspicuously, notoriously (and perhaps inevitably) absent from any "deontological" refinements of modal semantics (the one sketched earlier or any other) which are guidelines for "choices" of the hypothetical classes $\mathbf{W_T}$.

I will devote the rest of the section therefore to an impredicative effort to sketch a few minimal and incomplete criteria which "ought" to be met by such "realms", and argue on good skeptical grounds (and close alignment in this case with Kant) that more "complete" decision-procedures would be "inexpressible" in the theoretical contexts to which they applied.

Consider, for example, the following (rough) "marks" of "awareness", in ascending order of metatheoretic complexity.

23 It is a mark (indeed a characterisation) of a well-defined (idealised) Turing machine that it can implement object-theoretically feasible and well-demarcated "induction schemes" of inferences, and "know", in this sense, "how to go on" with them.

24 It is a mark of epistemic "intelligence" to "understand" metatheoretically (and perhaps even "know") that one does not "know how to go on" with respect to indefinitely extendible ranges of such (more and more complexly defined metatheoretic) schemes.

25 It is a deeper and more zetetic ("searching") mark of metatheoretic "insight" to search (perhaps in vain) for "reasons why" we do not "know how to go on"; make attempts to extend or modify such schemes (or "theories"); and propose new "characterisations" of them in other (more complexly schematic) "metatheories."

26 It is a still deeper and more zetetic mark of sentience, ethical insight and quasi-Kantian "reasonable faith" to seek to understand why and when "we" "should" and "should not" "be expected to" "know how to go on" with our attempts to query and refine current schemes; and formulate (in full awareness of our limited moral and conceptual capacities) quasi-Kantian practical *"Gebote"* ("commandments" or moral "injunctions").

The principle such *"Gebot"* might be called Pascal's imperative: to

27 value sentience and its preconditions, and respect "cognitive agents" enmeshed in its dilemmas;

and

28 attribute to such sentient agents "rights" and "responsibilities"—to the best of "our" abilities and "theirs" (cf. 24 above)—as (potentially) "reasonable beings" in need of mutual aid.

(We often seem to forget the obvious: not only do "we" *not* "*know*" how to go on"—"*we*" do not go on.)

If certain actions clearly seem to be incompatible with the "respect" and its "attributions" just invoked, we have ("autonomous") duties to abjure them, if nothing within our physical or conceptual horizon constrains us. And if—by parity of reasoning, so to speak—such forms of "respect" and their "attributions" clearly seem to entail certain actions or resorts to particular means, we have equally "autonomous" duties to carry them out.

Exactly because these deliberations are autonomous, however—"legislated" by an elusive and underdetermined "self"—they do not have sharp metatheoretic boundaries, and their conceptual horizons elude our view (cf. Leibniz' hypothesis that "[c]haque âme connait l'infini, connait tout, mais confusément ....", in the "Principles of Nature and of Grace", G VI, 604).

Metatheoretically nuanced efforts to interpret aspects of "experience" may extend them, and even "bring ... [them] closer to completeness" in this sense. But if they are intelligible—and in that sense communicable—they will not be able to determine "ultimate" boundary conditions of practical experience.

What endeavors to "prescribe", in short, cannot itself be "ultimately" "prescribed". What "it" can do (and "should" do), is (try to) enjoin itself to (try to) be more comprehensive (and *compréhensif*).

One could, of course, attempt dogmatically to assert the "existence" of an— essentially indefinable—*EA* notion of ("human") *consciousness*—a sort of intentional "last metatheory", of the sort mentioned earlier. But the arguments offered in prior sections suggest that such a "consciousness" could not express, much less "know" what [it itself] is, much less "know" with sovereign certainty that the thing behind the curtain is not [it].

(For a first indication that something like this might be inevitable, consider the "impredicative" nature of attempts to sketch criteria which "ought" to be met by "criteria"—such as the "W_T" sketched earlier—of what "ought" to be countenanced.)

There are, in short, meta-deontological counterparts of "autological" dilemmas, and there is no reason to believe that they can be finessed or ("ethically") suppressed.

Kant, for example—to his credit—essentially acknowledged, in the *Grundlegung der Metaphysik der Sitten*, that a "purely good will" is indefinable and indiscernible in human experience (cf., e.g., Gr 459–463). This seems to me methodologically right, on the analogies I have sketched. But the arguments of prior sections suggest (at least to me) that no metatheoretically stable Analogon of a "transcendental deduction" which would ensure a will's "existence"—even "regulatively"—could be formulated.

They also suggest another, somewhat more oblique analogy, between traditional theodicies and Kant's so-called ascription-problem—the ambiguities that arise when one tries to choose among various initial and boundary conditions for particular applications of "the" categorical imperative. For Kant felt compelled to acknowledge that the activities and "motives" of the good will, and its operative "Prinzip", the categorical imperative, may be unsurveyable and (internally) indefinable, so that he could postulate for it the utter detachment from merely empirical and hypothetical ("internal") ectypes that Leibniz sought in his "god".

This may sound less surprising, if one reformulates this putative analogy as follows.

Aristotle, Anselm, Aquinas, Leibniz and others wished to "ground" (or end) "merely" relational regresses—of origination, design and conception, for example—in the (allegedly) reflexive [self]-origination, [self]-design and [self]-conception of "god", along the lines sketched in prior sections. Kant, by contrast, wished to employ "practical" [self]-referential or fixed-point arguments to "ground" (end) certain other "merely" relational regresses—of actuation, purpose-seeking and other-directed *Heteronomie*—in the (alleged) [self]-actuation, [*Selbst*]*zweck* and [*Auto*]*nomie* of the noumenal *Wille*, and its equally noumenal *Prinzip*.

He sought, in other words, to provide in his ethics (among many other things) a "first cause"-argument for a "religion of humanity" (or "ultimate" "purpose than which no more ultimate purpose can be conceived").

This analogy may help explain why most of us find something numinous but thoroughly indeterminate in the resonant language of Kant's supposedly so affectless ethics. It might also help explain why Wöllner, Woltersdorff and the other opportunist and fundamentalist *Zensoren Seiner Durchlaucht Friedrich Wilhelms II* went after him with such ferocity in the 1790s, after he published *Religion innerhalb der Grenzen der bloßen Vernunft*.

Along with many others, I find the ideal of a Kantian *Reich der Zwecke*—and its more conceptually accessible Rawlsian counterpart—moving and "liminally" persuasive. But the culture bound rigidity and parochiality of the notorious four "applications" Kant offered in the *Grundlegung* for the first two versions of the categorical imperative—not to mention the even more notorious strains in his late essay "*Über ein vermeintliches Recht, aus Menschenliebe zu lügen*"—seem to require suspension of judgment, and a measure of (what might be called) noetic restraint.

For the very reflexivity of Kant's "*Selbstzweck(e)*", "*Selbstgesetzgebung*" and *reine Selbsttätigkeit* (pure self-activity or -actuation, the quality Aristotle attributed to *theoria*; cf. *Gr* 452) poses a (literal) dilemma: "practical" *Grenzideen*, on the analyses of the foregoing sections, are either

(i) indiscernible to themselves (and therefore of little "practical" use); or
(ii) plurally interpretable in ethically defensible "realms" of branching "*Zweckmäßigkeiten ohne* ('ultimate') *Zweck*"".

Such dilemmas would not be surprising if one construed Kant's "deduction" of practical *Autonomie* (self-lawgiving; cf., e.g., *Gr* 453–454) as an individualisation of traditional "first"- and "self-cause"-characterisations of "god", or at least

(iii) "[*etwas*] zu aller Reihen der Bedingungen *notwendig* ... *Unbedingtes, mithin auch eine sich gänzlich von selbst bestimmte Kausalität* ...." (*KdpV*, B 83–84).

As is the case with other would-be deductions of [self]-constituting metaphysical universality, this one may be (re)formulated as a (literally) self-referential paradox: (cf. 3.44.-3.45 above), that
29 Each "reasonable being" "must" be able to will that
    [each "reasonable being" "must" be able to will that
    [[...]]].

The indefinite "implosion" of "[[...]]"'s suggests a fault-line in Kant's attempts to assimilate a practical *Verstandeswelt* (*mundus intelligibilis*, or conceptual world; cf., e. g., *Gr* 458 and *KdpV*, 74) to the *Noumena* of his theoretical or speculative philosophy.

For the characteristic of the latter in the Dialectic is their conceptual underdetermination and openness to plurality of interpretation, at least "*in spekulativer Absicht*" ("in speculative intent"). (Compare the Antinomies, and Kant's incisive analyses of traditional "proofs" of the existence of god).

But this underdetermination and hermeneutic plurality contrasts sharply with the "regulative" "necessity" and unicity Kant postulated for his "realm of ends"—in (*Gr* 445), for example, when he wrote that
30 Wer also Sittlichkeit für Etwas und nicht für eine chimärische Idee ohne Wahrheit hält, muß das angeführte Prinzip derselben zugleich einräumen.

Who(soever) considers morality to be something, and not a chimerical idea without truth, must concede the principles we have advanced.

As a would-be "reasonable being", I would certainly "consider morality to be something". But I would also consider the dichotomy Kant imposes in this passage a groundless *petitio* of the doctrine I called "semantic monism": denial, in the case at hand, that
31 there could be any alternative to existence of "ultimate" speculative, practical and teleological interpretations and other "truths", except utter inconsistency ("eine chimärische Idee").

Such begged dichotomies and "first-cause"-analogies underlie the ambiguities that arise from Kant's "ascription problem", mentioned earlier, and in particular the notoriously uneven success of Kant's "examples" at *Gr* 422–424 and 429–430 mentioned earlier.

But they also suggest why there is something deeply, genuinely and recurrently correct as well as "problematic" in Kant's "transcendental" acknowledgment, at (Gr 419), that

32 es durch kein Beispiel, mithin empirisch auszumachen sei, ob es überall irgend einen dergleichen Imperativ gebe ....

it is not to be made out through any sort of example, and thus empirically, whether there is anywhere any imperative of the kind.

Since every moral dilemma we are likely to face comes trailing clouds of background conditions and boundary assumptions, it seems to me eminently "reasonable" (in Kant's own sense) to

33 abandon Kant's semantic monism, in the passage from *Gr* 445 quoted above, and interpret its implicit "if-then" assertion ("Whoever considers ... must concede ...."), as the hypothetical (rather than categorical) meta-imperative it seems in fact to be.

Once one does this, however, it also seems "reasonable" to grant him that

34 reflective—and ultimately reflexive and [self]-determining—contemplative activity and respect for such activity in all other "reasonable beings" are the recurrent—but "liminal" and essentially incomplete—"grounds" of moral action,

in keeping with

35 Aristotle's observations of children's desires (*orexeis*) to "know" and praises of *theoria* (or *energeia theoretike*) as the "highest" forms of thought, namely "thought about thought", and
36 Kant's injunction to "displace ourselves" ("*sich hineinversetzen*") into "intelligible" realms of ends (in his sense of "intelligible"),

conceived as an

37 empathetic imperative to "displace ourselves", however inadequately, into the sufferings and aspirations of other "reasonable beings" (Gustave Gilbert may have been right about the origins of "evil").[4]

---

[4] Based on his conversations with convicted Nazi war criminals, Gilbert suggested that the essence of evil consisted of "the absence of empathy." (*Nuremberg Diary*, New York: Farrar, Straus and Company, 1947). Ed.

I have already argued (in 27 and 28 above)
38 that respect for sentience, in all its gradations—construed as inchoate *energeiai theoretikai*—is the mark of Kant's elusive "good will", in all its gradations, and
39 that provisional respect (Kant's *Achtung*) for the emergence and possible presence of such *theoria* in others is an inherent constituent of that regulative ground, and basis for a more modest and heuristic categorical imperative.

This is the essence, for me, of Kant's alternative formulation of the imperative, as respect for the generative capacities of "[self]-legislating" *theoria* in other "reasonable beings", to be treated as ends (not only **in** [themselves], but **to** [themselves]), rather than (mere) "means".

What we should enjoin ourselves to respect, in other words—in a form of collective "self-legislation"—are "innocent" (non-"harming") forms of Aristotle's desire.

This respect, in a sense, might also be an appropriate dialectical/skeptical *elenchos*—"*in praktischer Absicht*"—of Descartes' rhetorical paranoid worries in Meditation II about the "hats and clothes under which might lie automata" ("*pileos et vestes sub quibus latere possent automata*", *AT* VII 32).

Who knows? But more to the point, I believe: what do "we" know about [ourselves], that entitles us to judge that other potential "reasonable beings"— Kant's phrase—are not up to "our" standards? Kant's explicitly latitudinarian view of the distribution of possible "reasonableness" (cf. *Gr*, 408 and 426) got this right, I think. And John Searle, in his "Chinese Room" parable ([Searle] 1984)—a fallacious "*EA*-argument" which exactly reverses Descartes' "worry" about the hats and clothes on springs—got it thoroughly and dogmatically wrong.

I believe, at any rate, that some of these insights may have inspired the "awe" (*Ehrfurcht*) Kant struggled to express, in his famous vision of the vault above him and moral law within him. For it is this processive "awe" itself that moves us: that we cherish in our children; that we want to foster and protect in each other, and in ourselves. The vault and the "law" are only placeholders for it, provisional emblems for an elusive and forever underdetermined *Erhabenheit* (sublimity) in Kant's "theoretical" vision—and ours.

6 Reflective Inquiry and "*Systeme von Zwecken*"

At *KdrV*, B 672, Kant characterised a "*System*" as "*eine gewisse kollektive Einheit zum Ziele der Verstandeshandlungen*", and he sketched suggestively "metatheoretical" and "hierarchical" roles for such "*Systeme*" in *KdrV*, B 692, quoted above.

"Experience", on the account I have offered, "is" (or may pragmatically be viewed as) an indefinitely ramified "*System*" of such "*Systeme*", which "individuate" and "objectify" in graduated ways as they "ascend".

"Stages" in such metatheoretic hierarchies, moreover, have an interesting sort of "distributive" or "*kollektive Einheit*", for

1   all Stone spaces (topological spaces of semantic interpretations of essentially incomplete first-order theories) are topologically and measure-theoretically "isomorphic" (cf. 1.3 and 7.29–30 below).

What distinguishes particular theories from other particular theories in this hierarchical myriad are their (metatheoretically "intended") formal vocabularies, of course;

but also, and more deeply,

2   their patterns of syntactical and semantic interpretation, which order and interrelate them; and
3   their (recursion-theoretic) complexities, which effectively determine the "isomorphisms".

In this section, I will argue

4   that Kant tacitly acknowledged in the third *Critique* that he had offered an artificially isolated and experimentally "controlled" structure of "experience", which could not (for example) even determine its allegedly fixed (Newtonian) boundary conditions.

(It was known, for example, in the late eighteenth century that Newton's analyses—which construed the planets as point-masses—could not cope with their mutual interactions, much less account for observed perturbations in their actual orbits.);

5   that Kant saw that some sort of metatheoretic ascent or *Regressus* might present itself if the "observational" base of his "*Anschauungen*" turned out not to "determine" more "heuristic" and nuanced aspects of scientific inquiry (the indefinite complexities of "the organic", for example);

and finally,

6   that he saw that underdetermination of that "intuitive" observational base might also be reflected in iterations *of KdrV*, B 692's conceptual register-shifts; and that such iterations might open the first *Critique*'s rigid two-stage architectonic, and blur or relativise its allegedly sharp "transcendental" demarcation of "*Verstand*" from "*Vernunft*".

One aspect of this recognition might be the oddly enhanced role given in the third *Critique* to *"reflectierende Urteilskraft"* ("reflective judgment"), characterised as a form of mediation (*"Mittelglied"*, *KdU*, iv) between "understanding" and "reason".

In the first and third *Critiques*, this *Vermögen* assumed the interestingly metatheoretic role of *"Überlegung"* ("reflection" or "deliberation"; literally, "overlaying"): a

> *Bewußtsein des Verhältnisses gegebener Vorstellungen zu unseren verschiedenen Erkenntnisquellen* (awareness of the relation of given representations to our various sources of knowledge, *KdrV*, B 316); and

> *Unterscheidung der Erkenntniskraft, wozu die gegenbenen Begriffe gehören* (*KdrV*, B 317) (discrimination of the capacity for knowledge to which the various concepts belong), [welche] *auf die Gegenstände selbst geht* ([which] applies to the objects themselves), (*KdrV*, B 319).

In this capacity, "reflective *Urteilskraft*" effectively

7   assigned orders or levels to (what might be called) "object"- as well as "concept-formation" ([welche] *"auf die Gegenstände ... geht"*);

and in so doing, it

8   ambiguated (and "transcended") the otherwise rigorously closed architectonic of "reason" in the first and second *Critiques*, in which *"Reflexion"* appeared only in the guise of an *"Amphibolie"* (cf. *KdrV*, B 316–349).

In metalogical analyses, counterparts of such (potentially higher-intentional) acts of "discernment"—"the" local distinction between theory and metatheory, for example—function as type- or register-raisers which ride the wavefronts of metatheoretical hierarchies, and locally "transcend" what they "discern".

It is plausible, therefore, that such a metaphysically "skeptical" (or at least inquisitive) *Vermögen* might have given Kant the conceptual instrument he needed to absorb and buffer the "awareness" I attributed to him in 4–6 above.

Similar preoccupations might also have led him to

9   give greater prominence to the incomplete, underdetermined and (in my analogy) type-straddling notion of a *"System"* which "merely" regulated emerging boundaries of "experiences" (cf., once again, *KdrV*, B 672 and *KdrV*, B 692); and

10   postulate evolving (and perhaps ramifying) forms of *"Reflexion"* about "natural" and "organic" *Systeme* of *"Zwecke"* and *"Zweckmäßigkeit"*, which would

(re)inform and (re)interpret extensions of the *Analytic*'s scientifically as well as "teleologically" inadequate *Anschauungen* and *Begriffe*.

And this, in turn, might have led Kant—or someone troubled by the equivocal ways in which he used words like "*Vollständigkeit*" in the *Analytic* and *Dialectic*—to

11 consider more latitudinarian (and more "counterfactual"-supporting) notions of "causality",
12 acknowledge at least tacitly that the "closure" of the *Aesthetic* and *Analytic* doesn't "work", and
13 explore "merely regulative "principles" and forms of "causality" in "heuristically" adequate notions of *Zweckmäßigkeit* (glossed at *KdU*, B 344 as "*die Gesetzmäßigkeit des Zufälligen*" ("the lawlikeness of the contingent"; cf. 18 and 19 below) *ohne* (a final, uniform) "*Zweck*" (purpose).

Kant's characterisation of "*Zweckmäßigkeit*" as a "merely regulative" "*Gesetzmäßigkeit des Zufälligen*", just cited, also raises methodological questions about the absence of "counterfactual support" for "lawlike" assertions in the *KdrV*):

14 do claims that "real" lawlikeness (*Gesetzmäßigkeit*) only holds for systems which are "necessary" in the sense that they are deterministic not beg the question of the Analytic's "underlying" ("Laplacian") determinism?

And if they do,

15 what are we to make of (ostensibly) metaphysical (as opposed to "merely" mathematical) assertions that "real" "experiential" causation—construed as a physical form of alethic modal entailment—"supports counterfactuals", if there are no counterfactuals to support?

The problems raised by these queries are oddly cognate to Hume's failure to distinguish between "factual" (indicative) and "counterfactual-supporting" (subjunctive) implications in his "definitions" of "Cause" in the first Enquiry (Hume, §60, 76). But they are hardly confined to the metaphysics of Hume and Kant. For they persist in cognate problems for any "causal" metaphysics in which we do not (and, I believe, cannot) know how to "choose" "ultimate" metatheoretic "boundary conditions" for "virtual" modal semantics of "worlds" and "accessibility"-relations which interpret their "causes".

Prompted by (what I have read as) Kant's tacit acknowledgment that "mediation" by "reflective judgment" might iterate "the" *Anschauung / Begriff / Idee*-hierarchy, I have therefore sought to

16 relativise that hierarchy;
17 "open" its rigid three-level Architektonik;
18 iterate such relativisations in ramified patterns of metatheoretical ascent; and finally
19 appeal to such patterns to formulate "conceptual" interpretations of ("theoretical", "practical" and "reflective") "experience" within shifting horizons of metatheoretic complexity.

Metalogical studies of such hierarchies suggest that essentially incomplete and underdetermined *Systeme* and their hierarchies may "regulate" emerging boundaries of experiences, and that notions of *Zweckmäßigkeit* may inform and (re)interpret such regulative systems and their notions of "causality". Within various stages of such "systems", for example, one can "rationally" as well as "reasonably" assimilate *Endzwecke* or *causae finales* to provisional forms of metalogical explanation or interpretation, but accept that no "ultimate" "finality" for them will ever come into view.

Such "principles" and forms of "causality", moreover, would be ("merely") "heuristic", "reflective", "regulative" and "problematic". In analogy with certain aspects of Kant's usage, they would be

20 "*heuristisch*", in that they ("merely") serve "'*den*' "*besonderen Gesetzen der Natur nachzuforschen .... ohne über 'die' Natur hinaus 'den' Grund 'der' Möglichkeit derselben zu suchen*" ("inquire into 'the' particular laws of nature ... without searching beyond nature [to inquire] about its ground", *KdU*, B 355).

(The single quotation marks in the quotation are obviously mine: we tend to forget that the original senses of "*physis*" and "*natura*" were processive as well as "organic".)

They would be

21 "*reflektiv*" in that they seek to discern theories' orders of complexity and relative interpretability, but do not define or determine "ultimately" "canonical" interpretations for them.

(Kant's "*reflective Urteilskraft*" sought "only" to "subsume" its intentional objects, whereas the "*Vermögen*" he called "understanding" "determined" them; cf., e.g., *KdU*, B 311–313 and B365).

They would be ("merely")

22 "regulative" in that they are "*nur ... Regel*[*n*]*, welche ... einen Regressus gebiete*[*n*]*, dem es niemals erlaubt ist, bei einem Schlechthinunbedingten*

stehen zu bleiben" ("only rule[s] which prescribe a regress in which one is never permitted to halt at an utterly unconditioned [limit]"; *KdrV*, B 536–537).

(Notice that Kant retained in this context Hume's distinction between "*Gesetzlichkeit*" and what Wittgenstein later called "*Regelbefolgung*"....)

And such "principles", "concepts" and "judgments" would finally be
23 "problematic" in that
(a) "*man das Bejahen oder Verneinen [derer] als bloß möglich (beliebig) annimmt*" ("one takes the affirmation or denial [of them] as possible (optional)", *KdrV*, B 100);

and
(b) they are "... *Begrenzung[en] gegebener Begriffe*" [welche] "*mit anderen Erkenntnissen zusammenhäng[en], dessen objektive Realität ... auf keine Weise erkannt werden kann*" ("boundaries of given concepts" [which] "fit together with other forms of knowledge whose objective reality cannot in any way be known", *KdrV*, B 310).

(Cf. also the remark that
(c) "[i]*ns Innere der Natur dringt Beobachtung und Zergliederung der Erscheinungen, und man kann nicht wissen, wie weit dieses mit der Zeit gehen werde*" ("observation and distinction of appearances penetrate into the interior of nature, and one cannot know how far this might go [on] over time", *KdU*, B 334)).

Some definitions and characterisations may clarify other aspects of the relativisations I have sketched. In the context of 16–19 above, for example,
24 a "system" may be any finitely (or finitarily) "overviewable" substructure of an eventually first-order metatheoretic hierarchy;
25 an act or instance of "reflection" (or "reflective judgment") locally identifies a given system's relative "position" in such a metatheoretic hierarchy (as above; cf. once again *KdrV*, B 316ff);
26 an "end" or "purpose" ("*Zweck*") (cf. 4.20) is once again a (syntactically or semantically) interpreted object-theoretic assertion (the German "*Sachverhalt*") at some stage in such a hierarchy;
27 a (regulative) "law" (or "merely" "*zweckmäßige*" "*Maxime*") is a modal implication which asserts that certain "means" or hypothetically contributory "causes" jointly entail certain "ends" in a metatheoretic environment in

which we do not and cannot know what ranges of "counterfactual" alternatives the implication might "support".

At a number of points in the third *Critique*, Kant argued for the claim that cognitive agents have a right to "impute" ("*ansinnen*") certain attributes or judgments to others, and such "imputation" is not confined to aesthetic judgments.

The more common noun-usage "*Ansinnen*" has many senses in Kant's and other German speakers' usage ("demand", "request" "expectation" "imposition" and "point of view" among them), and

28   an assertion that a given "reflective judgment" of mine is "ansinnbar" (an adjectival form Kant does not use) might therefore be construed as

29   a "liminal" assertion that whatever "interprets" "my experience" (in ways which may not be fully knowable to "me", or even expressible in "my" language) "should" interpret "my" "reflective judgment" or expression of that "experience" accordingly.

Such "interpretability"—which might be construed, as Kant suggests, as a form of communicability—would assure a form of relative consistency:

30   if "the world" "makes sense", so do "I", or at least "my" expression or reflection of a given aesthetic or "practical" aspect of it. (Or so "I" think).

Methodologically, at least, this metalogical reading of the *"subjektive Algemeingültigkeit"* of Kantian aesthetic judgments seems to me defensible.

More relevantly, it may clarify the conceptual asymmetry between aesthetic *"Ansinnbarkeit"* and moral "solidarity". For the former demands that others "appreciate" what I "appreciate", and the other that I try to "comprehend" the capacity to "appreciate" in others. The former may or may not be a "just" demand, but it is literally "egocentric" and "self-absorbed" in ways the latter is not.

Be that as it may, the principal burden of this section has been to make two closely related points:

31   that what is genuinely "regulative" of inquiry are attempts to consider, critique and appreciate "*Zwecke*" ("purposes"); and

32   that zetetic inquiry is not only regulative of "experience". It is also a form of generative *skepsis*, and a form—and perhaps prototype—of "*Zweckmäßigkeit ohne Zweck*".

Distributively, one might also characterise such "experiential" systems and their "dynamics" as graduated forms of "*Erfahrungsmäßigkeit ohne (schlechthinnige) Erfahrung*", and construe them as heuristic "regulative ideals" of such "experience" (or "*expérience(s)*").

Since uniform quasi-Kantian "preconditions" of such "experience" would not, in particular, be object-theoretically definable at any stage of theoretical inquiry, the underdetermination of such "systems'" boundary conditions and "counterfactual" environments might also be

33 compared with the "confusion" Leibniz attributed to attempts to understand *"notiones concretae"*, and
34 interpreted as a margin of the infinite (or at least indefinite) in the finite, or at least the conceptually finitary.

"Origins", "preconditions" and semantically "intended" interpretations are object-theoretically invisible in all the gradations of metatheoretical hierarchies they "condition" and "regulate".

We cannot, for example, define or measure "all" "conceivable …" / "conceptual …" definition or measurement. For the only way to remove the "'…'"s is to relativise one's inquiry to particular expressible theories in which one can begin to make sense of "conception", "definition" and "measurement". (Notice the parallels between these notions and mathematical "boundary conditions" and physical "experimental preparation.")

If the "cogito"-like responses Descartes, Berkeley and Kant offered to Section 2's criterial hierarchies "worked", in the presuppositionless ways their authors apparently intended, it would be to witness the *in*coherence of the (meta) theory in which they are formulated.

For "last" clauses in the stoic / skeptical dialogue which gave rise to "cogito"-like arguments will always reinstate skeptical doubt of skeptical doubt, in order to reinstate minimal stoic consistency-demands.

Fixed-point arguments, for example, would have little "constitutive" force for "the" unicity of "experience", but quite a bit of "regulative" value as "local" guidance for avoidance of certain conspicuous forms of "transcendental illusion".

If one could, for example, "define" "transcendence" in one of Section 1's "intelligible" theories, one could also formulate a "diagonal" assertion which would be equivalent to [its own] "transcendence". But this would undermine the "definition": for no syntactical interpretation of such a sentence would accurately "define" ("the" "preconditions" of) [its own] "immanence".

If Einstein had been "right", therefore, that
35 "the" world is "complete" (a word he often used in his debates with Niels Bohr) and epistemically accessible (since "der Alte", like Descartes' "god", does not "deceive" us),

he would also (on good Gödelian grounds) be "wrong",
36 for "it" would either be as incomplete as "we" are, or as epistemically inaccessible to "itself" as "it" is to "us".

On the evidence of such arguments, it might be a "transcendental illusion" that "experience" has a constitutive "*Grenzbestimmung*", and "reasonable beings" would be more "reasonable" to the extent they understand this.

This predicament might be regulative of Kant's "*Schicksal der Vernunft*" as well as "*expérience*" and "*compréhension*", in both French senses of these words. But it might also be one of reason's cradle-gifts: a source of
37 skeptical self-awareness that "ultimate" self-knowledge is little more than a semantic paradox; and of
38 "creativity" in that no such "ultimate" *Übersicht* bounds "reasonable beings"' conceptual event-horizons.

To see that this might be so, consider two recurrent arguments of the sort canvassed earlier:
39 a "neo-Kantian" one that some stage of metatheoretic ascent is closable as "experience"; and
40 a "Peircean" one that "the" entire ascent is linear and unique, and this "course of inquiry" is "experience."

In this context, all I have done is propose a nonlinear alternative to Peircean "limit(s)", in which convergence is plural and there is no "ultimate" hierarchy-internal "overview".

That "intelligible" theories cannot "intend" their own interpretation and that hierarchies of them are indefinitely iterable is a metalogical theorem.

That ramification of such hierarchies is limitless is a regulative ideal of experiential openness as well as contemplative "*Erhabenheit*" ("sublimity"), in ways I will try to explore in the next section.

That "we" can "go on" in such hierarchies is an article of "reasonable faith", sustained by liminal postulations of consistency and "intentional" interpretation which hover indefinitely at the margins of what they define.

7 Reflective Inquiry and Kant's Two "*Unermeßlichkeiten*"

At the end of the *Kritik der praktischen Vernunft*, Kant wrote the following resonant sentence, later inscribed on his cenotaph in Königsberg:

> Zwei Dinge erfüllen das Gemüt mit immer neuer und zunehmender Bewunderung und Ehrfurcht, je öfter und anhaltender sich das Nachdenken damit beschäftigt: der bestirnte Himmel über mir, und das moralische Gesetz in mir. (*KdpV*, 288)
>
> Two things fill the sensibility with ever new and growing awe, the more often and more persistently reflection occupies itself with them: the starred firmament above me, and the moral law within me.

To me at least, these lines suggest that for all his "critical" acceptance of Newtonian "Mechanism", Kant saw something comparably "*erhaben*" ("sublime") in "regulative ideals" of the *macrocosmos* "above" us and *microcosmos* "within" us. Prompted by this interpretation, I will appeal in this section to "liminal" as well as limitless aspects of the foregoing sections' interpretation of "experience" to

1. argue for an evenhandedly "regulative" interpretation of both these ideals; and
2. compare Kant's two "*Unermeßlichkeiten*" ("immeasurabilities"; cf. 7 below) with metalogical and metaphysical counterparts of Pascal's two *infinités* or *abîmes*".

Recall (or observe) first that Kant considered "*Achtung*" ("respect") and "*Ehrfurcht*" ("awe") intentional attitudes appropriate to (inadequate) attempts to contemplate two closely interrelated limiting ideals: "*Pflicht*" ("duty"), and "*das Erhabene*" ("the sublime").

"*Achtung*", for example, in the third *Critique* was the

3. "sense of the inadequacy of our capacity to attainment of an idea which obligates us" ("*Gefühl der Unangemessenheit unseres Vermögens zur Erreichung einer Idee, die für uns Gesetz ist*") (*KdU*, B 96).

When that "respect" is engendered by contemplation of

4. "the intellectual, in itself purposeful [and] (morally) good" ("*das intellektuelle, an sich selbst zweckmäßige (das Moralisch-)Gute*", cf. *KdU*, B 120),

then regard for this ideal is also "*erhaben*" ("sublime").
For it is

5. "not pleasure, but self-estimation (of the humanity in us), which raises us above our need [for pleasure]" ("*kein Vergnügen ist, sondern eine Selbstschätzung (der Menschheit in uns), die uns über das Bedürfnis desselben erhebt*") (*KdU*, B 228),

and its view or contemplation
6   "gives us just access to … the idea of a great system of natural ends" ("[*uns*] *zu der Idee eines großen Systems der Zwecke der Natur … berechtigt*"),

which we
7   "love as well as contemplate for its immeasurability, and find ourselves ennobled in that contemplation." ("*lieben, sowie ihrer Unermeßlichkeit wegen mit Achtung betrachten und uns selbst in dieser Betrachtung veredelt fühlen ….*") (*KdU*, B 303, emphasis mine).

Such "*Liebe*", finally, engenders a "higher" and more "autonomous" form of "humanist" respect, namely
8   "*Ehrfurcht*", or "respect for a ruler … which lies within us, [and therefore] awakens a sense of the sublime in our own [self-]determination which inspires us more than anything beautiful" ("*Achtung … gegen seinen Gebieter,* [*der*] *… in uns selbst liegt,* [*und daher*] *ein Gefühl des Erhabenen unserer eigenen Bestimmung erweckt, was uns mehr hinreißt als alles Schöne*") (*Die Religion innerhalb der Grenzen der bloßen Vernunft*, VI, 23–24fn).

In what follows, I will assimilate
9   Kant's *first* "*Unermeßlichkeit*" ("*der bestirnte Himmel über mir*") to the immeasurably ramifying metatheoretic "*Systeme*" of "intelligible" theories evoked above in Sections 1 through 3,

and
10 his second "*Unermeßlichkeit*" ("*das moralische Gesetz in mir*") to the correlatively immeasurable and ramifying "hypotheoretic" "*Systeme*" of variably encoded forms of awareness or consciousness which may be expressed "within" them,

respect for which I offered earlier as less prescriptive but (in my view at least) no less "*erhabene*" *Alternativen* to the various forms of Kant's categorical imperative.

These assimilations will carry with them cognate metalogical interpretations of two other "*Ideen*" in Kant's writings:
11 "physical" "Organism" (or "Organisation"), invoked in several contexts in the third *Critique*; and

12 "liminal" "systematic" "*Horizonte*", mentioned twice in the first *Critique*, and at considerably greater length in Kant's lectures on "*Logik*".

"*Physik*" (or "*das Physische*"), was emphatically not what we would call "(Newtonian) physics", but

13 ein Doctrinal-*System* empirischer Erken[n]tnis (*nicht ein empirisches System denn der Begriff von einem solchen enthält einen Wiederspruch*),

> a doctrinal system of empirical knowledge (*not an empirical system, for the concept of such a thing contains a contradiction*).

As the parenthetical remark suggests, such a "higher-order" *System* might require critical discernment of "empirical" theory from "transcendental" metatheory to avert (semantic) paradox ("*einen Wiederspruch*"), and have

14 zweyerley *Objecte*: 1) was überhaupt Gegenstand der Erfahrung ist 2.) dessen *Möglichkeit* selbst nicht anders als durch Erfahrung erkennbar ist, wovon also die *Wirklichkeit* vor der *Möglichkeit* nothwendig vorhergeht die also nicht *a priori* erkannt werden kann (*KW*, XXII, 398–399).

> *two sorts of objects*: 1) whatever is an object of experience 2.) [something] whose very *possibility* is not knowable other than through experience, the *reality* of which necessarily precedes therefore its *possibility*[, and] therefore which cannot be known *a priori*.

Quasi-Leibnizian "*notiones completae*", for example, might "regulate" Newtonian "Mechanism"'s reference-frames and boundary conditions, as well as instances of what Kant called "Organism", or "*das Organische*" (terms which do not appear in the third *Critique*, where he replaced them with more "dynamic" and "processive" usages such "*organisieren*" and "*Organisation*").

In his deconstruction of Swedenborgian mysticism, for example (*Träume eines Geistersehers*, *KW*, XVIII, 13), Kant observed in passing that "*das physische ist nicht pneumatisch, sondern organisch*", and he remarked in other non-"critical" texts that

15 [d]ie *dynamische Erklarungsart* ist entweder *mechanisch* durch Werkzeuge die selbst bewegender Kräfte zu ihrer Existenz bedürfen und wenn sie ihrer Natur nach Zwecke ihrer Bildung voraussetzen [o]rganisch vorgestellt werden. (*KW*, XXI, 233)

> the *dynamic mode of explanation* is either *mechanical*, by means of instruments which themselves require moving forces for their existence, [or] if they are represented *organically* in accordance with their nature [and] presuppose [aims or] ends for their formation.

and formulated a

16   ... *Definition eines organischen Körpers* ... daß er ein Körper ist dessen jeder Theil um des anderen willen (wechselseitig als *Zweck* und zugleich als *Mittel*) da ist.—Man sieht leicht daß dies *eine bloße Idee* ist der *a priori* die Realität (d.i. daß es ein solches Ding geben könne) nicht gesichert ist.

> Man kann die *Erklärung dieser Fiction* auch anders stellen: Er ist ein Körper an welchem die innere Form des Ganzen vor dem Begriffe der Composition aller seiner Theile ... in Ansehung ihrer gesammten bewegenden Kräfte vorhergeht (also *Zweck* und *Mittel zugleich* ist). (*KW*, XXI, 210)

> ... *definition of an organic body* ... that it is a body whose every part is there for the sake of [every] other (reciprocally as *end* and *means*).—One sees readily that this is *a mere idea*, whose reality (that is that such a thing could exist) is not *a priori* ensured. One can also put the *explanation of this fiction* differently: it is a body for which the inner form of the whole precedes the concept of the composition of all of its parts ... in view of their collective moving forces ([and] therefore is *at once end* and *means*).

(It may be worth observation that more precise analogues of Kant's struggle to "define" "the organic" may be found in mathematical-physical notions of an "interacting field").

Freed, in short, from the narrow confines of his "transcendental" rationales for Newton's differential equations, Kant acknowledged that his "*Physik*"'s "merely regulative" entailments might lead into "rational" but indefinitely complex realms of *Fiction(en)* and *bloße Ideen* (cf. Newton's notorious remark that "*hypotheses non fingo*"). One might, for example, "transcend" (*KW*, XXI, 233) deterministic "heteronomy" with appeals to prototypes of "complexity", "emergence" and "self-organisation", if such "*bloße Ideen*" were made (meta)mathematically precise in reasonably predictive ways.

"Skeptical" arguments, I believe, suggest that they can, and my modification of Galileo's famous dictum to suggest in 0.1 above that "the book of philosophy" might (and perhaps should) be written in the language(s) of metamathematics was partly animated by this conviction.

It was also guided by a belief that if such programmatic conjectures turn out to have regulative value, the fate of Carnap's *Aufbau* suggests that

17 they will have to be adaptive and provisional rather than prescriptive, and that
18 those who make them would be well advised to acknowledge their essential incompleteness and recurrently liminal metamathematical margins.

For it is quite conceivable that

19 interrelations between "the" "book of philosophy" and "language of (meta) mathematics" might themselves form "systems" of "intelligible" theories whose metametamathematical properties evolve and ramify in ways we cannot anticipate, much less dictate.

Still—as the Danish saying goes, "blind chickens find also a grain" ("*blinde høns finder også et korn*"), and an allegedly heuristic proposal should try to "find" ("*heurein*") something from time to time.

Here is a possible "*korn*". In one of 4.10's hierarchically organised "systems", answers to queries about object-theoretic consequences might serve as "means" to more complex metatheoretic "ends", and conversely. Finitary "intelligible systems" and their insights might therefore "evolve" in such hierarchies—in which higher-order metatheoretic "ends" in one "systematic" context became object-theoretic "means" in another—and stages of such evolution might be construed as

20 [d]as Materiale[,] in so fern es nur *problematisch gedacht* und eine Tendenz enthält es sich assertorisch als gegeben vorzustellen (Organisch, Unorganisch). (*KW*, XXII, 480)

[t]he material[,] in so far as it is only *thought problematically* and possesses a tendency to represent itself as given assertorily (Organic, Inorganic).

All of this, once again, would remain "merely regulative", "physically" underdetermined and metamathematically incomplete ... *ad indefinitum*. But such underdetermination might at least be *informative*, as well as compatible with several of Kant's more "skeptical" (or at least "merely regulative") remarks—that

21 [d]ie Mathematik wird durch Philos. indirect begründet ... (*KW*, XXII, 78), [m]athematics is indirectly grounded through philos[ophy],

for example; or that

22 ([d]er Begriff von *organisi[e]rten* Körpern gehört *zum Fortschreiten im **System** der Wa[h]rnehmungen des Subjects* das sich selbst affici[e]rt) (*KW*, XXII, 398);

[t]he concept of *organised* bodies belongs to *what is processive in the system of perceptions of the subject* which [ap]perceives itself;

or that

23 [m]an *fängt da nicht von Objecten an sondern von dem **System** der Möglichkeit sein eigenes denkendes Subject zu constitui[e]ren* und ist selbst Urheber seiner Denkkraft (*KW*, XXI, 79).

[o]ne *does not begin from objects*[,] *but from the system of the possibility to constitute one's own thinking subject* and [one] is the originator of one's capacity for thought;

or even that

24 die *zweyte* [*organische Ordnung der Natur, d.i. die Form derselben nach Regeln*] ist auf einer *Idee gegründet, die des einzelnen sich als Werkzeug zu einer Einrichtung bedient, die aus den einzelnen Naturdingen nach allgemeinen Gesetzen nicht entsprungen wäre*.) (*KW*, XVII, 418)

the *second* [*organic order of nature, i.e. the form of it in accordance with rules*] is based on an *idea, which makes use of the particular as* [*an*] *instrument for an arrangement which would not* [*otherwise*] *have arisen from the separate things of nature in accordance with general laws.*

My appeals to metatheoretical underdetermination might also be compatible with Kant's rare but carefully formulated (if "merely regulative") remarks about "*Horizonte*" in the first *Critique*:

25 Der Inbegriff aller möglichen Gegenstände für unsere Erkenntnis scheint uns eine ebene Fläche zu sein, die ihren scheinbaren *Horizont* hat, nämlich das, was den ganzen Umfang derselben befaßt, und von uns *der Vernunftbegriff der unbedingten Totalität* genannt worden. *Empirisch denselben zu erreichen, ist unmöglich, und nach einem gewissen Prinzip ihn a priori zu bestimmen, dazu sind alle Versuche vergeblich gewesen.* (*KdrV*, B 787)

The aggregate of all possible objects for our knowledge seems to us to be a plane surface, which has its apparent [or specious] horizon, namely that which includes the entire extent of that knowledge, and has been called by us the *concept-of-reason of unconditioned totality*. To reach this [horizon] empirically is impossible, and [as for efforts] *to determine* [or define] *it in accordance with a particular principle a priori, all* [such] *efforts have been in vain.*

I have also argued that "*der*" *Vernunftbegriff* "*der*" unbedingten "*Erfahrung*" is no more unique or "*a priori zu bestimmen*" than "*der*" *Vernunftbegriff* "*der*" unbedingten "*Totalität*"—an assimilation which also seems to me compatible with Kant's remark (*KdrV*, B 686–687, slightly abridged below) that

26 Man kann sich *die systematische Einheit* unter den drei logischen Principien auf folgende Art sinnlich machen. Man kann einen jeden Begriff als einen Punkt ansehen, der als der *Standpunkt* eines Zuschauers seinen *Horizont* hat, d.i. eine Menge von Dingen, die aus demselben können vorgestellt und

> gleichsam überschaut werden. Innerhalb diesem Horizonte muß *eine Menge von Punkten ins Unendliche angegeben werden können, deren jeder wiederum seinen engeren Gesichtskreis hat*; ... *und der logische Horizont besteht nur aus kleineren Horizonten ..., nicht aber aus Punkten, die keinen Umfang haben (Individuen)*. Aber zu verschiedenen Horizonten, d.i .... die aus eben so viel Begriffen bestimmt werden, läßt sich *ein gemeinschaftlicher Horizont*, daraus man sie insgesamt als aus einem Mittelpunkte überschaut, gezogen denken, ... bis endlich ... *der allgemeine und wahre Horizont ... aus dem Standpunkte des höchsten Begriffs bestimmt* wird und alle Mannigfaltigkeit als ... unter sich befaßt.

One can make *systematic unity* [more] graphic [sensory] under the three logical principles in the following way. One can see an arbitrary concept as a point which as the *standpoint* of a viewer has its *horizon*, i.e., a set of things which can be represented from it and surveyed, as it were. Within this horizon, *one must be able to posit a set of points in infinitum, each of which has its more limited field of vision*, ... *and the logical horizon consists only of smaller horizons*, ... *not of points, which have no extension (individuals)*. But to different horizons, ... which are determined from different [corresponding] concepts, *a common horizon*, from which one can survey them all as [if] from a center, can be traced out in thought, ... until finally ... *the general and true horizon is determined ... from the standpoint of the highest concept*, and the manifold comprehended ... under it.

Many other interesting if problematic things could be said about Kant's "*Sinnlichmachung*" in the metalogical framework of this essay. In the ellipses of this passage, for example, he interpreted "*Horizonte*" as "*Gattungen, Arten und Unterarten*" ..., whose most straightforward metalogical counterparts would be
27 (consistent and recursively defined) "types" or imbricated collections of unary predicates in the language of a given "intelligible" theory **T**.

One could, moreover, define
28 binary, ternary, *n-ary* and "nullary" "types" as well, and identify the latter with consistent, recursively axiomatisable theories which extend **T**.

For each such type, one could also define
29 a "canonical" boolean algebra (called its "Lindenbaum algebra"), as well as corresponding (topological) "Stone spaces"—mentioned earlier in 1.3 and 6.1–6.3—in which extensions of the given type determine closed subspaces.

"Complete" counterparts of such "types", finally, could also be defined (as elements of the aforementioned "Stone spaces"), but only in appropriate metatheories, and there would be uncountably many of them. The elements of such spaces —metalogical analogues of Kant's "points, which have no extension"—would be "*durchgängig bestimmt*", in Kant's language ("*notiones completae*", in Leibniz'), but almost all of them would fail to be "intelligible", in the metalogical sense sketched earlier.

One might try to assimilate Kant's "*allgemeine(r) und wahre(r) Horizont*" to an entire "Stone space" of such "points", and such spaces are indeed "universal" in various topological and metatheoretic senses. But they are also isomorphic to spaces of infinite sequences of "random" coin tosses, an observation which suggests that such "*Venunftideen*" would degrade rather than embody informational "truth".

For it is also known that every "intelligible" theory has the same Stone space, whose "generality" and "canonicity" would therefore render its "reference-frame" useless for discrimination of one "intelligible" theory from another.

To me at least, such observations call to mind two well-known predicaments in the history of philosophy:

30 the immanent white noise of Spinoza's transcendent *deus sive natura*; and
31 Kant's remark that "concepts" without "intuitions" are "blind" (in the sense that every intelligible theory may be said to furnish its own class of "intuitions").

But it also suggests that a form of "*Erhabenheit*" may be found in

32 limitless "universes" of "intelligible" theories' interpretations and capacities for "autological" [self]-reference,

and that this "sublimity" may be even more "disproportionate" than Kant's "*bestirnter Himmel*" and the "two abysses" evoked in one of the most resonant passages of Pascal's *Pensées*:

33 *Qui se considérera ... [l]es deux abîmes de l'infini et du néant, il tremblera dans la vue de ses merveilles, et je crois que sa curiosité se changeant en admiration, il sera plus disposé à les contempler en silence qu'à les rechercher avec présomption .... Quand on est instruit, on comprend que la nature ayant gravé son image ... dans toutes choses, elles tiennent presque toutes de sa double infinité: c'est ainsi que nous voyons que toutes les sciences sont infinies en l'étendue de leurs recherches .... Toutes choses sont sorties du néant et portées jusqu'a l'infini .... causées et causantes, aidées et aidantes, médiatement et immédiatement, ...* Voilà notre état véritable.

C'est ce qui nous rend incapables de savoir certainement et d'ignorer absolument.

Anyone who considers ... the two abysses of the infinite and the void, will tremble in the sight of its marvels, and I believe that—curiosity changing into admiration—will be more disposed to contemplate them in silence than explore them with presumption .... *When one is learned, one comprehends that since nature has graven its image in ... all things, almost all of them have something of this double infinity*: in this fashion we see that all forms of knowledge are infinite in the range[s] of their inquiries .... *All things have emerged from the void and are carried into the infinite, ... [and are] caused and causing, aided and aiding, mediately and immediately, ...*

Therein is our veritable condition. This is what renders us unable to know with certainty, and be ignorant absolutely.

In the essay's final section, I will
34 argue that the "underdetermination" of ramified "experiential" hierarchies—a fragile and inadequate form of "freedom"—has the dignity of one of its "preconditions";

and
35 compare the *"abîmes"* of its endless "ramifications" with the deltas and estuaries of Norman MacLean's "river," in which faint traces might linger long after we cease to "exist".[5]

8 Reflective Inquiry as a Form of Skeptical *"Theoria"*

In earlier chapters (1–3), I have
1 compared phenomenal underdetermination to a "transcendental" precondition of an activity I called "skeptical *theoria*" (a *"via negativa"*), and
2 assimilated such such *theoria* (or *"via contemplativa"*) to the *"reine Selbsttätigkeit"* ("pure self-actuation", *Gr*, 452) which Kant called *Freiheit*.

---

[5] The final sentences of Norman Maclean's *The River Runs Through It* (Chicago: University of Chicago Press, 1976) read: "Eventually, all things merge into one, and a river runs through it. The river was cut by the world's great flood and runs over rocks from the basement of time. On some of the rocks are timeless raindrops. Under the rocks are the words, and some of the words are theirs. I am haunted by waters." Ed.

I also observed that

3. the ancient skeptics attached a regulative (or "contemplative") value they found significant to such *epoche* and the search for it.

That "regulative" value, I also suggested (cf. 1.9 and 5.15 above), is Aristotle's "desire", not for "knowledge", but for quasi-Kantian forms of "disinterested" inquiry. For pyrrhonist and academic skeptics alike believed ("undogmatically", of course) that it conferred a kind of peace of mind they called "*ataraxia*" (roughly "equanimity", literally "non-perturbation"), a word one also finds in Stoic and Epicurean writings.

Also of interest, I believe, in the context of this essay's metalogical analogies, is the "problematic" "reflexivity" of such "ataractic" self-attribution and "practical" "self-actuation". For both introduce—as Kant observed—an implicit "critical" (or "metatheoretical") distinction between a "self" and what calms or "actuates" it.

In "Theory-Relative Skepticism," (1987), I also outlined a related analogy between skeptical *epoche* about "phenomenally" undecidable assertions, and (some of) the responses Immanuel Kant advocated to "transcendent", "antinomial" assertions of "pure reason", which are, by definition, undecidable within "the" *Bereich der Erscheinungen*.

I drew inspiration for this, in part, from Kant's own high valuation of the antinomies, despite his disparaging remarks about skepticism in the "*Transcendental Methodology*" (cf. *KdrV*, B 784–797), but also from the evidence canvassed in foregoing sections that

4. significant hermeneutic and alethic notions such as interpretability, definability, and (a fortiori) "truth" may indeed be Kantian "*Grenzideen*", rather than Kantian "*Begriffe*"—in metalogic and in metaphysics;

or at least that

5. such hermeneutic and alethic notions are relational notions: conjectural and schematic templates which our finite intellects are fortunate enough to be able to project for heuristic investigation and dialectical and practical inquiry.

"Disquotational" "truth" of a theory, for example—Tarski's "Convention T"—is a jejune second-order template. Nothing substantive emerges from it until we refine it to the relational metatheoretic notion of "truth in a structure".

Even then, moreover, "intelligible" first-order theories will propose infinitely many distinct candidates for such structures, which may be examined in infinitely many branching candidates for appropriate metatheories.

"Speculatively" as well as "practically", therefore, attempts to formulate [self]-grounding "intelligible" theories of the sort Descartes, Berkeley and Kant sought cannot escape a kind of complementarity between the intentional finitude of our conceptual horizons, and lack of closure of the concept formation(s) we undertake within them.

For "intelligible" theories face a choice between two "complementary" alternatives: undecidability and inexpressibility. This (literal) dilemma—a simple consequence, of Gödel's insights into [internal] diagonalisation—may be posed as follows.

Such a theory—of "the self", say; or "the world"; or Anselm's formula "that than which nothing greater can be conceived" (Kant's *"psychologische, kosmologische und theologische Ideen"*)—must itself be incomplete. It can formulate, but never prove, [its own] existence (consistency / interpretability).

By simple contraposition, then:

A complete limiting (meta)theory—not only of Kant's three *Ideen*, but also of *"the"* distinctions Kant tries to draw between speculative and practical, *Phaenomena* and *Noumena*, *Begriffe* and *Ideen*, *Vernunft* and *reflektierende Urteilskraft*—would not be "intelligible" and could not, *a fortiori*, formulate or express (encode) [its own] existence.

To put a mildly provocative point on it:

6   if a "canonical" intelligible theory **T** of "experience" (or "the god of the philosophers"; or *"der heilige Wille"*; or ....) were provably "complete" in some tacitly metatheoretic faithful extension **U** of **T** (i.e., if the extension **U** decided every assertion in the language of **T**), such a "proof" would bear witness to the inconsistency of **U** as well as **T**.

This recurrent complementarity—between intelligibility (or expressibility) and putative (metatheoretic) universality—seems to me to take many forms in the history of philosophy, as I've suggested in Section 2's examinations of the arguments of Descartes, Berkeley, and Kant and elsewhere. More precisely, I have argued that we cannot ensure, or even canonically express, certain ideals and *Grenzideen*, with anything like the constitutive *Vollständigkeit* and semantic unicity that most classical metaphysicians (and most analytic philosophers) have demanded of (and casually attributed to) "truth".

I have also tried to suggest

7   that such semantic *Vollständigkeit* (completeness) is what is really at issue—as Kant saw—in classical metaphysical examinations of the [self], and of "the god of the philosophers" and Kant's three *Ideen*; and

more controversially, I have argued

8   that one can trace a number of analogies between other metaphysicians' "probative" arguments made on behalf of these *Ideen* and Kant's tacit appeals to "transcendental" design arguments and the "Archimedean" leverage he attributed to his "critical" distinctions and "transcendental deductions"; and
9   that the "*durchgängige Bestimmung*" Kant claimed to have "deduced" for "the" "constitutive" structure of "experience" may be assimilated to the indeterminate "*Vollständigkeit der Bedingungen*" he (cogently) characterised in the *Dialectic* as a kind of underlying master-*Vernunftidee*.

In defense of the "locality" of "critical" theory / metatheory-distinctions, I have also argued that we have no "ultimate" decision procedure which would enable us once and for all to discern, in our theoretical and metatheoretical inquiries, proof from [proof] (existence from [existence]; consistency from [consistency]; ….).

Particular theories, for example—or at least their more feasible axiomatic approximations—may be "locally" immanent. But their "intended interpretations"—even "the" "intended" interpretations of theories as "obvious" as Peano arithmetic (an axiomatisation of the natural numbers)—are "locally" but recurrently *transzendent*.

Such "intended interpretations" are surely "abstractions" in some sense, if anything is. But such *abstracta*—which are not to be identified with the theories themselves, for this quickly gives rise to semantic paradoxes—require even more "abstract" metatheories in which their "intentions" can be grounded.

Even then, moreover, they are (I would suggest once again) either

10 theory-relative placeholders, which ride indefinite regresses of ascents of metatheoretic "forms"; or
11 "noumenal" notions which are inexpressible in [themselves] (the echo of Kant's usage is once again deliberate).

On the processive account of "experience" I have offered, the former are problematic but recurrently "immanent". The latter, by contrast, promote the notion of an "intended interpretation" to the level of its conceptual incompetence.

For nothing can (provably) "ground" or "interpret" what it cannot express—an observation which often seems to me to have eluded metaphysicians and philosophers of language who engage in relationally unqualified talk about "truth."

Correlatively, attempts to "ground" (end) metatheoretic ascent in [self]-interpretation—allegedly secured in the sort of "last" metatheory considered in Sections 1 and 2—seem to me

12 parade-examples of what Kant called *"transzendent(al)er Schein"* (cf. *KdrV*, B352 ff.), on a more or less equal basis with attempts to ground "the god of the philosophers" in [self]-causation, [self]-origination, [self]-organisation, or [self]-design.

Indeed, Kant's attempts to "intend" and secure a canonical extratheoretic interpretation of a collective (as well as subreptively uniform) [self], first, would seem to parallel rather closely Descartes' attempt to do this for a "psychological" [self], along the lines outlined above in Sections 1 and 2.

More generally, "secular" efforts to find last, "perfect", self-realising metatheories for such (putatively) "universal" notions aspire to "complete" [themselves], in ways which recall ontological, "cosmological", "physico-theological" and other "regress"-ending attempts to beg "theological" notions of ultimate "design".

Methodologically, for example,

13 claims that "all" of mathematical physics (say)—or "all" "real" knowledge of a language, for that matter—can be subsumed in some elusively "intended" or "intuitive" interpretation of "physics"—or what it "really" "is" to "speak a language"—

seem to me closely cognate to

14 Pascal's baffling embrace of Jansenist dogma; Spinoza's admirably stoic (but conceptually ill-posed) assertions about *"deus sive natura"*; Leibniz' more sympathetic (but equally ill-posed) *"principe de la raison suffisante"*; and Kant's *transzendentale Verstandeseinheit*, whose "deduction" he acknowledged would have to be a *Geschäft der Vernunft* in *KdrV*, B 692.

For such "global" and "universal" interpretations are all theory-marginal, in the sense that they would have to be "intended" and re-"intended" in ever-wider metatheories. But the more "global" one's theoretical aspirations—and more "holist" one's "contemplation" or *theoria*—the more open to metatheoretic ascent and emergent reinterpretation(s) such aspirations and *theoria* will have to be.

Such tensions already appeared in Gaunilo of Marmontiers' shrewd suggestion that limiting cases of the "conception" and "perfection" Anselm wanted us to "conceive" and mentally "perfect" might either be "trivial", or at least imperfect ("the most perfect island") or inconceivable.

They also recurred more strikingly in Hume's (or "Philo"'s) brilliantly formulated suggestions in the quotation from the *Dialogues Concerning Natural Religion* which began this essay—that "human" notions of [design] might provide very dubious templates for theological extrapolation.

In Kant's case, I have argued, there is also a sense in which he offered "cogito"-like "transcendental arguments" as attempts
15 to furnish (or beg) a collective counterpart of Descartes' individual (but allegedly generic) "existence"-proof, and
16 derive the "existence of the external world" (as well as a resolution of Descartes' "substantial-union"-problem) from the alleged exhaustiveness and universality of such a "complete" and "necessarily possible" collective "intentionality".

But if what is intentionally "possible" is collapsed in such ways to what is experientially "real"—a not entirely distortive thumbnail-sketch of Kant's "critical" rationale for Berkeleyan and Humean proposals—another variant of the recurrent dilemma I've tried to canvass presents itself: either
17 one opens "experience" to embrace indefinite hierarchies of provisional forms of semantic paradox; or
18 one acknowledges that "experience" is an inexpressible and almost neoplatonic "ideal", and its alleged "closure" a mystical as well as unintelligible "ultimate" semantic paradox.

One might construe this dichotomy as a sort of conceptual "zero-one-law", or counterpart of Niels Bohr's "*komplementarität*": if it is not the case that "everything intelligible" is "problematic", what "ensures" us that "it" is "unintelligible"?

Put in yet another form: we seem to want (or in our more judicious moods, "postulate") "complete" knowledge and "complete" (unique) interpretation (cf. Pascal's acknowledgment that "*le silence éternal de ces espaces infinis m'effraie*").

But all we can "intelligibly" (or "reasonably") attain is incomplete "understanding", conjectural as well as plurally interpretable clarification, and fleeting forms of *a-lētheia* ("non-ignorance"), rather than "ultimate" "truth". Postulations of "complete" knowledge, "objective" truth and "unique" interpretation, like postulations of "ultimate" subjective "enlightenment", may be little more than thinly veiled projections of intellectual vanity and vicarious aspirations to metaphysical immortality.

So much for what might be called the *via negativa* of skeptical *theoria* ("contemplation"), a dialectical interpellation which emerges—I have argued— from fairly straightforward forms of metalogical analysis.

Its *analogon*—partially and tantalisingly reflected in the remarks of writers such as Nicholas of Cusa—is the skeptical *via contemplativa* evoked in 1.9, at length in Section 4, *passim* in Section 6 and more briefly in 7.1–7.3 above. By way of transition to such a "way" or "path", consider first the

following remarks and conjectures about the differences between "skepticism" and "mysticism".
19 We tend to see more "simply" organised systems as "instruments" for "us" (or at least for Kant's "*empirisches ich*").
20 We are more reluctant (quite understandably) to regard "ourselves" as mere "means" or "instruments" for "others", whether or not they are more "complex" than "we" are.
21 We tend finally to beg the existence and uniqueness of "gods", theodicies and ineffably complex secular "designs" as *Wunschvorstellungen* of what might "confirm" or at least "(over)see" "us".

Intellectual vanity aside, there seems to me a *prima facie* tension between 20 and 21, just above, thrown into sharper relief by Hume's thumbnail critique of anthropomorphic "idealism" ("What particular privilege has this little Agitation of the Brain ....") in the *Dialogues Concerning Natural Religion*, cited earlier.

I have already argued, in effect, that such a tension defines that work's principle (and proto-"critical") *disputatio*—
22 whether the "deist" "Cleanthes" can define a tenable "critical" middleground between the "skeptic" "Philo"'s deconstructive counterexamples, and the quasi-"mystical" commonplaces of "Demea", a Christian strawman who stands in for the much more searching "fideist" musings of Blaise Pascal—and tried to offer suggestive metalogical evidence that he ("Cleanthes") cannot. To see why I find this evidence persuasive, consider first my suggestion that
23 "Demea"'s modest rhapsodies are indeed little more than faint echoes of much more eloquent passages from the *Pensées*, one of which I have already quoted, but
24 Pascal's evocations of our "disproportion" play a propaedeutic role in his larger dogmatic argument, as (at least arguably) did comparable invocations of ineffable infinities in the *Enneads* of Plotinus and *Ethica* of Spinoza.

In response to worthy defenders of these and other forms of "semantic monism" (animated by a "monistic pathos", in Lovejoy's words), I have endeavored to
25 interpret skeptical "criterial problems" and "lapse[s] into infinit[ies]" as generative structures worth further study;
26 argue that "ultimate" metaphysical "unities" "deduced"—by Spinoza, among others—from begged dichotomic premises are metalogically "unintelligible"; and

27 offer "locally transcendental" interpretations of a "pluralistic pathos" in which ramifying "intentional" hierarchies reflect Kant's "moral law within us", and ramifying "intensional" hierarchies reflect Kant's "starred heaven(s) above us".

This then is the skeptical counterpart of the dogmatic *"via contemplativa"*, mentioned earlier: a "reflective" "pathos" of limitless "critical" hierarchies for

28 Demea's (and Pascal's) metalogically untenable fideism, and Cleanthes' (and Kant's) metalogically untenable "middle way".

One indication of the latter's "untenability" may be found in Kant's "critically" calibrated claims to have to "deduced" the "necessary existence" and unique interpretability of "experience" from its "possibility".

For such formulations have appeared almost word for word in certain modal logicians' claims to "formalise" Anselm's "proof" of the "necessary existence" and unique interpretability of "god" from its "conceivability".

In §90 of the third *Critique*, moreover (which bears the interesting title *"Von der Art des Fürwahrhaltens in einem teleologischen Beweis des Dasein Gottes"*; "On the Manner of Holding-to-Be-True in a Teleological Proof of the Existence of God"), Kant struggled in 1790 to formulate

29 "analogical" notions of proof which might apply to "teleological" insights, desires for "higher" forms of justice and other "merely regulative" aspects of "reason"'s "fate".

In this chapter, I have tried to propose

30 "analogical" notions of proof in limitless metatheoretic and hypotheoretic hierarchies as "intelligible" media for the expression of "reason"'s "fate", as well as Kant's "secular humanist" ideals.

In accordance with the *"via negativa"* sketched earlier, much may be processively "hidden" and "uncovered" in such "experiential" hierarchies (*contra* Wittgenstein), in the sense that

31 no "intelligible" theoretical "horizon" is or will ever be "comprehensive", for example (much less "define" its own *"azimuth"*);
32 "merely regulative" postulation of certain forms and probabilistic extensions of metamathematical deduction are all the "inductive confirmation" we should anticipate;

and
33 Galileo's mathematical "book" and its "intelligible" metamathematical extensions are "physically" as well as "metaphysically" "precise" to the extent they are conceptually (and generatively) "heuristic" and "intelligible".

In keeping with such *Leitmotive*, I have also argued that Kant's "critical" attempts to discern "concepts" from "ideas of reason" led him to a valuable (and "heuristically" generative) insight:

34 that such methodological clarifications and "synthetic" realisations of otherwise uninterpreted syntactical evidence are—"locally", but "inherently"—semantic and metatheoretic rather than "immanent"; and
35 that his tacit acknowledgments in the *Dialektic* and third *Critique* that such metatheoretical undertakings might be—"locally" but recurrently—"*transzendent*" suggest two more skeptical, "neo-Kantian" conjectures: that
36 the very "intelligibility" of ("our") "experience" renders its "realisations" (its "*Verwirklichungen*" or "*Vergegenständlichungen*") provisional, plurally interpretable and inherently incomplete and "merely regulative"; and finally that
37 attempts to analyse and argue more precisely about such realisations may lead us into hierarchies—also local, but deeply heuristic as well as "merely" regulative—of "intended (re)interpretations" of "ourselves" as well as our "experiences".

"Local" and "heuristic" notions of "truth", "completeness" and "(intended) interpretation", after all—like their "absolute" dogmatic counterparts—are closely interrelated with (equally "local" and "heuristic") notions of "design".

And these, in turn, yield "critical" rationales for "essentially incomplete" notions of "knowledge", "structure" and "understanding"—"inductive understanding", for example, that one "knows" (provisionally, and in certain feasibly characterisable situations) "how to go on".

Like "design", "truth" and the other epistemic and metaphysical ideals just cited make sense for a given theory, in a particular metatheory. "Contemplatively" invaluable, they become delusive—and dogmatic as well as uninformative—only when they too are raised to the level of their incompetence in vain efforts to "diagonalise" them over "all" theories in "all" metatheories.

(We tend to lose sight, once again, of the facts that a "proof" was originally a "test" (a "*probatio*"), and "truth", in English, derived—as I remarked in an earlier footnote—from a common Germanic verb which meant to "(en)trust" or "believe".)

I have also tried to argue at various points that
38 iterated zetetic recourse to metatheoretic clarification and refinement may be heuristically "instrumental", where
39 "instruments" are ("locally") object-theoretic "means" to more "systematic" but counterfactual as well as potentially heuristic ("locally") metatheoretic "ends".

In particular, such "means" or "instruments" are object-theoretic with respect to what (conceptually or metatheoretically) "instrumentalises" them, and "experience" "organises" "means" and "ends" in metalogical hierarchies of "object"- and "concept"-formation which ramify without end.

(In this sense, what I wish to offer is a form of skeptical as well as "locally transcendental idealism", which hearkens back to Lucretius' "alphabet", as well as the "*logica magna*" Leibniz sought.)

In their efforts to explain "physical" "experience", for example, mathematicians and mathematical physicists—Galileo's successors—resorted to more and more elaborate metatheories to "define" (or posit "existence" of) object-theoretic "instruments" as "means" to more and more subtle (and conceptually "elegant") "physical" ends. Mindful of this, I have tried to argue that mathematically literate metaphysicians and epistemologists—Kant's successors—should search ("zetetically") for more elaborate "intelligible" metatheories to "define" (or posit "existence" of) object-theoretic "instruments" as "means" to subtler (and more conceptually "elegant") "intended" interpretations of
40 "cognitive" ends which clarify, at least provisionally, "liminal" notions of "identity", "awareness" and "individual consciousness", and
41 "ethical" and "deontological" ends which clarify, even more provisionally, collectively "liminal" notions of "equity", "dignity", "equanimity" and "collective moral sensibility".

If these analogies are tenable, one might also characterise
42 Hume's and Kant's "idealist" efforts to clarify "physical" patterns in terms of "cognitive" counterparts as an attempt to clarify *subtilia* in *subtiliora*, and
43 Kant's "*Primat des Praktischen*" as a conjecture that "ethical" and "deontological" ends or aims may be "deeper" and more conceptually complex than their "physical" and "cognitive" counterparts.

Kant's "counterfactual" and "nonconstructive" uses of "*Regel*" and "*regulativ*", for example, were as ambiguously suggestive as Hume's ambiguous attempts to define "Cause" in §60 of the first *Enquiry*. Perhaps it is "regulative" of broader forms of "experience" that we can "nonconstructively" "anticipate" schematic patterns we cannot "determinately" "understand"?

Graduated forms of "consciousness", in particular, might be attributed to Section 6's "systems" (or more accurately to metastable patterns of such systems' evolution in appropriate Stone spaces), to the extent they

44 range "freely" over "internal" interpretations of [themselves], and "external" interpretations of [what might be beyond themselves]; where
45 "freedom"—like its "physical" counterparts—is an inherently contextual and relational as well as graduated notion;
46 a system is "free" to the extent the range and complexity of its responses exceed that of any "internal" or "external" initial and boundary conditions it may "take into account" in formulation of such responses.

Admittedly, all these conjectures and analogies are impressionistic, and conceptually (much) more inchoate and rudimentary than I would wish them to be. But I believe they may be heuristically useful, and more particularly, that the indefinite extendibility of metatheoretic "ascent" and the non-"well-foundedness" of hypotheoretic "descent" are dual aspects of intelligibility which "mirror" each other in heuristically useful ways.

"Hypotheoretic" attributions of grades of "intention", "cognition" and "autonomy", for example, may "mirror" "metatheoretic" attributions of "interpretation", "verification" and "experimental isolation". ("The way up and the way down are one and the same.")

"Reflective inquiry", in this context, is inquiry in localisable stages of such hierarchies, and in their indefinitely iterable intentional ascents and intensional descents. As "we" "see" "simpler" conative and cognitive structures, so may more "complex" conative and cognitive structures "see" "us". ("Whatever ye do to the least of these ....")

Consider, by way of partial summary, the following assertions and (rhetorical) questions.

Kant's "*transzendentale Methode*" was arguably an attempt to make a certain sort of "sufficient ('synthetic *a priori*') reason" a fixed point for [itself].

One might construe the arguments of this essay as an attempt to make metalogical "inquiry" be an (endlessly transitional, but conceptually persistent) fixed point for [itself].

Contrary to Kant's express views in the *Analytic*, the more "*rein*" and "*a priori*" a theory (or "*Erkenntnis*") is, the more (metalogically) incomplete it is, and (therefore) the less "*vollständig*" it can be. Might this be a form of "practical" as well as conceptual "complementarity"?

The "*Vollständigkeit*" Kant attributed to "experience" was "constitutive", but that of ideas-of reason ("merely") "regulative". Might recognition that *both* forms of inquiry and concept formation are "essentially" incomplete, be a mark of John Keats' "negative capability" (cited above), as well as the *Würde* (dignity) Kant attributed to "reasonable beings"?

These questions are "rhetorical" in the sense that I find "good reasons" to believe (but cannot "prove") that the "contemplative" answer to each is "yes".

For it seems a mark of stoic epistemic virtue and awareness of our boundless limitations to continue to inquire and observe, knowing that "the" path(s) branch in liminally unobservable directions. It may even be a mark of some sort of epistemic or conceptual "existentialism" (in which the "essence" of "essential incompleteness" "precedes" "existence"). Indeed, an awareness of "reason's fate" and its Sisyphean nature may be part of that "fate", as well as a regulative ideal of Kant's "reasonable beings". But it might also (as I suggested earlier) be one of their more "enlightened" cradle-gifts. (To paraphrase Beckett: "I 'must' go on. I can't go on. I 'will' (to) go on ....")

Aristotle's "desire to know", for example, is not simply "sensory". It is also the desire to "value" and to "cherish" (This is a kind of "cognition"). And a broader and more "reflective" as well as "practical" form of this is the desire to "value" and "cherish" what can "value" and "cherish".

To "know" is to recognise what (apparently) "is". To "value" is to recognise (or "contemplate") what may be, or might have been.

To form a "purpose" and "evaluate" alternative "means" to attain it is to form a (usually counterfactual, and in all cases metatheoretic) "plan" in keeping with such "recognition" or "contemplation".

Such a plan, finally, is (practically) good to the extent it "values" the (indeterminate) plurality and range of other "reasonable beings'" "plans".

What is "just" is what "reaches out" (the original semantic sense of Aristotle"s "desire"), by analogy with what is "felt within" (the "golden rule"). This is "regulative" of equity and other "practically" significant forms of empathy and sympathy. An absence of both in a "cognitive system" is not "irrational". But it *is* (metatheoretically) "unreasonable". It marks the absence or suppression of a form of higher-order awareness.

The distinction between "irrational" and "unreasonable" is "practical", metatheoretical, inherently "liminal" and "merely regulative" at every stage of "experience". Refusal or inability to make this distinction is a dogmatic "cognitive" deficiency or lack of "sensibility" (or a deliberate attempt to dissemble, "blunt" or "coarsen" certain "natural" forms of cognitive awareness).

There is also a fairly clear parallel between Hume's *de facto* assimilation of "conceivable" to "consistent", Kant's "categorical imperative" and Kant's assimilation of *"gültig für uns"* = *allgemeingültig für jedermann* ("valid for us" = "universally valid for everyone"; *Pr*, Sects. 18–19). Both conflate "inconceivable to 'us'" with "inconceivable relative to a(n allegedly normalised but subreptively begged) conception of "us" (of what "we" are). (For comparison, cf. also Kant's remarks about *"nur komparative Allgemeingültigkeit"*—"merely comparative universal validity", *Pr*, Sects. 18–19).

Such conflation may also be construed as a begged assertion that "the" theory of ["the" world] is a faithful or conservative extension of "our" theory of [the world], whatever *Dinge an sich* it may hypothetically adjoin. If, moreover, "the world"'s interpretation were faithful, as in the analysis of "cogito"-like arguments sketched earlier, it would assure a not-so-desirable form of equiconsistency:

47 "the world" would be consistent if and only if "I" am (or in Kant's case, if and only if "we" are).

A "solipsist" metalogical consequence of this—that "the world" would cease to "exist" if and only if "I" cease to "exist"—would effectively nullify Kant's categorical imperative, as well as the "skeptical" variant I proposed above:

48 to treat others not as (mere) means, but as (vulnerable as well as "reasonable" beings in search of "higher") ends in (and beyond) themselves.

Given also that "strong" interpretations of "us" and "our" world-views are not "faithful" in the metalogical sense outlined earlier, the more relevant epistemic as well as "practical" question would not seem to me

49 whether "absolute" (ethical or epistemic) norms, standards or criteria "exist";

but rather—given the metalogical "unintelligibility" of "absolute" norms, standards and criteria in the senses outlined above—

50 whether (and when) it is reasonable for sentient beings to impose their norms, standards or criteria as such.

Along similar lines, it also seems to me that *dicta* that "everything is determined" would not imply that "everything is forgivable" (on the model of *"tout comprendre, c'est tout pardonner"*), but rather that

51 "practical" notions of "forgiveness" and "sympathy" as well as "responsibility" would have no meaning .... ; for

52 a "free" being which refused to admit its actions might be wrong would be an inconsistent cognitive as well as ethical agent, in the sense that

53   nothing could be "right"—"practically" or ethically "right"—to a being which could not admit—"wholeheartedly" admit—that it might be wrong.

A "rational" being is also "practical" as well as "reasonable" to the extent it acknowledges that it is neither unique nor self-sufficient, and undertakes to seek to respect cognate desires and forms of awareness in its (apparent) "*semblables*". At every stage of "reflective inquiry", this awareness is graduated, underdetermined and physically vulnerable. But its absence is a "cognitive" deficiency all the same, in the sense that it closes itself to broader forms of semantic awareness.

For zetetic inquiry is not only regulative of "experience". It is also, it seems to me, a "liminal" form—and perhaps prototype—of Kant's "*Zweckmäßigkeit ohne Zweck*" ("purposiveness without purpose") and other "nonconstructive" notions one might devise, such as

54   "*Gefühlsmäßigkeit ohne (eindeutig determinierte) Gefühl(e)*";
"*Regelmäßigkeit ohne (eindeutig determinierte) Regel(n)*;
"*Rechtmäßigkeit ohne (eindeutig determiniertes) Recht*";
"*Gesetzmäßigkeit ohne (eindeutig determiniertes) Gesetz*"; ....

To the extent such "*-mäßigkeiten*" urge respect for honest inquiry, they may express something like Kant's "purely good will" as well as Aristotle's "desire to 'know'". And respect for them enjoins us to value two forces which through the green fuse drive the flower: "desire" for "understanding" and "sympathy" "within" us, and "awe" for the "starred firmament" "above" us.

The "desire" in particular, takes many forms: a desire to (continue to be part of) (conscious) experience; a desire to "understand" the diapason of experience in ways which cannot be fulfilled; and a desire to seek, value and protect that fragile "understanding" as the "reason" of "reasonable beings".

Kant had a point, therefore, about the "intentional" and "self-referential" aspects of epistemic, "practical", aesthetic, and "reflective" "autonomy". But autonomous "selves"—essentially by definition—have no "ultimate" clarification of what it "is" they are supposed to "legislate", or the "ultimate" intelligible "aim" to which everything else is "reduced" or "interpreted".

For the self-referentiality of "intelligible" "*Selbstgesetzgebung*" and "*Selbstzwecke*" *must* be incomplete and "problematic", in the sense that they are assertible only in processively metatheoretic contexts which must provisionally be resolved in yet "higher" "coded" counterparts. Any conceptual act which claimed to "verify [its own] 'existence' (or '*Reinheit*', or 'experimental isolation', or ....") would therefore undo the "verification", by relativising the "existence" (or "*Reinheit*", or "experimental isolation", or ....).

"The" boundary line between Kantian "reason" and "experience" is also (in)determinate if and only if "the" boundary between Kantian "analytic" and "synthetic" is (in)determinate if and only if "the" boundaries between Kantian "*Spekulation*" and "*Praxis*" and "*Spekulation*" and "*Reflexion*" are (in)determinate.

The more global one's theoretical aspirations, therefore—and the more holist one's theory or *theoria*—the more open to metatheoretic ascent and emergent reinterpretation(s) such aspirations and *theoria* will have to be. The broader the *Rechtmäßigkeit* of *Deduktion* becomes, the less clear it is why it is not simply "reflective", "regulative" and "*problematisch*", and the more suggestive the thought that such breadth is a "good" thing.

Skeptical "theoria" ("contemplation") and the forms of *ataraxia* ("peace of mind") also have their own "light"-images—*skepsis* itself, for example, and *Horizonte* among them—which reflect (and reflect on) their more "dogmatic" and "hieratic" counterparts (Plato's *eidē*, Descartes' *intuitio*, Neo-Platonic "*emanations*," ....). Incomplete and inadequate as the skeptical images are, they are at least more open to further clarification, for they are exquisitely "relational", by definition as well as programmatic intent ("insight" against what backlight? "scrutiny" and "enlightenment" relative to what source? "escape" in which "direction", and toward which form(s) of greater "illumination"?)

Unlike Plato's (literally linear) "line" and Kant's "two truths", such images also evoke a distributive and graduated skeptical counterpart of Aristotle's "theoretical" *noesis noēseos* (thought about thought), whose ramifications suggest to me a great tree of conceptual life.

The "abstraction" characteristic of metalogical analysis is a counterpart of "egalitarianism" in ethics, and "experimental isolation" in "natural philosophy". (Cf. also Kant's "*rein*" and "*a priori*")

These views seem to me to reflect elusive egalitarian ideals of Kantian ethics which transcend his secular-pietist preoccupation with "absolute" forms of "duty", and enjoin us to "respect" fleeting nuances of "consciousness" which "reasonable beings" intermittently enjoy, and may (or may not) endeavor to "respect" in their fellows. Such genuine egalitarianism is a "merely regulative" ideal. And intelligible "*Zwecke*" exist only in provisional stages or levels of *Zweckmäßigkeit*. But the very openness and incompleteness of such ideals may offer a glimpse into comparably open aspects of moral eternity.

For whatever "particular privilege .... this little Agitation of the Brain" may have (in response to the opening remark from Hume's *Dialogues*), it is probably not particular to "us".

And complex "intentional" systems of inquiry which ramify with what they "discover" may be "regulative" of other (would-be) "reasonable beings", which may pose or have posed similar questions, in the past, in the future, or "in another galaxy, long ago and far away".

What rides the wavefronts of "for the sake of which" (so to speak) is ("reflective") "consciousness", which flourishes as it inquires, in unfinished forms of *"eudaimonia"* as well as *"theoria"*? And it is about *this* we should be generous and egalitarian (the "categorical imperative", like the "golden rule", is a moral "identity of indiscernibles" principle.)

Aspirations to understand deeper aspects of this ideal are "desirable in themselves", and collectively, they afford a kind of scheme of provisional fixed points for what I have called "reflective inquiry". The conviction that we can make "reflective" appeals to "merely" regulative ideals of such "schematic fixed points" is a regulative ideal "in itself". But it requires (as a "precondition", so to speak) that we accept (in simple terms) that

55 it's all "problematic", for expressible ("intelligible") "contemplation" must be incomplete.

(Compare William James' first "free" intentional "act", to "believe" in his "freedom" to act.) ([James], 1896) The skeptic's first "un(der)determined" intentional act is to "believe" in skeptical "un(der)determination"—as something "known to reason alone" *and* "known from (within) 'experience'".

None of this provides any "final" resolution for the questions which eternally recur. Which "systems", for example? "Consciousness" of "what"? Where will it "end"?

To interrogate ourselves in this way is to pose questions, once again (such as Kant's *drei Fragen* in *KdrV*, B 832–833 and *Logik*, IX, A 25),

> die sie ... nicht beantworten kann, denn sie übersteigen alles Vermögen der menschlichen Vernunft (*KdrV*, A VII, cited earlier)

> which one ... cannot answer, for they exceed every capacity of human reason ....

The same recurrent tensions also evoked one of Pascal's deepest and most "existential" (as well as "secular") insights in the following well-known passage, elliptically quoted earlier, and perhaps the most eloquent evocation ever written of the sort of courage Kierkegaard later attributed (dismissively) to his "knight of infinite resignation" (in the original Danish simply *"knight of infinity"*—*"Uendelighedens Ridder"*):

> La dernière démarche de la raison est de reconnaitre qu'il y a une infinité des choses qui la surpassent ....

> L'homme n'est qu'un roseau, le plus faible de la nature, mais c'est un roseau pensant. Il ne faut pas que l'univers entier s'arme pour l'écraser; une vapeur, une goutte d'eau suffit pur

le tuer. Mais quand l'univers l'écraserait, l'homme serait encore plus noble que ce qui le tue, puisqu'il sait qu'il meurt et l'avantage que l'univers a sur lui. L'univers n'en sait rien.

Toute notre dignité consiste donc en la pensée. C'est de là qu'il faut nous relever et non de l'espace et de la durée, que nous ne saurions remplir. Travaillons donc à bien penser: voilà le principe de la morale. (*Pensées*, 188 and 200)

The last step of reason is to recognise that there are an infinity of things that surpass It ….

A human being is a reed, the weakest of nature, but it is a thinking reed. It isn't necessary for the whole universe to take arms to crush it; a vapor, a waterdrop is enough to kill it. But when the universe crushes it, the human being is still nobler than what kills it, for it knows that it is dying and the advantage the universe has over it. The universe knows nothing.

All our dignity consists therefore in thought. It's from there we should take our orientation, and not from space and time, which we cannot fill. So let us work to think well: this is the principle of morals.

In the limitless contexts of reflective inquiry and skeptical "experience" (with apologies to Leonard Woolf), the journey not the arrival matters.

9 Metamathematical Interpretations of Free Will and Determinism

> I saw Eternity the other night,
> Like a ring of pure and endless light.
> All calm, as it was bright;
> And round beneath it, Time in hours, days, years
> Driv'n by the spheres,
> Like a vast shadow mov'd, in which the world
> And all its train were hurl'd ….
>
>   Henry Vaughan, "The World"

The principal purposes of this chapter (which might also be called "A Probabilistic Critique of Physical Determinism") are to

(i) extend Gödelian essential-incompletness arguments to initial and boundary conditions of "agency" and physical time-evolution;

(ii) argue that the more intelligible scientific analyses of physical time-evolution are, the more "ramified", "incomplete" and "indeterminate" the ranges of their predictions will be;

and

(iii) observe that almost all temporally evolving phenomena are not governed by continuity principles of the sort often found in early modern rationales for "deterministic" physical time evolution.

1 Three Venerable Quotations

The following passages expressed three closely interrelated classical views of physical "determinism", as well as (what I will call) mathematical "monism" (or "monist realism").

1.1 Gottfried Leibniz

(*Bruchstücke, die Scientia Generalis betreffend*, VI., K, *Die Philosophischen Schriften von Gottfried Wilhelm Leibniz*, VII, 118)

Die Mathematick oder wißkunst kan solche Dinge gar schöhn erleutern, denn alles ist in der Natur mit zahl, maaß und gewicht oder krafft gleichsam abgezirkelt. Wenn zum exempel eine kugel auf eine andere kugel in freyer lufft trifft, und man weiß ihre größe und ihre lini und lauff vor dem zusammentreffen, so kan man vorher sagen und ausrechnen, wie sie von einander prellen, und was sie vor einen lauff nach dem anstoß nehmen werden. Welches gar schöhne Regeln hat, so auch zutreffen, man nehme gleich der kugeln so viel als man wolle, oder man nehme gleich andere figuren als kugeln.

Hieraus siehet man nun, daß alles Mathematisch, daß ist ohnfehlbar zugehe in der ganzen weiten welt, so gar daß wenn einer eine gnugsame insicht in die inneren theile der dinge haben köndte, und dabey gedächtniß und verstand gnug hätte, umb alle umbstände vorzunehmen und in rechnung zu bringen, würde er ein Prophet seyn und in dem gegenwärtigen das zukünfftige sehen, gleichsam als in einem Spiegel.

Mathematics or the art of knowledge can clarify such [physical] matters beautifully, for everything in nature is as it were cordoned off with number, measure, weight or force. If for example, a sphere strikes another sphere in free air, and one knows their sizes, lines and courses before the strike, one can calculate and predict ahead of time how they strike each other, how they rebound and what sorts of courses they [will] follow after the strike. These events have rules so fine that they hold no matter how many spheres one takes, or solids other than spheres.

From this, one sees now that everything proceeds mathematically–that is, infallibly–in the whole wide world, so that if one could have sufficient insight into the inner parts of things, as well as sufficient memory and intelligence to consider all circumstances and take them into account, one would be a prophet and see the future in the present as it were in a mirror. (italics mine)

1.2 Roger Boscovich

Theoria Philosophiae Naturalis (Boscovich, 1763) (Latin-English Edition, 1922, §385, 280–84)

Sed licet ejusmodi problema vires omnes humanae mentis excedat; adhuc tamen unusquisque Geometra videbit facile, problema esse prorsus determinatum, & curvas ejusmodi fore omnes *continuas sine ullo saltu, si in lege virium nullus sit saltus* .....

Cognita autem lege virium & positione, ac velocitate, & directione punctorum omnium dato tempore, posset ejusmodi mens praevidere omnes futures necessarios motus, ac status, & omnia Naturae phenomena necessaria, ab iis utique pendentia, atque praedicere.

But if ever a problem of this kind should surpass all the forces of the human mind; even then any geometer would easily see that the problem is wholly determined, and its curves will all be *continuous without any jumps*[,] *if in the law of the forces there are no jumps* ....

> *If moreover the laws of all the points [and] the position, velocity and direction at a given time were known, the mind would be able in this fashion to foresee and predict all future necessary motions and all necessary phenomena of nature depending on them.*

1.3 Pierre-Simon Laplace

> (*Essai philosophique sur les probabilités*, 3-4, Sixième Édition, 1840)

Nous devons donc envisager l'état présent de l'univers comme l'effet de son état antérieur, et comme la cause de celui qui va suivre. Une intelligence qui pour un instant donné connaîtrait toutes les forces dont la nature est animée et la situation respective des êtres qui la composent, si d'ailleurs elle était assez vaste pour soumettre ces données à l'analyse, embrasserait dans la même formule les mouvements des plus grands corps de l'universe et ceux du plus léger atome; *rien ne serait incertain pour elle, et l'avenir comme le passé serait présent a ses yeux.*

We must therefore envisage the present state of the universe as the effect of its anterior state, and as the cause of the state which will follow. An intelligence which would know for a given instant all the forces by which nature is animated[,] and the respective situation[s] of the entities which make it up, and would moreover be sufficiently vast to submit these data to analysis, would embrace in [one] formula the motions of the greatest body in the universe and those of the lightest atom[.] [*N*]*othing would be uncertain for it, and the future as well as the past would be present to its eyes.*

2 Preliminary Observations

2.1 A first remark about these passages is the extent to which they drew on a long history of ancient and early modern "continuity principles"—tacitly in the passages from Leibniz and Laplace, and more explicitly in the italicised remarks by Boscovich.

2.2 A second is the extent to which their appeals to the implicit unicity of Leibniz' "god", Boscovich's "mind" and Laplace's "vast intelligence" were
(1) mathematically "Spinozan" (consider for example the tacitly "physical" implications of the latter's famous phrase "*deus sive natura*"); as well as
(2) metaphysically "Spinozan" (or perhaps in this case "Plotinian"), to the extent that they tacitly or explicitly drew on omniscient analogues of "the One" in the form of an ultimate synoptic metatheory.

2 Preliminary Observations

2.3 A third observation is the extent to which Laplace (who scrupulously acknowledged his debt to Leibniz a few lines before the well-known passage quoted above)
(1) begged metaphysical realist appeals to quasi-noumenal conceptions of mathematics, metamathematics and mathematical physics,

appeals which are difficult to reconcile (to put it mildly) with
(2) "secular" interpretations of his "scientific" rejoinder to Napolean's conventional invocation of *"le bon dieu"* (... *"Sire, je n'ai pas besoin de cette hypothèse"* / "Sire, I have no need of that hypothesis").

2.4 My aim in subsequent sections will be to propose alternative skeptical (or perhaps *zetetic*) "hypotheses", and outline a tenable mathematical framework in which
(1) temporal evolution of "agents" and physical systems in "globally hyperbolic" spacetimes[1] (or local counterparts of them), defined by
(2) "intelligible" first-order theoretical extensions of ZFC (Zermelo-Fraenkel set theory) (a notion defined more precisely in 5.2(2) and 5.2(3) below), may be
(3) partially but not linearly ordered, and ramify "forever" in theoretical hierarchies which are internally consistent but mutually inconsistent at every temporal instant;

and
(4) heuristic preconditions for (desirable) assumptions that data in such temporal hierarchies may be "intelligibly" encoded and "measured" suggest that their evolution is almost certainly not continuous.

2.5 The Physical Relevance of "Cadlag"[2] Temporal Processes

In what follows, I will work with four semiformal premises (or conjectures)—which I will attempt to make more formally rigorous in later sections—about physical measurements, classical as well as quantum-theoretic.

---

1 Roughly speaking, a spacetime is "globally hyperbolic" if it can be represented via a curvilinear "time" axis which is infinitesimally "orthogonal" to a temporal process of three dimensional Cauchy surfaces. Precise mathematical definitions of such properties may be found in Hawking and Ellis (1973, 200–206), and Wald (1984, 201–212).

2 *"Cadlag"* derives from the French phrase *continus à droite avec limites à gauche*, anglicised in some monographs as *"rcll"* (right continuous with left limits). Temporal characterisations of such processes appear in this section and the next, and references to their precise mathematical definition and basic properties may be found in the footnotes to section 6.4.

(1) The first is that one cannot give a mathematical account of physical measurement without (tacit or explicit) recourse to metamathematical ideas. For one cannot, I would argue, "measure" what one cannot—in some potentially formalisable sense—intelligibly "define" and "interpret".
(2) The second is that "outcomes" of measurements may ramify in time, as suggested above, in the sense that they range, "in prospect", over "temporally" parametrised "scales" of "values". (In philosophical language, they "support counterfactuals".) But each prior course of such "values" is "retrospectively" linear.
(3) The third is that such processes of values are temporally "conditioned" but not "determined", in the sense that topological as well as measure-theoretic "scales", readings and "distributions" of such values "exist", may be "sampled", and evolve in ways which permit epistemic access.
(4) The fourth, finally—compatible with the first three and prompted by criteria described in section 6 below—is that "almost all" such respectively linear but prospectively ramified temporal processes evolve in "cadlag" rather than "continuous fashion."

2.6 An informally "physical" rationale for the "cadlag"-nature of such processes' temporal evolution may briefly be sketched as follows:
(1) suppositions of continuity on the "right" (the future) reflect intuitions that "what has already happened" will be retrospectively well-defined, and
(2) suppositions of limit-existence on the "left" (the past) reflect assumptions that "systems" to be "measured" at any future time will have well-defined convergent antecedents.

3 Three Distinctions

3.1 In this section and the next, I will attempt to interrelate "limiting" constructions of the sort introduced in 2.5 and 2.6 with metaphysical and metamathematical distinctions between
(1) what is "infinite" and what is "indefinite";
(2) what is "intensional" and what is "intentional"; and
(3) what is (metalogically) "standard" and what is "nonstandard".

3.2 I have already suggested that
(1) physical "determinism" of the sort expressed by Leibniz, Boscovich and Laplace is itself an elusively indeterminate notion, in the sense that its

venerable advocates could not (and their metaphysical-realist heirs cannot) secure this "determination" without

(2) recourse to ill-defined dogmatic attributions of "metaphysical" "determinism" and appeals to quasi-mathematical forms of "monist realism".

3.3 In this section, I will attempt to

(1) critique such monism in the name of "local" or "feasible" recourses to logical inference and metalogically accessible consequence-relations; and

(2) interpret iterated recourses to provisional frameworks of intelligible syntactical representation as metamathematic variants of Kantian "regulative principles",

in opposition to

(3) "classically" authoritative (but question-begging) forms of "self evidence", and correspondingly "transcendent" (in Kant"s "critical" sense, in opposition to "transcendental") notions of mathematical "truth".

3.4 Colloquial Origins of Magisterial Usages

For purposes of pragmatic analysis, consider the "indefinite" (or "distributive", or plurally interpretable) historical antecedents of a number of ordinary English–language "abstractions", among them

(1) "evidence", derived from something that had been "seen" or "discerned";

(2) "consistency", from assertions that something was reliably "set or aligned together";

(3) "truth", from Scandinavian and older Germanic words for "believe", "belief" and "faith" (*jeg tror det ikke* ["I don't believe that"], or *den tro deler jeg ikke* ["I don't share that faith"], for example, in Danish);

(4) the Latin "verus", French "vrai" and German "wahr", which evolved from Indo-European designations for someone (or something) "worthy of respect";

and finally

(5) the Greek "*alethes*" ("un-ignored", "unconcealed", "unforgotten"), and ("*eteos*" and "*etymos*" ("real", "genuine", "trustworthy"), conjecturally thought to derive at least in part from *etoi* ("kinspeople", whom one could "trust"), and

(6) in part perhaps from "*etazein*" (to *test* or *examine*), and semantic parallels between such "examination" and the ancient pyrrhonists' "*zetesis*" (search, investigation") and "*skepsis*" ("insight" based on visual examination).

Whatever the colloquial origins of these metaphorical nuances, their distributive nature and un-canonical variety seem to me to suggest origins which are anything but "determinate", much less "timeless" or "apodictic".

3.5 Consider, for example, the semantic ramifications of the Greek plural *logoi* ("words", "expressions", "utterances"), from which we have of course the word "*logic*". It roughly referred on the one hand (*pace* Wittgenstein) to
(1) "interior" or mentally "private" communication, which neurophysiologists might assimilate to "intentional" equivalence classes of "synaptic firing patterns";

and on the other to
(2) "exterior" public or social communication, as plurally, distributively and indefinitely communicable intensional (but not intentional) placeholders.

In both senses, one might conclude that
(3) "logic" is a semantic intersection of many (metatheoretically consistent) syntactical *logoi*, rather than a union of them, in rough analogy with the metalogical observation that (formal)
(4) "theories" (from another "visual" Greek usage which Aristotle borrowed for his "contemplation(s)") are intersections, not unions, of the structures which may model or interpret them.

3.6 Prompted by such metalogical and linguistic reflections, I would therefore find it remarkable if
(1) rigorous mathematical-physical theories were "complete" (able "in principle" to decide "all" assertions expressible in their formal languages),

rather than
(2) "intelligible" but essentially incomplete (able "in principle" to propose, consider and perhaps accommodate pluralities of semantic interpretations in such languages)[3].

I would also argue that if
(3) "ordinary" inflected languages in which we talk informally but informatively about science and mathematics are plurally interpretable, in suggestive ways that may be nuanced *ad indefinitum*,
(4) we should be less shocked if "intelligible" mathematical and mathematical-physical theories—the theory of Peano arithmetic[4], for example—turned out to be plurally interpretable as well.

---

[3] For a sketch of Gödel's results for "intelligible" theories', cf. 5.2 and 5.6.
[4] Here and in the sequel, I will use the phrase "Peano arithmetic" to refer to the theory which includes all the usual Peano axioms except the one which bars negative integers, and replaces it with the axiom that asserts that for every nonzero integer a there exists an integer "$-a$" such that $a + (-a) = 0$

3.7 Two conjectural lines of inquiry may offer support for such arguments.

The first is that the formal arithmetic Peano studied and its "intelligible" extensions were the first "rigorous" theories of enumeration, in the sense that it supported
(1) carefully defined predicate-by-predicate metamathematical refinements of Humean "induction", in the form of
(2) "principles of mathematical induction" formulated explicitly as a distributive scheme[5] of axioms rather than a single finitary "induction axiom".

The second is that the theory of "rational numbers" (which "mildly" extends the theory of Peano's "natural numbers") includes
(3) dyadic (or binary) rational numbers (finite sums of fractions of the form 1/2), which Skolem and other parents of model theory exploited as follows:
(4) given a syntactical but metatheoretic notion of what "the" positive natural numbers ω are, and a "countable" enumeration of a theory **T**'s sentences,
(5) every metamathematical interpretation of **T** can be correlated with a metatheoretically complete dyadic sequence of zeros and ones,

which may in turn be interpreted on the one hand as
(6) an element of a Stone / Cantor space $2^\omega =_{Df} \{0, 1\}^\omega$ (cf. 6.2(1)–(3) below), named after Marshall Stone and Georg Cantor, in which "1" encodes "truth" and "0" "falsity",

and on the other as
(7) elements of a "primordial" probability-measure-space[6]—"primordial" in the sense of mathematically "universal"—in which all known aleatory arguments may be represented.

3.8 To me, these observations suggest again an analogical query, elaborated in more detail in the next section:
(1) if we respond heuristically but critically to schematic attempts to make mathematical sense of "Humean induction",

---

[5] A salient point of every serious exposition of "mathematical induction" (cf., e.g., Bell and Machover, 1977, 320–321) is that such induction is not an axiom (a single sentence in an given theory), but an infinite (or indefinite) *scheme* of such axioms (as, in effect, was Hume's "induction"). To me at least, the very word *inductio* (a "leading" or "bringing in") suggests something processive rather than fixed.

[6] Leibniz and Laplace anticipated the relevance of such spaces, and clear expositions of their properties may be found in introductions such as (Karr, 1922, 24 and 37–38).

(2) should we not respond heuristically but critically to cognate attempts to make mathematical sense of "volition", "physicality" and "intentionality"?

4 "Schematic Induction" and Metalogical "Nonstandardness"

4.1 In this brief section I will
(1) argue that distinctions between what is "infinite" and what is merely "indefinite" are intrinsically underdetermined;
(2) suggest that that such intrinsic underdetermination lies at the heart of "physical" counterparts of relevant "semantic paradoxes";
(3) interpret critical tolerance of underdetermined initial and boundary conditions as a regulative ideal of rational inquiry; and
(4) conjecture that it is rational to accept "schematic" "physical" refinements of Humean "induction", if (and only if) we practice such tolerance; and
(5) acknowledge that that the "scopes" or "ranges" of physical as well as metaphysical notions of inquiry may be intrinsically "indefinite", "intensional" and "nonstandard".

4.2 Consider once again, for example, Peano's careful attempt to "normalise" as well as axiomatise formal schemes of "mathematical induction" in ways which can be extended to their field- and set-theoretic counterparts.

Most mathematical realists readily accept that mathematical induction in formal as well as informal venues is a schematic principle. But they are less likely to accept that "definitions" and "semantic ranges" ("initial" and "boundary" conditions) of such "schemes" themselves—essentially metatheoretic notions—may be metametatheoretically indeterminate.

4.3 In opposition to rejections of the metatheoretical "ascent" sketched in the last paragraph, I would argue that
(1) it is a deep consequence of Gödel's work[7]—even deeper, I believe, than his famous conclusion that one cannot prove the consistency of arithmetic in arithmetic—that one cannot do any of the following:
(2) *schematically define* "the" *scope and limits* of *schemes*;
(3) *arithmetically "define"* "the" *scope and limits* of *arithmetic*;
(4) *physically "define"* "the" *scope and limits of what is "physical"*; or
(5) *inductively "define"* "the" *scope and limits* of (*mathematical*) *induction*.

---

[7] Anticipated, in part, by the work of Thoralf Skolem and closely related ancient as well as early-twentieth-century investigations of "self-referential paradoxes".

4.4 It is well known, for example, that one can *prove* in ZFC (construed as a metametatheory for "weaker" theories of intermediate "strength", such as Peano's arithmetic) that
(1) an initial copy of "the real" natural numbers, preceded and followed by rationally indexed copies of "the" strictly positive and negative "natural numbers",
(2) furnishes "canonical" order types for continuum-many "countable" nonstandard semantic interpretations of Peano's arithmetic[8];

and more unexpectedly, perhaps, that
(3) uncountable semantic interpretations of Peano's axioms also exist which have the property that uncountably many nonstandard "integers" are metatheoretically distinct but object-theoretically "indiscernible"[9] (a metatheoretical counterexample to Leibniz' "identity of indiscernibles").

4.5 To me at least—in borrowed Kantian language, and viewed from the ledges of particular set-theories—such observations suggest that
(1) things that are *"syntactically infinite"* may be *"semantically" "indefinite"*;
(2) things that are *"intensionally" "linear"* may be *"extensionally" ramified*;
(3) things that are *"intensionally" "countable"* may be *"extensionally" uncountable*;

and that
(4) the consistency of Gödel's theory "**T**-but-not-Con(**T**)"[10] for metatheoretically consistent intelligible theories **T** may have been the first of many rigorous signs and way-markers of potential noetic as well as physical "indeterminacy."

Confronted by such metalogically defensible "pathologies", I would also find no good reason to deny that they may be iterable and generalisable, in which case
(5) "subreptive" attempts (a nice Kantian usage) to impose unicity-assumptions on "the" structures of "the unique" natural numbers, "the" continuum and "the" "set-theoretic universe" may be instances of Kant's *transzendentaler Schein* ("transcendental illusion");

---

8 Cf. (Henkin, 1950, 91), where it is cited as a "simple result". A concise readable proof of this result may be found at http://math.stackexchange.com/questions/37418/non-standard-models-of-arithmetic-for-dummies. Accessed 2 March 2017.
9 An exposition of ways to "adjoin" such "indiscernibles" to semantic interpretations of Peano arithmetic and many other theories may be found in (Bell and Machover, 1977) 218–222.
10 Cf. Smorynsky, 1977, 828.

and
(6) indefinitely iterable but non-"well-founded" initial and boundary conditions may be assimilable to noetic, physical and mathematical forms of non-"standard" physics, as well as metaphysics and epistemology.

I will attempt to elaborate these conjectures in subsequent sections.

5 Gödelian "Incompleteness" and "Indeterminism"

5.1 The primary aims of this section are to
(1) consider more precisely well-studied "local" as well as "global" interpretations of "physical time-evolution" in relativistic contexts; and
(2) derive temporally ramified (and in that sense "anti-Laplacean") interpretations of physical time-evolution from the work of Kurt Gödel.

5.2 Linear Hierarchies of "Intelligible" Theories Defined on "Local" and "Global" Cauchy Surfaces

Consider a "globally hyperbolic" spacetime (M, γ) (or a "normal neighborhood"[11] in a larger spacetime (M', γ'), and assume metatheoretically (in keeping with Leibniz', Boscovich's and Laplace's "mathematisations") that properties of (M, γ)'s Cauchy surfaces are
(1) representable in intelligible (first-order) time-ordered theories T_ρ, where by "intelligibility" of the theories T_ρ, I mean that they are
(2) "parsable" (the languages of all the T_ρs are countable, and the metatheoretically defined Gödel-codes of their axioms are recursively generated); and
(3) "autological" (each T_ρ syntactically interprets theories of arithmetic (or finite sets) which are "*strong*" enough to permit recursively enumerable "encoding" of T_ρ's "intensional" proofs, consequence-relations and "internal" mathematical-induction schemes in T_ρ.

5.3 Analogies and Interpretations

(1) Parsability may be interpreted as an assumption that no inscrutable metatheoretic "oracles" (a word used in expositions of recursion theory) are needed to recognise the theories' "internal" (recursive) languages, premises and proof-structures

---

11 Cf. Wald, 1984, 42, 192, or Hawking and Ellis, 1973, 34, 41, 63, 280.

(an obvious scientific desideratum, as witnessed by Laplace's reply to Napoléon); and

(2) "autologicality" as an assumption that the theories can pose (but not, Gödel discovered, answer) "internally" coded counterparts of "intensionally" semantic questions about "themselves" ("Am 'I' 'consistent'"? "Do 'I' 'exist'"? "Are 'my' actions determined'"?).

The aim of the following definition (a generalisation of 5.2's linear hierarchies) is to provide an initial framework for Gödelian critiques of Leibniz', Boscovich's and Laplace's de facto continuity principles, and with them the sorts of determinism they attributed (*avant la lettre*) to "intelligible" scientific theories.

5.4 Ramified Hierarchies of "Intelligible" Theories

A temporal hierarchy $\langle \Theta_T, \prec \rangle$ of intelligible theoretical extensions of ZFC is ramified, if

(1) $\langle \Theta_T, \prec \rangle$ is a partially ordered "tree-structure" of intelligible first-order subtheories T_τ which extend ZFC,

where by "tree-structure", I mean in this case that

(2) the temporal predecessors with respect to $\prec$ of each theory T_τ forms a unique linearly ordered hierarchy of the form $\left(\cup_\rho (T_\rho | \rho < \tau \in Q) \right)$ for rationals $\rho < \tau$, but the successors, however, of any given T_ρ, will not in general be linearly ordered, but

(3) branches of branches of .... in a temporally indexed but partially ordered "tree structure" which will ramify "forever" into the "future" of the theory T_τ.

5.5 As in the linear hierarchies introduced in 1 above, I will also assume that

(1) each T_ρ is an proper subtheory of T_τ for $\rho < \tau$, and each T_ρ, T_τ and linear "limit"-subtheory $\left(\cup_\rho (T_\rho | \rho < \tau \in Q) \right)$ of the hierarchy $\langle \Theta_T, \prec \rangle$ (where Q is the rationals) is an intelligible first-order theory of "local" properties of the Cauchy surfaces S_ρ and S_τ.

For "physical" as well as metamathematical reasons sketched below, I will also assume that

(2) each theory T_τ and $T_\tau^\cup = {}_{Df} \left(\cup_\rho (T_\rho | \rho < \tau \in Q) \right)$ is faithfully interpretable[12] in every T_τ for each $\tau \succ \rho$, via "syntactical interpretations" $E_{\rho,\tau}$ of T_ρ in T_τ,

---

[12] Expositions of "faithful" syntactical translations and interpretations may be found in Enderton, 1972, 157–163 or Shoenfield, 1967, 61–62.

in the sense that

(3) syntactical translations (or interpretations) $E_{\rho,\tau}$ of formulae φ in T_ρ into counterparts $\varphi^{E_{\rho,\tau}}$ in T_τ exist such that [T_ρ proves φ] if and only if [T_ρ proves $\varphi^{E_{\rho,\tau}}$] in T_τ.

5.6 Remarks

(1) "Unfaithful" syntactical interpretations $E_{\rho,\tau}$ of T_ρ in T_τ— in which the "only if" clause in 5.5 (3) holds, but not the "if"—will have nothing to do with determinism or its denial.

But

(2) the Craig Interpretation Theorem[13] ensures that the $E_{\rho,\tau}$s are faithful whenever the time-indexed languages of the theories T_σs for Cauchy surfaces are mutually disjoint; and

(3) "faithfulness" of such $E_{\rho,\tau}$ ensures that later theories T_τ can interpret earlier events or assertions φ in T_ρ, but are unable to change them ("The past is another country.").

Given the framework just sketched—formulated in first-order set- or class-metatheories—one can formulate the following (mathematically straightforward)

5.6 Proposition

Let $\langle \Theta_\tau, \prec \rangle$ be an arbitrary ramified hierarchy as defined above, whose
(1) intelligible "nodes" T_τ, and their prior linear hierarchies $T_\tau^\cup =_{Df} (\cup_\rho (T_\rho | \rho < \tau \in Q))$ for rationals $\rho < \tau$, axiomatise
(2) the "physical" properties of "closed achronal surfaces[14] S_ρ, construed locally as "slices" of a "cylindrical" normal neighborhood in an ambient spacetime, or globally, as Cauchy surfaces S_ρ in the aforementioned globally hyperbolic spacetime (M, γ) (cf. Wald, 200–204).

Then each "prior" intelligible theory $T_\tau^\cup \subset T_\tau$, "prior" theoretical "agent" $A_\tau^\cup \subset T_\rho^\cup$, and "prior" "state of affairs" $S_\tau^\cup \subset T_\tau^\cup$, which is observable by $A_\tau^\cup$ at time $\tau \in R$ in $T_\tau^\cup$ will have

---

[13] Cf. Chang and Keisler, 1972, 84–87. The role of the theorem in this chapter is to ensure that "causal antecedents" of later events identified for the first time in T_t are entirely describable in theories T_s prior to T_t.

[14] The definition and relativistic properties of "closed achronal surfaces" may be found in Wald, 1984, 200–201.

(3) countably many internally consistent, but mutually inconsistent intelligible extensions $\mathbb{A}_\tau \supset \mathbb{A}_\tau^\cup$, "potential responses" $\mathbb{R}_\tau \subset \mathbf{T}_\tau$, and theoretical "outcomes" $\mathbb{O}_\tau \subset \mathbf{T}_\tau$.

Moreover
(4) Each $\mathbf{T}_\tau$'s "antecedent linear branch" $\mathbf{T}_\tau^\cup =_{Df} (\bigcup_\rho (\mathbf{T}_\rho | \rho < \tau \in Q))$ will be "*cadlag*", but not in general continuous1;

and
(5) (fleeting) "states", "agents", "responses" and "outcomes" $\mathbf{T}_\tau$, $\mathbb{A}_\tau$, $\mathbb{R}_\tau$, and $\mathbb{O}_\tau$ at a given time $\tau \in R$ will *not* in general be definable in their limiting temporal antecedents $\mathbf{T}_\tau^\cup$, $\mathbb{A}_\tau^\cup$ and $\mathbb{S}_\tau^\cup$, much less "determined" by them in $\mathbf{T}_\tau$.

5.7 Proof
All of these conclusions follow more or less directly from Gödel's (thoroughly non-trivial) arguments (cf., e.g., Smorynski, 1977, 827ff.), which require only "intelligibility" of the time-ordered theories involved, since
(i) all the intelligible theories $\mathbf{T}_\tau$ and intelligible subtheories $\mathbb{A}_\tau$, $\mathbb{S}_\tau$, $\mathbb{R}_\tau$, and $\mathbb{O}_\tau$, … . in each branch of the cadlag structure $\langle \Theta_\tau, \prec \rangle$ faithfully interpret their immediate temporal "pasts" $\mathbf{T}_\tau^\cup$, $\mathbb{A}_\tau^\cup$, $\mathbb{S}_\tau^\cup$, $\mathbb{R}_\tau^\cup$ and $\mathbb{O}_\tau^\cup$, … ;
(ii) the intelligibility of every T_τ in $\langle \Theta_\tau, \prec \rangle$ ensures the existence of countably many alternative "intelligible" extensions $\mathbf{T}_\tau^0 =_{Df} \mathbf{T}_\tau^\cup, \mathbf{T}_\tau^1, \mathbf{T}_\tau^2, \ldots$. of each (retrospectively linear) limiting "branch" $\mathbf{T}_\tau^\cup$ (or "antecedent linear hierarchy" $\mathbf{T}_\tau^\cup$), and
(iii) "local" aspects $\mathbb{A}_{\tau_1}$, $\mathbb{S}_{\tau_1}$, $\mathbb{R}_{\tau_1}$, and $\mathbb{O}_{\tau_1}$ of "local" "physical" phenomena $\mathbb{P}_{\tau_1}$ in $\mathbf{T}_{\tau_0} =_{Df} \mathbf{T}_\tau^\cup, \mathbf{T}_\tau^1, \mathbf{T}_\tau^2, \ldots$ of $\mathbf{T}_\tau^\cup$ will also be "conditioned" (but not "determined") by their cadlag antecedents $\mathbb{A}_\tau^\cup$, $\mathbb{S}_\tau^\cup$, $\mathbb{R}_\tau^\cup$ and $\mathbb{O}_\tau^\cup$.

5.8 It is also consistent in this framework (and indeed, I will argue in the next section, overwhelmingly likely) that
(1) agents $\mathbb{A}_{\tau_1}$ may have "free wills" $\mathbb{W}_\tau^\cup$ to initiate (conscious or semiconscious) "actions" $\mathbb{C}_\tau$, which may have origins in "nonstandard" or "indefinite" temporal "margins" between $\mathbf{T}_\tau^\cup$ and $\mathbf{T}_\tau$, and consequences in $\mathbf{T}_\tau$,

though
(2) such agents' volitional "freedoms" $\mathbb{F}_\tau$ may unfortunately be confined to the right to sleep under alternative counterfactual bridges $\mathbb{B}_\tau$, in intelligible but partially aleatory (or at least a priori uncertain) outcomes $\mathbb{O}_\tau$.

6 Mathematically Exiguous Interpretations of Physical "Continuity"

6.1 In this section, I will argue in greater topological and measure-theoretic detail that
(1) physical events in Cauchy surfaces described by intelligible theories T_τ in globally hyperbolic spacetimes

evolve in
(2) temporal "paths" or "branches" which are "almost certainly" ramified rather than linear, and *cadlag* rather than continuous.

More precisely, I will argue that
(3) continuity of the "linear" as well as "ramified" hierarchies defined above is a measure-theoretically "exiguous" (and in that sense an overwhelmingly rare) phenomenon.

To that end, it will be helpful to introduce several purely mathematical definitions and observations, among them a framework of

6.2 "Canonical" Polish Topologies and "Lebesgue" Measures on Temporally Evolving Spaces of Intelligible Theories

Observe that (metatheoretically)
(1) consistent (intelligible) theories T_τ of physically measurable properties of the Cauchy surface at given proper times τ

may be identified with
(2) closed sets $Cl(T_\tau)$ in a totally disconnected compact Stone space $St(T_\tau)$ (which is homeomorphic to the totally disconnected Stone / Cantor space 2^ω).

As is the case of the more familiar closed unit interval $[0, 1]$, moreover,
(3) one can impose a complete separable Fell metric[15] $\eta_{F,\tau}$ on $Cl(T_\tau)$ such that the subset $Cl(T_\tau)$ of all intelligible theories is a countable dense subset of $(Cl(T_\tau), \eta_{F,\tau})$.

A first observation in this framework is that $(Cl(T_\tau), \eta_{F,\tau})$ is more than measure-theoretically isomorphic to the totally disconnected compact separable metric space $(2^\omega, \varsigma_\tau)$.

---

[15] Cf., e.g., Kallenberg, 2001, 565–567.

6.3 Proposition
The space $(Cl(\mathbf{T}_\tau), \eta_{F,\tau})$ is homeomorphic to the Stone / Cantor space $(2^\omega, \varsigma_\tau)$.

Proof

Since $(Cl(\mathbf{T}_\tau), \eta_{F,\tau})$ is known to be a complete separable compact metric space for each τ (cf. Kallenberg, 2001, 565–66), it will suffice to show that it is totally disconnected.

To this end, observe in (again in Kallenberg, 2001, 565–66) that
(i) a countable basis for $(Cl(\mathbf{T}_\tau), \eta_{F,\tau})$ is given by the countably many sets $\{F_\kappa : F_\kappa \cap B_{S_n}^{r,m}\}$ and $\{F_\kappa : F_\kappa \cap \bar{B}_{S_n}^{r,m}\}$, where
(ii) F_κ in $(Cl(\mathbf{T}_\tau), \eta_{F,\tau})$ is the kth clopen base-set (or equivalently, the kth finitely axiomatisable theory) in the Stone-Cantor topology of $(2^\omega, \varsigma_\tau)$; and
(iii) $B_{S_n}^{r,m} = {}_{Df}\{t \in S : \rho(s_n, t) < r\}$ for each $s_n \in S$ and positive rational r is the clopen set corresponding to the mth finitely axiomatisable theory $T_{n,s}^{r,m}$.

Since there are only countably many such clopen basis-set pairs F_κ and $B_{n,s}^{r,m}$, the m-indexed product of them is
(iv) homeomorphic to a clopen base of the totally disconnected product space $(2^\omega)^\omega \approx S^\omega \approx 2^\omega$ (where "$\approx$" here means "is homeomorphic to"), so the space $(Cl(\mathbf{T}_\tau), \eta_{F,\tau})$ must also be homeomorphic to 2^ω.

It will also be necessary to introduce

6.4 "Canonical" Polish Topologies and "Lebesgue" Measures on Spaces of Continuous Linear Hierarchies
At this point one can also introduce
(1) a metric and "Lebesgue" (probability) measure μ_C on the space **C** of continuous "linear hierarchies" of
(2) functions $x(t)$ from $\bar{R} = {}_{Df}[\rho_1, \rho_2]$, into $(Cl(\mathbf{T}_\tau), \eta_{F,\tau})$, where $[\rho_1, \rho_2]$ is an arbitrary closed temporal segment,

defined by setting
(3) $m_C(x, y) = {}_{Df} \sup_t \eta_{F,\tau}[x(t), y(t)]$ (sometimes called the "uniform norm"), and
(4) μ_C on **C** to be the result of a (henceforth fixed) measure-theoretically isomorphic transfer of structure from the Banach space $(\mathbf{C}, m_C)$ to a G_δ subspace of Lebesgue measure on the Stone space 2^ω.[16]

---
[16] cf. Borkar, 1995, 2–3, and for a thorough account of such transfers, Levy, 2003, 234–243.

6.5 "Canonical" Polish Topologies and Lebesgue Measures on Spaces of "Cadlag" Hierarchies

A "canonical" complete separable Prohorov metric[17] m_{Cad} and corresponding Lebesgue (probability) measure μ_{Cad} on the space **Cad** of cadlag linear hierarchies from $\bar{R}$ into $(Cl(\mathbf{T}_\tau, \eta_{F,\tau}))$ may be defined by setting

(1) $m_{Cad}(x, y) =_{Df}$ the infimum of the set of $\varepsilon > 0$ such that both
(2) $\sup_t |x(t) - y(\lambda(t))|$ and $||\lambda|| =_{Df} \sup_{s \neq t} |\ln[(\lambda(t) - \lambda(s))/t - s)]|$ are less than or equal to ε,
(3) where $\lambda = \lambda(t)$ is an element of the set Λ of continuous and strictly increasing "scaling functions" $\lambda(t)$ from $\bar{R}$ into $(Cl(\mathbf{T}_\tau), \eta_{F,\tau})$; and
(4) the notions "$\lambda(t) - \lambda(s)$" and "increasing" make sense with respect to the homeomorphic transfer of structures from the Polish space $(Cl(\mathbf{T}_\tau), \eta_{F,\tau})$ to the Stone space 2^ω.

It is known that $m_{Cad}(x, y)$ defines a complete, compact and separable topology on **Cad**,[18] so one can introduce a quasi-"canonical" "Lebesgue" (probability) *measure*

(5) m_{Cad} on **Cad** as in 6.3(ii) above, once again via a measure-theoretic (but not homeomorphic) transfer of structure from the Polish space (**Cad**, m_{Cad}) to a $G_{\bar{o}}$ subset of the Stone / Cantor space 2^ω (cf. Borkar, 1995, 2–3).

In this framework, finally, one can make mathematical sense of

6.6 Definition

Measure-Theoretically "Exiguous" Sets in Polish Spaces of Cadlag Linear Hierarchies

(1) By a standard argument (cf. Oxtoby, 5), a subset M of the Polish space (**Cad**, m_{Cad}) has μ_{Cad}-measure zero (is a null set with respect to μ_{Cad}), if and only if
(2) for each $\varepsilon > 0$ there is a sequence of open sets O_n exists such that $M \subset \cup_n O_n$ and $\sum_n \mu_p(O_n) < \varepsilon$.

This longish string of definitions and observations makes it possible to formulate the following

---

[17] Cf. Borkar, 1995, 120–125, or Billingsley, 109–116.
[18] Cf. Borkar, 1995, 120–123.

6.7 Theorem

(1) The class **C** of continuous linear hierarchies into $Cl(\mathbf{T}_\tau)$ is dense in the cadlag metric space $(\mathbf{Cad}, m_{\mathbf{Cad}})$.
(2) The class **C** of continuous linear hierarchies into $Cl(\mathbf{T}_\tau)$ is also an $m_{\mathbf{Cad}}$-null-set in the cadlag measure space $(\mathbf{Cad}, \mu_{\mathbf{Cad}})$.

Proof of (1) (cf. Billingsley, 1968, 112, or Parasarathy, 1967, 248).

The simplest intuitive way to see this is to approximate an arbitrary element $x(t)$ of the space $(\mathbf{Cad}, m_{\mathbf{Cad}})$ more and more closely with a sequence of continuous elements $y(\lambda_{\kappa,n}(t))$, of $(\mathbf{C}, m_{\mathbf{C}})$ which

(i) "smooth the corners" of the countably many discrete "jumps"[19] of $x(t)$ at time-parameters t_k,
in the sense that
(ii) the scaling functions $\lambda_{\kappa,n}(t)$s are initially "flat" for "smaller" $t_{k-1} < t_{k,n} < t_k$, but rise later and more steeply as $t_{k,n}$ approaches each temporal "jump point" t_k, where
(iii) "corners", "flatness" and "steepness" make sense with respect to the homeomorphic transfer of structure from the Polish space $(Cl(\mathbf{T}_\tau), \eta_{F,\tau})$ to the Stone / Cantor space 2^ω.

Proof of (2)

Observe first that

(i) $m_{\mathbf{Cad}}(x, y)$ becomes $m_{\mathbf{C}}(x, y)$ when x and y are both continuous and $\lambda(t) = t$, since
(ii) $\|\lambda\| =_{Df} \sup_{s \neq t} |\ln[\lambda(t) - y(\lambda(s)/t - s)]|$ is then uniformly equal to zero, and

$$\sup_t |x(t) - y(\lambda(t))| = \sup_t |x(t) - y(t)|$$ at every temporal instant t.

The rest of the proof is a slight variant of a venerable argument invoked to disjoin the unit interval [0, 1] (or the Stone / Cantor space 2^ω) into a meager set and a null set (cf. again Oxtoby, 1970, 5).

In the case at hand, the schematic constructions in (i)–(ii) and (1) yield elements of y_i in **C** which are collectively countable as well as dense in the larger space **Cad**. One can therefore

(iii) define a doubly indexed collection of open "spheres" I_{ij} in $(\mathbf{Cad}, m_{\mathbf{Cad}})$ with centers y_i and radius 2^{i+j}, and

---

[19] Cf. Parthasarathy, 1967, 232.

(iv) for each integer j form an open set $G_j =_{Df} \bigcup_{i=1}^{\infty} I_{ij}$ in (**Cad**, $m_{\mathbf{Cad}}$) such that

$$\mathbf{C} \subseteq \bigcap_{j=1}^{\infty} G_j = \bigcap_{j=1}^{\infty} \bigcup_{i=1}^{\infty} I_{ij}$$

Then for every ε, however small, one can fix a sufficiently large j_ε so that

(v) $1/2^{j_\varepsilon} < \varepsilon$ and $\mathbf{C} \subseteq \bigcap_{j=1}^{j_\varepsilon} G_j = \bigcap_{j=1}^{j_\varepsilon} \bigcup_{i=1}^{\infty} I_{ij}$, so the right hand side has the property

that $\mu_{Cad}(\mathbf{C}) < \mu_{Cad}\left(\bigcap_{j=1}^{j_\varepsilon} \bigcup_{i=1}^{\infty} I_{ij}\right) \leq 1/2^j < \varepsilon$,

and observe that $\mu_{\mathbf{Cad}}(\mathbf{C})$ must therefore be 0, since ε was arbitrary and $\mathbf{C} \cap G_j$ is a (dense) subspace of every G_j.

Finally, one can draw upon this section's metamathematical arguments to examine two of Baruch Spinoza's "proofs" *more geometrico* in *Ethica*, *Pars I*.

6.7 "Id Quo Incomprehensibilius Nil Comprehendi Possit" ("That Than Which Nothing More Incomprehensible Could be Conceived")

Suppose that one interprets

intelligible temporally indexed theories (or threads), once again, as endlessly ramifying cadlag stages $\mathbf{T}_\tau$ of a quasi-Spinozan "*deus sive natura*",

Then the results of this section yield that

6.8 continuum-many mutually inconsistent theories or "threads" extend every temporal stage $\mathbf{T}_\tau^\cup$ (a conclusion incompatible with Spinoza's monism),

and

6.9 "almost all" such "threads" at every stage of temporal evolution will be metalogically "unintelligible" (a conclusion which is compatible with Spinoza's "*Neoplatonism*").

Proof

These conclusions follow directly from the arguments in 5.2 and 6.1–6.7.

7 The "Mystery" of Essentially Incomplete Inquiry

7.1 Fifty years ago I stumbled onto the following passage in *Religio Medici* (1643) by the essayist and physician Thomas Browne:

> There be not impossibilities enough in Religion for an active faith; [for] the deepest Mysteries ours contains have have not only been *illustrated*, but *maintained*, by Syllogism and the rule of Reason. I love to lose my self in a *mystery*, to pursue my Reason to an *O altitudo*! (*Religio Medici*, Part 1, Section 9, 14; italics mine).

A number of years later, prompted by desire to understand mathematical logic and critique hieratic qualities of assorted forms of "metaphysical monism" (or "realism"), it occurred to me that one might "lose one's selves" rather naturally in (critical as well as agnostic)
(1) "trees of inquiry" or "sentience" (analogues, perhaps, of traditional "trees of life"), marked by endlessly ramifying hierarchies of "intelligible" theories and their semantic interpretations.

This "tree-structure", in turn—influenced by the original senses of the Greek *"physis"* and its Latin cognate *"natura"*— suggested that one might
(2) interpret physical, metaphysical and epistemological conjectures in the language of metamathematics,

and this in turn that
(3) historically delimited flashes of clarity may ramify "forever" through dilemmatic paths of metatheoretic analysis.

7.2 By more or less general consensus, the most striking metalogical result of twentieth-century mathematical logic was
(1) Gödel's refined (*raffiniert*) but not malicious (*boshaft*) insight into the "essential incompleteness" of Peano arithmetic,

which later mathematicians generalised to an observation that
(2) every intelligible mathematical theory in which one can "calculate" has continuum-many mutually inconsistent semantic interpretations,

and therefore
(3) as many interpretations as there are "ramifications" through an infinite branching binary tree, in other words, the Stone / Cantor space 2^{ω}, ς.

7.3 Leibniz, Boscovich, and Laplace, heirs to the monist realism of Plato and Aristotle, were animated by the symmetries and uniformities they had found in (isolated) physical systems to assert that
(1) there "must" be a unique continuous linear hierarchy of conceptually "isolatable" physical events;

for
(2) alternative outcomes of such events "would not exist", since physical events with unique pasts "must" have mathematically predictable and therefore unique futures.

7.4 More abstractly, the aforementioned Leibniz, Boscovich, Laplace and most other researchers in the Platonic and Aristotelian tradition—"fellows of another college", as John Littlewood respectfully put it[20]—took as an unspoken *axioma* the

(1) existence of a unique continuous linear hierarchy of inquiry, a physical analogue of the ancients' belief that "the" highest form of *mathesis* (learning) would be determined by *noesis noeseos* (thought about thought), which Aristotle assimilated to a unique highest form of *theoria*,

and deduced from this *axioma* that

(2) alternative courses of theoretical hierarchies would be otiose, since the dignity of this uniquely empyrean theoria would require that physical, mathematical and philosophical inquiry converge to a single "canonical" (*or "hieratic"*) *limit*.

7.5 Close analogues of these views reappeared later in the nineteenth century in the

(1) unitarity of C. S. Peirce's "limit of inquiry", whose linearity and unicity he never questioned (to the best of my knowledge);

and (much) more dogmatically in the

(2) monism of Gottlieb Frege's assertions about "*das Wahre*" ("the True"), which led him to denounce plurality of mathematical interpretation in passages such as the following:

> Das Wort "*Deutung*" ist zu beanstanden; *denn ein Gedanke, richtig ausgedrückt, lässt für verschiedene Deutungen keinen Raum.*

> The word "*interpretation*" is to be admonished; *for a thought, correctly expressed, leaves no room for different interpretations.* (Frege, 1976, 384/301, italics mine)

(3) and in more explicitly mathematical form, his mistaken repudiation of the pluralities of non-Euclidean geometry:

> Niemand kann zwei Herren dienen. Man kann nicht der Wahrheit dienen und der Unwahrheit. Wenn die euklidische Geometrie *wahr* ist, so ist die nichteuklidische Geometrie *falsch*, und wenn die nichteuklidische *wahr* ist, so ist die euklidische Geometrie *falsch*.

---

20 Cf. Hardy, 81.

> *No one can serve two masters. One cannot serve truth and untruth.* If Euclidean geometry is *true*, non-Euclidean geometry is *false*, and if the non-Euclidean [geometry] is *true*, then Euclidean geometry is *false*. (Frege, 1969, 183, italics mine)

(Notice again the hieratic quality of Frege's invocation of Matthew 6:24 and Luke 16:13 in the second passage cited above, taken from the opening lines of *"Über Euklidische Geometrie"*.)

7.6 David Hilbert, by contrast—influenced perhaps by his careful study of non-Euclidean geometry and Felix Klein's *Erlanger Programm*[21]—may well have been the first mathematician who sought explicitly to conflate mathematical "existence" with logical consistency, in passages such as the following:

(1) Gelingt es ... zu beweisen, dass die dem Begriffe erteilten Merkmale *bei Anwendung einer endlichen Anzahl von logischen Schlüssen* niemals zu einem Widerspruche führen können, so sage ich, dass damit die *mathematische Existenz des Begriffes ... bewiesen worden ist.*

> If one succeeds ... in proving that the characteristics attached to the concept never can lead to a contradiction in the *application of a finite number of logical inferences*[22], I say that the *mathematical existence of the concept ... has thereby been proved.* (Hilbert (931/1935), 300–301); italics mine)

(2) Wenn sich ... willkürlich gesetzten Axiome nicht einander widersprechen *mit sämtlichen* [logischen] *Folgen,* so sind sie *wahr,* so *existieren* die durch die Axiome definierten Dinge. Das ist für mich *das Criterium der Wahrheit und der Existenz.*

> If ... arbitrarily posited axioms do not contradict each other *with all their [logical] consequences,* they are then *true,* the things defined by the axioms *exist.* This is for me *the criterion of truth and existence.* (Hilbert's letter to Frege in (Frege), 1976, 66), italics mine).

Notice also, however, the marginal "indefinition" in Hilbert's implicit recourses (replicated again and again in the philosophy of science), to

(3) hierarchies of metatheoretically (or metaphysically) initial and boundary conditions, in the first quotation's *"niemals ... können"* / "never can", and the second's *"... sämtlichen [metalogischen] Folgen"* / "...all their [metalogical] consequences").

---

[21] Cf. Hilbert, 1903.
[22] Notice that unicity of "finiteness" is exactly what a systematic willingness to take "indefiniteness" and "nonstandardness" calls into question.

7.7 Several decades later, Kurt Gödel and other logicians—respecters like many other mathematicians of Arthur Lovejoy's "monistic pathos"[23]—became aware of an implicit tension in such disclaimers and qualifications: that

(1) conjunctions of Hilbert's "existence"-criterion with Thoralf Skolem's "nonstandard models" and his own incompleteness results would eventuate in a dilemmatic or even polylemmatic "many-existence-theory".

Animated in part by "pluralistic" rather than "monistic" *pathe*, and in part by a search for a skeptical[24] *ataraxia* (tranquility), construed as a kind of overview of Browne's *"altitudo"*, I have tried in this chapter to defend the view that

(2) all forms of "intelligible" *mathesis* (learning) and *skepsis* (inquiry) are indefinitely extendible, and therefore plurally interpretable,

and will devote the rest of this section to an attempt at an analogical effort to vindicate this view.

7.8 I am not a mathematical historian, but I suspect a case could be made that geometry and arithmetic emerged hand over hand in the prehistory of mathematics. Whether or not this is correct, nineteenth-century "fellows of another college" established beyond a reasonable doubt that

(1) vast ranges of "nonstandard" geometries "exist" (a datum that may have influenced Hilbert's conflation of "existence" and "consistency"),

and their successors that

(2) some of these geometries (the aforementioned globally hyperbolic spacetimes, for example) might be more "physical" than others.

7.9 Later in the early and mid-twentieth-century, Skolem, Gödel and other spiritual heirs of Hilbert established beyond a reasonable doubt that

(1) "nonstandard" arithmetics as well as set-theories also "exist", in Hilbert's latitudinarian sense, in any metatheoretical realm in which one could make provisional sense of his *"Widerspruchsfreiheit"* ("freedom from contradiction").

---

[23] Cf. Lovejoy, 1965, 12–13.
[24] The obloquy attached to the word "skepticism" has always seemed to me revealing. *Skepsis*—as previously noted, derived from *skeptomai* ("inquire," "examine," "look carefully," or "(re)consider")—meant something very close to the English word "inquiry", itself derived from the Latin *quaerere* ("to *query* or *question*").

By parity of reasoning, therefore, it would seem to me logically and analogically tenable to conjecture that

(2) "nonstandard" mathematical-physical structures might have "physical" and "metaphysical" relevance, comparable to those of elliptical and hyperbolic geometries, among them the differential geometries of Lorentzian spacetimes.

I have obviously tried to argue in prior sections that such analogies are tenable, and will conclude with a heuristic suggestion that existence-criteria associated with semantic pluralities offer

(3) abstract metaphysical insights into mathematical and metamathematical forms of Shelleyan "intellectual beauty";
(4) concrete interpretations of other "physical" events (witnessed, perhaps, by quantum-theoretic "many-worlds"-theories);
(5) "regulative" interpretations of "agency" and "intentionality" (witnessed by liminal "agents" and their ephemeral "wills"); and
(6) "constitutive" evidence that "almost all" initial and boundary data of antecedent physical systems *condition*, but do not "*determine*", their consequent counterparts.

10 Time-Evolution in Random "Universes"

In this chapter, I endeavor to interpret (or represent)
(i) temporal "direction" (cf. 8.1. 8.4 and 8.5);
(ii) aspects of Bohr's "complementarity" (cf. 5.7 and 8.9);
(iii) "virtually" random "many worlds" semantics (cf. 7.1, 7.10 and 8.3);
(iv) "measured" quantum-theoretic systems in "observational" Hilbert spaces $\mathcal{H}$ (cf. 5.1–5.3, 6.1–6.7 and 6.10);
(v) "collapsed" states as Bochner integrals (or "state-valued expectations") in the "observational" spaces $\mathcal{H}$ (cf. 6.7);
(vi) "*cadlag*" time-evolution of "measured" quantum-theoretic systems as *continuous* time-evolution of counterparts defined in $\mathcal{H}$ (cf. 6.10);
(vii) heuristic physical conjectures that "almost all" time-evolving quantum-theoretic systems are neither "closed" nor "isolated" in $\mathcal{H}$ (cf. 9.1, 9.4 and 9.6); and
(viii) heuristic epistemological conjectures that "the physical" is an essentially incomplete notion, and that the semantic pluralities in (ii)–(iv) reflect this incompleteness (cf. 10.1–10.2).

1 Introduction

1.1 In "Metamathematical Quantum Theory: Random Ultrafilters as "Hidden Variables,"' (Boos, 1996), I interpreted "random" spectral values u of measure algebras determined by maximal extensions of self-adjoint operators as
(i) contextual, non-local and "theory-relative" "hidden variables" in a framework of stochastic "truth" for mathematical-physical assertions of von Neumann quantum theory, in the sense that
(ii) "random" or "virtual" spectral values u obey a theory-relativised functional calculus for Borel functions of a given set-theoretic universe M in "random extensions" $M(u)$ of M.

(The theory of such boolean and measure-algebraic structures, developed in successive waves by Scott, Vopenka, Solovay, Jech, Kunen and many others, may be found in Solovay 1970, 1–39, Jech, 2002, 201–283 and 511–541 and Bell, 2005, 149-153, as well as other more recent articles and monographs.)

1.2 Prompted in part by that paper's referee, I conjectured that (meta)mathematical settings might also exist in which

(i) one could correlate "time"-ordered "truth-values" for such assertions with elements of iterated boolean measure-algebras $\mathbb{U}_{sr}$ (where iteration for measure-algebras must be defined with some care—cf. Jech, 2002, 267–271 and 280–281),

and suggested that mathematical and metamathematical refinements of this framework might furnish precise representations for such heuristic notions as
(ii) time-evolution in the presence of intermittent "measurements",
(iii) "collapses" of "wave-functions" induced by such "measurements", and
(iv) "many worlds"-interpretations of observational processes, construed as "paths" through random (or stochastic) set-theoretic "universes".

"Virtual" widgets in measure-algebraically "random" extensions of set-theoretic universes V may be represented (roughly speaking) by stochastic widgets in V, so these *Ansätze* also suggested
(v) that "time" t might in some sense "be" (or be correlated with) a stochastic process T, which "appears" "*in retrospect*" to be a linear "parameter" with value 1 in measure-algebraic structures $V(u)$ (cf. Section 7).

1.3 Prompted by these heuristic analogies, and by pioneering work of Albeverio and Høegh-Krohn in the seventies, I developed a stochastic representation (Boos, 2007) of Richard Feynman's "sum over histories"-approach to quantum dynamics in which I observed that one can
(i) represent a "T-transformed" image of Feynman's "integrand" as an "iterated stochastic integral" σ with respect to Brownian motion, and
(ii) integrate this functional σ over the corresponding Wiener-space of non-differentiable "paths" to obtain time-evolved states φ_t of a given initial state φ under the influence of a given Hamiltonian $H = H + V$;

for a physically relevant class of potentials V.
 I also conjectured
(iii) that a class of "*cadlag*" (or "*rcll*"—"right-continuous with left limits") square-integrable martingales might represent processes of time-ordered "measurement" and "observation" (cf. 2.1(iv));
(iv) that continuous summands (in a well-studied decomposition) would represent "unmeasured" time-evolution of the "isolated" quantum systems considered in (vii) and (viii) above; and
(v) that their discrete summands (which are essentially generalisations of Poisson processes) might represent random temporal processes of "virtual" measurement of such systems.

1.4 In this chapter, I will attempt to bring together these heuristic conjectures in a more rigorous theory of

(i) càdlàg "time"-ordered "measurement"-processes, which "collapse" wavefunctions in branching "virtual" "worlds", interpreted as iterated extensions of "initial" set-theoretic universes,

and show that such a theory provides a common (meta)mathematical rationale for three recurrent but problematic notions in the foundations of quantum mechanics:

(ii) Feynman's heuristic conjectures about "sums over histories";

(iii) Schrödinger's ill-understood "*Schnitte*", or "collapse(s) of the wavefunction"; and

(iv) "many worlds" theories devised to reconcile discontinuous "measurements" with continuous quantum time-evolution.

1.5 A common thread in all these arguments will be a theory of integration over auxiliary spaces of "paths" or random "processes".

In 1.4(ii), for example—as construed in (Boos, 2007)—the integration is over "paths" associated with continuous random processes defined on (segments of) the (extended) real line with values in various sorts of "momentum-spaces" (the "continuous summands" of 1.3(iv)).

In 1.4(iii), the temporally continuous integrals of 1.4(ii) are randomly "perturbed" by Bochner integration (cf. 6.7) with respect to discrete point-processes of "martingale measures" over spectra, in accordance with a martingale-decomposition theorem (the "discrete summands" of 1.3(v)).

In 1.4(iv), the stochastic aspects of the constructions in 1.4(ii) and 1.4(iii) are transferred into a measure-algebraic semantics for interpretation of axiomatic set theories, developed by Solovay to prove a wide range of set-theoretic independent-results, and employed here to interpret 1.4(iv)'s "many worlds" theories.

1.6 In accordance with Everett's and Wheeler's original *Ansatz* (Everett and Wheeler, 1957), finally:

(i) the random integrals which implement 1.4(iii) and 1.4(iv) "appear" "almost surely" (with measure 1) to be "ordinary" integrals in the "virtual" ("graduated", "distributive") semantics of such "universes".

Put somewhat differently:

(ii) it is a "sure thing" (in such random "universes") that "*Einsteins "Alter" nicht zockt*" ("Einstein's 'old man' does not gamble") (cf. 6.16).

2 Four Heuristic Premises

2.1 In what follows, I will work with four semiformal premises.
(i) The first is that one cannot give a mathematical account of quantum-theoretic measurement without (tacit or explicit) recourse to metamathematical ideas. For one cannot, I would argue, "measure" what one cannot—in some potentially formal sense—"define".
(ii) The second is that outcomes of measurements ramify in time, in the sense that they range, "in prospect", over "temporally" parametrised "scales" of "values". (In philosophical terms, they "support counterfactuals".) But each course of such "values" is "retrospectively" linear.
(iii) The third is that something temporally "conditions" those values, though it does not "determine" them. I will attempt to interpret this conjecture mathematically as a stipulation that probability-measures on "scales" of such values "exist" at each temporal "instant", and stipulate that they evolve in ways which permit epistemic access, and may be "sampled".
(iv) The final premise is that these probability measures evolve temporally in a "cadlag" fashion with respect to a complete metric topology on the space of probability-measures. (cf., e.g., (Borkar, 1995, 19–30)); "cadlag", once again, is short for "*continue à droit, [avec] limites à gauche*" ("continuous on the right, [with] limits on the left").

(Suppositions of continuity on the right reflect intuitions that "what just happened" is retrospectively well-defined. Suppositions of limit-existence on the left reflect assumptions that "instantaneous" "systems" to be "measured" should have well-defined antecedents.)

The following sections offer "stochastic" and "metalogical" accounts of "measurement", alternative "worlds" and "wave-function-collapse" as formal realisations of these informal premises.

3 "Linguistic" and "Theoretical" Processes

3.1 Linguistic Processes

A ("universal") linguistic process is a "temporal" structure of the form
(i) $\langle L_r \rangle = \langle L_r | -\infty \leq r \leq \infty \rangle$, where the subscripts range over rational numbers $r \in Q$ and points at "$\pm \infty$", such that
(ii) $L_{-\infty}$ is the first-order language of ZFC and each L_r for $-\infty \leq r \leq \infty$ is a first-order language with equality (cf., e.g., Shoenfield, 14ff.);

(iii) $L_{-\infty} = L_r \cap L_s$ for all pairs (r, s) such that $-\infty \leq r < s \leq \infty$ (i.e., the "non-logical" vocabularies of L_r and L_s are disjoint except for "=" and "∈");
(iv) auxiliary "cumulative" languages $L_{\prec r}$ and $L_{\leq r}$ for $-\infty < r \leq \infty$ are defined by setting $L_{\prec r} = {}_{Df} \bigcup_{s < r} L_s$ and $L_{\leq r} = {}_{Df} \bigcup_{s \leq r} L_s - \infty \leq r \leq \infty$; and
(v) each L_r for $-\infty < r \leq \infty$ has countably many distinct m-ary predicate- and function-symbols for each integer m.

Such a linguistic process $\langle L_r | -\infty \leq r \leq \infty \rangle$ is standard if
(vi) a fixed recursive "Gödel-coding" scheme Γ of the logical and nonlogical symbols of the languages L_r for $r \in Q$ uniformly enumerates the sentences of each L_r in strictly increasing order.

3.2 Remarks

(i) It will be obvious in what follows that the "extended" interval $\{r| -\infty \leq r \leq \infty\}$ can be replaced by an arbitrary closed interval $\{r| c \leq r \leq d\}$ for c and d in R.
(ii) The process $\langle L_r | -\infty \leq r \leq \infty \rangle$ is "universal" in the sense that *every* Q-indexed process of countable first-order languages can be represented in its "limit" $L_{\leq \infty}$.

3.3 Theoretical Processes

A ("universal") theoretical process is a "temporal" structure of the form $\langle L_r, T_r \rangle = \langle (L_r, T_r) | -\infty \leq r \leq \infty \rangle$ such that
(i) $\langle L_r \rangle = \langle L_r | -\infty \leq r \leq \infty \rangle$ is a ("universal") cumulative linguistic process;
(ii) $T_{-\infty}$ is first-order Zermelo-Fraenkel set theory (cf. Shoenfield, 1967, 2392–40);
(iii) $T_{\leq \infty}$ is a consistent, essentially incomplete first-order theory in the language $L_{\leq \infty}$ (cf. 4.4(iii)); and
(iv) the theories T_r ($T_{< r}$, $T_{\leq r}$) for $-\infty \leq r < \infty$ are the sets of sentences in L_r ($L_{< r}$, $L_{\leq r}$) which are provable in the consistent "cumulative" *theory* $T_{\leq \infty}$.

3.4 Three immediate observations

(i) the theories $T_{\prec r}$ and $T_{\prec \infty, n}$ may well be proper subtheories of $T_{\leq r}$ and $T_{\leq \infty}$; and
(ii) each theoretical process $\langle L_r, T_r \rangle = \langle (L_r, T_r) | -\infty \leq r \leq \infty \rangle$ is uniquely determined by its "limiting" theory $T_{\leq \infty}$; and
(iii) the "limiting" theory $T_{\leq \infty}$ is a conservative extension of each T_r, $T_{\prec r}$ and $T_{\leq r}$ determined by the Q-indexed process $\langle T_r | -\infty \leq r \leq \infty \rangle$ (cf. e.g., Shoenfield,

1967, 61–65), i.e., every φ provable in $T_{\leq\infty}$ is provable in every "initial" theory $T_{\leq r}$ such that $\varphi \in L_{\leq r}$.

Well-known topological, measure-theoretic and metamathematical arguments also yield the following results for each such "theoretical process" $\langle T_r \rangle = \langle T_{\leq r} | -\infty \leq r \leq \infty \rangle$.

3.5 Lemma

(The "$_{(...)}$"s below range over the subscripts introduced earlier.)

If $\langle L_r | -\infty \leq r \leq \infty \rangle$ is standard, the Cantor-space $\Omega =_{Df} 2^\omega$ is "canonically" homeomorphic, for each $T_{\leq\infty}$ (via countable bases definable from the "coding" scheme Γ), to

(i) the Stone spaces $S_{\prec r} =_{Df} St(T_{\prec r}) (S_{\leq r} =_{Df} St(T_{\leq r}))$ of the cumulative theories $T_{\prec r}(T_{\leq r})$,
(cf., e.g., Bell and Machover, 1977, 143, 193, 203, and Levy, 2003, 253),

and these in turn to

(ii) the product-spaces $C_{\prec r} =_{Df} 2^{\omega \times Q_{<r}}$ $(C_{\leq r} =_{Df} 2^{\omega \times Q_{\leq r}})$, where $Q_{<r}(Q_{\leq r})$ is $\{s | s < r\}$ ($\{s | s \leq r\}$).

Moreover,

(iii) the Lindenbaum algebra $\mathbf{L}_{T(...)}$ of each theory $T_{(...)}$ may be "canonically" identified with the clopen algebra $\mathbf{C}_{(...)}$ of $C_{(...)}$ (or $2^{\omega \times Q_{(...)}}$, or 2^ω);

(iv) the Dedekind-MacNeille completion of each $\mathbf{L}_{T(...)T(...)}$ may be "canonically" identified with the regular open algebra of $\mathbf{R}_{T(...)}$ of $C_{(...)}$ (or $2^{\omega \times Q_{(...)}}$, or 2^ω)
(cf., e.g., Jech, 2002, 82–83);

and finally,

(v) every probability-measure μ on 2^ω yields a corresponding probability measure algebraic completion $\mathbf{M}_{(...)}$ of the clopen algebra of $C_{(...)}$ (or $2^{\omega \times Q_{(...)}}$, or 2^ω), and corresponding measure-algebras $\mathbf{M}_{T(...)}$ on the Stone space $St(T_{(...)})$ for each theory $T_{(...)}$.

By virtue of Marshall Stone's original analysis of the spaces named after him, one also has for each $r \in [-\infty, \infty]$

(vi) a continuous surjective function $s_{\infty, r}$ from $C_{\leq\infty}$ onto $C_{\leq r}$, such that

(vii) $s_{\infty, r}^{-1}$ implements the obvious embedding of the clopen algebra $\mathbf{C}_{\leq r}$ into its larger counterpart $\mathbf{C}_{\leq\infty}$, and completely embeds the regular-open and measure-algebras $\mathbf{R}_{\leq r}$ and $\mathbf{M}_{\leq r}$ into their larger homologues $\mathbf{R}_{\leq s}$ and $\mathbf{M}_{\leq s}$ for $r < s \leq \infty$.

In the case of the space(s) of theories $T_{\leq r}$, $T_{\leq s}$ is a conservative extension of $T_{\leq r}$ for $r < s$, as mentioned earlier, and each regular open completion $\mathbf{R}_{\leq r}$ of the clopen algebra $\mathbf{B}_{\leq r}$ on the Stone space $St(T_{\leq r})$ is therefore a complete subalgebra of $\mathbf{R}_{\leq \infty}$ on $St(T_{\leq \infty})$ for every $T_{\leq \infty} \in T_{\leq \infty}$. One can therefore define $s_{\infty, r}(\eta)$ for an ultrafilter $\eta = \eta_{\leq \infty} \in St(T_{\leq \infty})$ to be the restriction of η to its restricted counterpart $\eta_{\leq r}$ on $\mathbf{R}_{\leq \infty}$.

4 "Canonical" Measures and Topologies

4.1 The aim of this section will be to define (in an implicit metatheory T_Ω)
(i) a "canonical" Polish topology and "Lebesgue" probability measure $\pi_{\leq \infty}$ on the space $\mathcal{P}_{\leq \infty}$ of probability measures $\rho_{\leq \infty}$ on $2^{\omega \times Q_{\leq \infty}}$.
(ii) "canonical" Polish topologies and "Lebesgue" measures $\tau_{\leq \infty}$ ($\sigma_{\leq \infty}$) on the spaces $T_{\leq \infty}$ ($St(T_{\leq \infty})$) of essentially incomplete theories $T_{\leq \infty}$ in the "universal" language $L_{\leq \infty}$ (Stone spaces $St(T_{\leq \infty})$) of such theories.

4.2 A "canonical" Polish topology and "Lebesgue" measure $\pi_{\leq \infty}$ on the space $\mathcal{P}_{\leq \infty}$ can be defined as follows.

Prohorov defined a "canonical" complete separable metric on the space $\mathcal{P}_{\leq \infty}$ of probability measures on $\Omega =_{Df} 2^{\omega}$ (or $2^{\omega \times Q_{\leq \infty}}$) from a homeomorphism of $\mathcal{P}_{\leq \infty}$ onto a G_δ-subset of the product space $([0, 1])^\omega$, and a similar argument works for the product space $\Omega =_{Df} 2^{\omega}$ (cf. Borkar, 1995, 2–3, 19–20 and 26).

From a (definable) "natural" basis for Prohorov's topology, one can therefore define
(i) a "canonical" Borel isomorphism of 2^ω onto $\mathcal{P}_{\leq \infty}$, (cf. Takesaki, 1979, 376), and
(ii) a "canonical" "Lebesgue" probability measure $\pi_{\leq \infty}$ on $\mathcal{P}_{\leq \infty}$ as the image with respect to this bijection of "Lebesgue" measure on $2^{\omega \times Q_{\leq \infty}}$.

4.3 Turning to $T_{\leq \infty}$ and $\tau_{\leq \infty}$ (respectively $St(T_{\leq \infty})$ and $\sigma_{\leq \infty}$) in 4.1(ii) above, one can observe that
(i) well-studied results about countable products of Cantor spaces, and (for later use)
(ii) the fixed recursive "Gödel-coding" scheme Γ and uniform enumeration of the sentences of each L_r in 3.1(vi)

define
(iii) unique "Lebesgue" probability-measures $\lambda_{\leq \infty}$ on $C_{\leq \infty} =_{Df} 2^{\omega \times Q_{\leq \infty}}$ and $\lambda_{\leq r}$ on $C_{\leq r} =_{Df} 2^{\omega \times Q_{\leq r}}$ for each r such that $-\infty \leq r \leq \infty$.

One can define therefore "canonical" Polish topologies $\tau_{\leq\infty}$ ($\sigma^T_{\leq\infty}$) on the spaces $T_{\leq\infty}$ ($St(T_{\leq\infty})$) with the aid of the following

4.4 Lemma
Let
(i) $E_{\leq\infty}$ be the set of consistent extensions of the theory $T_{-\infty} = $ ZFC in $L_{\leq\infty}$;
(ii) $\tau_{\leq\infty}$ the *Michael-Vietoris* topology on the space $E_{\leq\infty}$ of such extensions (construed as closed subsets of $St(T_{-\infty})$; (cf., e.g., Kuratowski, 1966, 160–172); and
(iii) $T_{\leq\infty} \subseteq E_{\leq\infty}$ the subset of such consistent theories T which are essentially incomplete (i.e., no finite extension of a theory T in $T_{\leq\infty}$ decides every sentence in $L_{\leq\infty}$).

Then one can
(iv) fix a "canonical" sequence of $(G_n | n \in \omega)$ of open subsets of $E_{\leq\infty}$ such that $\bigcap_{n\in\omega} G_n = T_{\leq\infty}$;
(v) define from such a sequence a "Lebesgue" probability measure $\tau_{\leq\infty}$ on the Polish space $T_{\leq\infty}$ of essentially incomplete extensions of $T_{-\infty} = $ ZFC in the language $L_{\leq\infty}$,

and
(vi) define a corresponding "Lebesgue" probability measure $\sigma^T_{\leq\infty}$ on the Polish space $St(T_{\leq\infty})$ of Stone spaces $St(T_{\leq\infty})$ for $T_{\leq\infty} \in T_{\leq\infty}$, via "transfer of structure" from $T_{\leq\infty}$ to $St(T_{\leq\infty})$.

Proof.
It will suffice to verify (v), for one can "canonically" define
(a) a Polish topology τ_I on $T_{\leq\infty}$ from such a $(G_n | n \in \omega)$ (cf., e. g., Takesaki, 1979, 396).

For then
(b) a "Lebesgue" probability measure $\tau_{\leq\infty}$ on $T_{\leq\infty}$ can be defined as the image of "Lebesgue" measure on 2^ω with respect to the aforementioned definable Borel isomorphism from 2^ω to $T_{\leq\infty}$ (cf., e. g., Borkar, 1995, 16 and 26–30).

One can define (iv)'s sequence $(G_n | n \in \omega)$ such that $\bigcap_{n\in\omega} G_n = T_{\leq\infty}$ as follows. Given a particular theory T in $T_{\leq\infty}$, let
(c) $\langle \varphi_{T,i} | i \in \omega \rangle$ enumerate the sentences in $L_{\leq\infty}$ which T decides, and
(d) $\langle \psi_{T,i} | i \in \omega \rangle$ enumerate the sentences in $L_{\leq\infty}$ which T does not decide.

G_n may then be defined as the union

(e) $\bigcup_{T \in \mathcal{T}_{\leq \infty}} G_{T,n} =_{Df} \bigcup_{T \in \mathcal{T}_{\leq \infty}} (U_{T,n} \cap V_{T,n})$, where
(f) $U_{T,n}$ is the open set of theories $\bar{S}$ in $E_{\leq \infty}$ which decide the first n sentences in the list $\langle \varphi_{T,i} | i \in \omega \rangle$ in accordance with T; and
(g) $V_{T,n}$ is the open set of theories $\bar{\ }S$ in $E_{\leq \infty}$ which decide only finite collections of sentences which appear in the nth terminal segment $\langle \psi_{T,i} | i \geq n \rangle$ of the list $\langle \psi_{T,i} | i \in \omega \rangle$.

Then the intersection $\bigcap_{n \in \omega} G_{T,n}$ closes down on $\{T\}$ for each T in $\mathcal{T}_{\leq \infty} \subseteq E_{\leq \infty}$, so the intersection $\bigcap_{n \in \omega} \bigcup_{T \in I_{\infty}} G_{T,n}$ closes down on the set $\mathcal{T}_{\leq \infty}$.

4.5 Remarks

(i) The foregoing arguments can obviously be modified to work for appropriately defined "initial" Polish spaces $\mathcal{T}_{\leq r}(St(\mathcal{T}_{\leq \infty}))$ of essentially incomplete theories $\mathcal{T}_{\leq r}$ (Stone spaces of such $\mathcal{T}_{\leq r}$) in the language $L_{\leq r}$ for each r such that $-\infty \leq r \leq \infty$.
(ii) In keeping with the remarks about measurement and metatheoretic definability in Section 2, one might also consider "discrete" measures on
(iii) the countable class $\mathcal{R}_{\leq \infty}$ of consistent first-order theories $\mathcal{T}_{\leq \infty}$ in the language $L_{\leq \infty}$ which are recursively axiomatisable; or
(iv) the larger countable class $\mathcal{D}_{\leq \infty}$ of consistent essentially incomplete first-order theories in $L_{\leq \infty}$ which are pointwise definable in the shadow metatheory T_Ω,

"weighted" in each case by countably supported measures which are singular with respect to $\tau_{\leq \infty}$.

(In this case one would also have to keep in mind that $\mathcal{D}_{\leq \infty}$ itself is not definable in T_Ω, only in a "stronger" metametatheory $\bar{T}_\Omega$ for T_Ω.)

Still more generally, one could examine

(v) "unphysical" continuous measures $\sigma^s_{\leq \infty}$ on $\mathcal{T}_{\leq \infty}$ which are singular with respect to the "Lebesgue" measure $\tau_{\leq \infty}$.

I will consider briefly some of the interpretive possibilities of the constructions in (v) in Section 8.

5 "Observational Measures" and "Observational Frames"

5.1 Definition(s)

(i) An observational measure on $\mathcal{U}_{\leq \infty} =_{Df} \mathcal{T}_{\leq \infty} \times \mathcal{P}_{\leq \infty}$ is a $(\tau_{\leq \infty} \otimes \pi_{\leq \infty})$-absolutely continuous probability measure $v_{\leq \infty}$ on $\mathcal{U}_{\leq \infty}$,

where
(ii) $(\mathcal{T}_{\leq\infty}, \tau_{\leq\infty})$ and $(\mathcal{P}_{\leq\infty}, \pi_{\leq\infty})$ are the "canonical" Lebesgue measure spaces of essentially complete theories and probability measures on $C_{\leq\infty} =_{Df} 2^{\omega \times Q_{\leq\infty}}$ defined in prior sections.

The aim of this definition is to
(iii) represent "couplings" of theories $T = T_{\leq\infty} \in \mathcal{T}_{\leq\infty}$ with probability measure spaces $\rho = \rho_{\leq\infty} \in \mathcal{P}_{\leq\infty}$,

where such measure spaces represent
(iv) (continuous) "scales" or "gauges" $\mathcal{A}_{\leq\infty}$ of (equivalence-classes of) "measurement-apparatus" $A = A_{\leq\infty} \in \mathcal{A}_{\leq\infty}$ which may be "read" in accordance with "instructions" formulated in the theories $T = T_{\leq\infty} \in \mathcal{T}_{\leq\infty}$.

In the sequel, let
(v) $\theta = \theta_{\leq\infty} \in \mathcal{T}_{\leq\infty}$ be an alternative notation for elements of $\mathcal{T}_{\leq\infty}$,
(vi) $\rho = \rho_{\leq\infty} \in \mathcal{P}_{\leq\infty}$ be an alternative notation for elements of $\mathcal{P}_{\leq\infty}$,

and
(vii) $\tilde{\upsilon}_{\leq\infty}$ the $\tau_{\leq\infty}$-absolutely continuous "projection" of the measure $\upsilon_{\leq\infty}$ on $\mathcal{T}_{\leq\infty}$.

5.2 Definition
The ("theory-based") "observational frame" associated with the observational measure $\upsilon_{\leq\infty}$ (cf. 5.6) is
(i) the (complex) Hilbert space $L^2(\mathcal{U}_{\leq\infty}, \upsilon_{\leq,\theta})$, or equivalently (varying the proof of the Fubini theorem), the direct-integral Hilbert space $\mathcal{K}_{\leq\infty} =_{Df}$
$$\int^{\oplus}_{\theta_{\leq\infty} \in (\mathcal{T}_{\leq\infty}, \tilde{\upsilon}_{\leq\infty})} L^2(\mathcal{P}_{\leq\infty}, \mathsf{v}_{\leq,\theta});$$

where
(ii) $\mathsf{v}_{\leq\infty,\theta} = \mathsf{v}_{\leq\infty}(\theta_{\leq\infty}, \cdot)$ is the $\tilde{\upsilon}_{\leq\infty}$-random probability measure on $\mathcal{P}_{\leq\infty}$ such that $\upsilon_{\leq\infty}$ on $\mathcal{U}_{\leq\infty} =_{Df} \mathcal{T}_{\leq\infty} \times \mathcal{P}_{\leq\infty}$ is the "disintegrational" product of $\tilde{\upsilon}_{\leq\infty}$ and the random measure $\mathsf{v}_{\leq\infty,\theta}$ (cf., e.g., Borkar, 1995, 41).

5.3 As before, each observational measure $\upsilon_{\leq\infty}$, projection $\tilde{\upsilon}_{\leq\infty}$, and $\tilde{\upsilon}_{\leq\infty}$-random measure $\mathsf{v}_{\leq\infty}(\theta_{\leq\infty}, \cdot)$ also determines, for rational numbers r in R,
(i) initial measures $\upsilon_{\leq r}$ on $\mathcal{U}_{\leq r}$ and $\tilde{\upsilon}_{\leq r}$ on $\mathcal{T}_{\leq r}$,
(ii) $\tilde{\upsilon}_{\leq r}$-random initial measures $\mathsf{v}_{\leq r}(\theta_{\leq r}, \cdot)$ on $\mathcal{P}_{\leq r}$, and
(iii) initial observational frames $L^2(\mathcal{U}_{\leq r}, \upsilon_{\leq,r}) = \int^{\oplus}_{\theta_{\leq r} \in (\mathcal{T}_{\leq\infty}, \tilde{\upsilon}_{\leq\infty})} L^2(\mathcal{P}_{\leq r}, \mathsf{v}_{\leq r}(\theta_{\leq r}))$,

as well as

(iv) "initial" complete measure subalgebras $\mathbf{U}_{\leq r}$ of the "limit" measure algebra $\mathbf{U}_{\leq \infty}$ determined by the measure space $(\mathcal{U}_{\leq \infty}, \nu_{s,\theta})$ (about which more in Section 7).

At this point, one can formulate the cardinal definition of this section.

5.4 Definition

The measurement process or measurement martingale $\dot{\mathcal{M}}_r = \dot{\mathcal{M}}_r(u) = \dot{\mathcal{M}}_r(u_{\leq \infty}) = \dot{\mathcal{M}}_r(\theta_{\leq \infty}, \rho_{\leq \infty})$ determined by $(\mathcal{U}_{\leq \infty}, \nu_{s,\theta})$ is

(i) the uniformly integrable martingale $\dot{\mathcal{M}}_r(\theta_{\leq \infty}, \rho_{\leq \infty})$ of conditionalisations of the square root of the Radon-Nikodym derivative $\dot{\mathcal{M}}_\infty(\theta_{\leq \infty}, \rho_{\leq \infty})$ of $\nu_{\leq \infty}$ with respect to the Lebesgue measure $(\tau_{\leq \infty} \otimes \pi_{\leq \infty})$ for rational numbers r in R (cf. Kallenberg, 2001, 67–69 and 109).

Notice first that

(ii) the martingale $\dot{\mathcal{M}}_r = \dot{\mathcal{M}}_r(u) = \dot{\mathcal{M}}_r(\theta, \rho) = \dot{\mathcal{M}}_r(\theta_{\leq \infty}, \rho_{\leq \infty})$ is square-integrable in the sense of (Ikeda and Watanabe, 1981, 47), since the square root of $\mathcal{M}_\infty$ is square-integrable in the usual sense as a function from the measure space $(\mathcal{U}_{\leq \infty}, \nu_{\leq \infty})$ into $R \subseteq C$.

Translated into the present context, theorems of Doob and other pioneers of stochastic analysis also ensure that

(iii) every uniformly square-integrable martingale $\dot{\mathcal{M}}_r = \dot{\mathcal{M}}_r(u)$, defined above for rational r such that $-\infty < r < \infty$, has a $\lambda_{\leq \infty}$-almost surely unique cadlag extension $\dot{\mathcal{M}}_r = \dot{\mathcal{M}}_r(u)$ to the entire extended real line $\bar{R} = [-\infty, +\infty]$ (cf., e. g., Kallenberg, 2001, 130 and 134–135);

that

(iv) the processes of $(\tau_{\leq \infty} \otimes \pi_{\leq \infty})$-absolutely continuous probability measures $\nu_{\leq r}$ and their corresponding $\dot{\mathcal{M}}_r = \dot{\mathcal{M}}_r(u)$ for rational r such that $-\infty \leq r < \infty$, may be extended to unique cadlag versions $\nu_{\leq r}$ and $\dot{\mathcal{M}}_r = \dot{\mathcal{M}}_r(u)$ for all $r \in R$ (cf. Borkar, 1995, 26–30);

and that

(v) every such uniformly integrable martingale $\dot{\mathcal{M}}_r = \dot{\mathcal{M}}_r(u)$ of $(\tau_{\leq \infty} \otimes \pi_{\leq \infty})$-absolutely continuous probability measures $\nu_{\leq r}$ for r such that $-\infty \leq r < \infty$ is determined by such a $(\tau_{\leq \infty} \otimes \pi_{\leq \infty})$-absolutely continuous random measure $\nu_{\leq \infty}$ on $\mathcal{U}_{\leq \infty, \theta}$ (cf. Kallenberg, 2001, 129).

5.5 Remark

Recall that the original definition of an "observational measure" $\nu_{\leq \infty}$ on $\mathcal{U}_{\leq \infty} =_{Df} \mathcal{T}_{\leq \infty} \times \mathcal{P}_{\leq \infty}$ in 5.1 above was symmetrical with respect to interchanges of elements $\rho_{\leq\infty}$ of $\mathcal{P}_{\leq\infty}$ and $\tau_{\leq\infty}$ of $\mathcal{T}_{\leq\infty}$. So also, therefore, are the definitions and results which followed (*mutatis mutandis*), in the sense that one can formulate from the measure $\upsilon_{\leq\infty}$ on $\mathcal{U}_{\leq\infty}$ the following alternative

5.6 Definition
A (measure-based) "observational frame" associated with the observational measure $\upsilon_{\leq\infty}$ (cf. 5.2) is
(i) the (complex) Hilbert space $L^2(\mathcal{U}_{\leq\infty}, \upsilon_{\leq,\theta})$, or equivalently, the direct-integral Hilbert space $\mathcal{K}_{\leq\infty} =_{Df} \int^{\oplus}_{\rho_{\leq\infty} \in \mathcal{P}_{\leq\infty}} L^2(\mathcal{T}_{\leq\infty}, \mu_{\leq\infty,\rho})$;

where
(ii) $\underline{\upsilon}_{\leq\infty}$ is the $\pi_{\leq\infty}$-absolutely continuous "projection" of the measure $\upsilon_{\leq\infty}$ on the space of probability measures $\mathcal{P}_{\leq\infty}$, and
(iii) $\mu_{\leq\infty,\rho} = \mu_{\leq\infty}(\rho_{\leq\infty}, \cdot)$ is the $\underline{\upsilon}_{\leq\infty}$-random probability measure on $\mathcal{P}_{\leq\infty}$ such that $\upsilon_{\leq\infty}$ on $\mathcal{U}_{\leq\infty} =_{Df} \mathcal{T}_{\leq\infty} \times \mathcal{P}_{\leq\infty}$ is the "disintegrational" *product* of $\underline{\upsilon}_{\leq\infty}$ and the random measure $\mu_{\leq\infty,\rho} = \mu_{\leq\infty}(\rho_{\leq\infty}, \cdot)$ (cf., e.g., Borkar, 1995, 41).

5.7 Remark
One can make a case that the symmetries of such "frames" offer straightforward interpretations of some of Niels Bohr's suggestive but elusive heuristic remarks about "complementarity" (cf. also 8.9).

For in (quasi-) Bohrian language, one might well construe such "disintegration-analyses" as reflections of a pattern of inextricable "duality" (or "complementarity"), in which
(i) neither the "theoretical systems" $\mathcal{T}_{\leq\infty}$ nor the "scales" $\mathcal{P}_{\leq\infty}$ and "apparatus" $\mathcal{A}_{\leq\infty}$ they "observe" have any intrinsic conceptual "priority",

but
(ii) "composite" probability measures $\upsilon_{\leq\infty}$ on $\mathcal{U}_{\leq\infty}$ $\mathcal{U}_{\leq\infty} =_{Df} \mathcal{T}_{\leq\infty} \times \mathcal{P}_{\leq\infty}$ are "complementary" products of the "projections" $\tilde{\upsilon}_{\leq\infty}$ of $\upsilon_{\leq\infty}$ on $\mathcal{T}_{\leq\infty}$ (respectively, $\underline{\upsilon}_{\leq\infty}$ on $\mathcal{P}_{\leq\infty}$) with a $\tilde{\upsilon}_{\leq\infty}$-random measures $\nu_{\leq,\theta}$ on $\mathcal{P}_{\leq\infty}$ (respectively, $\underline{\upsilon}_{\leq\infty}$-random measures $\mu_{\leq\infty,\rho}$ on $\mathcal{T}_{\leq\infty}$).

Initial heuristic attempts to describe such notions in "ordinary language" might well seem equivocal, but so, arguably, is the notion of "ordinary language". It has long been a regulative ideal of mathematical inquiry to clarify "equivocal" notions with the aid of precise structural definitions.

Consider, for example, the ambiguity (or "absurdity") of "infinite" charges which "concentrate" at "infinitesimal" points.

6 A Continuous Representation of Wave-Function "Collapse"

My aim in this section will be

6.1 to "embed" "wave-function collapse" of "measured time-evolution" in the "observational frame" $L^2(\mathcal{U}_{\leq\infty}, \upsilon_{\leq\infty}) = \int^{\oplus}_{\tilde{\upsilon}_{\leq\infty} \in (\mathcal{T}_{\leq\infty}, \tilde{\upsilon}_{\leq\infty})} L^2(\mathcal{P}_{\leq\infty}, \nu_{\leq\infty,\theta}) = \int^{\oplus}_{\rho_{\leq\infty} \in \mathcal{P}_{\leq\infty}} L^2(\mathcal{T}_{\leq\infty}, \mu_{\leq\infty,\rho})$

via a construction which
(i) "absorbs" discontinuities associated with "measurements" expressible in terms of the $(\tau_{\leq r} \otimes \pi_{\leq r})$-absolutely continuous "observational measures" $\upsilon_{\leq r}$ at temporal stages r,

where (cf., e. g., Kallenberg, 2001, 492, 499)
(ii) the "measurement process" (or martingale) $\dot{\mathcal{M}}_r = \dot{\mathcal{M}}_r(u) = \dot{\mathcal{M}}_r(u_{\leq r})$ "canonically" associated with the process of measures $\upsilon_{\leq r}$ may have "accessible jumps" (cf. 6.4(i)).

6.2 More precisely, I will
(i) "embed" wave-function collapse in continuous time-evolution defined on such "frames",

and
(ii) outline two syntactically distinct but semantically related ways to interpret "measured" time-evolution of a classical n-dimensional quantum system:
(a) "metalogically", in a "virtual" many worlds theory (considered in more detail in later sections), and
(b) "stochastically", in ways which apply the apparatus of stochastic analysis to the martingale $\mathcal{M}_r = \mathcal{M}_r(u)$ (considered in the present section).

To that end, consider
(iii) an arbitrary self-adjoint quantum-theoretic Hamiltonian $H = H_0 + V$ on $L^2(R^d)$, and in particular
(iv) a Hamiltonian $H = H_0 + V$ of the sort considered in (Boos, 2007), such that V is polynomially bounded and $H = H_0 + V$ is essentially self-adjoint on the Schwartz space $\mathcal{S}(R^d)$.

A suggestive way-marker for this section's arguments will be the following

6 A Continuous Representation of Wave-Function "Collapse" — 419

6.3 Decomposition Theorem (cf. Gihman and Skorohod III, 1979, 62, 88 and 90)

Any complex uniformly (square-)integrable *cadlag* $(\mathcal{U}_{\leq\infty}, \mathcal{v}_{\leq\infty})$-martingale $\mathcal{M}_s$ for $-\infty < s \leq \infty$ can be decomposed into a sum $\mathcal{M}_{c,s} + \mathcal{M}_{d,s}$ such that
(i) $\mathcal{M}_{c,s}$ is a continuous martingale; and
(ii) $\mathcal{M}_{d,s}$ is a discrete martingale $\int_{2^\omega} \sigma \cdot (d\mu_r(\sigma))$; where
(iii) σ varies over the regular measure-space $S = 2^\omega$, and
(iv) μ_r is a $v_{\leq\infty}$ random martingale measure on $S = 2^\omega$ (cf. 6.8 and Gihman and Skorohod, 1979, 62, 88);

(that is,
(v) $\mu_s = \mu_s(u) = \mu_s(u_{\leq\infty})$ is a point-process of "time-indexed" measures on S;
(vi) $\mu_s(A)$ is a cadlag square-integrable $v_{\leq\infty}$-martingale for each measurable $A \subseteq S = 2^\omega$;

and
(vii) $\mu_s(A) \cdot \mu_s(B)$ is also a cadlag $v_{\leq\infty}$-martingale for each measurable pair A and B such that $A \cap B = 0$).

6.4 The discrete martingale $\dot{\mathcal{M}}_{d,s} = \int_{2^\omega} \sigma \cdot (d\mu_r(\sigma))$ of 6.3(ii) above makes a non-null contribution at s such that
(i) the cadlag martingale $\mathcal{M}_s(u)$ has an "accessible jump" at s, i.e. (cf. Kallenberg, 2001, 499);
(ii) $\{u | \dot{\mathcal{M}}_s(u) \neq \dot{\mathcal{M}}_{s-}(u)\}$ has $v_{\leq\infty}$-measure greater than 0, where $\dot{\mathcal{M}}_{s-}(u)$ is the limit in $v_{\leq\infty}$-probability of the random variables $\dot{\mathcal{M}}_{s'}(u)$ for rational $s' < s$;

and
(iii) a predictable $v_{\leq\infty}$-random time $\varsigma = \varsigma(u)$ exists such that $\dot{\mathcal{M}}_{\varsigma-} \neq \dot{\mathcal{M}}_\varsigma$ on a set of positive $\tau_{\leq\infty}$-measure, where
(iv) ς is predictable if it is "announced" by a series of $\tau_{\leq\infty}$-optional times ς_n which are bounded $v_{\leq\infty}$-almost surely below s.

6.5 Heuristic Remark

The optional times ς_ns and "predictable" time ς they "announce" in 6.4(iv) may offer stochastic analogues of
(i) theoretical "preparations" (formulated in a theory $T_{<r}$) which permit "observation" and "measurement" of "U-governed time-evolution" of a "system" S,

in such a fashion that

(ii) time-evolution of the "incoming" system S is "measured" at the "instant" r by an "apparatus" ρ_r whose "coupling" with S is defined in the "limiting" theory $T_{\prec r}$).

In addition, one can make the following

6.6 Definition(s)

Given an arbitrary Hamiltonian H on $L^2(R^d)$, let

(i) $\dot{M}_s(u)$ ("M" for "measurement") be 5.4(i)'s square-integrable cadlag martingale of square roots of Radon-Nikodym derivatives of the conditionalisations $v_{\leq r}$ of the "observational measure" $v_{\leq \infty}$.

Further, let

(ii) $\mathcal{H}$ be the observational frame $L^2(\mathcal{U}_{\leq \infty}, v_{\leq \infty})$, or equivalently, the Hilbert space of square-integrable martingales $\mathcal{N}_s(u)$, and

(iii) $\mathcal{N}$ an element of $L^2(\mathcal{U}_{\leq \infty}, v_{\leq \infty})$ (respectively $\mathcal{N}_s$ an element of $L^2(\mathcal{U}_{\leq s}, v_{\leq s})$) determined by "canonical" measure-isomorphism(s) $e_{\leq \infty}$ (respectively $e_{\leq s}$) from the Gaussian probability space (R^d, γ_d) onto $(\mathcal{U}_{\leq \infty}, v_{\leq \infty})$ (respectively $(\mathcal{U}_{\leq s}, v_{\leq s})$).

Then

(iv) $U^{isol}_{H,s}(\mathcal{N}) = \mathcal{N}^{isol}_{H,s}$ in $L^2(\mathcal{U}_{\leq \infty}, v_{\leq \infty})$ (respectively $U^{isol}_{H,r,s}(\mathcal{N}_s) = \mathcal{N}^{isol}_{H,r,s}$ in $L^2(\mathcal{U}_{\leq r}, v_{\leq r})$) is $(\exp(-isH)(\mathcal{N}))$ (respectively $(\exp(-isH_r)(\mathcal{N}_r))$) for $s \in \bar{R} =_{Df} ([-\infty, \infty])$,

where "isol" is short for "isolated", and

(v) H on $L^2(\mathcal{U}_{\leq \infty}, v_{\leq \infty})$ (respectively $H_{r,s}$ on $L^2(\mathcal{U}_{\leq r}, v_{\leq r})$) is (abusive notation for) the spatial image(s) of H on $L^2(R^d, \lambda_d)$ via the "canonical" measure-isomorphism $e_{\leq \infty}$ (respectively $e_{\leq r}$) in (iii) above;

or equivalently, for the class of Hamiltonians H considered in (Boos, 2007),

(v) $U^{isol}_{H,s}(\mathcal{N}) = \mathcal{N}^{isol}_{H,s}$ in $L^2(\mathcal{U}_{\leq \infty}, v_{\leq \infty})$ is $(\exp(-isH)(\mathcal{N})) = \int_W \sigma^{\mathcal{N}}_s(w)\, dw$, where $\sigma^{\mathcal{N}}_s(w)$ is the corresponding $e_{\leq \infty}$-image of functional on the space W of Wiener paths such that $\exp(-isH)\varphi(x) = \int_W \sigma^x_s(w)\, dw$ in Theorem 1.1 (4) of (Boos, 2007) (and similarly for $U^{isol}_{H,r,s}(\mathcal{N}_s) = \mathcal{N}^{isol}_{H,r,s}$ in $L^2(\mathcal{U}_{\leq s}, v_{\leq s})$).

It is immediate, of course, that $U^{isol}_{H,s}$ (respectively $U^{isol}_{H,r,s}$) is strongly continuous on $L^2(\mathcal{U}_{\leq \infty}, v_{\leq \infty})$ (respectively $L^2(\mathcal{U}_{\leq r}, v_{\leq r})$), since $\exp(-isH)$ is strongly continuous on $L^2(R^d, \lambda_d)$.

In an earlier effort to "measure" such "isolated states" and their time-evolution, I construed the "ordinary" expectation value $\int_R u \cdot (d\mu_{H,r,sp}(u))$ as a "discrete" real-valued term $M_{d,r} = \int_{2w} \sigma \cdot (d\mu_r(\sigma))$ of the sort considered in 6.3(ii).

Such an assimilation might be a first step toward a "classical" study of "cadlag Markov operators" acting on distribution functions of the expectation values of quantum-theoretic states. But a quantum-theoretic expectation value is not a quantum-theoretic state (cf. Blank, Exner and Havlicek, 275 and 6.8).

A more reasonable *Ansatz* therefore for reconstruction of Schrödinger's "Schnitt" (Schrödinger, 1926), (anglicised as "collapse" of the "wave function"), would be to interpret

6.7 The "Measured" State as a "State-Valued Expectation" (or Bochner Integral).

Let the "measured" unitary group $U_{H,s}^{meas}$ be defined on each of the conditionalisations $L^2(\mathcal{U}_{\leq r}, \upsilon_{\leq r})$ of $L^2(\mathcal{U}_{\leq \infty}, \upsilon_{\leq \infty})$ for $s \in \bar{R} =_{Df} (-\infty, \infty]$ by setting

(i) $U_{H,r,s}^{meas}(\mathcal{N}_r) = \mathcal{N}_{H,r,s}^{meas} = \mathcal{N}_{H,r,s}^{isol} =_{Df} U_{H,s}^{isol}(\mathcal{N}_r)$ for each $\mathcal{N}_r \in L^2(\mathcal{U}_{\leq r}, \upsilon_{\leq r})$ if $\dot{M}_r(u)$ does not have an "accessible jump" at $r \in \bar{R}$ (cf. 6.3(ii) and 6.4(i)).

For each $r \in \bar{R} =_{Df} (-\infty, \infty]$ at which $\dot{M}_r(u)$ does have such a "jump", let

(ii) $\dot{M}_{r-}(u)$ be the limit in $\upsilon_{\leq \infty}$-probability of the random variables $\dot{M}_{s'_r}(u)$ for rational $r' < r$ (cf. the definition in 6.4);

(iii) $\upsilon_{\leq r}^{J_r}$ be the restriction of the measure $\upsilon_{\leq r}$ to J_r, where $J_r \subseteq \mathcal{U}_{\leq \infty}$ is the set of u such that $\{u | \dot{M}_r(u) \neq \dot{M}_{r-}(u)\}$ has $\upsilon_{\leq \infty}$-measure greater than 0;

(iv) $\mathcal{N}_{H,r,s}^{perturbed}$ be the projection of $\mathcal{N}_{H,r,s}^{isol}$ onto $L^2(J_r, \upsilon_{\leq \infty}^{J_r}) \subseteq L^2(\mathcal{U}_{\leq \infty}, \upsilon_{\leq \infty})$; and

(v) $\mathcal{N}_{H,r,s}^{unperturbed}$ the projection of $\mathcal{N}_{H,r}^{isol}$ onto $L^2(J_r, \upsilon_{\leq \infty}^{J_r})$'s orthogonal complement $L^2(\mathcal{U}_{\leq \infty} - J_r, \upsilon_{\leq \infty}^{J_r}) \subseteq L^2(\mathcal{U}_{\leq \infty}, \upsilon_{\leq \infty})$.

Then if

(vi) $\mu_{H,r,s}^{meas-}$ is the spectral measure determined for such r by $\mathcal{N}_{H,r,s}^{meas} =_{Df} \mathcal{N}_{H,r,s}^{perturbed-}$ on the spectrum of H,

let

(vii) $\mathcal{N}_{H,r,s}^{measured}$ be the Bochner integral $\int_R^{\mathcal{B}} \mathcal{N}_{H,r,s} \cdot (d\mu_{H,r,s}(u))$, where $\mathcal{N}_{H.r.s}$ in $L^2(\mathcal{U}_{\leq \infty}, \upsilon_{\leq \infty})$ is the limit of a sequence of finite unit-normalised approximations of the form $\left(\sum_{k \leq n} \left(P_H(A_k) \mathcal{N}_{H,r,s,k}^{meas-}\right)\right)$ in $L^2(\mathcal{U}_{\leq \infty}, \upsilon_{\leq \infty})$ (cf. Yosida, 1974, 132–136), where $P_H(A_k)$ is the spectral projection given by H and $A_k \subseteq R$, and for $r \in \bar{R} =_{Df} (-\infty, \infty]$ at which $\dot{\mathcal{M}}_r(u)$ has an accessible "jump" as in (ii) above, let

(viii) $U_{H,r,s}^{meas}(\mathcal{N}_s) = \mathcal{N}_{H,r,s}^{meas}$ be the orthogonal sum $\mathcal{N}_{H,r,s}^{unperturbed} + \mathcal{N}_{H,r,s}^{measured}$ in $L^2(\mathcal{U}_{\leq s}, \upsilon_{\leq s})$.

6.8 Remark

$\int_{R}^{\circledast} \mathcal{N}_{H,r,s} \cdot (d\mu_{H,r,s}(u))$ in 6.7(vii) above—a "wave-functional" analogue of 6.3 (iv)'s ordinary integral (or expectation value) $\int_{2\omega} \sigma \cdot (d\mu_r(\sigma))$—implements and elaborates heuristic assertions one can find in the literature—in (Blank et al. 1994, 257), for example, that "if ... the measured value is contained in the set $P_H(A_k)$ ... the system after the measurement will be in the [normalised] state ... $(P_H(A_k)\mathcal{N}_{H,r,k}^{meas-})/\|P_H(A_k)\mathcal{N}_{H,r,k}^{meas-}\|$". (The italics here are mine).

6.9 Remark

"Disintegration of measures" offers a slightly different view of the random measure $\mu_{H,r,\mathcal{N},s}(u)$'s role in the definition of $\mathcal{N}_{H,r,s}^{meas}$ in 6.7(ii) above, as well as its relevance to 'wave-function collapse' at the "instant" r.

For let

(i) $i_{\mathcal{U},\mathcal{N},r}$ be the "canonical" embedding of the measure-space $(\mathcal{U}_{\prec r} \times Sp(H), \upsilon_{\prec r} \otimes \tilde{\mu}_{H,\mathcal{N},r,s}(u))$ into $(\mathcal{U}_{\leq r}, \upsilon_{\leq r})$,

given by

(ii) the "canonical" surjection of $(\mathcal{U}_{\leq r}, \upsilon_{\leq r})$ onto $(\mathcal{U}_{\prec r} \times \mathcal{U}_r, \upsilon_{\prec r}(u) \otimes \mu_{H,r,\mathcal{N},s}(u))$, where $\tilde{\mu}_{H,r,\mathcal{N},s}(u)$ is a random image of Lebesgue measure on $\mathcal{U}_r$.

Then the obvious homeomorphism between $\mathcal{U}_{\prec r} \times \mathcal{U}_r$ and $\mathcal{U}_{\leq r}$, and the random measure $\upsilon_{\prec r}(u)$ of $\tilde{\mu}_{H,r,\mathcal{N},s,d}(u)$ on $\mathcal{U}_r$ give rise to

(iii) a "random product" $\mathcal{N}_{H,r}^{meas}$ of $\mathcal{N}_{H,\prec r}(u)$ and a $\upsilon_{\prec r}$-random Bochner integral on the $\mathcal{N}_{H,\prec r}(u)$-measured spectrum of H.

The foregoing definitions and observations provide a theoretical scaffolding for the following

6.10 Theorem

In the framework of 6.3 through 6.9 above,

(i) each $U_{H,r,s}^{meas}$ for $s \in \bar{R} =_{Df} (-\infty, \infty]$ is a unitary operator from $\mathcal{N} \in L^2(\mathcal{U}_{\leq r}, \upsilon_{\leq r})$ onto $U_{H,r,s}^{meas}(\mathcal{N}) \in L^2(\mathcal{U}_{\leq r}, \upsilon_{\leq r})$.

(ii) The construction of the operators $U^{meas}_{H,r,s}$ for $s \in \bar{R} =_{Df} (-\infty, \infty]$ can be extended to arbitrary self-adjoint operators H' which act on L^2-functions from an arbitrary Polish measure-space (S, ρ) into C.

(iii) The group of unitary operators $U^{meas}_{H,s} = U^{meas}_{H,\infty,s}$ on the Hilbert space $L^2(\mathcal{U}_{\leq\infty}, \upsilon_{\leq\infty})$ is strongly continuous, though its arguments $\mathcal{N}$ and images $\mathcal{N}^{meas}_{H,r} = U^{meas}_{H,r}(\mathcal{N})$ are cadlag (but not necessarily continuous) martingales in $L^2(\mathcal{U}_{\leq\infty}, \upsilon_{\leq\infty})$.

(iv) The infinitesimal generator $\mathfrak{H}_{H,\mathcal{M}} =_{Df} H_\mathcal{M}$ of the group $U^{meas}_{H,s} = U^{meas}_{H,\infty,s}$ (cf. 9.5) is unitarily equivalent to the self-adjoint operator H on $L^2(\mathcal{U}_{\leq\infty}, \upsilon_{\leq\infty})$ (abusively identified in 6.6(iii) and 6.6(v) with the Hamiltonian H on $L^2(R^d, \lambda_d))$ if and only if the real martingale $\dot{\mathcal{M}}_r(u)$ defined from the "observational measure" $\upsilon_{\leq\infty}$ in 6.6(i) is continuous.

Proof

The basic idea, roughly speaking, is that time displacements in each $U_{H,r,s}$ "ride out" temporal discontinuities in the "measurement"-processes which condition their domains.

(i) Isometry and surjectivity of the operators $U^{meas}_{H,r,s} : L^2(\mathcal{U}_{\leq r}, \upsilon_{\leq r}) \to L^2(\mathcal{U}_{\leq r}, \upsilon_{\leq r})$ for each $s \in \bar{R} =_{Df} (-\infty, \infty]$ are immediate from the definitions.

(ii) For (ii), it suffices to observe that every Polish measure-space (S, ρ) can be "canonically" embedded in the Lebesgue measure space $(2^\omega, \lambda)$.

(iii) For (ii) above, strong continuity in s of each $U^{meas}_{H,r,s}$ for $s \in \bar{R} =_{Df} (-\infty, \infty]$ which is at the limit in $\upsilon_{\leq\infty}$-probability of the $\dot{\mathcal{M}}_{r'}(u)$ for rational $r' < r$ is also immediate from the definitions.

For in $\dot{\mathcal{M}}_{r'}(u)$ at which an accessible jump occurs, Johan von Neumann's remarkable sufficiency-theorem of (Theorem VIII.9, Reed and Simon I, 1980, 267–268) ensures that

(iv) strong continuity in s of each $U^{meas}_{H,r,s} = U^{meas}_{\mathcal{M},H,r,s}$ for each s in $J_{\dot{\mathcal{M}}}(u) =_{Df} \{r | \{u | \dot{\mathcal{M}}_r(u) \neq \dot{\mathcal{M}}_{r-}(u)\}$ has positive $\upsilon_{\leq\infty}$-measure$\}$

follows from

(v) strong continuity in s of each $U^{meas}_{H,r,s}$ for $s \in \bar{R} =_{Df} (-\infty, \infty]$ for each $\dot{\mathcal{M}}$ such that $J_{\dot{\mathcal{M}}}(u) =_{Df} \{r | \{u | \dot{\mathcal{M}}_r(u) \neq \dot{\mathcal{M}}_{r-}(u)\}$ has positive $\upsilon_{\leq\infty}$-measure$\}$ is finite.

For the set of such $\dot{\mathcal{M}}$ is dense in the space of all $\dot{\mathcal{M}}$ defined from the Polish space measures $\upsilon_{s,\theta}$ in 5.4 (cf. also 6.15).

But (v) follows from

(vi) strong continuity in s of each $U^{meas}_{H,r,s}$ for $s \in \bar{R} =_{Df} (-\infty, \infty]$ for each $\dot{\mathcal{M}}$ such that $J_{\dot{\mathcal{M}}}(u) =_{Df} \{r | \{u | \dot{\mathcal{M}}_r(u) \neq \dot{\mathcal{M}}_{r-}(u)\}$ has positive $\nu_{\leq\infty}$-measure$\}$ is a singleton $\{s\}$.

And this, finally, is ensured by the fact that $U^{meas}_{H,r,s}(\mathcal{N}_{\leq r}) = U^{meas}_{\dot{\mathcal{M}},H,r,s}(\mathcal{N}_{\leq r})$ for such $\dot{\mathcal{M}}$ is determined by

(vii) a continuously evolving s-indexed sequence of Lebesgue-Stjeltes spectral measures $\mu^{meas-}_{H,r,s}$ (as defined in 6.7(vi)),

and

(viii) a corresponding continuously evolving s-indexed sequence of *Bochner integrals* $\mathcal{N}^{measured}_{H,r} =_{Df} \int_R^{\mathcal{B}} N_{H,r,s} \cdot (d\mu_{H,r,s}(u))$ (as defined in 6.7(vii)).

The assertions in clause (iv) of 6.10, finally, follow immediately from the definition of the unitary group $U^{meas}_{H,s} = U^{meas}_{H,\infty,s}$ in 6.6 and 6.7, and the proof of its strong continuity with respect to the "temporal" parameter s.

These arguments complete the proof of Theorem 6.10.

6.11 Two Glosses of the Conclusions of 6.10

(i) The first is that Theorem 6.10 as well as Theorem 1.1(4) of (Boos, 2007), can be generalised to globally hyperbolic spacetimes, since the latter are well-known to be homeomorphic to $R^1 \times R^3$ (cf. Hawking and Ellis, 1973, 212).

(ii) The second is that

(a) the Hilbert space $L^2(\mathcal{U}_{\leq r}, \nu_{\leq r})$ is spatially isomorphic to $L^2(R^d, \lambda_d)$ via definable Borel isomorphisms of the sort exploited in Sections 3 and 4 above,

so that

(b) the infinitesimal generator $H^{meas}_{H,r,s}$ of the unitary group $U^{meas}_{H,r,s}$ acting on $L^2(\mathcal{U}_{\leq r}, \nu_{\leq r})$ in Theorem 6.10 may "canonically" be correlated with

(c) the generator $\tilde{H}^{meas}_{H,r,s}$ of a corresponding group $\tilde{U}^{meas}_{H,r,s}$ which acts continuously on the Lebesgue space $L^2(R^d, \lambda_d)$.

The conclusions of Theorem 6.10 therefore furnish an interpretation of "measurement" of a given quantum-theoretic system as a replacement of

(d) "unmeasured" Hamiltons H on $L^2(R^d, \lambda_d)$ with

(e) "measured" counterparts $\tilde{H}^{meas}_{H,r,s}$ on $L^2(R^d, \lambda_d)$ "perturbed" (and tacitly parametrised) by a class of "observational measures" $\nu_{\leq r}$ on $\mathcal{U}_{\leq\infty} =_{Df} \mathcal{T}_{\leq\infty} \times \mathcal{P}_{\leq\infty}$.

6.12 Three Extensions

Theorem 6.10 and the essentials of its proof may be extended to
(i) "measured" time-varying Hamiltonians H_t, and doubly time indexed "unitary propagators" $U_{s,t}$, whenever such propagators $U_{s,t}$ are well-defined (cf., e.g., Reed and Simon II, 1980, 282–292, and Blank et al., 1994, 317–321);
(ii) "measured" unitary time-evolution $U_{A,t}$ generated (in the "Heisenberg picture", for example) by the commutator $[H, A_t]$ of an arbitrary observable A on $L^2(R^2)$ under the action of $U_t = {}_{Df} \exp(-itH)$ (cf., e.g., Blank et al., 1994, 323–324); and
(iii) "measured" unitary time-evolution governed by "superoperators" $\mathcal{U}_{s,t}$ (cf., e.g., Prugovecki, 1971, 398), which map states Σ_s in "Liouville space" $\mathcal{L}^2_C(S, \rho)$ (the space of complex Hilbert-Schmidt operators on $L^2(S, \rho)$) unitarily to "time-evolved" counterparts Σ_t for $s < t$.

One can also apply the constructions of this section to Hamiltonians of the sort studied in (Boos, 2007) to formulate a "Feynmanian"

6.13 "Tensor-Product"-Representation of "Measured" Schrödinger Time-Evolution

Recall once again the formula
(i) $\exp(-isH)\varphi(x) = \varphi_s(x) = \int_W \sigma_s^x(\omega)\, d\omega$,

(cf. Theorem 1.1(4) in (Boos, 2007 and 6.2).

Varying the program of 6.2(ii)(b) above, one can interpret the integrand $\sigma_r^x(\omega) = \sigma(r, x, \omega)$ of (Boos, 2007) (in effect) as
(ii) a Wiener space parameterised functional $\sigma_\omega(s, x)$, rather than
(iii) a "space-time" parameterised integrand $\sigma_s^x(\omega)$, defined on the measure space (W, μ) of Wiener paths ω from $[0, t]$ into d-dimensional momentum-space.

For H as in Theorem 1.1(4) of (Boos, 2007), one can vary 6.10's constructions slightly to
(iv) identify $x \in L_2(R^d)$ with $u \in L_2(\mathcal{U}_{\leq \infty}, \mathcal{V}_{\leq \infty})$, and rewrite the "space-time"-parametrised integrand $\sigma_s^x(u, \omega)$ as a "measurement parametrised" integrand $\sigma_s^u(\omega) = \sigma(u, s, \omega)$,

and develop a "tensor-product"-representation of "measured" Schrödinger time-evolution as follows.

6.14 Given H as in 6.2, and
(i) an "observational frame" $L^2(\mathcal{U}_{\leq\infty}, v_{s,\theta})$ as in 5.2 and 5.3(iii) above (or equivalently, its associated direct-integral Hilbert space $\mathcal{K}_{\leq\infty} =_{Df} L^2(\mathcal{U}_{\leq\infty}, v_{s,\theta}) = \int_{\theta_{\leq\infty} \in (\mathcal{T}_{\leq\infty}, \tilde{v}_{\leq\infty})}^{\oplus} L^2(\mathcal{P}_{\leq\infty}, v_{s,\theta}))$,

one can define
(ii) a Wiener measure $\mu_{\leq\infty}$ on the space $\mathcal{W}_{\leq\infty}$ of continuous paths from $\bar{R} = [-\infty, \infty]$ into R^d is an appropriate Borel-isomorphic image of Wiener measure $\mu = d\omega$ on the space W of continuous paths from $[0, 1]$ into R^d,

and form the tensor-product space(s)
(iii) $L^2\left(\mathcal{W}_{\leq\infty} \times \mathcal{U}_{\leq\infty}, \mu_{s,\theta} \otimes v_{s,\theta}\right) \cong L^2(\mathcal{W}_{\leq\infty}, \mu_{\leq\infty}) \otimes L^2(\mathcal{U}_{\leq\infty}, v_{\leq\infty}) \cong L^2((\mathcal{W}_{\leq\infty}, \mu_{\leq\infty}), L^2(\mathcal{U}_{\leq\infty}, v_{\leq\infty})) \cong L^2((\mathcal{W}_{\leq\infty}, \mu_{\leq\infty}), \int_{\theta_{\leq\infty} \in (\mathcal{T}_{\leq\infty}, \tilde{v}_{\leq\infty})}^{\oplus} L^2(\mathcal{P}_{\leq\infty}, v_{\leq\infty}))$.

Since $\sigma_r^u(\omega) = \sigma(r, u, \omega)$ (construed as a u-parametrised functional from $(\mathcal{W}_{\leq\infty}, \mu_{\leq\infty})$ into C), is also square-integrable (cf. Theorem 1.1(6) of Boos, 2007),
(iv) a "randomisation" of the process $\mathcal{N} \in L^2(\mathcal{U}_{\leq\infty}, v_{\leq\infty})$ of 6.6(iv) can be represented as an element $\int_{\mathcal{W}_{\leq\infty}} \sigma_{H,s}(u, \omega) \mu_{\leq\infty}$ of $L^2((\mathcal{W}_{\leq\infty}, \mu_{s,\theta}) \otimes L^2(\mathcal{U}_{\leq\infty}, v_{s,\theta}), C)$,

and
(v) the "continuous parts" of the unitary operators $U_{H,r,s}^{meas}(\mathcal{N})$ for $r \in R$ can be represented as acting on the tensor product space $L^2(\mathcal{W}_{\leq\infty} \times \mathcal{U}_{\leq\infty}, \mu_{s,\theta} \otimes v_{s,\theta}) \cong L^2((\mathcal{W}_{\leq\infty}, \mu_{s,\theta}), L^2(\mathcal{U}_{\leq\infty}, v_{s,\theta}))$.

6.15 A "Parametric" Generalisation of Theorem 6.10

One can iterate the constructions of Theorem 6.10 if one construes the measures $v_{\leq\infty}$ on $\mathcal{U}_{\leq\infty} =_{Df} \mathcal{T}_{\leq\infty} \times \mathcal{P}_{\leq\infty}$ as "new" stochastic parameters in their turn (cf. also 9.6–9.7).

For
(i) the measure space $(\mathcal{T}_{\leq\infty} \times \mathcal{P}_{\leq\infty}, \tau_{\leq\infty} \otimes \pi_{\leq\infty})$ is Borel-isomorphic to the measure space $(C_{\leq\infty} \times C_{\leq\infty}, \lambda_{\leq\infty} \otimes \lambda_{\leq\infty})$, and
(ii) the collection $\mathcal{V}_{\leq\infty}$ of $(\tau_{\leq\infty} \otimes \pi_{\leq\infty})$-absolutely continuous measures $v_{\leq\infty}$ on $\mathcal{U}_{\leq\infty} =_{Df} \mathcal{T}_{\leq\infty} \times \mathcal{P}_{\leq\infty}$ is therefore in one-to-one correspondence with the Polish space $\mathcal{L}_{\leq\infty}^1$ of $\lambda_{\leq\infty} \otimes \lambda_{\leq\infty}$-almost surely strictly positive elements of $L^1(C_{\leq\infty} \times C_{\leq\infty}, \lambda_{\leq\infty} \otimes \lambda_{\leq\infty})$.

A Lebesgue measure $\lambda^1_{\leq\infty}$ on $\mathcal{L}^1_{\leq\infty}$ (and corresponding measure-algebra $\mathcal{L}_{\leq\infty}$) may therefore be defined on the Polish space $\mathcal{L}^1_{\leq\infty}$ as in Sections 3 and 4 above, and instead of working with

(iii) elements $u = u_{\leq\infty}$ of $\mathcal{U}_{\leq\infty} =_{Df} \mathcal{T}_{\leq\infty} \times \mathcal{P}_{\leq\infty}$, uniformly integrable martingales $\mathcal{N}_r(u)$ and a fixed $(\tau_{\leq\infty} \otimes \pi_{\leq\infty})$-absolutely continuous measure $\upsilon_{\leq\infty}$ on $\mathcal{U}_{\leq\infty}$ as in 6.10, one can work with

(iv) elements $z = z_{\leq\infty}$ of $\mathcal{Z}_{\leq\infty} =_{Df} \mathcal{L}^1_{\leq\infty} \times \mathcal{T}_{\leq\infty} \times \mathcal{P}_{\leq\infty}$, uniformly integrable martingales $\mathcal{N}_r(z)$ and a fixed $(\lambda^1_{\leq\infty} \otimes \tau_{\leq\infty} \otimes \pi_{\leq\infty})$-absolutely continuous measure $\zeta_{\leq\infty}$ on $\mathcal{Z}_{\leq\infty}$.

The proofs of corresponding generalisations of 6.10 go through in this wider context.

6.16 Historical Remark

Feynman was well-aware that heuristic as well as rigorous "path-theoretic" arguments ((Boos, 2007)'s transform-based interpretation of Feynman "integration", for example) might furnish quasi-"classical" *Ansätze* for Einsteinian reconstructions of quantum mechanics as

(i) a "statistical" theory, in which time-evolution is determined by an underlying class of "hidden parameters" (the paths).

As remarked earlier in 1.5,

(ii) the extended theory developed in this section can be construed in part at least as such a "statistical theory", in the sense that its "discrete part"

(iii) the Bochner integral $\int_R^\mathcal{B} \mathcal{N}_{H.r.s} \cdot (d\mu_{H,r,sp}(u))$

is defined in terms of

(iv) ("path"-indexed) "integration" over stochastic spectral measures $\mu_{H,r,s}(u)$.

In Sections 7 and 8, I will argue that such "statistical" reconstructions do *not* support Einstein's other great hope: that "*der Alte würfele nicht*" ("the old man doesn't "gamble"") (cf. 1.6).
 For

(vi) "*der Alte*" **selbst** ((die "*Semantik des Kasinos*", sozusagen), könnte sich etwa als $\upsilon_{\leq\infty}$-*wertige Zufallsgröße entpuppen* ...

("the old man" **himself** (the "semantics of the casino", so to speak) might turn out to be a $\upsilon_{\leq\infty}$-valued random variable ....")

7 "Virtual" Set-Theoretic Universes

7.1 The aims of this section will be
(i) to set forth in brief outline a "virtual" (measure-algebraically random) semantics for "physical" assertions which parallels the last section's stochastic analysis;
(ii) to show that this framework gives rise to a correspondingly "virtual" (measure-algebraically random) "many-worlds-interpretation" of quantum time-evolution; and
(iii) to vary and iterate such "virtuality" in ways which ambiguate the "metaphysical" "bivalence" ("sharpness", "two-valuedness") Einstein was determined to "save" (cf. 8.5(i)).

More precisely, I will describe and employ a sense in which "many worlds" or "many universes" interpretations emerge when one does not "integrate out" the random parameters of the last three sections' martingales and stochastically parametrised "integrals".

I will also show that one can implement this *Ansatz* at ascending "levels", which can be varied and iterated *ad libitum* (cf. 6.15 and 9.5 and 9.6).

7.2 At the first "level", one can construe
(i) the last section's space $\mathcal{H}$ of martingales $\mathcal{N}_r(u)$, as a dense subset $\tilde{\mathcal{H}}$ of the $\mathbb{U}_{\leq\infty}$-valued Hilbert-space $\mathcal{L}^2(D) = \mathcal{L}^2(D_{\bar{R},C})$ of square-integrable functions from the Polish space $D = D_{\bar{R},C}$ of cadlag functions (cf. Borkar, 1995, 119–126) into C.

7.3 To make this somewhat more precise, consider
(i) the "measure-algebraic" "model" $\mathbf{V}^{\mathbb{U}_{\leq r}}_{(\ldots)} = \mathbf{V}^{\mathbb{T}_{\leq r} \otimes \mathbb{N}_{\leq r}}_{(\ldots)}$ of ZFC, where
(ii) $\mathbb{U}_{\leq\infty}$ is the complete measure algebra associated with the measure space $(\mathcal{U}_{\leq\infty}, \nu_{\leq\infty})$;
(iii) $\mathbb{T}_{\leq\infty}$ is the complete measure algebra associated with the measure space $(\mathcal{T}_{\leq\infty}, \tilde{\nu}_{\leq\infty})$;
(iv) $\mathbb{P}^{\tilde{\nu}}_{\leq\infty}$ is the $\mathbb{T}_{\leq\infty}$-valued complete measure-algebra defined with value one in $\mathbb{T}_{\leq\infty}$ by the measure $\nu_{\leq\infty, \theta}$ on the $\mathbb{T}_{\leq\infty}$-valued completion $\tilde{\mathcal{P}}_{\leq\infty}$ of $\mathcal{P}_{\leq\infty}$; and
(v) $\mathbb{T}_{\leq\infty} \otimes \mathbb{P}^{\tilde{\nu}}_{\leq\infty}$ is the complete "iterated" measure-algebra "canonically" isomorphic to $\mathbb{U}_{\leq\infty}$ such that $\mathbf{V}^{\mathbb{U}_{\leq r}}_{(\ldots)}$ is indeed isomorphic to $\mathbf{V}^{\mathbb{T}_{\leq\infty} \otimes \mathbb{P}^{\tilde{\nu}}_{\leq\infty}}_{(\ldots)}$.

In this context, one has the following

7.4 Theorem

Let $D = D_{\bar{R}, C}$ be the Polish space of cadlag functions from the extended real line $\bar{R}$ into C (cf. Borkar, 1995, 120–126), and $\mathbf{V}^{U_{\leq r}}_{(\ldots)} = \mathbf{V}^{T_{\leq r}}_{(\ldots)} \otimes N_{\leq r}$ the measure-algebraic structure just defined.

Then the Hilbert space $\mathcal{H}$ of square-integrable martingales $\mathcal{N}_r(u)$ (or equivalently, of their limits $\mathcal{N}_{\leq \infty}(u)$) "is" (corresponds to)

(i) a $\mathbb{U}_{\leq \infty}$-valued subset $\tilde{\mathcal{H}}$ of the $\mathbb{U}_{\leq \infty}$-valued Hilbert-space $\mathcal{L}^2(D_{\bar{R}, C})$ of corresponding square-integrable cadlag functions $n(r)$ from the Polish space D into C such that $\| \tilde{\mathcal{H}}$ is dense in $L^2(D_{\bar{R}, C}) \|_{\mathbb{U}_{\leq \infty}} = 1$.

"Almost surely", moreover (with value 1 in the measure-algebra $\mathbb{U}_{\leq \infty}$),

(ii) the $\mathbb{U}_{\leq \infty}$-valued operators $\tilde{\mathcal{U}}^{meas}_{H, r, s}$ from $\tilde{\mathcal{H}}$ to $\tilde{\mathcal{H}}$ which map $n(r) \approx \mathcal{N}_r(u)$ to $\tilde{\mathcal{U}}^{meas}_{H, r, s}(n_r) \approx U_{H, s}(\mathcal{N}_r)$ for $s \in R$ (cf. 6.3) are extendible to a unitary group of operators $\tilde{\mathcal{U}}^{meas}_{H, r, s}$ on the Hilbert space $\mathcal{L}^2(D)$ with value 1 in $\mathbb{U}_{\leq \infty}$.

Proof

Observe first that $\mathcal{N} = \mathcal{N}(u) = \mathcal{N}_{\leq \infty}(u)$ can be construed as a square-integrable random function $\left(\tilde{\mathcal{N}}(u)\right)(r) =_{Df} \mathcal{N}_r(u) = \mathcal{N}(r, u)$ from (D, δ) into C (cf. Borkar, 1995, 126), where D has the Skorohod (Polish) topology defined in Borkar, 1995, 120–124, and δ is defined as in Section 4.

Any such $\left(\tilde{\mathcal{N}}(u)\right)(r)$, moreover, defines a $\mathbb{U}_{\leq \infty}$-valued measurable function n from (D, δ) into C. To see this, one can stand on the shoulders of

(a) "Borel-coding" arguments transferred from $(2^\omega, \lambda)$ to (D, δ) (cf. Solovay, 1970, 24–33, Jech, 2002, 504–507 and Kanamori, 1994, 137–142), and

(b) results of boolean-valued analysis applied to $L^2(\mathcal{U}_{\leq \infty}, \upsilon_{\leq \infty})$ and $\mathcal{L}^2(D)$ (cf., e.g., Takeuti, 1978, 56–68, particularly 62 and 64),

to see that

(c) n is an element of $\mathcal{L}^2(D)$ with value 1 in $\mathbb{U}_{\leq \infty}$ whenever $\int_R |\mathcal{N}(u)|^2$ is finite for each r and almost all u in $(\mathcal{U}_{\leq \infty}, \upsilon_{\leq \infty})$,

and this is clearly the case for $\mathcal{N}(u)$ such that $\int_{(\mathcal{U}_{\leq \infty}, \upsilon_{\leq \infty})} |\mathcal{N}(u)|^2 < \infty$.

But the set-theoretic results just mentioned also establish that the set $\tilde{\mathcal{H}}$ defined in (i) above is a proper subset of $\mathcal{L}^2(D)$ with value 1 in $\mathbb{U}_{\leq \infty}$. For the integral of $|n(r, u)|^2$ in $\mathcal{L}^2(D)$ is a $\upsilon_{\leq \infty}$-random function from u in $\mathcal{U}_{\leq \infty}$ to $\int_D |n(r, u)|^2$ in R (cf. Takeuti, 1978, 64), and there is no reason to assume that this measurable function will be $\upsilon_{\leq \infty}$-integrable.

$\tilde{\mathcal{H}}$ does, however, constitute a dense subset of $\mathcal{L}^2(D)$ with value 1 in $\mathbb{U}_{\leq\infty}$, and this will suffice to extend the isometric operators $\tilde{\mathcal{U}}^{meas}_{H,r,s}$ from $\tilde{\mathcal{H}}$ to all of $\mathcal{L}^2(D)$ with value 1 in $\mathbf{V}^{\mathbb{U}_{\leq\infty}}$.

To see this, consider an $n'(r,u)$ such that $[\![n'(r,u) \in \mathbf{L}^2(D)]\!] \geq \mathbf{b}$ for some nonzero $\mathbf{b} \in \mathbb{U}_{\leq\infty}$. Since $v_{\leq\infty}$ is a probability-measure on $\mathcal{U}_{\leq\infty}$, one can define

(d) a sequence of Borel subsets J_k of $\mathcal{U}_{\leq\infty}$, characteristic functions $\chi_{J_k}(u)$ and square integrable $\mathcal{S}_k(u) =_{Df} \mathcal{N}'(u) \cdot \chi_{J_k}(u)$ in $L^2(\mathcal{U}_{\leq\infty}, v_{\leq\infty})$, and from these "standard" functions $\mathcal{S}_k(u)$

(e) a corresponding sequence of $\mathbb{U}_{\leq\infty}$-valued functions $s_k(d)$ in $\mathcal{L}^2(D)$ such that ($[\![$all the $s_k(r)$s are in $\check{H}]\!] \wedge [\![$the sequence $s_k(r)$converges to n' in $\mathrm{L}^2(D)]\!]) \geq \mathbf{b}$.

This construction secures the first assertion in the theorem.

To see the second, observe that unitarity of the operators $\mathcal{U}_{H,r,s}$ on $\mathcal{H}$ ensures $\mathbb{U}_{\leq\infty}$-valued isometry of the operator $\tilde{\mathcal{U}}_{H,r,s}$ on the dense subset $\tilde{\mathcal{H}}$ of $\mathcal{L}^2(D)$ with value 1.

For suppose not. Then a D-valued random variable $\left(\tilde{\mathcal{N}}(u)\right)(r) =_{Df} \mathcal{N}_r(u) = \mathcal{N}(r,u)$ exists such that $\|\tilde{\mathcal{U}}_{H,r}(n(r,u))\| \neq \|n(r,u)\| \geq \mathbf{b}$ for a nonzero $\mathbf{b}$ in $\mathbb{U}_{\leq\infty}$. But this would contradict the last section's conclusion that $\|U^{meas}_{H,r,s}(\mathcal{N}_r(u))\| = \|\mathcal{N}_{H,r,s}(u)\|$.

7.5 In the $\mathbb{U}_{\leq\infty}$-valued semantics just outlined, the "accessible jumps" of the last section correspond to "instants" r at which

(i) the $\mathbb{U}_{\leq\infty}$-value $[\![\dot{\mathbf{M}}_r(u) = \dot{\mathbf{M}}(u,r) \neq \lim_{r'<r} \dot{\mathbf{M}}(u,r') = \dot{\mathbf{M}}_{r-}(u)]\!]^{\mathbb{U}_{\leq\infty}} > \mathbf{0}$, or in stochastic terms, at which

(ii) the $v_{\leq\infty}$-likelihood of the "random" assertion that "$\dot{\mathcal{M}}_r(u) \neq \dot{\mathcal{M}}_{r-}(u)$" is greater than 0.

7.6 In the "stochastic" context of the last section, moreover,

(i) the values of the "wave functions" $\mathcal{N}_{H,r-,s}(u)$ (for a given Hamiltonian H), as well as

(ii) the values of the $v_{\prec(r+s)}$-random spectral measures $\mu_{U,r,s} = \mu_{U,r,s}(u)$ on H's spectrum $Sp(H)$, and

(iii) the "measured wave functions"

$$U^{meas}_{H,r,s}(\mathcal{N}) =_{Df} \mathcal{N}^{meas}_{H,r,s}(u) = \int_R^B \mathcal{N}_{H,r-,s}(u) \cdot \left(d\mu_{H,r,sp}(u)\right)$$

"ramify" at the random "instants" r at which accessible jumps occur.

In the measure-algebraic context, standard measure-algebraic arguments (cf. Jech, 2002, 243, 513) show that these "ramifying" objects are determined with $\mathbb{U}_{\leq\infty}$-value **1** by

(iv) $\mathbb{U}_{\leq\infty}$-valued "Solovay-random" ultrafilters $\mathbf{u}_{\leq\infty}$, which generate "ramifying" $\mathbb{U}_{\leq\infty}$-valued structures $V^{\mathbb{U}_{\leq\infty}}(\mathbf{u}_{\leq\infty})$ with value 1 in $V^{\mathbb{U}_{\leq\infty}}$, and ("virtually") *determine* "virtual" counterparts $n_{H,r,s}^{meas} = \mathcal{N}_{H,r,s}^{meas\prime}(\mathbf{u}_{\leq\infty})$ of $\mathcal{N}_{H,r}^{meas}(\mathbf{u}_{\leq\infty})$, where $\mathcal{N}_{H,r}^{meas\prime}$ in $V^{\mathbb{U}_{\leq\infty}}$ is the identically Borel-coded $\mathbb{U}_{\leq\infty}$-valued counterpart of $\mathcal{N}_{H,r,s}^{meas}$ in V (cf., e.g., Boos, 1996, 105).

These set-theoretic observations prompt the following

7.7 Definition
(i) A ($\mathbb{U}_{\leq\infty}$-) "virtual world" (or "universe") is a $\mathbb{U}_{\leq\infty}$-valued structure $V^{\mathbb{U}_{\leq\infty}}(\mathbf{u}_{\leq\infty})$, or equivalently, the

(ii) $\mathbb{U}_{\leq\infty}$-"Solovay-random" ultrafilter $\mathbf{u}_{\leq\infty}$, which generates $V^{\mathbb{U}_{\leq\infty}}(\mathbf{u}_{\leq\infty})$ with value 1 in $V^{\mathbb{U}_{\leq\infty}}$ (cf. Solovay, 1970, 24–40, Jech, 2002, e.g., 211–218, 223–224, 243–244 and 511–515).

7.8 Remark
Well-studied "iteration"- analyses (cf. Jech, 2002, 267–71, and Bell, 1985, 104 and 146–147) also yield that each such "virtual world" $V^{\mathbb{U}_{\leq\infty}}(\mathbf{u}_{\leq\infty})$ also has a corresponding iterated representation

(i) $V^{\mathbb{T}_{\leq\infty} \otimes \tilde{\mathbb{P}}_{\leq\infty}}(\mathbf{u}_{\leq\infty, T} \otimes \mathbf{u}_{\leq\infty, P})$, where

(ii) $\mathbf{u}_{\leq\infty, T}$ is a $\mathbb{T}_{\leq\infty}$-random ultrafilter in the measure-algebra $\mathbb{T}_{\leq\infty}$, and

(iii) $\mathbf{u}_{\leq\infty, P}$ is a $\mathbb{T}_{\leq\infty}$-valued random ultrafilter in the $\mathbb{T}_{\leq\infty}$-valued counterpart $\tilde{\mathbb{P}}_{\leq\infty}$ of the measure space $\mathbb{P}_{\leq\infty}$ in $V^{\mathbb{T}_{\leq\infty}}$.

In this context, one can also formulate the following

7.9 Definition
A $\mathbb{U}_{\leq\infty}$-valued quantum-theoretic system is a $\mathbb{U}_{\leq\infty}$-valued structure $\mathcal{S} = (\mathcal{H}, \varphi, H)$ such that (with value 1 in $\mathbb{U}_{\leq\infty}$)

(i) $\mathcal{H}$ is a L^2-space $\mathcal{L}^2(S, \rho)$;
(ii) φ is a "pure state" in the Hilbert space $\mathcal{H}$;
(iii) H is a $\mathbb{U}_{\leq\infty}$-valued Hamiltonian on $\mathcal{L}^2(\mathcal{U}_{\leq\infty}, \nu_{\leq\infty, \theta})$; and, finally,
(iv) all these mathematical objects are definable from S, ρ with value one in the "universal" $\mathbb{U}_{\leq\infty}$-valued random ultrafilter $\mathbf{u}_{\leq\infty}$.

These observations and the metamathematical results which underlie them eventuate in the following

7.10 Theorem

Let $S = (\mathcal{H}, \varphi, H)$ an arbitrary $\mathbb{U}_{\leq\infty}$-valued quantum-theoretic system as in 7.6. Then (cf. Jech, 2002, 201–224 and Bell, 1985, 88–108)

(i) each $\mathbb{U}_{\leq\infty}$-valued "world" $V^{\mathbb{U}_{\leq\infty}}(\mathbf{u}_{\leq\infty}) = V^{\hat{\mathbb{T}}_{\leq\infty}} \otimes \hat{\mathbb{P}}_{\leq\infty}(\mathbf{u}_{\leq\infty, T} \otimes \mathbf{u}_{\leq\infty, P})$ determines (with value **1** in $V^{\mathbb{U}_{\leq\infty}}$) a unique "temporal" evolution of this system S; and such "virtual" determination is also faithful, in the sense that

(ii) for each pair of "virtual" ultrafilters $\mathbf{u}_{\leq\infty}$ and $\mathbf{u}'_{\leq\infty}$ and $p > 0$ in $\mathbb{U}_{\leq\infty}$ such that

(a) $[\![\mathbf{u}_{\leq\infty} \neq \mathbf{u}'_{\leq\infty}]\!] \geq p$,

a $\mathbb{U}_{\leq\infty}$-valued $S = (S, H, \varphi)$ exists such that

(b) $\left[\!\left[\begin{pmatrix} (\mathbf{u}_{\leq\infty} - \text{determined time-evolution of the system } S) \neq \\ (\mathbf{u}'_{\leq\infty} - \text{determined time-evolution of the system } S). \end{pmatrix} \right]\!\right] \geq p$.

Finally,

(iii) semantic distinctions in the "worlds" $V^{\mathbb{U}_{\leq\infty}}(\mathbf{u}_{\leq\infty})$ "appear" (with $\mathbb{U}_{\leq\infty}$-value 1) to be "sharp", in the sense that $[\![$(syntactically well-defined assertions that have no 'intermediate' truth-values)$]\!]^{\mathbb{U}_{\leq\infty}} = 1$.

Sketch of the Proof

The essential observation is that each $\mathbb{U}_{\leq\infty}$-valued ultrafilter $\mathbf{u}_{\leq\infty}$ is interdefinable with a $\mathbb{U}_{\leq\infty}$-valued homomorphism $\mathbf{h}_{\mathbf{u}_{\leq\infty}}$ from $\mathbb{U}_{\leq\infty}$ to **2** with value 1 in $V^{\mathbb{U}_{\leq\infty}}$ (cf. Solovay, 1970, 35–36 and Jech, 2002, 215–218). In effect, $\mathbf{h}_{\mathbf{u}_{\leq\infty}}$ "collapses" $\mathbb{U}_{\leq\infty}$-valued semantics to **2**-valued semantics, but only "with $\mathbb{U}_{\leq\infty}$-value 1", since $\mathbf{u}_{\leq\infty}$ and $\mathbf{h}_{\mathbf{u}_{\leq\infty}}$ only exist with $\mathbb{U}_{\leq\infty}$-value 1. These observations also underlie the "apparent" $\mathbb{U}_{\leq\infty}$-valued "sharpness" (or "two-valuedness) in $V^{\mathbb{U}_{\leq\infty}}(\mathbf{u}_{\leq\infty})$ in (iii), as well as the interdefinability of $\mathbf{u}_{\leq\infty}$ and $V^{\mathbb{U}_{\leq\infty}}(\mathbf{u}_{\leq\infty})$.

Syntactical definition of "systems" $S = (\mathcal{H}, H, \varphi)$ in one or another $T_{\leq\infty} \in \mathcal{T}_{\leq\infty}$, finally, is "absolute" enough to guarantee $\mathbb{U}_{\leq\infty}$-valued "faithfulness" of their time-evolution in (ii).

7.11 Remark

The stochastic generalisation of Theorem 6.10 as outlined in 6.16 has a corresponding metamathematical analogue which generalises the constructions in 7.8, namely

7.12 Theorem 7.10

Given

(i) an $(\lambda^1_{\leq\infty} \otimes \tau_{\leq\infty} \otimes \pi_{\leq\infty})$-absolutely continuous measure $\zeta_{\leq\infty}$ on $\mathcal{Z}_{\leq\infty} =_{Df} \mathcal{L}^1_{\leq\infty} \times \mathcal{T}_{\leq\infty} \times \mathcal{P}_{\leq\infty}$ and $\tilde{\zeta}_{\leq\infty}$ the "projection" of $\zeta_{\leq\infty}$ on $\mathcal{L}^1_{\leq\infty}$ (cf. 6.14(iv)),

let
(ii) $\mathbb{Z}_{\leq\infty}$ be the measure-algebra associated with the measure space $(\mathbb{Z}_{\leq\infty}, \zeta_{\leq\infty})$;
(iii) $\tilde{\mathbb{Z}}_{\leq\infty}$ the measure-algebra associated with the measure space $\left(\mathcal{L}^1_{\leq\infty}, \tilde{\zeta}_{\leq\infty}\right)$; and
(iv) $\tilde{\mathbb{U}}_{\leq\infty}$ the $\tilde{\mathbb{Z}}_{\leq\infty}$-valued measure-algebra associated with the random measure-space $(\mathcal{U}_{\leq\infty}, \upsilon_{\leq\infty, z})$ with $\tilde{\mathbb{Z}}_{\leq\infty}$-value 1 in $V^{\tilde{\mathbb{Z}}_{\leq\infty}}$, such that $V^{\tilde{\mathbb{Z}}_{\leq\infty}} \otimes \tilde{\mathbb{P}}_{\leq\infty} = V^{\tilde{\mathbb{Z}}_{\leq\infty}}$.

If one substitutes $\mathbb{Z}_{\leq\infty}$ for $\mathbb{U}_{\leq\infty}$, $\tilde{\mathbb{Z}}_{\leq\infty}$ for $\tilde{\mathbb{T}}_{\leq\infty}$ and $\tilde{\mathbb{U}}_{\leq\infty}$ for $\tilde{\mathbb{P}}_{\leq\infty}$ and makes other notational changes, all the assertions in Theorem 7.10 and its proof go through *mutatis mutandis*.

In particular, one has an analogous "many worlds" interpretation of $\mathbb{Z}_{\leq\infty}$-valued quantum-theoretic systems $S = (\mathcal{H}, \varphi, \mathrm{H})$ and its
(v) iterated counterpart $V^{\tilde{\mathbb{Z}}_{\leq\infty}} \otimes \tilde{\mathbb{U}}_{\leq\infty} (\mathbf{u}_{\leq\infty, Z} \otimes \mathbf{u}_{\leq\infty, U})$, where
(vi) $\mathbf{u}_{\leq\infty, Z}$ is a $\tilde{\mathbb{Z}}_{\leq\infty}$-random ultrafilter in the measure-algebra $\tilde{\mathbb{Z}}_{\leq\infty}$, and
(vii) $\mathbf{u}_{\leq\infty, U}$ is a $\mathbb{Z}_{\leq\infty}$-valued random ultrafilter in the $\mathbb{Z}_{\leq\infty}$-valued counterpart $\tilde{\mathbb{U}}_{\leq\infty}$ of the measure-space $(\mathcal{U}_{\leq\infty}, \upsilon_{\leq\infty, z})$ in $V^{\tilde{\mathbb{Z}}_{\leq\infty}}$.

I will explore methodological implications of such "iterated randomness" in the next two sections.

8 The "Direction" of "Time"

8.1 In this section, I will exploit the metamathematical properties of the measure-algebras $\mathbb{U}_{\leq\infty}$ (or $\mathbb{Z}_{\leq\infty}$) to argue that "temporal" processes of prior sections' "temporally" parametrised
(i) ("... measurements of couplings of measurements of couplings of ..."),

which one might compare with "von Neumann's ladder", offer rigorous stochastic and metamathematical representations which align with "directionality" of "time", in the sense that
(ii) what is "random" and "underdetermined" in prospect becomes "sharp" and "determined" in retrospect.

8.2 In the stochastic interpretation sketched earlier, for example, one can identify
(i) "temporally" parametrised

"..... measurements of systems of measurements of systems of measurements of ...
..... measurements of systems of measurements of systems of measurements of ...."

with
(ii) "irreversible" stochastic processes of stochastically parametrised "point-values" of a right- but not necessarily left-continuous ("cadlag") martingale.

8.3 In the metamathematical interpretation, one can define (e.g.)
(i) "irreversible" semantic processes of measure-algebraic extensions $\mathbf{V}_{\leq r}^{\mathbb{U}_{\leq r}}$ of a "ground model" $\mathbf{V}_{-\infty}^{\mathbb{U}_{-\infty}}$ of ZFC
(for extended reals r such that $-\infty \leq r \leq \infty$), as
(ii) the "worlds" (or "universes") of a "many worlds" interpretation of quantum-mechanical measurement and time-evolution.

8.4 As in the last section (7), the "universes" $\mathbf{V}_{\leq r}^{\mathbb{U}_{\leq r}}$ are
(i) generated by "Solovay-random" ultrafilters $\mathbf{u}_{\leq r}^{\mathbb{U}_{\leq r}}$ in the "temporally indexed" measure-algebras $\mathbb{U}_{\leq r}$ (again, cf. Jech, 2002, 243 and 513),

and
(ii) particular realisations of these "universes" $\mathbf{V}_{\leq r}^{\mathbb{U}_{\leq r}}$ and their Solovay-random ultrafilters $\mathbf{u}_{\leq r}^{\mathbb{U}_{\leq r}}$ ramify in ways which are governed by alternative "paths" which have well-defined temporal "directions".

As in Section 5, finally,
(iii) the measure-algebras $\mathbb{U}_{\leq r}$ are co-determined by "physically" enriched extensions $\theta_{\leq r} = T_{\leq r}$ of ZFC which "grow" "temporally" with the parameter r, and "simultaneous" ("scale-reading") $\tau_{\leq r}$-random measures $v_{\leq r, \theta} \in \mathcal{P}_{\leq r}$.

These observations prompt the following

8.5 Proposal
The "arrow of time" orients
(i) the appearance of "bivalence" ("sharpness", "two-valuedness") which Einstein was determined to "save" (cf. 7.1(iii)), in the sense that the "arrow"'s "trajectories" mirror the "temporal asymmetries" of
(ii) the "cadlagness" of the martingales $\dot{\mathcal{M}}_{\leq r}$ and $\mathcal{N}_{\leq r}$;

(iii) the "growth" of the "scope" of the theories $\theta_{\leq r} = T_{\leq r}$ and measures $\nu_{\leq r, \theta} \in \mathcal{P}_{\leq r}$; and finally (construing "semantic randomness" at stage r as "$\mathbb{U}_{\leq r}$-valuedness"),

(iv) the "collapses" of "semantic randomness", effected by the Solovay random ultrafilters $\mathbf{u}_{\leq r}^{\mathbb{U}_{\leq r}}$, in their dual roles as

(a) "virtual" elements $\mathbf{u}_{\leq r}^{\mathbb{U}_{\leq r}}$ of the Stone spaces $St(\mathbb{U}_{\leq r})$ which "generate" the "universes" $\mathbf{V}_{\leq r}^{\mathbb{U}_{\leq r}}$,

and

(b) (complete) "virtual" homomorphisms $\mathbf{h}_{\leq r}^{\mathbb{U}_{\leq r}}$ of the measure-algebras $\mathbb{U}_{\leq r}$ onto the simplest "measure-algebra" **2** which "collapse" $\mathbf{V}_{\leq r}^{\mathbb{U}_{\leq r}}$ to the "phenomenally bivalent" structure $\mathbf{V}_{\leq r}^{\mathbb{U}_{\leq r}}\left(\mathbf{u}_{\leq r}^{\mathbb{U}_{\leq r}}\right)$ (cf. 8.6(ii)).

8.6 We cannot "see" these "collapses" in the "virtual universes" $\mathbf{V}_{\leq r}^{\mathbb{U}_{\leq r}}$—not only because "we" ramify with them, as suggested in the last section; but because

(i) "we" "collapse" with them as well.

For the standard arguments in the study of measure–algebraic semantics just recapitulated clarify a sense in which

(ii) $\mathbb{U}_{\leq r}$-valued "truth" in $\mathbf{V}_{\leq r}^{\mathbb{U}_{\leq r}}$ is "collapsed" to two-valued "truth" by "virtual" homomorphisms $\mathbf{h}_{\leq r}^{\mathbb{U}_{\leq r}}$ "almost surely" (with $\mathbb{U}_{\leq r}$-value **1**) in $\mathbf{V}_{\leq r}^{\mathbb{U}_{\leq r}}$ (cf., e.g., Jech, 2002, 215ff).

8.7 Taken together, the "stochastic" and the "metamathematical" interpretations of "measured time evolution" sketched offer rationales for:

(i) the randomness of "prospective" limits of the evolution of "unperturbed" "initial" systems, and

(ii) the well-definition of "retrospective" limits of "instantaneous" "couplings" of such systems with ambient apparatus.

8.8 They also offer stochastic as well as metalogical rationales for

(i) the retrospective "linearity" and prospective underdetermination of "measured time-evolution".

For in notationally different ways, each of these reconstructions gives rise to an irreversible partial ordering, in which "processes of measurement" are linear in "retrospect," but ramify in "prospect".

More precisely,

(ii) stochastic irreversibility is secured by the almost-surely cadlag underdetermination of its martingales (or "processes of observation"); and
(iii) metamathematical "irreversibility" is secured by the "retrospective" nature of "sharp" individuation in a "prospective" sea of randomness and potentiality.

A final remark: the (essentially metamathematical) observations of the last two sections suggest

8.9 A Methodological "Complementarity"-Principle:

The "sharper" one's "measurement-grids", the more "random" (or "virtual") will be their "interpretations" (cf. also 10.2).

Proof (or Explication)

A random ultrafilter $\mathbf{u} = \mathbf{u}_{\leq r}^{\mathbb{U}_{sr'}} \neq$ on $\mathbb{U}_{sr}$ is identifiable with a "random real" $r = r_\mathbf{u}$ in $\mathbf{V}_{\leq r}^{\mathbb{U}_{sr'}}$, characterised by the requirement that it be "ideally sharp":
  i.e., that it
(i) "respect" (with value $\mathbb{U}_{sr}$-value1) every "ordinary" partition (or "measurement-grid") $\mathcal{A}$ of 2^ω into countably many disjoint Borel sets A_n which have positive Lebesgue measure $v_{\leq\infty}(A_n)$ in $\mathbf{V}_{\leq r}^{\mathbb{U}_{sr'}}$ (lie, that is, in exactly one element A_n of $\mathcal{A}$) (cf. Jech, 2002, 513). ("Ordinary" here means that the grid $\mathcal{A}$ lies in some $\mathbf{V}_{\leq r'}^{\mathbb{U}_{sr'}}$ for $r' < r$.)

It follows from this characterisation that no "ideally sharp" $\mathbf{u} \in \mathbf{V}_{\leq r}^{\mathbb{U}_{sr'}}$ can lie in any of the "initial" universes $\mathbf{V}_{\leq r'}^{\mathbb{U}_{sr'}}$ for $r' < r$. One can, however, "coarse-grain" the "sharpness" of u in $\mathbf{V}_{\leq r}^{\mathbb{U}_{sr'}}$ by relaxing the requirement that u "respect" a fixed countable set of such "measurement"-grids' $\mathcal{A}$ in $\mathbf{V}_{\leq r'}^{\mathbb{U}_{sr'}}$ for $r' < r$.

Such "pragmatic" or "phenomenal" sharpness of "ordinary" $u \in \mathbf{V}_{\leq r}^{\mathbb{U}_{sr'}}$ suffices (indeed has to suffice) for most purposes, since the set of "ordinary" A_n-random reals $u_\mathcal{A}$ for any given countable set of such "grids" A_n has $\lambda_{\leq\infty}$-measure 1 (an instance of the Rasiowa-Sikorski Lemma: cf. Bell, 1985, 10, or Kunen, 1980, 243).

We might therefore be inclined to think that "noumenal sharpness" of $\mathbf{V}_{\leq r'}^{\mathbb{U}_{sr'}}$ for $r' < r$ is attainable "in principle", or at least at some "ideal limit". But this would be a "transcendental illusion".

For an "ideally sharp" or "precise" $\mathbf{u} = \mathbf{u}_{\leq r}^{\mathbb{U}_{sr'}}$ can only "exist" as a $\mathbb{U}_{sr}$-valued ultrafilter in $\mathbf{V}_{\leq r}^{\mathbb{U}_{sr'}}$, represented by a thoroughly un-"sharp" $(\mathcal{U}_{\leq\infty}, v_{\leq\infty})$- (or $(\mathcal{Z}_{\leq\infty}, \zeta_{\leq\infty})$-, ...)-random element of the space $(2^\omega, \lambda)$ *of countably many tosses of an honest coin.*

9 The Elusive Ideals of a "Closed" and "Isolated System"

9.1 Given a "quantum system" $S = (\mathcal{H}, \varphi, H)$, define
(i) "closure" of such a "system" as freedom from "unobserved boundary-conditions",

and
(ii) "isolation" of such a "system" as absence of perturbative "measurements" of it.

Then
(iii) "closure" is mathematically as well as metamathematically unstable, and "isolation" is topologically negligible.

9.2 Remarks
(i) By "unstable" in (iii), I mean that random "perturbations" may impinge on the system's time-evolution in assorted mathematically or metamathematically unobservable ways.
(ii) "Closure", "stability", "observation" and systematic "isolation" are loosely interpretable notions, and "closure" and "isolation" are often identified. Mindful of this, I offer the following arguments for the assertions in 9.1(ii) and 9.1(iii) as (heuristic) *"scholia"*.

9.3 "Isolation"
Consider
(i) the real Hilbert space $\mathcal{H}$ of martingales generated by real-valued uniformly square-integrable "measurement-processes" $\dot{\mathcal{M}}_r = \dot{\mathcal{M}}_r(u) = \dot{\mathcal{M}}_r(u_{\leq\infty})$ of the sort introduced in Section 5; and
(ii) the subspace $\mathcal{H}_c$ of such processes generated by processes $\dot{\mathcal{M}}_r = \dot{\mathcal{M}}_r(u) = \dot{\mathcal{M}}_r(u_{\leq\infty})$ whose "sample paths" are almost everywhere continuous in r for $-\infty \leq r \leq \infty$,

i.e.,
(iii) those processes generated by elements $\dot{\mathcal{M}}_r(u)$ of $\mathcal{H}$ whose "discrete parts" $\dot{\mathcal{M}}_{d,r}$ and martingale measures $\mu_r(\sigma)$ vanish $v_{\leq\infty}$-almost everywhere (and whose potential "measurements" are "never implemented").

Then
(iv) $\mathcal{H}_c$ is a "rare" (i.e., closed, nowhere-dense) subset of $\mathcal{H}$.

Proof

Verification of (iii) is straightforward, for $\mathcal{H}_c \subseteq \mathcal{H}$ is a closed proper subspace of the real Hilbert space $\mathcal{H}$ (cf., e.g., Ikeda and Watanabe, 1981, 47), and every such subspace is nowhere dense in $\mathcal{H}$.

For if $\mathcal{N}_r(u)$ is an arbitrary element of $\mathcal{H}_c$ and $O_\varepsilon = O_\varepsilon(\mathcal{N}_r(u))$ is an arbitrarily small ε-neighborhood of $\mathcal{N}_r(u)$, one can use the martingale decomposition theorem above (or Kallenberg, 2001, 527) to find a cadlag but non-continuous square-integrable martingale $\mathcal{N}'_r(u)$ in the orthogonal complement of $\mathcal{H}_c$. One can then "shrink" this $\mathcal{N}'_r(u)$ to make the non-continuous sum $\mathcal{N}_r(u) + \mathcal{N}'_r(u)$ an element of O_ε, and define a smaller neighborhood $O_{\varepsilon'}$ of $\mathcal{N}_r(u) + \mathcal{N}'_r(u)$ which is disjoint from $\mathcal{H}_c$.

9.4 "Closure"

By 9.3 and the arguments of prior sections, reconstructions of "stochastic" and "virtual" representations of measured time-evolution have in common that

(i) "bare" evolution of a quantum system is "almost always" mediated by "measurements";

that

(ii) the outcomes and readings of such systems' evolution, as suggested earlier, are aleatory in prospect, and appear to be "sharp" only in retrospect;

and that

(iii) Hilbert spaces $\mathcal{H}$ of "random" and "virtual" processes which represent such measurements are relatively indifferent to particular representations of the "bare" systems they "measure",

a pattern exemplified by the enormous variety of L^2-spaces which may be represented as spatially isomorphic images of subspaces of $L^2(\mathcal{U}_{\leq\infty}, \mathcal{V}_{\leq\infty})$.

9.5 With these preliminary observations in mind, consider therefore the following argument.

If $\mathfrak{H}_H$ is the infinitesimal generator of Theorem 6.10"s continuous unitary group $U_{H,s}$ on the Hilbert space $\mathcal{H} \cong L^2(\mathcal{U}_{\leq\infty}, \mathcal{V}_{\leq\infty})$, time-evolution governed by $\mathfrak{H}_H$ can be "measured" by

(i) "new" Hilbert spaces such as $L^2(\mathcal{Z}_{\leq\infty\leq\infty}, \zeta_{\leq\infty\leq\infty})$ (cf. 6.16), where $\zeta_{\leq\infty}$ is a $(\lambda^1_{\leq\infty} \otimes \tau_{\leq\infty} \otimes \pi_{\leq\infty})$-absolutely continuous measure on $\mathcal{Z}_{\leq\infty} =_{Df} \mathcal{L}^1_{\leq\infty} \times \mathcal{T}_{\leq\infty} \times \mathcal{P}_{\leq\infty}$.

More generally, unitary groups generated by self-adjoint operators in arbitrary "observational frames" $L^2(S_{z \leq \infty}, \sigma_{\leq \infty})$ can be

(ii) "remeasured" (cf. 6.15) in hierarchies of "higher-order randomisations" $(S_{z \leq \infty}, \sigma_{\leq \infty})$ and their measure-algebras $\mathbb{S}_{\leq \infty}$.

9.6 Such iterations of "random boundary-conditions" might introduce (for example)

(i) "new" classes $\mathcal{O}_{\leq \infty}$ of theories $\Theta_{\leq \infty}$ which prove the consistency of corresponding theories $T_{\leq \infty}$ in $\mathcal{T}_{\leq \infty}$, and are in that sense "hidden" to such $T_{\leq \infty}$; or they might involve

(ii) "new" classes $\mathcal{Q}_{\leq \infty}$ of measures $\varsigma_{\leq \infty}$ which are singular with respect to the "canonical" Lebesgue measures $\pi_{\leq \infty}$ (respectively $(\tau_{\leq \infty} \otimes \pi_{\leq \infty})$, $(\lambda^1_{\leq \infty} \otimes \tau_{\leq \infty} \otimes \pi_{\leq \infty}))$ on $\mathcal{P}_{\leq \infty}$ (respectively $(\mathcal{U}_{\leq \infty}, \mathcal{Z}_{\leq \infty})$);

or . . . .

9.7 Such thought-experimental iterations give rise to

(i) indefinitely ascending sequences of observational frames $L^2(\mathcal{U}_{1, \leq \infty}, \upsilon_{1, \leq \infty})$, $L^2(\mathcal{U}_{2, \leq \infty}, \upsilon_{2, \leq \infty})$, ..., $L^2(\mathcal{U}_{n, \leq \infty}, \upsilon_{n, \leq \infty})$, ..., in what might be called

(ii) "Wigner's ladder", or "von Neumann's hierarchy" (cf., e.g., Wigner, 1971, 16–17),
whose frames $L^2(\mathcal{U}_{n, \leq \infty}, \upsilon_{n, \leq \infty})$ would be

(iii) indefinitely re-representable, in the sense that they would all be spatially isomorphic to the "canonical" Hilbert space $L^2(2^\omega, \lambda)$.

Historically, the underlying mathematical rationale for the "homogeneity" in 9.7(iii)—the "universality of the Cantor space of binary expansions' 2^ω"— was known to Leibniz, who drew on it to outline the basic idea of a propositional Stone space (cf. Leibniz, 1966, 1–3, and the brief glosses of these passages in "Virtual Modality," Boos, 2003b, 283).

Leibniz might therefore have found partial confirmation in the foregoing argument for one of his more striking conjectures: that

> chaque Monade est un miroir . . . doué d'action interne, représentatif de l'univers, suivant son point de vue, et aussi reglé que l'univers lui-même.

> every monad is a mirror ... endowed with an internal action, representative of the universe, following its point of view, and as regulated as the universe itself.
> (Leibniz, 1978, VI, 599)

10 A Promissory Note and Postscript

10.1 Promissory Note

I hope to draw on the chapter's measure-theoretic apparatus and "virtual" constructions to consider

(i) mutually singular (and in that sense epistemically "inaccessible") "observational measures" (as potential sources of "dark energy", perhaps—cf. 9.6(ii));

and on measure-algebraic counterparts of (Boos, 2003b)'s Boolean-valued semantic interpretation of quantified modal logic to offer "virtual" reconstructions of

(ii) "temporally" indexed "contextual hidden variables"; and
(iii) Bas van Fraassen's (diodorean) "modal" interpretation of nonrelativistic quantum mechanics (cf. van Fraassen, 1991, 299–330).

Here

(iv) "mutual singularity" in (i) has its usual measure-theoretic sense, given by Lebesgue-Radon- Nikodym analyses of real and complex measures, and
(v) "diodorean" (after Diodorus Cronus) refers in (iii) to a first-order modal logic whose "necessity"-operator asserts a sentence is "necessary" if and only if "it is and always will be the case".

10.2 Postscript

This chapter's underlying Ansätze have been

(i) that the "book of [natural] philosophy", as Galileo observed, is written "in the language of mathematics";
(ii) that critiques and interpretations of what is "written" in this (immeasurably vast) "book" should therefore be undertaken (in part at least) in "the language(s) of metamathematics";

and finally

(iii) that heuristic efforts to postulate the "physical existence" of "worlds" which "randomly" evolve and ramify "forever" almost cry out for such metamathematical clarification and interpretation.

To me at least, these Ansätze also suggest the following "regulative" conjectures (in metamathematical variants of the senses in which Immanuel Kant often employed the word "*regulativ*"—to refer to "principles" which are

(iv) "only rule[s] which prescribe a regress in which one is never permitted to halt at an utterly unconditioned [limit]"
("*nur ... Regel[n], welche ... einen Regressus gebiete[n], dem es niemals erlaubt ist bei einem Schlechthinunbedingten stehen zu bleiben*") (Kant, *KdrV*, B 536–537):

namely that
(v) "everything is processively random";
(vi) "the physical" is an essentially incomplete notion;
(vii) the "intrinsically probabilistic" aspects of quantum-theoretic "ontology" reflect the semantic pluralities and potentialities which follow from such incompleteness;

and a bit more precisely (cf. 8.9)), that heuristic analogies might exist between
(viii) hierarchies of

[.... random interpretations of "sharp" observations of ....]
[.... random interpretations of "sharp" observations of ....],

and
(ix) hierarchies of

[.... metatheoretic interpretations of theoretical assertions about ....]
[.... metatheoretic interpretations of theoretical assertions about ....];

and that these analogies offer confirmation for yet another conjecture—epistemological this time: that
(x) incompleteness-phenomena reflect "intrinsic" dilemmas in any "mathematical principles of natural philosophy" "critical" enough to examine "questions ...

[they] cannot decide" (cf. Kant's "fate of reason" (or "*Schicksal der Vernunft*"), in Kant, *KdrV*, AVII).

Bibliography

Abelard, Peter. *Peter Abelards Philosophische Schriften*, B. Geyer, ed., *Beitrage zur Geschichte der Philosophie des Mittelalters*, 21. Münster: Achendorff, 1933, 7–30.
Åberg, Claus. "Relativity Phenomena in Set Theory," *Synthese*, 27 (1974), 189–98.
Aczel, Peter. "Saturated Intuitionistic Theories," H. A. Schmidt et al. eds., *Contributions to Mathematical Logic*. Amsterdam: North Holland, 1968. 1–13.
Annas, Julia and Jonathan Barnes. *The Modes of Skepticism, Ancient Texts and Modern Interpretations*. Cambridge: Cambridge University Press, 1985.
Anselm of Canterbury. *Fides Quaerens Intellectum*, ed. with a French translation by Alexandre Koyré. Paris: Vrin, 1978.
Anselm of Canterbury. *Monologion* and *Proslogion*, in *A New Interpretive Translation of St. Anselm's Monologion and Proslogion*, ed. with an English translation by Jasper Hopkins. Minneapolis: Banning, 1986.
Anselm of Canterbury. *St. Anselms Proslogion, with a Reply on Behalf of the Fool by Gaunilo and the Author's Reply to Gaunilo*. Trans'. M.J. Charlesworth. Notre Dame, Indiana: University of Notre Dame Press, 1979. Gaunilo's "Reply" is 83–89.
Aquinas, Thomas. *Summa Theologiae*. Madrid: Biblioteca de Autores Cristianos, 1951.
Aristotle. *Metaphysics*, ed. with an English translation by Hugh Tredennick. Cambridge: Harvard University Press, 1975 and 1977.
Aristotle. *Nicomachean Ethics*, ed. with an English translation by H. Rackham. Cambridge: Harvard University Press, 1975.
Barcan, R. C. "A Functional Calculus of First Order based on Strict Implication," *Journal of Symbolic Logic*, 11 (1946), 1–16.
Bartlett, James. "Funktion und Gegenstand. Eine Untersuchung in her Logik von Gottlob Frege," Ph. D. Dissertation, University of Munich, 1961.
Barwise, Jon, ed. *Handbook of Mathematical Logic*. Amsterdam: North Holland, 1977.
Baumgartner, James. "Iterated Forcing," A. R. D. Mathias, ed., *Surveys in Set Theory*. Cambridge: Cambridge University Press, 1983.
Bell, John. *Boolean-Valued Models and Independence Proofs in Set Theory*, 2nd edn. Oxford: Clarendon Press, 1985.
Bell, John and Moshe Machover. *A Course in Mathematical Logic*. Amsterdam: North Holland, 1977.
Bell, John and A. B. Slomson. *Models and Ultraproducts*. Amsterdam: North Holland, 1969.
Beller, Aaron, Ronald Jensen, and Philip Welch. *Coding the Universe*. Cambridge: Cambridge University Press, 1982.
Berkeley, George. *The Works of George Berkeley, Bishop of Cloyne*, ed. A. A. Luce and T. E. Jessop. 9 vols. London and Edinburgh: Thomas Nelson, 1948–57; reprint, Nendeln: Kraus Reprint, 1979.
Billingsley, Patrick. *Convergence of Probability Measures*. London: Wiley, 1968.
Blaedel, Niels. *Harmoni og Enhed, Niels Bohr: En Biografi*. Kobenhavn: Rhodos, 1985.
Blank, Jiri, Pavel Exner, and Miloslav Havlicek. *Hilbert Space Operators in Quantum Physics*. New York: AIP Press, 1994.
Bolton, Martha. "Berkeley's Objection to Abstract Ideas and Unconceived Objects", E. Sosa. ed., *Essays on the Philosophy of George Berkeley*. Dordrecht: Reidel, 1987, 61–81.

Bolzano, Bernard. *Bernard Bolzano's Grundlegung der Logik, Ausgewählte Paragraphen aus der Wissenschaftslehre*, Band I and II, ed. Friedrich Kambartel. Hamburg: Felix Meiner Verlag, 1975. (*Wissenschaftslehre* first published in 1837).

Bolzano, Bernard. *Pardoxien des Unendlichen*. Hamburg: Felix Meiner Verlag, 1975.

Boolos, George. "A New Proof of the Gödel Incompleteness Theorem," *Notices of the American Mathematical Society*, 36.4 (1989), 388–90.

Boolos, George, *The Unprovability of Consistency*. Cambridge: Cambridge University Press, 1979.

Boos, William. "Consistency and *Konsistenz*," *Erkenntnis*, 26 (1987a), 1–43.

Boos, William. *Lectures on Large Cardinal Axioms*. Berlin: Springer Verlag, 1974.

Boos, William. "Limits of Inquiry," *Erkenntnis*, 20 (1983a), 150–94.

Boos, William. "A Metalogical Critique of Wittgenstein's 'Phenomenology,'" Dan Kolak and J. Symons, eds., *Quantifiers, Questions and Quantum Physics: Essays on the Philosophy of Jaakko Hintikka*. New York: Springer, 2004, 75–99.

Boos, William. "Metamathematical Quantum Theory: Random Ultrafilters as 'Hidden Variables'", *Synthese*, 107 (1996), 83–143.

Boos, William. "'*Parfaits miroirs de l'univers*': A 'Virtual' Interpretation of Leibnizian Metaphysics", *Synthese*, 136 (2003a), 281–304.

Boos, William. "A Self-referential 'Cogito'", *Philosophical Studies*, 44 (1983b), 269–90.

Boos, William. "Stochastic Representations of Feynman Integration", *Journal of Mathematical Physics*, 48.12 (2007) 122106-1–122106-18.

Boos, William. "Theory-relative Skepticism", *Dialectica*, 41.3 (1987b), 175–207.

Boos, William. "Thoralf Skolem, Hermann Weyl and 'Das Gefühl der Welt als Begenztes Ganzes,'" Jaako Hintikka, ed., *Essays on the Development of the Foundations of Mathematics*. Dordrecht: Reidel, 1995, 283–329.

Boos, William. "The *Transzendenz* of Metamathematical 'Experience,'" *Synthese*, 114 (1998), 49–99.

Boos, William. "'The True' in Gottlob Frege's 'Über die Grundlagen der Geometrie,'" *Archive for History of Exact Sciences*, 34 (1985), 141–92.

Boos, William. "Virtual Modality," *Synthese*, 136 (2003b), 435–91.

Boos, William. "The World, the Flesh, and the Argument from Design," *Synthese*, 101 (1994), 15–52.

Borkar, Vivek. *Probability Theory: An Advanced Course*. New York: Springer, 1995.

Boscovich, Rogerius Josephus. *Theoria Philosophiae Naturalis*. Venice, 1763.

Boswell, James. *The Life of Samuel Johnson*. London: Dent, 1976.

Brook, Richard. *Berkeley's Philosophy of Science*. The Hague: Nijhoff, 1973.

Browne, Thomas. "Religio Medici", Geoffrey Keynes, ed., *Selected Writings*. Chicago: University of Chicago Press, 1968.

Bull, Robert and Krister Segerberg. "Basic Modal Logic," D. Gabbay and F. Guenther, eds., *Handbook of Philosophical Logic, Volume II, Extensions of Classical Logic*. Dordrecht: Reidel, 1984, 1–88.

Burnyeat, Myles. "Can the Skeptic Live His Skepticism?," Malcolm Schofield, Myles Burnyeat and Jonathan Barnes, eds., *Doubt and Dogmatism*. Oxford: Clarendon Press, 1980, 20–54.

Burnyeat, Myles, ed. *The Skeptical Tradition*. Berkeley: University of California Press, 1983.

Cantini, Andrea. "Paradoxes and Contemporary Logic," *Stanford Encyclopedia of Philosophy* (Fall 2014 edition), ed. Edward N. Zalta, http://plato.stanford.edu/archives/fall 2014/entries/paradoxes-contemporary-logic

Cantor, Georg. *Gesammelte Abhandlungen*. Berlin: Springer Verlag, 1933; reprint Hildesheim: Olms Verlag, 1966.
Capaldi, Nicholas, James King and Donald Livingston. "The Hume Literature of the 1970's," *Philosophical Topics*, 12.3 (1981), 167–92.
Cassirer, Ernst. *Das Erkenntnisproblem in der Philosophie und Wissenschaft der neureren Zeit*. New Haven: Yale University Press, 1922; reprint Darmstadt: Wissenschftliches Buchgesellschaft, 1974.
Chaitin, Gregory. "Randomness and Mathematical Proof", *Scientific American* (May, 1975), 47–52.
Chang, C. C. and H. J. Keisler, *Model Theory*. Amsterdam: North Holland, 1972.
Cicero. *De Natura Deorum, Academica*, with an English translation by H. Rackham. Cambridge: Harvard University Press, 1972.
Cusanus, Nicholas. *Opera*. Paris: Iod. Badium Ascensium, 1514.
Davidson, Donald. *Inquiries into Truth and Interpretation*. Oxford: Oxford University Press, 2001.
Dedekind, Richard. *Was sind und was sollen die Zahlen*. 10th ed. Braunschweig: Friedrich Vieweg und Sohn, 1969.
Descartes, René. *Discours de la méthode, texte et commentaire*, Etienne Gilson, ed. Paris: J. Vrin, 1976.
Descartes, René. *Oeuvres de Descartes*, Charles Adam et Paul Tannery, ed., Nouvelle Présentation. Paris: Librairie Philosophique, J. Vrin, 1973–1978.
Devlin, Keith. "Constructibility," in Barwise, *Handbook of Mathematical Logic*, 453–92.
Devlin, Keith. *Constructibility*. Berlin: Springer, 1984.
Diels, Hermann and Walter Kranz. *Fragmente der Vorsokratiker*. Berlin: Weidmannsche Verlagsbuchhandlung, 1951.
Duhem, Pierre. *La Thèorie physique, son object, sa structure*. Paris: Marvel Rivière et Cie, 1974.
Dummett, Michael. "Frege on the Consistency of Mathematical Theories," in [Schirn 1975], *Studies on Frege*, ed. Matthias Schirn, 229–42.
Dodd, A. J. *The Core Model*. Cambridge: Cambridge University Press, 1982.
Dummett, Michael. *Intuitionism*. Oxford: Clarendon Press, 1977.
Enderton, Herbert. *A Mathematical Introduction to Logic*. New York: Academic Press, 1972.
Ehrenfeucht, Andrzei and Andrzei Mostowski. "Models of Axiomatic Theories Admitting Automorphisms," *Fundamenta Mathematica*, 43 (1956), 50–68.
Enderton, Herbert. *A Mathematical Introduction to Logic*. New York: Academic Press, 1972.
Everett, Hugh [with John A. Wheeler]. "'Relative State' Formulation of Quantum Mechanics," *Reviews of Quantum Physics*, 29 (1957): 454–62.
Feferman, Solomon. "Theories of Finite Type Related to Mathematical Practice," in Barwise, *Handbook of Mathmatical Logic*, 913–72.
Fisch, Max. *Peirce, Semeiotic and Pragmatism*. Bloomington: Indiana, 1986.
Flage, Daniel. *Berkeley's Doctrine of Notions: A Reconstruction based on His Theory of Meaning*. New York, St. Matins, 1987.
Frede, Michael. "Des Skeptikers Meinungen," *Neue Hefte fur Philosophie*, 15.16 (1979), 102–29.
Frege, Gottlob. *Begriffsschrift, eine der arithmetischen nachgebildete Formelsprache des reinen Denkens*. Halle: L. Nebert, 1879; reprint *Begriffsschrift und andere Aufsätze*, ed. Ignacio Angelelli. Hildeshiem: Olms Verlag, 1964.
Frege, Gottlob. *Funktion, Begriff, Bedeutung, Fünf Logische Studien*, ed. Günther Patzig. Göttingen: Vandenhoeck und Ruprecht, 1962.
Frege, Gottlob. *Grundegesetze der Arithmetik*. Band I. Jena: Pohle Verlag, 1893; Band II. Jena: Pohle Verlag, 1903; Olms Verlag, 1962.

Frege, Gottlob. *Die Grundlagen der Aritmetik*. Breslau: Verlag von Wilhelm Koebner, 1884; reprint translated J. L. Austin, Oxford: Basil Blackwell, 1974.
Frege, Gottlob. *Die Grundlagen der Arithmetik*. Breslau: Verlag von Wilhelm Koebner, 1884; reprint Hildesheim: Olms Verlag, 1977.
Frege, Gottlob. *Kleine Schriften*, ed. Ignacio Angelelli. Hildesheim: Olms Verlag, 1967.
Frege, Gottlob. "Kritische Beleuchtung einiger Punkte in E. Schröders Vorlesungen über die Algebra der Logik," *Archiv für Systematische Philosophie*, 1, 433–56. Reprinted in [Frege 1960].
Frege, Gottlob. *Logische Untersuchungen*, ed. Günther Patzig. Göttingen: Vandenhoeck und Ruprecht, 1966.
Frege, Gottlob. *Nachgelassene Schriften*, ed. Hans Hermes, Friedrich Kambartel und Friedrich Kaulbach, Hamburg: Felix Meiner, 1969.
Frege Gottlob. "Über die Grundlagen der Geometrie," I und II. *Jahresberichte der Deutschen Mathematikvereinigung*, XII (1903), 319–29, 368–75.
Frege Gottlob. "Über die Grundlagen der Geometrie," I, II und III. *Jahresberichte der Deutschen Mathematikvereinigung*, XV (1906), 293–309, 377–403, and 423–30.
Frege, Gottlob. *Wissenschaftlicher Briefwechsel*, ed. Gottfired Gabriel et al. Hamburg: Felix Meiner Verlag, 1976.
Freudenthal, Hans. "The Main Trends in the Foundations of Geometry in the 19th Century," in [Nagel, Suppes, Tarski], *Logic, Methodology and Philosophy of Science*, ed. Ernest Nagel, Patrick Suppes and Alfred Tarski. Stanford: Stanford University Press, 1962, 613–21.
Friedman, Sy. "Strong Coding," *Annals of Pure and Applied Logic*, 35 (1987), 1–98.
Gadamer, Hans Georg. *Wahrheit und Methode*. Tübingen: Mohr, 1660.
Galilei, Galileo. *The Assayer. Opere*, 6. Napoli: F. Rossi, 1970.
Garber, Daniel. *Leibniz: Body, Substance, Monad*. Oxford: Oxford University Press, 2009.
Garey, Michael and David Johnson. *Computers and Intractability, A Guide to the Theory of NP-Completeness*. San Francisco: Freeman, 1979.
Garson, James. "Quantification in Modal Logic," D. Gabbay and F. Guenther, eds., *Handbook of Philosophical Logic, Volume II, Extensions of Classical Logic*. Dordrecht: Reidel, 1984, 249–308.
Gaunilo of Marmontiers. "An Answer to Anselm," http://philosophy.lander.edu/intro/articles/gaunilo-a.pdf
Gaunilo of Marmontiers, "In Praise of the fool." See Anselm, *Prologion, with a Reply on Behalf of the Fool*.
Gihman, Iosif and Skorohod, Anatoli. *The Theory of Stochastic Processes*, Vol. III. Berlin: Springer, 1979.
Gilson, Etienne, ed., *Descartes, René: Discours de la méthode, texte et commentarie*. Paris: J. Vrin, 1976.
Gödel, Kurt, *Collected Works, Volumes I–III*, eds. Solomon Feferman, John Dawson et al. Oxford: Oxford University Press, 1986–1995.
Goldblatt, Robert. *The Mathematics of Modality*. Stanford: Center for the Study of Information and Language, 1993.
Goldfarb, Warren. "Logic in the Twenties: The Nature of the Quantifier", *Journal of Symbolic Logic*, 44.3 (1979): 351–68.
Grattan-Guiness, Ivor. "How Bertrand Russell Discovered his Paradox," *Historia Mathematica*, 5 (1978), 127–37.
Grayling, A. C. *Berkeley: The Central Arguments*. Lasalle: Open Court, 1986.

Gueroult, Martial. *Descartes selon l'ordre des raisons*. Paris: Aubier, 1953.
Hadamard, Jacques. *An Essay on the Psychology of Invention in the Mathematical Field*. Princeton: Princeton University Press, 1954.
Hardy, G. H. *A Mathematician's Apology*. Cambridge: Cambridge University Press, 1967
Hawking, S. W. and G. F. R. Ellis. *The Large Scale Structure of Spacetime*. Cambridge: Cambridge University Press, 1973.
Hegel, Georg. *Theorie Werkausgabe*, Band 5: Wissenschaft der Logik I. Frankfurt: Suhrkamp Verlag, 1969.
Henkin, Leon. "The Completeness of the First-Order Functional Calculus", *Journal of Symbolic Logic*, 14 (1949), 159–66.
Henkin, Leon. "Completeness in the Theory of Types", *Journal of Symbolic Logic*, 15 (1950), 81–91.
Henkin, Leon, Patrick Suppes and Alfred Tarski. *The Axiomatic Method*. Amsterdam: North Holland, 1958.
Hilbert, David. "Axiomatisches Denken," *Mathematische Annalen*, 78 (1918), 405–15; reprinted in Hilbert, *Gesammelte Abhandlungen*, 1935, 146–56.
Hilbert, David. *Gesammelte Abhandlungen*. Berlin: Springer, 1935.
Hilbert, David, *Grundlagen der Geometrie*, 2nd ed. Leipzig: Teubner Verlag, 1903; 7th ed. Teubner, 1930.
Hilbert, David. "Mathematische Probleme. Vortrag, gehalten auf dem internationalen Mathematiker-Kongress zu Paris 1900," *Archiv der Mathematik und Physik*, 3rd series, 1 (1901), 44–63, 213–37; reprinted in Hilbert, David, *Gesammelte Abhandlungen*, Berlin: Springer Verlag, 1935, 290–329.
Hilbert, David, "Über die Grundlagen der Logik und der Arithmetik", *Verhandlungen des Dritten Internationalen Mathematiker-Kongresses in Heidelberg vom 8. bis 13. August 1904*. Leibzig: Teubner, 1905.
Hintikka, Jaakko. "Cogito Ergo Sum: Inference or Performance?" *Philosophical Review* 71, 3–32; reprinted in Willis Doney, ed., *Descartes, A Collection of Critical Essays*. New York: Anchor. 1967.
Hintikka, Merrill and Jaakko Hintikka. *Investigating Wittgenstein*. Oxford: Blackwell, 1986.
Hintikka, Jaakko. *Language, Truth and Logic in Mathematics (Jaakko Hintikka Selected Papers, Volume III)*. Dordrecht: Kluwer, 1998.
Hintikka, Jaakko. *Ludwig Wittgenstein, Half-Truths and One-and-a-Half Truths (Jaakko Hintikka Selected Papers, Volume I)*. Dordrecht: Kluwer, 1996.
Hintikka, Jaakko. "Modality and Quantification," *Theoria*, 27 (1961), 119–28.
Hintikka, Jaakko. *Models for Modalities: Selected Essays*. Dordrecht: Reidel, 1969.
Hintikka, Jaakko. *Quantifiers in Deontic Logic*, Vol. 12. Helsinfors: Societas Scientarum Fennica, 1957.
Hintikka, Jaakko. *Wittgenstein*. Belmont: Wadsworth, 2000.
Hughes, G. E. and M. J. Cresswell. *An Introduction to Modal Logic*. London: Methuen, 1972.
Hume, David. *Dialogues Concerning Natural Religion*, ed. J. Price. Oxford: Clarendon Press, 1976.
Hume, David. *Enquiries Concerning Human Understanding and Concerning the Principles of Morals*, ed. P. H. Nidditch. Oxford: Clarendon Press, 1975.
Hume, David. *Essays Moral, Political and Literary*. Oxford: Oxford University Press, 1963.
Hume, David. *A Treatise of Human Nature*, ed. P. H. Nidditch. Oxford: Oxford University Press, 1978.
Husserl, Edmund. *Cartesianische Meditationen*. Hamburg: Meiner, 1977.

Husserl, Edmund. *Die Krisis der europäischen Wissenshaften und die transzendentale Phänomenologie*, Hamburg: Meiner, 1977.
Husserl, Edmund. *Logische Untersuchungen II/1*. Tübingen: Niemeyer, 1913.
Ikeda, Nobuyuki and Shinzo Watanabe. *Stochastic Differential Equations and Diffusion Processes*. Amsterdam: North Holland, 1981.
James, William. "The Will to Believe," *New World*, June, 1896.
Jech, Thomas. *Set Theory*. New York: Academic Press, 1978; Berlin: Springer, 2002.
Jesseph, Douglas. *Berkeley's Philosophy of Mathematics*. Chicago: University of Chicago Press, 1993.
Jonsson, Bjarni and Alfred Tarski. "Boolean Algebras with Operators, Part I," *American Journal of Mathematics*, 73 (1951), 891–939.
Kallenberg. Olav. *Foundations of Modern Probability*. Second Edition, New York: Springer, 2001.
Kambartel, Friedrich. "Frege und die aximatische Methode, Zur Kritik mathematikhistorischer Legimationsversuch der formalistischen Ideologie," Matthias Schirn, ed., *Studies in Frege*, Stuttgart/Bad Constalt: Fromann-Holzboog-Verlag, 1975, 215–28.
Kanamori, Akihiro. *The Higher Infinite*. Berlin: Springer, 1994.
Kanger, Stig. "The Morning Star Paradox," *Theoria*, 23 (1957), 1–11.
Kanger, Stig. New Foundations for Ethical Theory, Stockholm, 1971, reprinted in Hilpinen, R., ed., *Deontic Logic: Introductory and Systematic Readings*. Dordrecht: Reidel, 1971.
Kanger, Stig. "A Note on Quantification and Modalities," *Theoria*, 13 (1957), 131–34.
Kanger, Stig. *Provability in Logic*. Ph. D. dissertation, Stockholm, 1971.
Kant, Immanuel. *Grundlegung der Metaphysik der Sitten*, ed. Karl Vorländer. Hamburg: Meiner, 1965.
Kant, Immanuel. *Kritik der prakitschen Vernunft*. Meiner: Hamburg, 1974.
Kant, Immanuel. *Kritik der reinen Vernunft*, ed. Raymund Schmidt. Hamburg: Meiner, 1956.
Kant, Immanuel, *Kritik der Urteilskraft*, ed. Karl Vorländer. Hamburg: Meiner, 1974.
Kant, Immanuel. *Prolegomena zu einer jeden künftigen Metaphysik*, ed. Karl Vorländer. Hamburg: Meiner, 1976.
Kant, Immanuel. *Werke, Briefwechsel und Nachlaß auf CD Rom*. Berlin: Karsten Worm, 2003 (*KW*; cited roman numerals refer to the 1905 *Akademie-Ausgabe*).
Karr, Alan. *Probability*. New York: Springer, 1992.
Kaye, Richard. *Models of Peano Arithmetic*. Oxford: Clarendon Press, 1991.
Keats, John, *Selected Poems and Letters*, ed. Douglas Bush. Boston: Houghton Mifflin, 1959.
Kleene, Stephen. *Introduction to Metamathematics*. Amsterdam: North-Holland, 1974.
Kline, Morris. *Mathematical Thought from Ancient to Modern Times*. Oxford: Oxford University Press, 1972.
Kolmogorov, Andrey. *Foundations of the Theory of Probability*. 2nd ed. New York: Chelsea, 1956.
König, Julius. "Zum Kontinuum-Problem," *Mathematische Annalen*, 60 (1905), 117–80, in *From Frege to Gödel*, ed. Jan van Heijenoort, 1967, 145–49.
Koppelberg, Sabine, J. D. Monk, and R. Bonnet, eds., *Handbook of Boolean Algebras*, vol. 1. Amsterdam: North Holland, 1989.
Korselt, Alwin. "Paradoxien in der Mengenlehre," *Jahresberichte der Deutschen Mathematikervereinigung*, 15 (1906), 215–19.
Korselt, Alwin. "Über die Grundlagen der Geometrie," *Jahrsberichte der Deutschen Mathematikervereinigung*, 12 (1903), 402–07.
Korselt, Alwin. "Über die Gundlagen der Mathematik," *Jahrsberichte der Deutschen Mathematikervereinigung*, 14 (1905), 365–89.

Korselt, Alwin. "Über die Logik der Geometrie," *Jahresberichte der Deutschen Mathematikerveringung*, 17 (1908), 98–124.
Kotlarski, H., S. Krajewski, and A. H. Lachlan. "Construction of Satisfaction-Classes for Nonstandard Models," *Canadian Mathematical Bulletin*, 24 (1981), 283–93.
Kripke, Saul. "A Completeness Theorem in Modal Logic," *Journal of Symbolic Logic*, 24 (1959), 1–14.
Kripke, Saul. "Semantic Considerations on Modal Logic," *Acta Philosophica Fennica*, 16 (1963), 83–94.
Kripke, Saul. "Semantical Analysis of Modal Logic I: Normal Propositional Calculi," *Zeitschrift fur Mathematische Logik und Grundlagenforschung*, 9 (1963), 67–96.
Kripke, Saul. "Semantical Analysis of Modal Logic II: Non-Normal Propositional Calculi," J. W. Addison et al., eds. *The Theory of Models*. Amsterdam: North Holland, 206–20.
Kunen, Kenneth, *Set Theory, An Introduction to Independence Proofs*. Amsterdam: North Holland, 1980.
Kuratowski, Kasimir. *Topology*, Vol. I. New York: Academic Press, 1966.
Lachlan, A. H. "Full Satisfaction Classes and Recursive Saturation," *Canadian Mathematical Bulletin*, 24 (1981), 295–97.
Laplace, Pierre Simon. *Essai philosophique sur les probabilités*, Sixième Édition. Paris: Bachelier, 1840.
Largeault, Jean. *Logique et philosophie chez Frege*. Publications de la Faculté des Lettres et Sciences de Paris-Sorbonne. Série « Recherches » tome 50. Louvain: Editions Nauwelaerts, 1970.
Leibniz, Gottfried. *Die philosophische Schriften von Gottfried Wilhelm von Leibniz*, ed. C. Gerhardt. Hildesheim: Olms, 1978.
Leibniz, Gottfried. *Opuscules et fragments inédits*, ed. Louis Couturat. Hildesheim: Olms, 1966.
Leibniz, Gottfried. *Nouvelles lettres et opuscules inédits de Leibniz*, ed. A. Foucher de Careil. Hildesheim: Olms, 1971.
Leibniz, Gottfried. *Philosophical Essays*, ed. and trans. by Daniel Garber and Roger Ariew. Indianapolis: Hackett, 1989.
Leibniz, Gottfried. *Philosophical Papers and Letters*, ed. and trans. by Leroy Loemker. 2nd ed. Dordrecht: Reidel, 1969.
Leibniz, Gottfried and Christian Wolff. *Briefwechsel zwischen Leibniz und Christian Wolff*, ed. C. Gerhardt. Hildesheim: Olms, 1963.
Levy, Azriel. *Basic Set Theory*. Berlin: Springer, 1979; Mineola: Dover, 2003.
Lindström, Per. "On Extensions of Elementary Logic," *Theoria*, 35 (1969), 1–11.
Locke, John. *An Essay concerning Human Understanding*, ed. Peter Nidditch. Oxford: Clarendon Press, 1975.
Loewer, Barry. "Descartes' Skeptical and Antiskeptical Arguments," *Philosophical Studies*, 39 (1981), 163–82.
Lovejoy, Arthur. *The Great Chain of Being*. New York: Harper, 1965.
Luria, Alexander. *The Mind of a Mnemonist*, trans. Lynn Solataraff. New York: Basic Books, 1968.
MacIntosh. J. "Leibniz and Berkeley", *Proceedings of the Aristotelian Society* (1970–1971), 147–63.
Mates, Benson. *The Philosophy of Leibniz*. Oxford: Oxford University Press, 1986.
McKinsey, J. C. C. and A. Tarski. "The Algebra of Topology," *Annals of Mathematics*, 45 (1944), 141–91.
Monk, J. D. *Mathematical Logic*. Berlin: Springer, 1976.
Moore, Gregory. *Zermelo's Axiom of Choice*. Berlin: Springer, 1980.

Moschovakis, Yianis. *Descriptive Set Theory*. Amsterdam: North Holland, 1980.
Muehlmann, Robert, *Berkeley's Ontology*. Indianapolis: Hackett, 1992.
Murray, Michael and Michael Cannon Rea. *An Introduction to the Philosophy of Religion*. Cambridge: Cambridge University Press, 2008.
Myhill, John and Dana Scott. "Ordinal Definability," Dana Scott, ed., *Axiomatic Set Theory*. Providence: American Mathematical Society, 1971, 271–78.
Nagel, Ernest, Patrick Suppes and Alfred Tarski, eds., *Logic, Methodology and Philosophy of Sciences*. Stanford: Stanford University Press, 1962.
Neurath, Otto. "Protokollsätze," *Erkenntnis*, 3 (1932–33), 204–14.
Neurath, Otto. "Radikaler Physikalismus und 'wirliche Welt,'" *Erkenntnis*, 4 (1934), 346–62.
Oxtoby, John. *Measure and Category*. New York: Springer, 1970.
Paqué, Ruprecht. *Das Pariser Nominalistenstatut: Zur Enstehung des Realitatsbegriffs der Neuzeitlichen Naturwissenschaft*. Berlin: de Gruyer, 1970.
Parthasarathy, K. R. *Probability Measures on Metric Spaces*. New York: Academic Press, 1967.
Pascal, Blaise. "De l'espirit géométrique," *Oeuvres Completes*, ed. Louis Lafuma. Paris: Editions du Seuil, 1963, 348–55 (written before 1658, first published in full in 1844).
Pascal, Blaise. *Pensées*, in *Oeuvres Complètes*, ed. Louis Lafuma. Paris: Éditions du Seuil, 1963.
Plantinga, Alvin. *God and Other Minds*. Ithaca: Cornell University Press, 1967; rev. ed. 1990.
Pederson, Gert. *Analysis Now*. Berlin: Springer, 1988.
Pierce, Charles. *Collected Papers of Charles Sanders Peirce*, ed. Charles Hartshorne and Paul Weiss. Vols. 5 and 6. Cambridge: Harvard University Press; reprinted in single vol., Cambridge: Belknap Press, 1974.
Planck, Marx. "Positivismus und reale Aussenwelt," Vortrag gehalten am 12. November 1930 im Harnack Haus der Kaiser-Wilhelm-Gesellschaft zur Forderung der Wissenschaften, Vortrage und Erinnerungen. Darmstadt: Wissenschaftliche Buchgesellschaft, 1979.
Plato. *Apology and Phaedo*, ed. with a translation by H. N. Fowler. Cambridge: Harvard University Press, 1977.
Plato. *Parmenides*, ed. with a translation by H. N. Fowler. Cambridge: Harvard University Press, 1977.
Plato. *Republic*, ed. with a translation by Paul Shorey. Cambridge: Harvard University Press, 1969.
Plato. *Theaetetus* and *Sophist*, ed. with a translation by H. N. Fowler. Cambridge: Harvard University Press, 1977.
Popkin, Richard. *The History of Skepticism*. Berkeley: University of California Press, 1979.
Poincaré, Henri. *La Science et l'hypothese*. Paris: Flammarion; reprint, ed. J. Vuillemin, 1968.
Prugovecki Eduard. *Quantum Mechanics in Hilbert Space*. Cambridge Massachusetts: Academic Press, 1971.
Putnam, Hilary. *Meaning and the Moral Sciences*. London: Routledge and Kegan Paul, 1978.
Putnam, Hilary. "Models and Reality," *Journal of Symbolic Logic*, 45 (1980), 464–82.
Putnam, Hilary. *Reason, Truth and History*. London: Cambridge University Press, 1981.
Putnam, Hilary. "What is 'Realism'?" *Proceedings of the Aristotelian Society* (1975–76), 177–94.
Quine, Willard van Orman. *Word and Object*. Cambridge, MA: MIT Press, 1960.
Rasiowa, Helena and Roman Sikorski. *The Mathematics of Metamathematics*. Warsaw: P. W. N., 1963.
Ratke Heinrich. *Systematisches Handlexikon zu Kants Kritik der Reinen Vernunft*. Hamburg: Meiner, 1991.
Reed, Michael and Barry Simon. *Methods of Modern Analysis*. Volumes I and II. New York: Academic Press, 1975; 2nd ed., 1980.

Reinhardt, William. "Remarks on Reflection Principles, Large Cardinals, and Elementary Embeddings," *Axiomatic Set Theory, Proceedings of Symposia in Pure Mathematics*, ed. Thomas Jech, 13.2 (1974). Providence: American Mathematical Society, 189–205.
Resnik, Michael. "The Frege-Hilbert Controversy," *Philosophy and Phenomenological Research*, 34 (1973–74), 386–403.
Rucker, Rudolph v. B. *Infinity and the Mind*. Princeton: Princeton University Press, 1995.
Rudin, Walter. *Functional Analysis*. New York: McGraw Hill, 1973.
Rudin, Walter. *Real and Complex Analysis*. New York: McGraw Hill, 1974.
Russell, Bertrand. "Les Paradoxes de la logique," *Révue de métaphysique et de morale*, 14 (1906), 627–50.
Russell, Bertrand. *Principles of Mathematics*. Cambridge: Cambridge University Press, 1903; reprint, New York: W. W. Norton, 1938.
Russell, Bertrand. "Recent Work on the Principles of Mathematics," *International Monthly*, 4 (1901), 83–101.
Rutherford, Donald. *Leibniz and the Rational Order of Nature*. Cambridge: Cambridge, 1995. (Rutherford 1995a)
Rutherford, Donald. "Metaphysics: The Late Period," N. Jolley, ed., *The Cambridge Companion to Leibniz*, Cambridge: Cambridge, 1995. (Rutherford 1995b)
Sambursky, Samuel. *Physics of the Stoics*. Princeton: Princeton University Press, 1956, rpt. 2014.
Sacks, Oliver. *The Man Who Mistook His Wife for a Hat*. New York: Harper, 1990.
Schirn, Matthias, ed. *Studies in Frege*. Stuttgart/Bad Constalt: Fromann-Holzboog-Verlag, 1975.
Schlick, Moritz. "Positivismus und Realismus," *Erkenntnis*, 3 (1932–33), 1–31.
Schlick, Moritz. "Über das Fundament der Erkenntnis," *Erkenntnis*, 4 (1943), 70–99.
Schmidt, H. A., ed., *Contributions to Mathematical Logic*. Amsterdam: North Holland, 1968.
Schneider, Martin. *Analysis und Synthese bei Leibniz*. Ph. D. Dissertation, University of Bonn, 1974.
Schröder, Ernst. *Vorlesungen über die Algebra der Logik* (Exakte Logik). Leipzig: B. G. Teubner Verlag, 1890.
Schofield, Malcolm, Myles Burnyeat, and Jonathan Barnes, eds. *Doubt and Dogmatism*. Oxford: Oxford University Press, 1980.
Schrödinger, Ervin. "An Undulatory Theory of the Mechanics of Atoms and Molecules," *Physical Review*, 28.6 (1926), 1049–70.
Searle, John. *Minds, Brains, and Science*. Cambridge: Harvard University Press, 1984.
Sextus Empiricus. Vols. I–IV, ed. with an English translation by R. G. Bury. Cambridge: Harvard University Press, 1976. Vol. I, *Pyrroneion Hypotyposeon (Outlines of Pyrrhonism)* (PH I, PH II); Vol. II, *Adversus Mathematicos (Against the Teachers)* (M VII, M VIII).
Shoenfield, Joseph. *Mathematical Logic*. Reading: Addison Wesley, 1967.
Skolem, Thoraf. "A Critical Remark on Foundational Research," in Skolem 1970, 581–86.
Skolem, Thoralf. "Einige Bemerkungen zur axiomatischen Begrundung der Mengenlehre," in Skolem 1970, 144–52.
Skolem, Thoralf. "Review of Weyl 1910 (signed 'Sk')," *Jahrbuch der mathematicschen Fortschritte*, 41 (1910), 89–90.
Skolem, Thoralf. *Selected Works in Logic*. Oslo: Universitetsforlaget, 1970.
Skolem, Thoralf. "Some Remarks on the Foundation of Set Theory," in Skolem, 1970, 519–28.
Skolem, Thoralf. "Sur la Portée du Thèoréme de Löwenheim-Skolem," in Skolem 1970, 457–82.
Skolem, Thoralf. "Über die Gundlagendiskussionen in der Mathamatik," in Skolem 1970, 207–25.

Skolem, Thoralf. "Über die Nichtcharakterisierbarkeit mittels endlich oder abzählbar unendlich vieler Aussagen mit asusschliesslich Zahlenvariablen," *Fundamenta Mathematicae*, 23 (1934), 150–61.

Skolem, Thoralf. "Une relativisation des notions mathmatiques fondamentales," in Skolem 1970, 633–38.

Slaattelid, Rasmus. *Horizons of the Mind, An Aporetic Approach to Concepts of Mental Horizons*. Ph. D. Dissertation, University of Bergen, 1999.

Slezak, Peter. "Descartes' Diagonal Deduction," *British Journal of the Philosophy of Science*, 34 (1983), 13–36.

Sluga, Hans. *Gottlob Frege*. London: Routledge and Kegan Paul, 1980.

Smorynski, Craig. "The Incompleteness Theorems", John Barwise, ed., *Handbook of Mathematical Logic*. Amsterdam: North Holland, 1977, 821–65.

Solovay, Robert. "A Model of Set Theory in Which Every Set of Reals is Lebesgue Measurable", *Annals of Mathematics*, 92 (1970), 1–56.

Solovay, Robert and Stanley Tennenbaum. "Iterated Cohen Extensions and Souslin's Problem," *Annals of Mathematics*, 94 (1971), 201–45.

Spinoza, Baruch. *Werke I and II (Tractatus de Intellectus Emendatione and Ethica)*. Darmstadt: Wissenschaftliche Buchgesellschaft, 1967.

Takesaki, Masamichi. *Theory of Operator Algebras I*. New York: Springer, 1979.

Takeuti, Gaisi. *Two Applications of Logic to Mathematics*. Princeton: Princeton University Press, 1978.

Takeuti, Gaisi and Wilson Zaring. *Axiomatic Set Theory*. Berlin: Springer, 1973.

Tarski, Alfred. "Der Wahrheitsbegriff in den formalisierten Sprachen," *Studia Philosophica*, 1 (1935), 261–405.

Thiel, Christian. *Sinn und Bedeutung in der Logik Gottlob Freges*. Meisenhelm am Glan: Verlag Anton Hain, 1965.

Thrane, Gary. "The Spaces of Berkeley's World", Colin Turbayne, ed., *Berkeley: Critical and Interpretative Essays*. Minneapolis: Minnesota, 1982, 127–47.

Tipton, Ian. "Berkeley's Imagination", Ernest Sosa ed., *Essays on the Philosophy of George Berkeley*, Dordrecht: Reidel, 1987, 85–102.

Vaihinger, Hans. *Die Philosophie des Als Ob*. Leipzig: Felix Meiner, 1922.

Van Benthem, Johan. "Correspondence Theory," D. Gabbey and F. Guenther, eds., *Handbook of Philosophical Logic, Vol. II, Extensions of Classical Logic*. Dordrecht: Reidel, 167–248.

van Fraassen, Bas. *Quantum Mechanics: An Empiricist View*. Oxford: Clarendon Press, 1991.

Van Heijenoort, Jan, ed., *From Frege to Gödel, A Source Book in Mathematical Logic, 1879–1931*. Cambridge: Harvard University Press, 1967.

Vesley, Richard. "Boolos' Nonconstructive Proof of Gödel's Incompleteness Theorem," *Kurt Gödel: Proceedings of a Symposium Held in Neuchatel*. Neuchâtel: Neuchâtel, 1992.

von Arnim, Joannes, ed., *Stoicorum Veterum Fragmenta*, vols. I–IV. Stuttart: Teubner, 1968.

Von Kues, Nikolaus. *Philosophisch-Theologische Schriften*, ed. Leo Gabriel. Vienna: Verlag Herder, 1967.

Von Kutschera, Franz. *Die Antinomien der Logik*. Freiburg im Breisgau: Karl Alber Verlag, 1965.

Wald, Robert. *General Relativity*. Chicago: University of Chicago Press, 1984.

Wang, Hao. *Reflections on Kurt Gödel*. Cambridge, MA: MIT Press, 1987.

Wagon, Stanley. "The Collatz Problem," *The Mathematical Intelligencer* 7.1 (1985), 72–76.

Weyl, Hermann. "Der ciculus vitiosus in der heutigen Begründung der Mathematik," in Weyl 1968, II, 43–50.

Weyl, Hermann. "Die heutige Erkenntnislage in der Mathematik, " in Weyl 1968, II, 511–42.
Weyl, Hermann. *Gesammelte Abhandlungen*, I–IV, herausgegeben von K. Chandrasekharan. Berlin: Springer, 1968.
Weyl, Hermann. *The Open World. Three Lectures on the Metaphysical Implications of Science*. New Haven: Yale University Press, 1932.
Weyl, Hermann. *Philosophie der Mathematik und Naturwissenschaften*. München: Oldenbourg, 1927.
Weyl, Hermann. *Philosophy of Mathematics and Natural Science*. Princeton: Princeton University Press, 1949.
Weyl, Hermann. *Die Stufen des Unendlichen*. Jena: Gustav Fischer, 1931.
Weyl, Hermann. "Uber die Definitionen der mathematischen Grundbegriffe," in Weyl 1968, II, 299–304.
Weyl, Hermann. "Über die neue Gundlagenkrise der Mathematik," in Weyl 1968, II, 511–42.
Wheeler, John. *A Journey into Gravity and Spacetime*. New York: Freeman, 1990.
Wigner, Eugene. "The Subject of Our Discussions", Bernard d'Espagnat ed., *Foundations of Quantum Mechanics* New York: Academic, 1971, 1–19.
Wigner, Eugene. "The Unreasonable Effectiveness of Mathematics in the Natural Sciences," *Communications on Pure and Applied Mathematics*, 13 (1959), 1–14.
Wilson, Margaret. "The Phenomenalisms of Leibniz and Berkeley," Ernest Sosa, ed., *Essays on the Philosophy of George Berkeley*. Dordrecht: Reidel, 1987, 2–22.
Winkler, Kenneth. *Berkeley: An Interpretation*. Oxford: Clarendon Press, 1994.
Wittgenstein, Ludwig. *Bemerkungen über die Grundlagen der Mathematik*, in *Werkausgabe in Acht Bänden*. Frankfurt am Main: Suhrkamp, 1984.
Wittgenstein, Ludwig. *Logisch-Philosophische Abhandlung* and *Philosophische Untersuchungen*, in *Werkausgabe in Acht Bänden* (Band I). Frankfurt am Main: Suhrkamp, 1984.
Wittgenstein, Ludwig. *Nachlaß*. Oxford: Oxford (CD Rom Edition), 2000.
Wittgenstein, Ludwig. *Philosophische Bemerkungen*, in *Werkausgabe in Acht Bänden* (Band II). Frankfurt am Main: Suhrkamp, 1984.
Wittgenstein, Ludwig. *Philosophische Grammatik*, in *Werkausgabe in Acht Bänden* (Band IV). Frankfurt am Main: Suhrkamp, 1984.
Wittgenstein, Ludwig. *Remarks on the Philosophy of Psychology*, G. E. M. Anscombe and G. H. Von Wright, eds. 2 vols. Chicago: University of Chicago Press, 1980.
Wittgeinstein, Ludwig. *Schriften* I. Frankfurt: Suhrkamp, 1969.
Wittgenstein, Ludwig. *Über Gewißheit*, in *Werkausgabe in Acht Bänden*. Frankfurt am Main: Suhrkamp, 1987.
Wright, John. *The Skeptical Realism of David Hume*. Minneapolis: University of Minnesota Press, 1983.
Yosida, Kosaku. *Functional Analysis*. New York: Springer, 1974.

Index of Names

Abelard 10
Adam (Biblical) 258, 281, 292, 293
Albeverio, Sergio 407
Anaxagoras 34
Anselm of Canterbury / Bec 6, 8, 26, 30, 31, 61, 83, 84, 96, 101; Chapter 4: 131–159; 184, 187, 228, 323, 345, 367, 369, 372
Antisthenes 120
Aquinas, Thomas 6, 31, 155, 189, 323, 345
Arcesilaus 95
Aristotle 1, 2, 4, 6, 7, 22, 27, 29–31, 59, 63; Chapter 2: 68–104, esp. 68, 70, 71, 76, 77, 81–85, 96–101; 115, 128, 135, 149, 154–156, 159, 170, 184, 221, 227, 231, 237, 293, 310, 314, 341, 345, 347, 348, 366–379, 388, 401, 304
Arnim, Ludwig Achim von 106
Augustine, of Hippo 26, 30, 91–95, 105–107, 113, 114, 131, 187, 310, 320, 323
Autrecourt, Nicolas, of 102

Bachofen, Johann 221
Bacon, Roger 29
Barwise, Kenneth Jon 200
Beckett, Samuel 159, 376
Berkeley, George 32–42, 44–46, 62, 63, 74, 131, 152–154 Chapter 6: 186–232; 236, 238–240, 243, 246–255, 257–259, 261–265, 269–271, 276, 278, 279, 282, 287, 293, 301, 307, 310, 312, 313, 320, 322, 323, 325, 326, 328, 331, 336, 341, 355, 367, 370
Bernays, Paul 86, 87, 141
Berry, G. G. 12, 16, 38, 74, 110, 112, 141, 200–203, 266
Boethius 35, 190
Bohr, Niels 25, 194, 355, 370, 406, 417
Bolzano, Bernard 172
Boscovich, Roger 383, 384, 386, 392, 393, 401, 402
Boswell, James 212, 235
Browne, Thomas 274, 400, 404

Caesar, Gaius Julius 282
Cantor, George 70, 71, 74, 90, 134, 141–145, 157, 162, 227, 228, 309, 310, 316, 327, 333, 389, 397
Carnap, Rudolf 242, 247, 272, 274, 360
Carneades 15, 16, 95, 112
Carroll, Lewis 19, 142
Cassirer, Ernst 25, 64, 307, 308
Chaitin, Gregory 62, 170, 180, 182, 200
Chrysippus 16, 113, 115–120, 123, 184
Cicero, Marcus Tullius 116, 117, 119–122, 238
Clement of Alexandria 30, 91, 105, 106
Cohen, Hermann 307
Copernicus, Nicolaus 44, 283, 321
Cusa, Nicholas of 101, 142, 145, 181, 370

Davidson, Donald 125
Dedekind, Richard 141, 142, 160, 411
Democritus 22, 28, 123
Descartes, René 6, 7, 14, 15, 26, 27, 29–33, 46, 63, 70, 88, 91–96, 101, 105–107, 112–114, 131, 136–138, 140, 157, 158, 170, 179, 184, 187–190, 194, 211, 218, 220, 236, 241, 243, 247, 250, 251, 256, 269, 279, 291, 299, 302, 307, 310, 312, 320–326, 328, 331, 336, 341, 348, 355, 367, 369, 370, 379
Diodorus Cronus 48, 127, 150, 188, 189, 440
Diogenes 120
Diogenes Laertius 121
Doob, Joseph L. 416
Duhem, Pierre 140, 308

Eco, Umberto 210
Einstein, Albert 170, 171, 221, 308, 355, 408, 427, 428, 434
Empedocles 22, 28, 123
Epictetus 120
Epicurean 29, 32, 39, 116, 338, 366
Euclid 46, 47, 277, 402, 403
Euler, Leonhard 308
Everett, Hugh 408

Feynman, Richard 407, 408, 425, 427
Forman, Milos 249
Fourier, Joseph 181
Fraassen, Bas van 440
Fränkel, Adolf 141, 385, 410
Frege, Gottlob 89, 90, 171, 191, 194, 196, 308, 402, 403

Gadamer, Hans-Georg 59
Galileo 32, 238, 307, 308, 312, 319, 360, 373, 374, 440
Gassendi, Pierre 102, 247, 268
Gauß, Carl Friedrich 307, 420
Gilbert, Gustave 347
Gödel, Kurt 3, 12–15, 17, 23, 27, 31, 37, 38, 48, 50, 51, 62, 68, 70, 72, 73, 77, 82, 86, 89, 90, 92, 93, 110, 111, 113, 115, 131, 133, 141, 143–145, 153, 170, 171, 176, 182, 184, 194, 195, 199–201, 203, 204, 206, 210, 222, 227, 229, 232, 242, 243, 263, 266, 270–273, 301, 304, 307, 308, 315, 318, 319, 330, 333, 335, 356, 367, 382, 388, 390–393, 305, 401, 404, 410, 412
Grelling, Kurt 12, 110, 266

Halle, Georg von 141
Hamilton, William 407, 418, 420, 423, 425, 430, 431
Hardy, Godfrey Harold 12, 110, 402
Heisenberg, Werner 305, 425
Helmholtz, Hermann von 308
Henkin, Leon 36–38, 61, 160, 161, 167–169, 172, 176, 177, 184, 198, 200, 333, 336, 391
Heraclitus 22, 28, 123, 146
Hertz, Heinrich 308
Hilbert, David 31, 71–73, 89, 142, 143, 178, 184, 196, 198, 227, 228, 242, 307, 308, 327, 403, 404, 425
Hintikka, Jaakko 75
Hoegh-Krohn, Raphael 407
Hume, David 16, 19, 25, 26, 37, 39–46, 56–58, 61–63, 66, 102, 103, 118, 129, 131, 135, 136, 138, 149, 188, 198, 208, 209, 215, 216, 219–221, 224; Chapter 7: 233–304; 305, 306, 312, 320, 334, 351, 353, 369–371, 374, 377, 379, 389, 390

James, William 66, 139, 380
Jech, Thomas 160, 406, 407, 411, 429, 431, 432, 434–436

Kant, Immanuel 1–6, 10, 15, 17, 18, 22, 25–27, 29, 44–49, 51–60, 62–64, 71, 73, 75–77, 82, 84, 89, 90, 96, 98–101, 103, 106, 108, 112, 119–124, 126–129, 131–134, 136, 138, 140, 143, 144, 146–148, 151–157, 159, 167, 170, 188, 190, 197, 214, 218, 220–225, 227, 228, 230–232, 234, 235, 238, 242, 245, 246, 248, 249, 252, 254, 260, 278, 279, 281, 283, 285–291, 293–295, 297, 298, 301, 302, 304, 305; Chapter 8: 306–381; 387, 391, 440, 441
Kaplan, Wilfred 14
Keats, John 306, 337, 376
Kepler, Johannes 44
Kierkegaard, Soren 102, 380
Klein, Felix 403
Kolmogorov, Andrei 62
Kraut, Edgar A. 79
Kripke, Saul Aaron 75
Kronecker, Leopold 58
Kuhn, Thomas 25, 66
Kunen, Kenneth 406, 436

Laplace, Pierre-Simon 44, 118, 128, 172, 283, 330, 384–386, 389, 392, 393, 401, 402
Leibniz, Gottfried Wilhelm 5, 7, 12, 22, 26, 27, 32–37, 39, 46, 57, 63, 83, 85, 100, 101, 110, 116–120, 123, 139, 140, 147, 151; Chapter 5 160–185; 187, 189, 190, 196, 213, 216, 217, 221, 224, 231, 237, 241, 242, 245, 282, 299, 308, 309, 312, 325, 326, 344, 345, 355, 359, 364, 369, 374, 382, 384–386, 389, 391–393, 401, 402, 439
Leucippus 22, 28, 123
Lindström, Per 332
Littlewood, John 12, 110, 402
Lobachevsky, Nikolai 308
Locke, John 22, 186, 196
Löb, Martin 304, 334
Lorentz, Hendrick 308, 309, 405

Lovejoy, Arthur 5, 6, 8, 52, 70, 98, 100, 126, 139, 188, 231, 371, 404
Lucretius 374
Luria, Alexander 248, 249

MacLean, Norman 365
MacNeille, Holbrook Mann 160, 411
Malebranche, Nicolas 34, 175
Marmoutier, Gaunilo of 134, 135
Marriner, Neville 249
Mill, John Stuart 21
Montague, James 14
Montaigne, Michel de 102
Mozart, Wolfgang 249

Napoléon Bonaparte 385, 393
Neumann, John von 3, 406, 423, 433, 439
Newton, Isaac 32, 39, 44, 46, 66, 96, 116, 117, 219, 283, 290, 296, 298, 300, 305, 307, 308, 312, 320, 321, 323–325, 331, 349, 357, 359, 360

Ockham, Wilhelm 29, 176

Parmenides 69, 74, 79, 82, 85, 100, 101, 147, 167, 169, 176
Pascal, Blaise 5, 21, 32, 52, 98, 99, 102, 103, 159, 187, 236, 239, 283, 337, 343, 357, 364, 369–372, 380
Paul of Tarsus 246
Peano, Giuseppe 71, 83, 145, 171, 179, 182, 308, 335, 368, 388–391, 401
Peirce, Charles Sanders 5, 6, 25, 61, 85, 188–191, 356, 402
Plantinga, Alvin 149
Plato 6, 8, 11, 22, 26, 28–30, 60, 61, 63; Chapter 2: 68–104, esp. 68, 69, 71, 74–76, 78–80, 84–90, 93, 96–101, 104; 109, 114–116, 123, 140, 147, 157, 167, 176, 184, 197, 246, 302, 315, 318, 340, 379, 401, 402
Plotinus 21, 32, 85, 86, 100, 116–120, 135, 371, 384
Plutarch 118, 135
Poincaré, Henri 308

Protagoras 8, 22, 29, 69, 89, 92, 94, 113, 315, 319
Putnam, Hilary 91
Pythagoras 8, 78, 140

Quine, Willard van Orman 169, 237, 239, 272, 274, 275, 333

Robinson, Abraham 200
Russell, Bertrand 30, 31, 61, 74, 76, 110, 111, 135, 171, 194, 201, 201, 205, 217, 227, 264–266, 279, 308, 333

Salieri, Antonio 249
Sambursky, Samuel 116
Sanches, Francisco 102
Schrödinger, Erwin 408, 421, 425
Scott, Dana 308, 406
Searle, John 348
Sextus Empiricus 14, 63, 97, 116, 208, 234, 238, 239, 242, 289, 302, 339
Shaffer, Peter 249
Shelley, Michael 405
Skolem, Albert Thoralf 3, 12, 27, 37, 38, 63, 68, 72, 83, 87, 90, 110, 170, 171, 184, 194, 199, 200, 204, 266, 270, 271, 308, 333, 389, 390, 404
Snow, Charles Percy 309
Socrates 22, 34, 69, 81, 86, 88–92, 94, 95, 101, 238, 315, 319
Solovay, Robert M. 62, 180–182, 308, 406, 408, 429, 431, 432, 434, 435
Spinoza, Baruch 6, 7, 20, 22, 23, 25, 26, 31–33, 63, 79, 84, 100, 105, 116, 117, 120, 140–142, 145, 182, 184, 187, 225, 364, 369, 371, 384, 400
Stone, Marshall 79, 80, 160, 167, 171, 176, 183, 309, 314, 316, 389, 397, 411

Tarski, Alfred 3, 12, 14, 15, 75, 76, 110, 167, 182, 194, 200, 308, 366
Tertullian 107

Vaihinger, Hans 338
Voltaire 34, 100, 172
Vopenka, Petr 406

Walker, William 96
Weierstrass, Karl 172
Weil, Simone 246, 302
Weyl, Hermann 63
Wheeler, John 193, 408
Wigner, Eugene 193, 439
Wittgenstein, Ludwig 16, 17, 19, 25, 49, 51, 60, 63, 72, 73, 104, 112, 121, 128, 133, 157, 158, 162, 173, 211, 212, 230, 231, 238, 239, 265, 301, 334, 339, 353, 372, 388
Woolf, Leonard 104, 381
Wright, John 286, 295

Xenophanes 22, 28, 122, 136

Zeno of Elea 101, 116, 121
Zermelo, Ernst 71, 141, 308, 385, 410

Main Index

This list contains English terms used in the text. For French, German, Greek and Latin expressions, see also the foreign words index.

absolute space 44, 283, 290
abstract, abstraction 18, 74, 83, 85, 133, 134, 150, 152, 171, 176, 190–196, 208, 213, 214, 224, 229, 233, 241, 246, 248, 249, 257, 289, 303, 316, 331, 368, 379, 397, 402, 405
absurd, absurdity 10, 11, 21, 26, 37, 38, 59, 61, 62, 73, 107, 108, 109, 114, 125, 133, 154, 155, 187, 195, 198–201, 206, 207, 212–214, 219, 241, 244, 247, 327, 417
action / *praxis* 19, 54, 57, 97, 122–124, 126, 129, 158, 163–165, 229, 250, 251, 296, 297, 299, 317, 323, 342, 343, 347, 377, 379, 393, 395, 425, 439
AE-sentences (universal-existential) 89, 314, 315
agency, agent 2, 47, 53, 128, 136, 152, 158, 189, 209, 225, 255, 326, 343, 354, 377, 382, 385, 394, 395, 405
agnosticism, agnostic 59, 102, 206, 216, 255, 320, 401
algebra 80, 150, 151, 160, 183, 363, 406–408, 411, 412, 416, 427–429, 431, 433–435, 439, 440
anachronistic, anachronism 33, 78, 89, 96, 116, 121, 142, 155, 183, 192, 207, 232, 242, 310
analytic 4, 17, 23, 27, 29, 31, 45, 47, 48, 58, 59, 77, 91, 109, 148, 154, 194, 197, 202, 214, 239, 242, 272, 295, 308, 322, 330, 336, 351, 367, 375, 379
anthropomorphy 101, 285, 371
apodicticity 30, 42, 44, 54, 100, 113, 126, 305, 320, 387
appearance 42, 46, 71, 88, 97, 228, 231, 237, 238, 284, 286, 289, 290, 306, 312, 320, 353, 434
application, self-referential 14, 15, 26, 72, 73, 93, 94, 104, 114
archetype 5, 6, 31, 42, 68, 86, 161, 206, 226, 228, 230

argument, argumentation 3, 6, 8, 10–13, 15–19, 21–33, 35–39, 42–51, 57–65, 69, 72–74, 76–79, 81, 82, 84, 88–92, 94, 96, 99–103, 105–112, 124, 125, 127, 131, 132, 134–140, 142, 146, 148–157, 160, 169, 172, 176–179, 182, 186–192, 194–216, 218, 220–222, 225, 226, 228, 231–234, 236, 238–241, 244, 246, 248, 250, 252, 254–258, 260–262, 265, 267, 268, 271–276, 279, 282–286, 288, 289, 291–294, 297, 299, 300, 302, 304, 305, 307, 309, 310, 312–318, 320–332, 335–337, 339–341, 344, 345, 348, 355, 356, 360, 367, 368, 370, 371, 375, 377, 382, 389, 395, 398–400, 408, 411, 412, 414, 418, 423, 424, 427, 429, 431, 435, 437–439
– arguments to or from design 24, 25, 58, 153, 182, 291
– chain(s) of arguments 43, 282, 288, 302
– cogito-argument 30, 31, 33, 48, 90–92, 94, 96, 105, 107, 124, 125, 131, 222, 307, 310, 312, 322, 324, 325, 336
– cogito-like argument 315, 316, 321, 324, 326, 328, 329, 331, 335, 355, 370, 377
– EA(=existential-universal) argument 89, 314, 315, 348
– fixed-point argument 15, 26, 31, 105, 107, 152, 153, 154, 310, 313, 320, 327, 345, 355
– master argument 36–39, 62, 74, 125, 131, 153, 186–188, 191, 192, 194, 196–201, 205, 207, 212, 214, 222, 225, 231, 232, 307, 310, 322, 325, 336, 368
– negative argument 69, 88
– pari-ty-of-reasoning argument 13, 138, 315, 343, 405
– physico-theological (= design-based) argument 136, 137, 146, 318, 369
– *reductio*-argument 48, 74, 114, 135, 136, 206, 207, 222, 270, 281
– *tritos-anthropos* argument 22, 29, 82, 83, 169

- uniqueness of maximality and determination argument 314, 318, 341
arithmetic 12, 46, 71, 83, 110, 143, 145, 171, 179, 200, 202, 304, 318, 335, 368, 388–392, 401, 404
artificial intelligence 284
atom, atomic 41, 116, 181, 236, 237, 246, 248, 259, 260, 279, 302, 384
atomism 29, 241, 244, 246, 248, 255, 264, 298
autologicality 72, 143, 318, 344, 364, 392, 393
autonomy 56, 57, 103, 105, 111, 118, 125, 127, 128, 130, 337, 341, 343, 344, 345, 348, 375, 378
axiological 54, 126
axiom, axiomatic, axiomatisable 6, 19, 24, 29, 38, 46, 54, 64, 71–73, 140, 141, 143, 156, 160, 168, 171, 176, 179, 184, 194, 200, 200, 203, 227, 228, 236, 241, 242, 258, 266, 267, 271, 272, 309, 310, 316, 318, 332, 363, 368, 388–392, 394, 397, 402, 403, 408, 414

believe, belief 11, 20, 60, 109, 123, 149, 151, 211, 213, 216, 220, 230, 236, 243, 248–250, 254, 282, 291, 307, 318, 360, 366, 373, 376, 380, 387, 402
binary, binarity 80, 138, 162, 168, 180, 203, 363, 389, 401, 439
bivalent, bivalence 15, 64, 88, 89, 117, 120, 183, 194, 428, 434, 435
Bochner integral 406, 408, 421, 422, 424, 427
Boolean (-valued, algebra, analysis) 80, 120, 150, 160, 162, 178, 342, 363, 406, 407, 429, 440
Borel (coding, functions, images, isomorphism, set) 181, 406, 412, 413, 424, 426, 429–431, 436
Borgesian (cartography, maps, theory) 58, 61–63
boundary assumptions 347
boundary, boundary conditions / demarcations 33, 40, 41, 56, 65–67, 101, 108, 118, 119, 144, 151, 158, 159, 170, 172, 173, 179, 182, 223, 229, 235, 237, 271, 289, 301, 311, 327, 328, 330, 334, 341, 344, 349, 351, 355, 359, 375, 379, 382, 390, 392, 403, 405, 437, 439
bounded, boundedness 67, 115, 418, 419
Brownian motion 407

cadlag 385, 386, 395, 396, 398–400, 406–409, 416, 419–421, 423, 428, 429, 434, 436, 438
canonical 8, 38, 62, 77, 81, 84, 88, 105, 150, 152, 161, 162, 199, 200, 273, 309, 313, 330, 331, 340, 352, 363, 364, 367, 369, 387, 391, 396–398, 402, 411–413, 415, 418, 420, 422, 424, 428, 439
cardinal, cardinality 41, 67, 147, 236, 246, 273, 299, 416
Cartesian 7, 14, 29, 31, 33, 70, 88, 91, 93, 94, 107, 113, 114, 131, 137, 157, 187, 194, 211, 218, 241, 250, 251, 269, 299, 302, 320, 322, 324–326
categorical imperative 56, 344, 345, 347, 348, 358, 377, 380
categories of the understanding 46, 155, 260, 290, 293, 298, 304, 320, 323, 324, 329
category-mistakes 155, 336
cauchy surfaces 385, 392–394, 396
causation 39–41, 84, 119, 135, 149, 234, 237, 254, 276, 351, 369
cause 6, 41, 42, 59, 83, 103, 107, 116–120, 149, 152, 181, 185, 213, 215, 219, 233, 237, 240, 253, 254–258, 260, 263–265, 273–275, 282, 282, 284, 285, 287, 288, 292–294, 296, 297, 299, 311, 314, 338, 345, 346, 351, 353, 364, 365, 374, 384
cave 11, 28, 68, 74, 75, 78, 90, 99, 100, 104, 123, 124, 264, 302, 340
chaos 219, 244, 245, 277
circle, Berkeley's 220
circle, Cartesian 7, 31, 105, 137, 140, 157, 323–326
circle, Hume's 303, 304
circle, Kant's 304
classical 9, 77, 121, 127, 155, 169, 191, 208, 226, 234, 238, 247, 252, 263, 272, 282, 298, 314, 315, 324, 367, 382, 385, 387, 418, 421, 427
clinamen 32 see also swerve

Main Index — 461

clopen 397, 411, 412
code, coding 12, 23, 27, 72, 110, 171, 203, 204, 303, 315, 410–412, 429 *see also* encode, encoding
cognition 318, 320, 326, 327, 338, 339, 375, 376
cognitive 53, 105, 200, 222, 225, 320, 326, 343, 354, 374–378
coherent, coherence 14, 19, 33–35, 47, 82, 86, 90, 93, 94, 115, 125, 127, 132, 170, 199, 212, 215, 220, 222, 238, 250, 260, 277, 291, 305, 309
coherentist, coherentism 32, 34, 35, 39, 46, 186, 192, 207, 216, 219, 223, 231, 325
coin-toss 70, 162, 227, 364, 436
compossibility 165, 168, 169, 173
comprehend, comprehensible 28, 122, 130, 157, 210, 246, 269, 280, 354, 363, 365, 400
comprehension 30, 159, 215, 222, 225, 229, 231, 232, 256, 269, 279, 297, 316, 356
comprehensive 13, 32, 79, 82, 83, 90, 107, 113, 122–124, 149, 270, 272, 307, 313, 316, 324, 341, 344, 372
conceivable, conceivability 14, 37, 41, 74, 99, 156, 185, 198, 199, 201, 204–207, 221, 225, 229, 236, 242, 244–246, 250, 252, 261–266, 269–273, 278, 281, 294, 301, 303, 321, 336, 355, 361, 372, 377
conceive 4, 5, 18, 30, 37, 50, 51, 63, 66, 74, 77, 84, 101, 124, 129, 131–131, 134–139, 145, 149, 156, 157, 159, 184, 185, 187, 194, 198, 202, 210, 211, 213, 218–220, 222, 225, 226, 228–230, 235, 240, 241, 243–245, 247–250, 253, 255, 257, 258, 261, 265, 268–271, 273, 277, 279, 281, 322, 325, 328, 345, 347, 367, 369, 400
concept 59, 146, 191, 223, 231, 237, 250, 293, 295, 307, 311, 350, 353, 359–364, 373, 403
concept-formation 6, 13, 60, 64, 66, 74, 76, 111, 122, 137, 138, 143, 144, 149, 186, 192, 201, 211, 213, 217–219, 224, 226, 227, 230, 244, 262, 264, 268, 276, 279, 281, 291, 298, 308, 309, 337, 339, 350, 367, 374, 376
confutation 8, 17, 94, 95, 106, 131, 152, 326

conjecture 2–4, 21, 23, 34, 35, 46, 52, 55–57, 64, 65, 69, 75, 86, 96, 98, 101, 116, 132, 135, 136, 144, 150, 154, 183, 215, 220, 224, 225, 231, 252, 274, 290, 307, 332, 333, 360, 371, 373–375, 385, 390, 392, 401, 405–409, 439–441
consciousness 4, 55, 57, 129, 231, 159, 193, 224, 231, 251, 285, 297, 344, 358, 374, 374, 379, 380
consequence-relations 18, 20, 21, 143, 194, 318, 331, 332, 334, 387, 392 *see also* Prohorov metric/topology, proof-schemes
consistency 19, 31, 38, 40, 41, 47, 48, 51, 54, 67, 72, 73, 77, 94, 115, 118, 122, 126, 152, 155, 156, 159, 170, 171, 179, 182, 184, 198, 199, 201, 203, 205–207, 209, 210, 221, 222, 228, 235, 242, 252, 257, 261–263, 265, 269, 271–273, 300, 301, 318, 329, 330, 333, 334, 336, 354–356, 367, 368, 387, 390, 391, 403, 404, 439
constitutive 4, 30, 40, 44, 46–48, 55, 77, 98, 103, 122, 131, 147, 148, 153, 167, 170, 171, 186, 189, 219, 222–224, 285, 291, 293–295, 300, 305, 309, 311, 313, 314, 323, 327, 330, 338, 339, 355, 356, 367, 368, 376, 405
constructive 2 20, 26, 30, 53, 87, 92, 113, 125, 128, 199, 211, 221, 222, 226
contemplate, contemplation 43, 70, 96, 97, 158, 159, 240, 293, 357, 358, 369, 370, 376, 379, 380, 388
contextual, contextualism 11, 59, 66, 74, 75, 89, 118, 150, 186, 197, 202, 206, 207, 210, 262, 302, 315, 375, 406, 440
contingency 42, 50, 51, 54, 66, 98, 126, 162, 173, 180
continuity 156, 177, 183, 382, 384, 386, 393, 396, 397, 409, 423, 424
continuum 45, 67, 70, 145, 150, 177, 227, 319, 391, 400, 401
contradiction 10, 41, 108, 133, 141, 144, 152, 165, 170, 203, 204, 219, 226, 235, 241–244, 245, 258, 263, 271, 277, 286, 288, 325, 335, 336, 359, 403, 404, 430
contraposition 77, 95, 205, 316, 367
contrapositive 243, 267, 315, 317
convergence 7, 58, 59, 69, 154, 188, 189, 356

cosmological 84, 136, 137, 140, 146, 152, 220, 369
cosmos 105, 115–117, 357 see also world in main index and ho kosmos in foreign words index
countable 71, 143, 181, 227, 318, 389, 391, 392, 396, 397, 399, 410–412, 414, 436
counterfactual 42, 43, 105, 129, 183, 217, 219, 237, 239, 243, 248, 249, 252, 276, 277, 278, 327, 341, 342, 354, 355, 374, 376, 395
counterfactual concept-formation 149, 186, 192, 217
countermodel 136, 161, 181, 270
criteria of truth 9–12, 19, 23, 26, 89, 107–109, 113, 155, 156, 170, 230, 245, 299, 300, 314, 324, 403
criterial ascent 238, 263, 273, 274, 300, 301
criterion, criteria 3, 8, 9, 10–14, 17, 23, 26, 34, 55, 64, 70, 88, 91, 107, 109, 113, 129, 138, 139, 148, 177, 238, 255, 273, 275, 281, 282, 300, 337, 342, 344, 377, 386, 404, 405
criterion-problem / problem of the criterion 12, 17, 19, 23, 26, 32, 85, 109, 137, 251, 326
custom 16, 25, 39, 40, 42–46, 103, 220, 238–240, 245, 246, 257, 260, 269, 275, 208–286, 288, 290–300, 303, 304, 312, 320, 344

decompose, decomposition 329, 407, 408, 419, 438
deconstruct, deconstructive, deconstruction 4, 22, 49, 146, 213, 230, 271, 259, 359, 371
deduction 44, 45, 47, 48, 58, 103, 125, 131, 132, 152, 153, 222, 304, 307, 310, 311, 315, 318, 320–324, 326, 328, 329, 331, 334, 336, 344–346, 368, 369, 372
definability, pointwise 169, 204, 264, 266, 414
deist 59, 189, 212, 371
deontic, deontology 4, 138, 237, 342, 344, 374
design 3, 5, 7, 8, 21–25, 27, 28, 32, 34, 35, 40, 42, 58–61, 80, 84, 102, 107, 121–123, 136, 140, 153, 155, 176, 205, 209, 213, 214, 219, 220, 225, 257, 318, 341, 345, 369 see also arguments to or from design
– hieratic design 35, 105
– transcendental design 32, 44, 368
– secular design 5, 23, 371, 373
determinism, deterministic 33, 105, 117–119, 121, 127, 172, 173, 183, 351, 360, 382, 386, 387, 393, 394
diagonal 12, 17, 18, 74, 89, 92, 110–114, 135, 222, 243, 323, 355
diagonal lemma 13, 14, 70, 72, 94, 203, 232, 243, 263, 301
diagonalise, diagonalisation 12, 29, 48, 77, 92–94, 109, 135, 158, 203, 227, 271, 313, 320, 367, 373
dialectic, dialectical 2, 4, 7–10, 12, 21, 23, 26, 27, 45, 47, 52–54, 62, 71, 75, 88, 90, 92, 100–102, 104–107, 108, 110, 111, 14, 122, 123, 126, 130, 131, 134, 148, 186, 190, 192, 205–207, 209, 212–214, 218, 219, 221–224, 230, 231, 237, 252, 262, 266, 271, 280, 288, 295, 298, 301, 308, 310, 316, 319, 320, 326, 328, 330, 338–340, 346, 348, 351, 366, 369, 370
dichotomy, dichotomisation 25, 30, 54, 69, 77, 79, 84, 116, 118, 119, 212, 302, 305, 339, 346, 370, 371
dilemma 4, 7, 21–23, 77, 86, 95, 103, 118, 133, 134, 142, 156, 191, 224, 226, 230, 235, 236, 239, 251, 258, 262, 270, 273, 274, 292, 294, 295, 303, 343–345, 347, 367, 370, 401, 401, 441
directedness see also linearity 61, 188
disambiguation 169, 176, 195, 262
distributive, distributivity 150, 151, 178, 181, 182, 309, 349, 354, 379, 387–389, 408
divine, divinity 7, 33, 35, 40, 41, 71, 77, 84, 97, 100, 101, 118, 135, 136, 138, 139, 155, 157–159, 171, 172, 184, 212, 217, 324
dogma, dogmatic 8, 22, 24, 26, 35, 39, 41–44, 46, 89, 90, 92, 94, 95, 103, 111, 130, 135, 148, 154, 186–189, 206–209, 212, 214, 217, 219, 223, 223, 232, 235–239, 241, 246, 250–254, 256, 257, 269, 278, 280, 290, 299, 302, 312, 318,

324, 339, 344, 348, 369, 371–373, 376, 379, 387, 402
doubt 13, 17, 26, 28, 29, 30, 39, 49, 51, 53, 70, 84, 85, 92–95, 113–115, 125, 131, 136, 154, 208, 223, 229, 233, 244, 255, 261, 274, 281, 285, 288, 299, 306, 322, 326–328, 355, 404

ectypes 5, 79, 86, 161, 226, 228, 230, 345
empiricist, empiricism 4, 33, 35, 39, 41–44, 131, 148, 208, 230, 235–239, 246, 250, 261, 268, 275, 300, 302, 303, 308, 309
Empyrean 117, 218, 402
encode, encoding 19, 20, 49, 71, 72, 77, 92, 95, 99, 101, 110, 113, 124, 141, 143, 145, 154, 177, 200–203, 205, 206, 270, 272, 273, 304, 318, 323, 332, 358, 367, 385, 389, 392 *see also* code, coding
epistemological, epistemology 3–5, 63, 88, 152, 179, 211, 230, 239, 266, 268, 337, 340, 374, 392, 401, 406, 441
eternity 35, 75, 140, 190, 229, 289, 379, 382
ethics 6, 29, 56, 57, 63, 97, 100, 120, 231, 306, 345, 379
evidence 43, 46, 49, 50, 55, 62, 63, 65, 67, 73, 81, 91, 90, 98, 113, 118, 122, 124, 125, 136, 142, 150, 154, 155, 157, 172, 179, 184, 188, 202, 206, 208, 225, 233, 239, 271, 275, 277, 282, 285, 288, 295, 206, 330, 333, 334, 356, 366, 371, 373, 387, 405
evident, evidential 8, 42, 66, 208, 115, 234, 238, 242, 244, 247, 248, 254, 258, 277 *see also* order, first
existence assertion / assumption / claim 19, 20, 21, 38, 71, 81, 171, 186, 187, 198–200, 205, 207, 214, 262, 315, 316
existential 38, 81, 139, 184, 266, 380
experience 33, 34, 46, 241, 242, 281, 286, 303
experientiality 23, 27, 28, 42–44, 50, 57, 58, 60, 63, 64, 123, 144, 155, 172, 190, 237, 241, 259, 273, 281–283, 285, 298, 302, 307, 311, 313, 324, 327, 328, 330, 336, 341, 351, 354, 356, 365, 370, 372
expressible, expressibility 4, 8, 12, 15, 17, 20, 27, 28, 39, 41, 48, 50, 60, 62, 72, 74, 75, 77, 85, 86, 93, 99, 110, 115, 122–125, 127, 144, 169, 171, 172, 176, 178, 181, 182, 184, 191, 194, 195, 202–204, 206, 221, 224–227, 229, 266, 272, 325, 339, 354, 355, 367, 380, 388, 418
extension, extensional 37, 38, 55, 62, 71, 72, 75, 93, 119, 120, 149, 156, 160, 161, 168, 169, 176–178, 181, 195, 196, 198, 199, 204, 207, 227, 228, 243, 248, 260, 265, 275, 308, 319, 329, 333, 338, 339, 342, 351, 363, 364, 367, 372, 373, 377, 385, 389, 391, 393, 395, 406–408, 410, 412, 413, 416, 425, 434

faith, faithful 35, 59, 127, 131, 132, 134, 140, 148, 151, 311, 328, 335, 336, 339, 343, 356, 367, 377, 387, 393–395, 400, 432, 436
Fell metric 396
fideist 32, 35, 103, 179, 189, 190, 205, 207, 219, 220, 230, 238, 371, 372
form 2, 8, 11–14, 16, 26, 27, 30, 37, 46, 49, 68, 69, 77, 79–83, 96, 111, 157, 178, 199–201, 204, 206, 207, 221, 222, 243, 245, 263, 314, 362, 365, 393, 400, 409, 410, 421
form of the Good 76, 77, 79, 97, 217, 340
formalise, formalisation 13, 15, 73, 114, 141, 177, 200, 202–204, 263, 272, 308, 330, 372
formula 134, 167, 178, 202, 203, 238, 243, 367, 384, 425
foundational, foundationalism 35, 189, 226, 228, 241, 244, 252, 280, 294
France, the present king of 303, 264, 265, 278, 279
free will 382, 395
functional similarity / equivalent / resemblance 45, 245, 260, 286, 295
functions 293, 397, 411, 416, 420, 421, 423–425, 428–430
– scaling function 398, 399

geometry 6, 29, 46, 89, 171, 244, 307, 383, 400, 402–405
god 6, 7, 28, 30, 31, 34, 35, 42, 43, 45, 58, 76, 84, 96, 117, 123, 124, 128, 130, 135, 136, 138, 140, 142, 144, 147, 148, 151–154,

158, 159, 164–166, 170, 172, 173, 175–177, 179, 182, 184, 188–191, 197, 210, 212, 217, 218, 220, 221, 225, 228, 229, 240, 249, 269, 293, 295, 306, 310, 312, 313, 320, 322–325, 338, 345, 346, 355, 367, 369, 371, 372, 384
– night watchman 35, 189, 217
– Supreme Being 285, 288, 292
god of the fish 22, 28, 122, 136
god of the philosophers 58, 140, 144, 367, 369
golden mountain 241, 278, 279
good will 53, 56, 57, 103, 121, 344, 348
great chain of being 126, 139
guilt / moral responsibility 128, 267

habit 16, 25, 39, 40, 42, 44, 45, 95, 122, 238, 257, 284–286, 290, 299, 312, 320
harmony 39, 43, 103, 174, 240, 260, 282, 283, 292–294
hermeneutic 2, 3, 10, 11, 35, 37, 59, 66, 75, 87, 99, 108, 109, 126, 148, 150, 152, 159, 189–191, 194, 198, 207, 208, 217, 280, 307, 308, 312, 323, 337, 340, 346, 366
heuristic 3, 9, 25, 33, 45, 63, 66, 67, 82, 101, 123, 132, 142, 182, 191, 200, 216, 259, 309, 312, 329, 331, 333, 337, 340, 341, 348, 349, 351, 352, 354, 361, 366, 373–375, 385, 389, 390, 405–409, 417, 419, 422, 427, 437, 440, 441
hierarchic 6, 51, 187, 193, 316, 348, 349, 361
hierarchies 5–7, 23, 24, 26, 28, 30, 31, 40, 48, 50–53, 60, 62–64, 74, 83, 85, 118, 119, 122, 124, 125, 135, 145, 128, 131, 132, 134, 137, 139, 142, 144, 148–151, 156, 158, 159, 169, 179, 187, 188, 193, 194, 197, 207, 209, 224, 246, 300–302, 309, 311, 313–315, 319, 320, 328, 329, 331–333, 336, 339, 340, 349–353, 355, 356, 361, 365, 370, 372–375, 385, 392–399, 401–403, 439, 441 see also metatheoretic hierarchies
– ramified / ramifying hierarchies 28, 48, 51, 63, 74, 134, 139, 144, 156, 158, 197, 209, 331, 336, 365, 393, 394, 396, 401
– type-hierarchies 194, 308, 333

hieratic 4–6, 8, 10, 15, 22–26, 28–32, 35, 40, 43, 45, 49, 54, 59, 60, 64, 105, 108, 111, 122, 123, 125, 126, 218, 340, 379, 401–403
holistic, holism 100, 101, 161, 163, 180, 272, 369, 379
human nature 42, 239, 290, 296, 299, 300
Hylas 187, 190, 192, 209–214, 218, 219, 235, 257, 261, 262
hypostase 69, 82, 122, 132, 150, 317, 318
hypothesise, hypotheses 3, 4, 20, 33, 44, 55, 57, 74, 75, 79, 88, 128, 165, 173, 183, 190, 194, 220, 230, 238, 253, 254, 259, 275–278, 282–284, 330, 331, 335, 337, 344, 360, 385

ideas 6–8, 35–37, 40–43, 63, 75, 81, 82, 84, 88, 96, 97, 100, 110, 125, 131, 141, 146, 147, 153, 155, 157, 171, 176, 183, 184, 186, 187, 189, 190–192, 194–196, 198, 202, 204–206, 208, 210, 211, 213–216, 218, 223, 229, 230, 233, 235–237, 240–244, 246–260, 262, 266, 267, 269, 271–281, 285, 288, 290–295, 300, 302, 303, 305–307, 322, 328, 338, 346, 357, 358, 360, 362, 373, 376, 386, 409, 423, 439
ideal 3, 5, 6, 13, 22, 24, 25, 29, 32, 33, 35, 45, 46, 48, 49, 52, 55, 58–60, 65, 77, 96, 100, 106, 121, 123, 127–130, 139–141, 151, 153, 157, 158, 270, 307, 310, 330, 332, 337, 341, 345, 354, 356, 357, 367, 370, 372, 373, 376, 379, 380, 390, 417, 436, 437
idealistic, idealism 40, 44, 60, 116, 234, 269, 286, 312, 325, 374
identity, personal identity 39, 57, 119, 234, 250, 251
ignorance 14, 57, 87, 131, 268, 289, 294, 370
illusion 54, 101, 118, 126, 191, 289 see also transcendental illusion
image / likeness 11, 22, 46, 78, 81, 101, 181, 248, 286, 324, 339, 351, 364, 365, 379, 407, 412, 413, 420, 422, 423, 426, 438 see also eikasia in foreign words index
immanent, immanence 17, 27, 49, 51, 67, 87, 101, 124, 141, 229, 338–340, 355, 364, 368, 373

immaterialism, immaterialist 40, 152, 186, 187, 206, 207, 212, 214, 231, 232, 234, 238, 252, 259, 270
impredicativity 131, 133, 137, 141, 145, 157, 243, 244
impression, impressive 37, 40–42, 136, 151, 166, 198, 235–237, 241–244, 246, 248, 250, 251, 253–256, 259, 267, 269, 276, 278–282, 285, 290–292, 301–303
incoherent, incoherence 14, 21, 49, 70, 86, 90, 93, 94, 96, 101, 114, 122, 185, 212, 265, 316, 355
incomplete, incompleteness 2, 13, 16, 21, 33, 36, 38, 40, 45, 47, 48, 50, 52–57, 63, 66, 73, 77, 83, 97–99, 106, 112, 123, 125–129, 151, 156, 158, 160, 162, 163, 170, 172, 180, 188, 195, 197, 200, 214, 267, 307, 309, 313, 328, 335, 337, 339, 341, 342, 347, 349, 350, 352, 356, 360, 361, 367, 370, 373, 375, 376, 378–380, 382, 388, 392, 400, 401, 404, 406, 410, 412–414, 441
incompleteness result 13, 16, 50, 112, 200, 335, 404
inconceivable, inconceivability 14, 38, 41, 125, 187, 199, 201, 218, 231, 235, 237, 243, 245, 246, 254, 257, 262–267, 270, 279, 283, 302, 305, 369, 377
inconsistence see *Inkonsistenz* in foreign words index
indefinable, indefinability, undefinability 11, 14, 16, 39, 41, 62, 74, 103, 108, 112, 142, 180, 198, 202, 203, 205, 218, 251, 257, 266, 270, 279, 344, 432
indefinite regress 20, 207, 154, 155, 188, 236, 238, 243, 292
indefinition 57, 135, 251, 313, 403
indeterminate, indeterminacy, indeterminism 28, 29, 98, 118, 183, 194, 207, 222, 225, 239, 275, 301, 311, 345, 368, 376, 382, 386, 390–393, 395
individuate, individuation 8, 17, 20, 21, 27, 38, 56, 67, 75, 80, 117, 120, 124, 137–140, 148, 154, 169, 172, 193, 198, 200, 204, 206, 221, 225, 245, 247, 259, 260, 277, 349, 436
induction 21, 49, 62, 143, 206, 268, 298, 303, 318, 334, 343, 389, 390, 392

inductive 19, 21, 62, 72, 111–113, 154, 177, 182, 200, 211, 226, 267, 268, 270, 298, 303, 332, 335, 372, 373, 390
inductiveness 19, 21, 49, 62, 72
ineffability 17, 28, 48, 86, 100, 144, 218, 221, 228, 371
ineluctable, ineluctability 5, 7, 21, 22, 25, 51, 62, 121, 133, 229, 230, 331, 332
inevident 42, 208, 234, 237, 238, 252 see also order, second
inexpressible, inexpressibility 16, 36, 60, 69, 76, 81, 84, 85, 92, 95, 112, 113, 122, 140, 141, 187, 189, 190, 203, 205, 206, 221, 226, 328, 341, 342, 367, 368, 370
inference 57, 58, 107, 150, 173, 208, 210, 213, 220, 221, 234, 236, 251, 253, 255–259, 268, 277, 281, 282, 284, 292, 303, 314, 317, 318, 328, 343, 387, 403
infinitary 61, 62, 83, 84, 146, 147, 220, 239, 267
infinite 7, 20, 34, 38, 70, 75, 80, 85, 87, 95, 102, 107, 115, 122, 124, 135, 141, 142, 151, 154, 155, 162, 164, 166, 176, 178, 181, 188, 201, 211, 227, 236, 238, 239, 243, 244, 248, 264, 267, 282, 292, 294, 313, 333, 355, 364–366, 380, 386, 389–391, 401, 417
infinitesimal 132, 167, 227, 325, 385, 417, 423, 424, 428
infinity 87, 100, 102, 107, 110, 135, 157, 166, 181, 207, 259, 292, 294, 365, 371, 380, 381
inquiry 2–8, 11, 20, 24–27, 31, 32, 36, 40, 41, 48–53, 55–58, 60, 61, 63–66, 68, 70, 75, 82, 83, 85, 86, 96–99, 105, 106, 115, 121–123, 126, 127, 129, 130, 132, 134, 139, 144, 150, 157–159, 186, 188–190, 192, 211, 221, 225, 230, 260, 274, 275, 281, 307, 308, 312, 319, 320, 328, 333, 337, 338, 341, 348, 349, 354–356, 365, 366, 368, 375, 376, 378, 379, 380, 381, 389, 390, 400–402, 404, 417
– mathematic inquiry 186, 192, 417
– philosophical inquiry 24, 54, 115, 123, 402
– reflective inquiry 50, 132, 308, 312, 320, 328, 338, 348, 356, 365, 375, 378, 380, 381
– scientific inquiry 24, 65, 66, 186, 192, 307, 349, 385, 393

integers 58, 388, 391
integrals 22, 406–408, 415, 417, 421, 422, 424, 426–429
integrand 407, 425
intellect 75, 82, 132, 246, 302, 366 see also understanding
intellectual 51, 101, 246, 248, 275, 279, 298, 302, 357, 370, 371, 405
intelligible, intelligibility 3, 8, 13, 15, 16, 21, 24, 27, 31, 33, 49, 50, 59, 61, 62, 64, 67, 71–73, 82, 119, 122, 123, 128, 140, 141, 143–145, 147, 149, 151–153, 155–159, 174, 175, 178, 184, 185, 190, 220, 221, 223, 225, 231, 241, 253, 261, 271, 301, 307, 309, 311, 318, 319, 330, 332, 337, 338, 341, 342, 344, 346, 347, 355, 356, 358, 361, 363, 364, 366, 367, 370–375, 378–380, 382, 385–389, 391–396, 400, 401, 404
intensional 13, 37, 70, 71, 95, 110, 114, 160, 198, 227, 228, 264, 294, 333, 372, 375, 386, 388, 390–393
intensions 71, 72, 92, 110, 113, 160, 161, 181, 227, 230, 294
intentional, intentionality 16, 31, 34, 35, 40, 42, 51, 62, 71, 76, 83, 97, 110, 114, 125, 136, 137, 140, 158, 186–188, 190, 193, 205–207, 214, 227, 236, 239, 241, 248, 252, 259–263, 273, 275, 279, 303, 312, 313, 320, 325, 327, 334, 338, 344, 350, 352, 356, 357, 367, 368, 370, 372, 375, 378–380, 386, 388, 390, 405
interpretable, interpretability 20–22, 27, 28, 34, 37, 39, 46, 59, 62, 72–75, 77, 85, 92–94, 99, 101, 113, 114, 129, 140, 141, 144, 145, 155–157, 170, 190, 194, 196, 198, 199, 201, 204–276, 210, 221, 222, 226, 228, 232, 322, 323, 332, 334, 339, 341, 345, 352, 354, 366, 367, 370, 372, 373, 387, 393, 404, 347
intersubjective 40, 43, 47, 89, 153, 197, 206, 216, 217, 220, 223, 223, 225, 232, 282, 293, 299, 315, 325, 326
intersubjective uniformity 40, 43, 153, 197, 206, 216, 217, 220, 225, 232, 282, 293, 299

intuition 4, 82, 96, 155, 199, 207, 211, 222, 225, 229, 246, 260, 264, 279, 290, 293, 298, 304, 307, 320, 323, 329, 349, 364, 369, 379, 386, 399, 409
intuitionism 87, 88
irrelationalism 40, 60, 234, 238, 252, 253, 255, 257, 259, 270, 280, 302
isomorphic, isomorphism 46, 61, 75, 81, 122, 177, 320, 329, 349, 364, 396, 397, 412, 413, 420, 424, 426, 428, 438, 439
iterate, iteration 14, 38, 59, 68, 85, 90, 94, 99, 104, 110, 137, 154, 187–189, 192–194, 197, 221, 280, 309, 313, 315, 327, 329, 333, 340, 351, 352, 374, 387, 407, 408, 426, 428, 431, 433, 439

Jansenist dogma 369

King of Glory 265, 268
knowable, knowability 59, 91–95, 106, 113, 114, 354, 359

language of (meta)mathematics 272, 308, 360, 361, 401, 440
Latitudinarian 25, 54, 126, 128, 177, 187, 196, 275, 329, 348, 351, 404
law 175, 215, 287, 296, 306, 340, 348, 353, 362, 370, 378, 384 see also Gesetz in foreign words index
law of nature 215, 216, 219, 287, 297, 352
Lebesgue measures 396–398, 412–416, 422, 423, 427, 429
Lebesgue-Stjeltes spectral measures 424
lemma 13, 258, 273, 404, 411, 314 see also diagonal lemma
lemma, fixed-point 37
lemma, Rasiowa-Sikorski-Lemma 436
liminality 57, 170, 180, 183
limit 4–9, 25, 27, 30–32, 35, 36, 43, 53, 61, 63, 83–85, 101, 122, 132, 133, 135–137, 146, 149, 152–154, 157–159, 185, 188, 190, 193, 219, 226, 228, 230, 240, 252, 260, 278, 307, 310, 312, 313, 318, 319, 320, 331–333, 343, 353, 356, 363, 385, 386, 390, 393, 402, 407, 409, 410, 416, 419, 421, 423, 429, 435, 426, 441

Lindenbaum algebra 160, 363, 411
line, Plato's line 68, 78, 84, 90, 101, 340, 379
line, real line 408, 416, 429
linear, linearity 8, 23, 25, 61, 78, 100, 135, 139, 188, 189, 333, 340, 356, 379, 385, 386, 391–399, 401, 402, 407, 409, 435 *see also* directedness
localize 9, 33, 55
logic 3, 9, 10, 12, 26, 30, 86, 87, 108, 109, 120, 150, 156, 191, 194, 232, 258, 265, 275, 278, 300, 308, 314, 331, 388, 401, 440
logical positivist 4, 302, 308, 333

Manichean 136
manifold 306, 309, 363 *see also* many-existence-theory
many-universes 165, 428
many-worlds 165, 405–408, 418, 428, 433, 434
Markov operators 421
martingale 407, 408, 416, 418–420 423, 427–429, 434, 436–438
mathematical 4, 7, 15, 29, 31, 33, 35, 46, 68, 71, 81–83, 85–88, 116, 118, 119, 143, 170, 172, 186, 188, 191–194, 207, 243, 276, 289, 294, 307, 311, 323, 314, 315, 318, 330, 330, 339, 351, 355, 373, 374, 382–390, 392, 394, 396, 398, 401–409, 417, 431, 437, 439, 331
mathematical logic 3, 68, 70, 80, 87, 98, 142, 199, 227, 300, 401
mathematical physics 4, 46, 119, 307, 330, 339, 360, 369, 374, 385, 388, 405, 406
mathematics 4, 59, 78, 158, 307, 308, 312, 330, 339, 361, 383, 385, 388, 404, 440
mathematics, philosophy of 67, 68, 83, 86–88, 196, 210, 262, 263, 294
matters of fact 40, 44, 131, 216, 224, 236, 241–243, 246, 258–260, 263, 271–275, 281, 283, 290, 292, 300, 301–303, 305
maxim / *Maxime* 201, 207, 236, 239, 240, 241, 244, 246, 252, 253, 274, 275, 296, 299, 305, 342, 353
maximal, maximality 6, 8, 139, 157, 159, 160, 166–168, 177, 179, 184, 188, 307, 310, 313, 314, 318, 319, 341, 406

maximum 83, 136, 175, 278, 306, 307
measure, probability-measure 389, 409, 412, 415–417
measure space 309, 389, 399, 415, 416, 419, 422, 423, 425, 426, 428, 431, 433
measure-algebra 406–408, 411, 416, 427–429, 431, 433–435, 439, 440
measurement 65, 183, 355, 385, 386, 407–409, 414, 415, 418–420, 422–425, 433, 434, 436, 437, 438
metalogical turn / *metalogische Wende* 337
metalogical, metalogic 2–5, 7, 9, 10, 13, 18, 19, 21, 22–24, 26, 31, 33, 35, 36, 38, 40, 43, 49, 55, 58, 59, 63, 65, 66, 68–70, 76, 79, 84, 88, 90–92, 96, 98, 100, 101, 103, 105, 113, 117–119, 122–125, 128, 131–134, 136, 137, 139, 142–148, 150, 153, 154, 156, 157–163, 166, 168, 169–171, 173, 175–180, 183, 184, 186, 189–201, 204–208, 210–212, 214, 222, 226, 228, 230, 232, 234, 235, 237, 242, 245, 258, 259, 263, 265, 266, 270, 271, 272, 274, 275, 279, 300, 304, 307–311, 313, 316, 318, 321, 322, 324, 327, 328, 331, 333, 334, 336, 337, 339, 340, 350, 352, 354, 356–358, 363, 364, 366, 370–372, 374, 375, 377, 379, 386–388, 390, 391, 400, 401, 403, 409, 418, 435
metamathematical 4, 14, 31, 68–70, 78, 80, 82, 85–88, 132, 146, 161, 162, 170, 183, 191, 192, 226, 272, 279, 294, 308, 311, 321, 360, 361, 372, 373, 382, 386, 389, 393, 400, 405–407, 409, 411, 431–437, 440
metamathematics 66, 69, 154, 206, 207, 258, 307, 308, 360, 385, 401, 440
metametatheoretic, metametatheory 20, 21, 38, 73, 86, 263, 266, 270, 274, 276, 300, 343, 390, 391, 414
metaphilosophical 225, 276
metaphysics, metaphysicians 1, 3–6, 9–11, 15, 17, 22, 26, 29, 33, 35, 36, 39–43, 45, 48, 49, 52, 54, 55, 58, 59, 63, 66, 69, 71, 76–78, 81–83, 85, 86, 88, 91, 97, 99, 100, 103, 107, 109, 111, 113, 115, 119, 123, 125, 126, 129, 131, 132, 139, 142, 152, 160, 161, 163–167, 169, 171–173, 175, 179,

182, 186, 188, 189, 191, 192, 194, 197, 206, 207, 209, 211, 214, 226, 228, 230, 233–236, 239–241, 244, 246, 247, 252, 257, 258, 261, 266, 268, 270, 271, 273, 295, 279, 280, 281, 286, 293, 295, 298–303, 305–312, 314, 315, 318, 320, 322, 324, 327, 329, 330, 334, 337, 338, 340, 346, 350, 351, 357, 366–368, 370, 371, 373, 374, 384–387, 390, 392, 401, 403, 405, 428
metatheorem 6, 38, 161, 182, 195, 199, 202, 206, 207, 242, 270, 271, 304, 394, 399, 408, 416, 419, 420, 422, 424–426, 429, 430, 432, 433, 438
metatheoretic ascent 2, 10, 17–23, 25, 26, 28, 29, 31, 35, 36, 40, 42, 43, 45, 48–50, 53, 55, 58, 60–62, 65, 66, 68, 76, 78, 81, 83–85, 88–90, 94, 95, 101, 108, 121, 122, 124, 128, 131, 142, 169, 192, 194, 197, 206, 207, 212, 217, 224, 226–229, 266, 303, 315, 322, 328, 349, 356, 368, 369, 379
metatheoretic hierarchies 30, 53, 60, 152, 169, 179, 301, 302, 329, 333, 349, 353 *see also* hierarchies
metatheory, metatheoretic 3, 5, 8, 11, 18–21, 28, 30, 33–38, 42, 48, 49–51, 53, 55, 60, 65, 67–71, 73–77, 81–86, 88–90, 93, 99, 101, 108–110, 114, 115, 117–122, 124, 125, 127, 131–135, 144, 145, 147, 148, 154, 155, 157, 158, 162, 167, 168, 171, 176, 178, 179, 181–186, 189, 190, 193–200, 202–207, 209, 211, 212, 214, 217, 221, 224–226, 228, 230, 235, 237–239, 243, 257, 258, 261–264, 266, 268–270, 272–275, 278, 300, 301, 303, 307, 309, 311, 314–319, 322–325, 329, 330, 331, 333, 334, 336, 339, 340, 342–344, 350–353, 358, 359, 361, 364, 366–368, 368, 372–376, 378, 384, 389–391, 394, 401, 412, 414, 441
methodology, methodological 7, 9, 10, 17, 29, 37, 39, 40, 58, 60, 63, 66, 70, 98, 103, 105, 109, 111, 126, 131, 136, 148, 152, 171, 198, 209, 211, 219, 220, 235, 236, 239, 244, 250, 251, 252, 255, 261, 262, 267, 275, 278, 280, 281, 294, 295, 298, 301, 302, 305, 309, 336, 344, 351, 354, 366, 369, 373, 433, 436

Michael-Vietoris topology 413
microscope, Hume's microscope 41, 236, 241, 247, 250, 255, 256, 259–261, 261, 265, 267, 271, 273, 275, 278, 280, 290, 302, 303, 305
miniature 15, 112, 162, 183, 191, 207, 230, 316
miniature, formal 1, 17, 19, 26, 54, 59, 78, 110, 191
mirror, mirroring 120, 164, 177, 303, 328, 375, 383, 434, 439
mirroring principle 57, 161, 163, 168, 169
mite´s foot 41, 190, 236, 264, 322
modal, modality 42, 43, 134, 147, 149–151, 161, 169, 171, 172, 177–179, 182, 183, 193, 224, 238, 243, 354, 365, 270, 280, 315, 317, 342, 351, 353, 372, 440 *see also* virtual modality
modal extensions 119, 120, 149, 160, 181
monad(s) / *monades* / monadic, monadology 5, 34, 57, 88, 100, 120, 147, 163, 165, 166, 170, 173, 174, 179, 180, 216, 231, 439
monism, monist 6, 8, 23, 52, 97–100, 105, 115, 117, 118, 126, 131, 184, 188, 194, 316, 339, 382, 387, 400, 401
– semantic monism 69, 70, 76, 85, 88–90, 98, 99, 298, 313, 315, 316, 346, 347, 371
monistic pathos 6, 52, 70, 98, 100, 126, 139, 188, 371, 404
monotheism, monotheist 28, 58, 64, 177, 188
moral, morality 42–44, 52, 54, 55, 102–105, 118, 126, 128, 173, 190, 230, 239, 240, 260, 284, 298–300, 317, 341–343, 346, 347, 354, 357, 374, 379–381 *see also* moralisches Gesetz *in foreign words index*
moral identity 56, 57, 121, 277
moral science 39, 44, 103, 219, 283, 296, 299, 302, 305, 320
mysticism 101, 125, 359, 371

necessary connexion 41, 46, 216, 237, 254, 255, 261, 273, 274, 280, 286, 295
necessary existence 126, 149–151, 155, 314, 372
necessary-precondition-for-the-possibility 35, 50, 189, 320, 334

Newton of the moral sciences, the 39, 44, 219, 283, 305, 320 *see also* Hume *in names index*
noetic 24, 75, 78, 83, 86, 98, 100, 104, 132, 138, 139, 194, 340, 345, 392, 192
nominal, nominalist, nominalism 5, 23, 31, 50, 61, 81, 87, 123, 163, 164, 186, 187, 196, 200, 221, 227, 228, 232, 243, 340
nonconstructive 74, 133, 140, 157, 207, 210, 211, 213, 214, 217, 224, 226, 229, 230, 262–265, 268, 269, 279, 374, 375, 378
normative 1, 5, 29, 32–34, 45, 52, 54, 55, 54, 64, 87, 123, 126, 129, 140, 148
number 49, 96, 112, 164, 192, 202, 203, 222, 264, 289, 319, 383
– natural number 70, 74, 202, 227, 368, 389, 391
– nuptial number 78
– rational number 46, 389, 409, 415, 416
– real number 202, 264, 319

objective, objectivity 6, 31, 45, 89, 135, 137, 179, 188, 223, 286, 293, 315, 353, 370
Occam´s razor 176, 265, 331
one-over-many problem 83, 194
ontological 2, 11, 19, 30, 33, 45, 54, 58, 68, 76, 78, 83, 84, 87, 90, 96, 100, 109, 116, 126, 132, 135–138, 140, 142, 146, 148, 150, 152, 153, 163, 169, 171, 175–179, 187, 188, 194, 201, 206, 209, 212, 213, 217, 220, 221, 225, 241, 321, 322, 324, 332–334, 340, 342, 344, 369, 374, 441
order, first order 16, 18, 35, 36, 40–42, 73, 119, 143, 150, 156, 160, 166, 170, 176, 182, 189, 190, 192, 194, 195, 198–203, 226, 234, 236, 246, 255, 257–259, 270, 276, 308–310, 318, 319, 328, 329, 331–333, 335, 336, 342, 349, 353, 366, 385, 392–394, 409, 410, 414, 440 *see also* evident
– higher order 18, 35, 38, 40, 42, 43, 117, 121, 122, 182, 214, 225, 234, 239, 267, 270, 311, 359, 361, 376, 410, 439 *see also* metatheoretic
– lower order 117 *see also* object-theoretic
order, ordered 6, 8, 15, 23, 46, 49, 61, 70, 71, 101, 107, 115, 117, 120, 123, 135, 136, 165, 215, 218, 219, 227, 228, 260, 285, 287, 292, 300, 310, 314, 317, 319, 330, 343, 349, 350, 362, 391, 393, 395, 407, 408, 435
order, second order 16, 35, 39, 40, 42, 45, 60, 75, 90, 112, 191, 225, 233, 234, 236, 237, 264, 252–261, 270, 286, 287, 316, 325, 340, 366 *see also* inevident

parable 6, 11, 22, 61, 78, 99, 101, 103, 104, 340
paradox
– Berry paradox 12, 16, 74, 110, 112, 141, 200–203
– doubter's paradox 92, 113, 131
– heap(er) paradox 12, 110
– liar paradox 11, 12, 15, 92, 106, 108, 109, 113, 203
– paradox of the knower 14
– Richard paradox 200, 201, 271, 279
– Russell's paradox 74, 76, 110, 111, 227, 333
– Skolem's paradox 110, 390
particularity 251, 264, 265, 275, 280
particularity-doctrine 243, 247–250, 278
perceive, perception 34, 36–38, 40, 41, 74, 151–153, 163, 164, 166, 173, 181, 187, 193, 193, 197, 198, 201, 205, 206, 212, 208, 212–214, 217, 218, 220, 224, 232, 233, 235–237, 241, 244–248, 250–257, 259–261, 269, 271, 276, 277, 280, 285, 298, 299, 302, 325, 327, 329, 330, 361
perfect, perfection 8, 30, 61, 84, 96, 119, 120–122, 124, 137–139, 141, 150, 157–159, 161, 165, 166, 173–180, 184, 219, 247, 248, 276, 289, 314, 323, 324, 369
philosophers 6, 21, 38, 52, 58, 66, 75–77, 100, 123, 135, 140, 154, 164, 214, 215, 219, 247, 284, 300, 309
– analytic philosophers 4, 59, 77, 91, 194, 197, 202, 367
– philosophers of language 4, 75, 192, 265, 368
– philosopher of mathematics 4, 68, 87, 196, 263, 294
philosophical logic / logicians 4, 10, 86–88, 108, 110, 117, 156, 265, 270
philosophy / *philosophia* 2, 8, 26, 29, 54, 63, 64, 66, 77, 100, 104, 135, 209, 220, 239,

273, 289, 292, 297, 307, 308, 312, 322, 339, 340, 346, 360, 361, 364, 167, 379, 440, 441
- first philosophy 2, 135, 226, 228, 235
- moral philosophy 54, 126
- philosophy of mathematics 86–88, 210, 262
- philosophy of science 66, 192, 214, 216, 223, 281, 403
- Western philosophy 6, 25, 35, 45, 59, 78, 100, 160
physics 4, 29, 35, 67, 105, 120, 126, 307, 330, 339, 359, 369, 385, 392
plenitude 161, 173, 175, 177, 178, 180
pluralist, pluralism 3, 8, 52, 53, 70, 88, 98, 104, 125, 139, 219, 225, 231, 232, 295, 339, 372, 404
Poisson processes 407
Polish space 398, 399, 413, 414, 423, 426–429
Polish topologies 396–398, 412, 413, 429
positivism, positivist 4, 230, 239, 302
possibility 7, 14, 35, 86, 99, 118, 149, 150, 153, 165, 172, 173, 175, 178, 187, 244, 245, 261, 265, 269, 281, 303, 310, 320, 328, 334, 359, 362, 372, 414
potential, potentiality 30, 34, 43, 50, 52, 55, 57, 69, 74, 80, 92, 97, 104, 110, 117, 129, 130, 150, 159, 163, 173, 182, 193, 214, 217, 218, 223, 224, 231, 236, 245, 267, 305, 332, 333, 341, 343, 348, 350, 374, 386, 391, 395, 407, 409, 436, 437, 440, 441
predestinarianism 121
predicate 10, 12, 13–15, 38, 39, 74, 75, 92, 93, 108, 109, 111–114, 122, 141, 143, 144, 146–149, 155, 161, 163, 164, 166–168, 181, 187, 194, 202, 203, 250, 266, 272, 279, 327, 363, 389, 410
prime mover 6, 185 see also aitia in foreign words index
principles
- clarity-and-distinctness principle 14, 15, 41, 235, 242, 251, 264, 169, 278, 300
- continuity-principle 183, 382, 384, 393
- first principle 321, 322
- invisible intelligent principle 291
- predicate-in-notion principle 161, 163

- principle of sufficient reason 120, 169–171
- principle of the identity-of-indiscernibles 41, 161, 163, 168, 235, 380, 391
- regulative principle 305, 308, 338, 387
probabilistic, probabilism 11, 109, 121, 182, 372, 382
probability-measure(-space) 389, 397, 398, 409, 411–417, 420, 430
problem, substantial-union problem 70
Prohorov metric / topology 398, 412
proof 3, 6, 8, 13, 16, 18, 21, 27, 28, 31, 33, 34, 38, 40, 43, 48, 49, 48, 93, 96, 101, 112, 115, 122, 132, 133, 137–141, 152, 157, 159, 171, 176, 179, 202–204, 208, 211, 233, 235, 238, 239, 242, 258, 259, 267, 273, 297, 298, 304, 305, 318, 324, 326–328, 330, 334–336, 346, 367, 368, 370, 372, 373, 391, 392, 395, 397, 399, 400, 413, 415, 423–425, 427, 429, 432, 433, 436, 438
- proof-schemes 18 see also consequence-relations
- proof-structure 318
propositional logic 9, 30
propositions 15, 163, 163, 242–245, 253, 271, 279, 288, 299, 321, 335, 336, 394, 397, 439
Protagorean relativism 8, 22, 29, 92, 94, 113
proto- / transcendental argument 16, 17, 25–30, 47, 88, 99, 103, 112, 226, 300, 317, 322, 328, 332, 336, 370
prototype 21, 22, 30, 31, 36, 117, 170, 181, 188, 197, 200, 209, 218, 232, 235, 237, 240, 252, 310, 322, 338, 354, 360, 378
psychology, psychologist, psychologism 44, 47, 48, 58, 66, 77, 83, 84, 123, 191, 219, 248, 284, 367, 369
Pyrrhonist, pyrrhonian 7, 9, 11, 17, 22, 42, 55, 91, 92, 101, 106, 108, 113, 126, 138, 184, 187, 205, 208, 211, 219, 234, 236–239, 245, 263, 269, 273, 278, 282, 286, 290, 299, 301, 321, 339, 366, 387

quantification 9, 51, 124, 147, 150, 178, 270, 279, 280, 308
quantifier-alternation 59, 156, 314
quantifier-error 70, 313, 31

quantifier-exportation 265, 266
quantifier-mistake 59, 90, 107, 137, 155, 257, 313
quantifier-shift 43, 257, 282, 283
quantifier-structure 89, 314, 317

random, randomness 62, 157, 160–163, 169, 180–183, 215, 309, 364, 406–408, 415–417, 419, 421, 422, 424, 426–431, 433–441
ratiocinative, ratiocination 18, 40, 44, 197, 267, 268
rational, rationality 2, 18, 20, 21, 23, 24, 46, 54, 64, 70, 107, 113, 116, 162, 205, 307, 332, 360, 378, 390, 397, 416, 419, 421, 423
rationalism, rationalist 44, 148, 283, 300, 325, 326
rationalisation 296, 298
Rawlsian 342, 345
rcll / right continuous with left limits 385, 407
real / reality 6, 18, 20, 23, 31, 52, 63, 79, 83, 135–137, 139, 146, 157, 174, 175, 179, 188, 193, 196, 213, 216, 217, 219, 229, 236, 242–244, 250, 251, 255–257, 261, 271, 282, 289, 301, 333, 338, 342, 351, 353, 359, 360, 369, 370, 387, 391, 408, 416, 421, 423, 429, 436–438, 440
realist 6, 8, 25, 59, 82, 83, 85, 126, 188, 286, 295, 298, 336, 385, 387, 390
reason, reasoning 3, 5, 8, 13, 15, 18, 19, 27, 33, 44–47, 50–54, 57, 63, 81, 82, 84, 98, 100, 102, 120–123, 129, 133, 134, 138, 144, 146, 147, 161, 162, 169–171, 173, 173, 178, 181, 197, 200, 208, 211, 212, 213, 217, 225, 229, 230, 231, 233, 236, 241, 244, 253, 354, 258, 260, 270, 271, 274, 275, 278, 280, 282, 284, 285, 287–292, 295, 303, 306, 311, 314, 315, 323, 325–330, 336–338, 340–344, 350, 356, 362, 366, 372, 373, 375–381, 393, 400, 405, 441 *see also* understanding *in main index and* intellectus / Vernunft / Verstand *in foreign words index*
recursive, recursion 93, 110, 117, 142, 156, 308, 363, 392, 410, 412

recursively axiomatisable 24, 71, 72, 156, 160, 176, 179, 200, 227, 228, 309, 310, 318, 332, 363, 414
reflect, reflective, reflections 2, 4, 11, 19, 20, 41, 42, 50, 54, 58, 71, 78, 82, 96, 97, 99, 103, 123, 126, 128, 129, 133, 147, 151, 154, 155, 157, 158, 162, 166, 177, 178, 188, 191, 197, 199, 211, 224, 228, 230, 231, 236, 237, 243, 265, 274, 274, 285, 288, 292, 298, 303, 304, 307, 317, 329, 338, 340–342, 347, 348, 350–354, 357, 372, 376, 378–370, 386, 417, 441
reflexion, idea of reflexion 42, 208, 233, 237, 302
reflexion, reflexive, reflexivity 10, 11, 23, 30, 76, 97, 99, 108, 109, 154, 254, 279, 299, 310, 330, 345, 347, 350, 366, 379
regress, regressive 20, 26, 29, 59, 81, 83–85, 88, 94, 107, 189, 213, 237, 238, 291–293, 301, 303, 313–315, 326, 327, 345, 349, 352, 353, 368, 369, 441
relation of ideas 40, 131, 242, 243, 246, 259, 260, 266, 271–275, 277, 278, 291, 292, 294, 300, 302, 303, 305
relational, relationality 10, 11, 20, 29, 44, 59, 60, 62, 75, 104, 108, 109, 136, 137, 138, 142, 248, 150, 163, 168, 169, 182, 193, 205, 259, 260, 283, 313
relativity, semantic relativity 89, 315
representations 8, 19, 20, 36, 80, 113, 118, 193, 223, 294, 350, 387, 407, 418, 425, 431, 433, 438
reverends and right reverends 233, 238, 259
rule, golden rule 55, 376, 380
rules 19, 46, 49, 55, 146, 215, 217, 223, 231, 286, 294, 311, 353, 362, 376, 380, 383, 400, 441

savants, sages 8, 75, 105, 123, 140, 249
scheme / schemata 18, 19, 27, 49, 60, 87, 90, 94, 101, 125, 141, 143, 149, 171, 178, 203, 241, 267, 306, 307, 318, 320, 325, 343, 380, 390, 390
Schlick-Neurath dispute 302
scholastic 6, 61, 135, 142, 188, 190, 218, 287
science / *scientia* 29, 63, 66, 67, 103, 143, 219, 244, 258, 274, 285, 289, 296,

298–300, 302, 308, 312, 324, 325, 331, 364, 388
scientific, Newtonian-scientific 308, 312, 323
secular, secularity 5, 23, 24, 31, 32, 40, 123, 153, 188, 221, 225, 232, 312, 323, 324, 326, 330, 340, 369, 371, 372, 379, 380, 385
secularise, secularisation 24, 44, 65, 170
self 19, 56, 61, 62, 76, 77, 84, 96, 98, 99, 105, 106, 130, 133, 150, 154, 156, 184, 251, 252, 277, 320, 331, 337, 341, 344, 345, 354, 356, 357, 366, 367, 369, 378, 387, 400, 401, 406, 418, 423, 439
self-actuation 76, 127, 345, 365, 366
self-awareness 2, 5, 70
self-causation 84, 345, 369
self-constituting 76, 83, 96, 295, 346
self-determination 98, 127, 347, 358
self-doubt 70, 92–95, 113–115
self-interpretation 35, 61, 77, 84, 118, 149, 189, 368
self-legislation 125, 128, 348
self-organisation 84, 360, 369
self-origination 84, 345, 369
self-reference 4, 14, 15, 26, 56, 72, 73, 93, 96, 97, 111, 114, 128, 231, 280, 300, 341, 346
self-refuting 90, 92–94, 113, 115, 316
semantic paradox 3, 4, 10, 12, 36, 38, 49, 54, 71, 72, 108–110, 125, 131–133, 135, 144, 154, 164, 170, 186, 187, 191, 194, 197, 200, 205–209, 213, 220, 222, 223, 226, 230, 232, 233, 262, 263, 271, 273, 300–302, 308, 327, 331, 337, 346, 356, 359, 368, 370, 390
semantic pluralism / plurality 70, 85, 88, 98, 339, 405, 406, 441
semantic relativism / relativity 69, 70, 89, 90, 315
semantics 10, 16, 17, 24, 73, 76, 112, 119, 144, 147, 151, 154, 160, 163, 166, 169, 172, 178–181, 191, 201, 226, 312, 330, 342, 351, 406, 408, 427, 428, 430, 432, 435
sensation 41, 236, 250, 251, 255
sensory 1, 276, 291, 306, 324, 363, 376
sentient, sentience 1, 2, 5, 50, 55, 98, 124, 158, 170, 214, 224, 230, 231, 309, 341, 343, 348, 377, 401

skeptical (self) doubt 17, 26, 28–30, 49, 51, 53, 70, 84, 92–95, 113, 114, 125, 154, 261, 355
skeptical, anti- 29, 154, 236, 237, 252
skepticism 26, 30, 68, 92, 94, 97, 105, 106, 114, 132, 152, 208, 233, 238, 243, 244, 252, 259, 263, 286, 298–300, 305, 320, 366, 371, 404
skeptics 9–11, 13, 15, 16, 22, 23, 28, 29, 32, 25, 38–41, 93, 95, 97, 101, 103, 105, 108, 111, 112, 114, 122, 123, 131, 135, 137, 138, 140, 187, 188, 190, 208, 212, 219–221, 230, 234, 235, 238, 243, 246, 252, 275, 282, 299, 303, 307, 312, 313, 326, 336, 337, 366, 371
Socratic, pre- 22, 28, 122
Socratic, Socratics 95, 101, 106, 340
soul 107, 116, 121, 151, 166, 181, 224, 231, 256, 260, 277, 284
space 44, 69, 79, 102, 120, 160, 162, 171, 172, 283, 290, 309, 320, 349, 363, 364, 370, 381, 389, 396–400, 406–409, 411–420, 422–429, 431, 433, 436–439
– Cantor space 80, 162, 389, 396–399, 401, 412, 439
– Hilbert space 406, 415, 417, 420, 423, 424, 426, 428, 429, 431, 437–439
– Stone space 69, 71, 70, 80, 120, 147, 150, 160, 162, 163, 170, 180, 195, 349, 363, 364, 375, 389, 396–399, 401, 411–414, 439
spacetime 46, 307, 320, 385, 392, 394, 396, 397, 404, 405, 424, 425, 435
spectral, spectra 406, 408, 421, 424, 427, 430
spirit, spiritual 1, 7, 35–37, 40, 104, 121, 127, 153, 169, 178, 186, 187, 198, 205, 209–213, 215–218, 220, 250, 251, 280, 285, 293, 322, 325
square-integrable 407, 416, 419, 420, 426, 428–430, 437, 438
standard, standardness 9, 22, 46, 65, 72, 79, 80, 87, 89, 97, 107, 119, 131, 162, 228, 263, 272, 314, 315, 327, 348, 377, 386, 398, 410, 411, 430, 431, 435

Stoa, stoic, stoics 9, 10, 12–16, 22, 25, 26, 29–32, 39, 55, 56, 73, 75, 77, 88, 91–95, 98, 103, 105–109, 112–123, 126–128, 130–132, 148, 170, 187, 194, 229, 230, 234, 236, 269, 290, 302, 310, 313, 318, 355, 366, 369, 376
stochastic 119, 406–409, 416, 418, 419, 426–428, 430, 432–436, 438
subjective, subjectivity 206, 271, 370
Swedenborgian mysticism 359
swerve 116, 338 *see also* clinamen
symbol 25, 72, 202, 203, 314, 410
sympathy / sympathetic 57, 116, 130, 159, 231, 369, 376, 377, 378
synthetic, synthesis 4, 34, 44–46, 48–50, 64, 131, 144, 222, 239, 242, 252, 272, 283, 286, 293, 294, 298, 305, 308, 310, 312, 326, 327, 373, 375, 379

teleological 5, 25, 138, 159, 218, 224, 308, 310, 317, 319, 339–341, 346, 351, 372
tensor-product 425, 426
theodicy 105, 123, 136, 177, 219, 344, 271
theological 5, 23, 28, 29, 39, 58, 59, 77, 84, 123, 136, 137, 140, 142, 146, 152, 153, 171, 177, 188, 190, 219, 224, 232, 292, 317, 318, 330, 367, 369
theorem   *see also* metatheorem
– Beth's theorem 202
– completeness theorem 36, 38, 73, 176, 184, 198, 242, 258, 263, 265, 271, 336
– Craig's theorem 316, 394
– Fubini theorem 415
– incompleteness theorem 38, 73, 200, 204, 263, 319
– Lindström's theorem 331, 332, 356
– martingale-decomposition theorem 408
– sufficiency theorem 423
theory / metatheory distinction 11, 49, 65, 118, 119, 131, 148, 182, 190, 194, 263, 300, 309, 322, 329, 331, 333, 350, 368
theory / *theoria* 3–6, 12, 15, 16, 18, 20, 37, 38, 61, 64, 65, 67, 69, 70, 72, 73, 75–77, 79, 81, 82, 83, 90, 92–101, 109, 111–114, 118, 122, 125, 130, 131, 140, 143–145, 147–149, 153, 156, 159–162, 167, 168, 170, 172, 176–178, 181, 182, 184, 190, 194–203, 221, 226, 227, 231, 239, 249, 257, 263, 264, 266, 267, 270, 271, 288, 300, 302–304, 309, 310, 313, 316, 318, 319, 322, 324, 328–330, 333, 335, 345, 342, 347, 348, 355, 359, 363–370, 373, 375, 377, 379, 380, 383, 388, 389, 391–394, 397, 401, 402, 406, 408, 410, 411, 413, 415, 418–420, 427
– design-theoretic, design-theory 158, 224, 318
– extratheoretic 68, 80–84, 89, 96, 156, 197, 314, 369
– information-theoretic / information-theory 47, 72, 95, 225
– many-existence-theory 404
– measure-theory, measure-theoretic 147, 182, 349, 386, 396–398, 411, 440
– model theory 24, 62, 70, 87, 227, 389
– number theory 68, 70–72, 171, 182, 227
– object-theory, object-theoretic 8, 9, 19, 33, 36, 38, 48, 51, 65, 67, 68, 73, 76, 77, 84, 94, 95, 110, 114, 117, 199, 132, 155, 157, 182, 192, 193, 199, 200, 202, 203, 205, 206, 211, 230, 243, 257, 261, 265, 271–274, 276, 300, 311, 313, 317, 328, 341–343, 353, 355, 361, 374, 391
– probability theory 29
– proof-theory 21, 58, 147, 179, 304, 334
– quantum-theory, quantum theoretic 56, 385, 406, 409, 418, 421, 424, 431–433, 441
– recursion theory 318, 349, 392
– redundancy theory 17, 60, 75, 249
– set-theory, set-theoretic 46, 70–72, 84, 87, 120, 141, 150, 167, 177, 179, 181, 227, 228, 264, 308, 309, 333, 342, 385, 390, 391, 404, 406–408, 410, 428, 429, 431
– theory-formation 190, 194
– theory-relativity, theory-relative 4, 15, 28, 50, 51, 53, 57, 62, 66, 71, 74, 76, 86, 87, 89, 92, 110, 112, 124, 132, 133, 137, 162, 171, 182, 188, 206, 229, 232, 262, 263, 270, 271, 301, 304, 305, 313, 330, 366, 368, 406
– type theory 31, 35, 189, 300, 308
thought, entities-of thought 208, 233
time 16, 19, 28, 65, 66, 102, 104, 166, 193, 207, 215, 219, 222, 231, 244, 276, 277,

309, 320, 333, 353, 365, 381, 382, 385, 386, 387, 394–396, 399, 407–409, 419, 423, 425, 432–434 *see also* spacetime
time-evolution 28, 118, 167, 172, 382, 392, 406–408, 418–421, 425, 427, 428, 432, 434, 435, 437, 438
topological, topology 79, 147, 154, 160, 167, 181, 182, 209, 343, 363, 364, 386, 396–398, 409, 411–413, 437
transcend, transcendent, (proto-)transcendental, (proto-)transcendence 2, 15–18, 21, 24–33, 35, 36, 44, 46–49, 51, 54, 58, 60–63, 67–70, 86, 88, 90, 100, 101, 103, 111, 112, 119, 124–126, 131–134, 141, 147, 148, 151–153, 167, 170, 183, 214, 220–224, 229, 232, 286, 295, 298, 304, 307–312, 314, 316, 317, 320, 321, 323, 326–331, 334, 337–339, 347, 349, 350, 355, 359, 360, 364, 364, 366, 372, 379, 387
transcendental illusion 2, 21, 25, 49, 54, 126, 183, 337, 355, 356, 391, 436
triangle 195, 243, 269, 277
true, truth 8, 11, 12, 20, 22, 31, 39, 60, 75, 81, 82, 91, 106, 109, 117, 138, 152, 161, 164, 165, 225, 245, 250, 251, 253, 270, 284, 300, 363, 372, 402, 403
truth in a structure 24, 75, 366
Turing machine 343
Tychism 188
type 65, 90, 135, 143, 144, 147, 148, 160, 166, 168, 169, 173, 177, 194, 317, 333, 350, 363, 364, 391 *see also* archetype, ectype, prototype
type-level 65, 236

uncountable 38, 87, 145, 181, 364, 391
undecidability 3, 50, 65, 66, 76, 80, 92, 96, 98, 99, 101, 113, 173, 183, 203, 204, 222, 272, 295, 300, 301, 366, 367
undefinability 16, 112, 142, 203, 300
underdetermination 3, 45, 55, 66, 109, 118, 147, 155, 162, 163, 180, 182, 191, 202, 204, 206, 207, 212, 216, 221, 222, 226, 230, 231, 237, 250–252, 259, 265, 274, 281, 295, 310, 311, 325, 328, 337, 344, 346, 348–350, 352, 355, 361, 362, 365, 378, 390, 433, 435, 436

understanding 3, 5, 13, 46, 53, 59, 60, 64, 82, 122, 123, 128, 129, 132, 146, 147, 153, 155, 158, 191, 222, 225, 229, 231, 260, 267, 276, 287, 287, 288, 290, 293, 294, 298, 299, 301, 304, 306, 307, 311, 320, 323, 329, 341, 350, 352, 370, 373, 378 *see also* reason *in main index and* intellectus / Vernunft / Verstand *in foreign words index*
undogmatic 11, 109, 186, 208, 234, 238, 290, 312, 339, 366
unintelligible, unintelligibility 32, 85, 141, 247, 248, 277, 280, 282, 284, 285, 307, 370, 371, 377, 400
uninterpretable 73, 74, 87, 93, 114, 144, 198, 199, 206, 210, 373
uniqueness 100, 152, 314, 318, 341, 371
universal 12, 13, 24, 28, 35, 47, 60–62, 76, 82–84, 122, 171–173, 183, 184, 189, 216, 275, 288, 290, 296, 305, 309, 313, 314, 364, 369, 377, 389, 409, 410, 412, 431
universality 13, 32, 40, 77, 100, 126, 184, 234, 271, 275, 297, 298, 324, 346, 367, 370, 439
universe 5, 25, 28, 31, 59, 71, 84, 100, 102, 103, 114, 120, 123, 129, 151, 164–166, 170, 181, 212, 216, 219, 221, 227, 242, 244, 247, 276–278, 306, 364, 381, 384, 391, 406–408, 410, 428, 431, 434–436, 439
unknowable 43, 59, 93, 94, 114, 240, 253, 297
unprovability 92, 113, 161, 203–206

valid, validity 12, 13, 15, 47, 111, 136, 147, 177, 201, 223, 273, 293, 318, 377
validate, validation 15, 148, 159, 320, 324, 327, 330, 331, 334, 337
veracity 9, 107, 170, 323
virtual modality 119, 150, 160, 161, 169, 180, 192, 342, 351, 439 *see also* modality
virtue 121, 171, 194, 297, 376
volition 216, 243, 290, 295

wave 116, 120, 151, 158, 166, 181, 287, 350, 3
wave-function 407, 408, 409, 418, 421, 422, 430

Wiener (measure, paths, space) 407, 420, 425, 426
wisdom 1, 2, 6, 98, 121, 123, 215, 339
world 1, 23–25, 33, 34, 35, 44, 104, 107, 115–117, 120, 164, 165, 168, 173, 181, 182, 189, 216, 219, 231, 265, 283, 289, 292, 319, 342, 346, 351, 354, 355, 365, 367, 370, 377, 382, 383, 405–409, 418, 428, 431–434, 440 *see also* cosmos *in main index and* ho kosmos *in foreign words index*

Foreign Words Index

This list contains French, German, Greek and Latin terms used in the text. References to English words appear in the main index.

abîme (abyss) 288, 357, 364, 365
Achtung (respect) 2, 19, 53, 57–59, 337, 343, 348, 353, 357, 358, 379, 387, 436
aitia (first cause) 6, 83, 116, 119, 314
akatalepsia (lack of comprehension) 15, 73, 112
alethes (the True) 22, 387, 402
alethic (of or relating to truth, modalities of truth) 10, 11, 14, 22, 32, 75, 82, 85, 92, 104, 106, 108, 109, 122, 138, 149, 169, 170, 182, 224, 237, 238, 245, 270, 342, 351, 366, 387
Amphibolien (equivocations) 42, 59, 74, 99, 138, 152, 195, 229, 237, 265
Anschauung (intuition) 96, 224, 245, 306, 349, 351
Antinomien / antinomies (incompatible laws) 111, 45, 52, 108, 123, 222, 223, 232, 260, 262, 288, 289, 308, 314, 346, 366
anupotheon (free from hypothesis) 88
apodeixis, apodeictic, apodicity (self-evident, incontrovertibly true or false) 30, 113, 320
aporia, aporiai, aporetic (internal contradiction or logical disjunction) 10–12, 17, 27, 39, 61, 63, 81, 85, 104–106, 108–110, 120, 125, 129, 130, 141, 149, 191, 207, 224, 226, 228, 266, 303
arche (principle) 9, 13, 27, 33, 39, 44, 43, 57, 59, 68, 102, 103, 108, 116, 117, 120, 149, 160, 161, 163, 168–171, 176, 179, 183, 202, 207, 208, 209, 234, 235–237, 239–241, 243–248, 250, 252, 253, 255–263, 268, 269, 273–277, 280–282, 284, 286–288, 290–296, 298–302, 304, 305–310, 321, 325, 331, 338, 343, 346, 351–353, 362, 363, 371, 380–382, 384, 387–390, 393, 436, 440, 441
ataraxia (tranquility, equanimity, non-perturbation, peace of mind) 55, 70, 92, 96, 113, 123, 126, 128, 130, 307, 366, 379, 404

Begriff des Ganzen (concept of the whole) 59
Begriff, Inbegriff (concept, totality, aggregate) 3, 46, 59, 71, 142, 145–147, 160, 223, 225, 227, 230, 231, 250, 286, 293, 295, 306, 311, 326, 329, 350, 351, 353, 359, 360–363, 366, 367, 403
Bereich der Erfahrung (realm of experience) 17, 25, 49, 60, 123, 309

characteristica magna (universal characteristic) 27, 33, 36, 179
conceptio quid= conceptio de re (conception of the nature of [an entity]) 74, 262, 265, 266, 268, 269, 278, 279
conceptio quod = conceptio de dicto (conception that [an entity] exists) 262, 265, 268, 269, 278, 279

daimon (fate, power, god) 101
de dicto (a claim that [an entity] exists, lit., of the word) 37, 74, 262–266, 268, 270, 272, 273, 278
de re (a claim about the nature of [an entity. lit., of the thing) 38, 39, 74, 262, 264–268, 270, 272, 278, 303
demiourgos (demiurge) 7, 34, 77, 124, 139, 177, 185
der Alte würfelt (nicht) (the old man does (not) gamble) 171, 180, 355, 427
deus sive natura (god within nature) 25, 32, 100, 117, 140, 141, 182, 190, 221, 225, 313, 364, 369, 384, 400
dianoia (discursive thinking, as opposed to immediate apprehension) 71, 82
Dinge an sich (things in themselves) 51, 124, 197, 223–225, 287, 312, 377
dioikesthei (administration) 115–117
doxa (common belief, popular opinion) 97, 98, 208, 270, 339
durchgängig bestimmt (thoroughgoingly determined) 47, 49, 146, 246, 298, 364

durchgängige Bestimmung (thoroughgoing definition / determination) 84, 120, 122, 147, 154, 155, 248, 290, 368

Ehrfurcht (awe) 103, 348, 357, 358
eidos auton (the Very Idea) 88, 125
eidos, eide, eidein, eidetic (form, appearance) 1, 2, 7, 8, 29, 37, 60, 68, 82, 88, 96, 97, 100, 101, 147, 197, 198, 247, 379
eikasia (likeness) 11, 22, 61, 78, 80, 98, 101, 103, 109
ekptosis eis apeiron (lapse into (the) infinite (s)) 38, 85, 169, 211, 230, 236, 282, 371
elenchos, elenchoi (refutation) 69, 88, 91, 95, 106, 186, 188, 192, 201, 248
elenctic, (serving to refute) 8, 22, 26, 29, 30, 37, 86, 95, 100, 135, 149, 154, 191, 198, 208, 209, 212, 223, 238, 245, 262, 268, 295, 315
Eliploken (superposition) 117, 120, 173, 181, 310
energeia (active power) 70, 76, 97, 98, 130, 347, 348
epoche (suspension of judgment) 4, 15, 16, 19, 21, 25, 28, 42, 44, 48–50, 64, 65, 70, 73, 89, 90–92, 96, 98, 100, 106, 112–114, 121, 124, 134, 157, 158, 181, 187, 211, 212, 216, 217, 220, 225, 229, 240, 244, 259, 284, 287, 306, 310–312, 315, 317, 331, 332, 345, 348, 350, 351, 353, 354
Erfahrung (experience) 17, 25, 44, 46, 49, 58, 60, 100, 123, 147, 148, 170, 197, 221, 223–225, 281, 283, 286, 307, 309, 321, 327, 330, 338, 354, 359, 362
Erfahrung überhaupt (absolute experience) 44, 46, 283, 302
Erhabenheit (sublime, sublimity) 218, 348, 356–358, 364
eristic (characterized by debate, argument, esp. one aimed primarily at refutation) 9, 16, 17, 108, 112, 233, 336
Erscheinung / phenomenon (appearance) 46, 167, 231, 266, 286, 287, 289, 306, 313, 314, 318, 353, 366, 396
esse (est) percipi (to be is to be perceived) 36, 197, 199

eteos / etymos (true, real, actual) 387
eudaimonia (enlightened, enlightenment) 55, 78, 99, 100, 104, 121, 340, 370, 376, 379

Fliegenglas (fly bottle) 121, 128
Freiheit (freedom) 2, 53, 54, 56, 57, 98, 99, 125–127, 130, 156, 337, 338, 365, 375, 380, 395, 404, 437

Geschäft der Vernunft (affair of reason) 298, 306, 323, 369
Gesetz (law) 175, 215, 287, 296, 306, 340, 348, 353, 362, 370, 378, 384
Gesetz, moralisches (moral law) 2, 46, 103, 106, 218, 221, 286, 337, 342, 348, 351, 357, 358, 372
Grenzbestimmung (boundary demarcation) 101, 223, 229, 230, 356
Grenzidee (regulative/ limiting idea / ideal) 6, 25, 45, 48, 49, 52, 55, 58, 71, 75, 77, 131, 157, 228, 307, 310, 317, 337, 345, 354, 356, 357, 366, 367, 376, 379, 380, 390, 417

Heimarmene (fate) 105, 116
hen (the One) 100, 115, 34
henades (ideal unities) 88, 100
ho kosmos, kosmoi, kosmou / die Welt (order, world, cosmos, universe) 2, 28, 105–106, 107, 115–117, 120, 122, 130, 136, 354, 355, 367, 377
holon (the Whole / the Universe) 5, 29, 59, 115, 123, 170, 221
homunciones (human-lings) 132, 138

id quod maius (that than which greater) 6, 30, 132–134, 142, 145
Ideen / ideai / idein (idea, ideas) 6–8, 35–37, 40–43, 63, 75, 81, 82, 84, 88, 96, 97, 100, 110, 125, 131, 141, 146, 147, 153, 155, 157, 171, 1176, 83, 184, 186, 187, 189, 190–192, 194–196, 198, 202, 204–206, 208, 210, 211, 213–216, 218, 223, 229, 230, 233, 235–237, 240–244, 246–260, 262, 266, 267, 269, 271–281, 285, 288, 290–295, 300, 302, 303, 305–307, 322,

328, 338, 346, 357, 358, 360, 362, 373, 376, 386, 409, 423, 439
ignoratio(nes) elenchi (logical fallacy, irrelevant argument) 87, 131, 268
Inbegriff des Denkbaren (aggregate / totality of everything thinkable) 71, 141, 142, 145, 227, 230
incomplétude (incompleteness) 40, 52, 53, 56, 129
Inkonsistenz (inconsistence, inconsistency, inconsistent) 28, 37, 41, 47, 48, 54, 72, 73, 90, 99, 124, 126, 133, 141, 142, 145, 157, 200, 201, 203, 207, 208, 210, 212, 214, 233, 252, 254, 262, 263, 270, 301, 316, 346, 367, 377, 385, 395, 400, 401
insipient, insipientes (foolish) 136, 138
intellectus/ Verstand (reason, understanding) 3, 5, 13, 46, 53, 59, 60, 64, 82, 122, 123, 128, 129, 132, 146, 147, 153, 155, 158, 191, 222, 225, 229, 231, 260, 267, 276, 287, 287, 288, 290, 293, 294, 298, 299, 301, 304, 306, 307, 311, 320, 323, 329, 341, 350, 352, 370, 373, 378

katalepsis / katalepsia / kataleptike (comprehension, apprehension) 12, 14, 15, 30, 73, 77, 111–113, 229, 230
Komplementarität (complementarity) 31, 45, 66, 75–77, 118, 149, 173, 176–179, 188, 194, 362, 367, 370, 375, 406, 417, 436
kopernikanische Wende (Copernican turn) 44, 45, 52, 286

L´Homme Machine (Man a Machine [materialist 1748 treatise]) 297, 298
le bon dieu (the good god) 7, 35, 39, 40, 189, 215, 216, 219, 221, 223, 313
logike / logismos (art of reasoning) 8, 9, 12, 109, 118
logismos (argumentation) 8, 10, 35, 46, 47, 108
logos, logoi (logical argument) 8, 388

metalogische Wende (metalogical turn) 337
more geometrico (in the manner of (Euclidean) geometry) 6, 171, 400

Mystisches (mystical) 21, 24, 61, 95, 125, 145, 221, 222, 307, 370, 371

natura formaliter spectata / materialiter spectata (the form of nature; a reality of sensations) 46, 286
nisus, animal (animal efforts) 216, 281, 285, 291
noesis noeseos (thought about thought) 6, 96, 97, 99, 128, 135, 154, 1159, 84, 227, 228, 239, 347, 379, 402
notwendige Allgemeingültigkeit (necessary universal validity) 45, 47, 232, 293
Notwendigkeit / necessité / ananke (necessity) 40, 42, 45, 105, 115, 118, 119, 120, 150, 161, 165, 169, 170, 172, 173, 175, 182, 189, 216, 234, 240, 254–256, 260, 265, 273, 281, 284, 286, 294–296, 304, 310, 320, 334, 346, 440
noumen / noumena (the unknowable, unknowable existence) 20, 21, 42, 124, 135, 136, 144, 151, 153, 155, 156–159, 208, 229, 233, 234, 237, 239, 242, 255, 278, 287, 302, 345, 346, 367, 368, 385, 436
Numen Lumen (God, our light) 158

obscura per obscuriora (explaining the obscure by the more obscure) 68, 69, 243

pan / pantou (the All) 116, 117
parfaits miroirs de l'univers (perfect mirrors of the universe) 119, 120, 160
pathos, pathe (suffering, emotion) 6, 8, 45, 52–54, 70, 97, 98, 100, 125, 139, 188, 231, 236, 244, 371, 372, 404
petitio, petitiones (a logical fallacy in which what is to be proved is implicitly assumed, begged question) 59, 90, 128, 129, 131, 141, 152, 157, 187, 207, 214, 221, 223, 304, 316, 320, 331, 364
Pflicht (duty) 100, 121, 122, 133, 340, 357, 379
phainomena (the appearances) 42, 46, 88, 237, 238, 310, 312, 320
phantasia / fantasia (imagination, process of receiving images) 14, 113

physis / natura (nature) 22, 29, 31, 32, 39, 42, 43, 46, 56, 57, 102, 111, 115–117, 119, 121–123, 127, 128, 137, 151, 163–166, 181, 186, 215–217, 219–221, 225, 238, 239, 240, 244, 258, 260, 277, 280, 283–287, 289, 290, 296, 297, 299, 300, 304, 307, 344, 352, 353, 362, 364, 365, 380, 381, 383, 394, 401

pou sto (fixed-point) 13, 15, 17, 26, 27, 29–32, 27, 48, 51, 92–94, 112–114, 135, 138, 145, 146, 152, 153, 154, 159, 172, 229, 320, 325, 326, 334, 375, 380

Primat des Praktischen (primacy of the practical) 1, 2, 52, 55, 103, 128, 129, 341, 342, 374

principium individuationis (principle of identity) 277

raison suffisante (sufficient reason) 161, 180, 369

Reich der Zwecke (realm of ends) 338 see also Systeme von Zwecken

roseau pensant (thinking reed) 52, 102, 159, 380, 381

Schicksal der Vernunft (fate of reason) 5, 45, 51, 53, 54, 230, 306, 311, 328, 329, 340, 441

Schweigen (silence) 1, 17, 104, 112, 133, 157, 238, 364, 365, 370

Selbstgesetzgebung (self-legislation) 12, 77, 345, 378

semeion (sign) 1, 7, 37, 195, 197, 198, 200, 206, 209, 213, 215, 217, 218, 232, 257, 325, 391

skepsis / skeptesthai / skeptikoi (observation, examination; reluctance to believe) 7, 65, 97, 307, 354, 379, 387, 404

sophos (wise, skilled) 77, 98, 105, 123

sorites, sorites, soritical (paradox of the heap) 12, 16, 110, 112, 211, 278

spectrum, spectra (appearance, image) 406, 408, 421, 422, 424, 427, 430

Sprachspiele (language games) 4, 25

Systeme von Zwecken (systems of ends) 348 see also Reich der Zwecke

tropos, tropoi (modes) 26, 28, 40, 50, 69, 85, 169, 205, 226, 230, 238, 313

uendelighedens ridder (knight of infinite resignation) 102, 380

Unermesslichkeiten (immeasurabilities) 356–358

Unschärfe (fuzziness) 179, 298, 305

Unsinn, Unsinnigkeit (nonsense) 16, 17, 49, 73, 112

Vermögen (capacity, faculty) 5, 19, 44–46, 58, 63, 74, 133, 159, 190, 200, 213, 229, 231, 249, 283, 286, 289, 290, 306, 312, 327, 343, 348, 350, 352, 354, 357, 362, 364, 380

Vermögen, verknüpfendes (connecting faculty) 45, 286

Vernunft, vernünftig (reason, reasonable) 3, 5, 44, 45, 48, 51, 63, 101, 133, 134, 156, 223, 224, 230, 231, 232, 281, 290, 298, 304, 306, 311, 323, 328, 329, 337, 338, 345, 349, 356, 358, 362, 367, 369, 380, 441

Vernunftideen (ideas of (pure) reason) 51, 84, 147, 155, 167, 225, 230, 288, 295, 302, 323, 327, 368, 373, 376

Verstand, Verstehen (reason, understanding) 3, 5, 48, 59, 147, 197, 225, 231, 287, 295, 298, 304, 306, 311, 323, 346, 348, 349, 363, 383

Verstandesbegriffe (concepts of the understanding) 3, 147, 225, 295, 306, 307, 311

verus / vrai / wahr (true) 387

via contemplativa (the contemplative way) 365, 370, 372

via negativa (description stating what something is not, esp. in theology with reference to attributes of God) 365, 370, 372

Vollständigkeit (completeness) 2, 21, 31, 36, 38, 46, 48, 77, 84, 117, 124, 126, 128, 147, 155, 157, 161, 167, 168, 176, 194, 195, 197, 198, 200, 209, 248, 258, 272, 273, 288, 313, 320, 323, 324, 329, 338, 339, 344, 351, 367, 368, 373, 375, 376

Wissenschaft (science) 64, 66
Würde (dignity) 2, 5, 53, 54, 57, 58, 102–104, 106, 121, 126, 130, 133, 140, 193, 156, 337, 365, 374, 376, 381, 383, 402

zetein, zetesis, zetetic, zetetikoi (seek for, inquire into, proceeding by inquiry) 4, 7–11, 24, 26, 29, 30, 34, 35, 39, 42, 45, 48, 49, 51–55, 57, 60, 62, 63–65, 97, 123–129, 148, 307, 343, 354, 374, 378, 385, 387

www.ingramcontent.com/pod-product-compliance
Lightning Source LLC
Chambersburg PA
CBHW030234250426
43668CB00047BA/245